Planetary Geologic Mapping and Remote Sensing

Planetary Geologic Mapping and Remote Sensing

Guest Editors

Kaichang Di
Long Xiao
Jessica Flahaut

MDPI

Basel • Beijing • Wuhan • Barcelona • Belgrade • Novi Sad • Cluj • Manchester

Guest Editors
Kaichang Di
State Key Laboratory of Remote Sensing and Digital Earth
Chinese Academy of Sciences
Beijing
China

Long Xiao
School of Earth Science
China University of Geosciences
Wuhan
China

Jessica Flahaut
CRPG
CNRS/University of Lorraine
Vandoeuvre-les-Nancy
France

Editorial Office
MDPI AG
Grosspeteranlage 5
4052 Basel, Switzerland

This is a reprint of the Special Issue, published open access by the journal *Remote Sensing* (ISSN 2072-4292), freely accessible at: https://www.mdpi.com/journal/remotesensing/special_issues/Planetary_Geo.

For citation purposes, cite each article independently as indicated on the article page online and as indicated below:

Lastname, A.A.; Lastname, B.B. Article Title. *Journal Name* **Year**, *Volume Number*, Page Range.

ISBN 978-3-7258-7851-2 (Hbk)
ISBN 978-3-7258-7852-9 (PDF)
https://doi.org/10.3390/books978-3-7258-7852-9

Contents

About the Editors

Kaichang Di

Kaichang Di is a Professor at Aerospace Information Research Institute, Chinese Academy of Sciences, Beijing, China, where he is also the founding director of the Planetary Mapping and Remote Sensing Lab (PMRS Lab) at the institute. He was the chair of Inter-Commission Working Group III/II "Planetary Remote Sensing and Mapping" of International Society for Photogrammetry and Remote Sensing from 2016 to 2022. His research interests include planetary photogrammetry and remote sensing, visual localization and navigation, and planetary science. He is a core member of the scientific research teams for Chang'e-3 and Chang'e-4 lunar missions and Tianwen-1 Mars mission, and a member landing site selection working group for China's crewed lunar exploration. He led the PMRS Lab and developed technologies and software of high-precision topographic mapping and visual localization, which have been operationally applied to the Chang'e-3, 4, 5, 6, and Tianwen-1 missions, and greatly contributed to landing site mapping and evaluation, lander localization, rover localization and navigation in these missions. He has also made a series of research achievements in lunar control network and global mapping, lunar and Mars target recognition, topographic and geomorphologic analysis, mineral inversion, and geological evolution. He has authored 2 academic books, 196 academic journal papers, and 128 conference papers. His achievements have been recognized through more than ten international and national awards.

Long Xiao

Long Xiao is a distinguished professor at the China University of Geosciences (Wuhan), a leading scholar in planetary geology, and the chief editor of the Planetary Geology textbook for the national "101 Plan". His research focuses on planetary volcanism, lunar and Martian geological evolution, impact processes, and terrestrial analogs for planetary exploration, with core contributions to deep space exploration missions including China's Chang'E and Tianwen projects. He has led and participated in groundbreaking studies on young lunar mare volcanism, Martian sedimentary and volcanic features, and lunar soil geochemistry, publishing over 200 high-impact papers in top journals such as *Science, Nature Astronomy, Nature Geoscience,* and *Geology*. His work on Chang'E-3/4/5/6 lunar mission data and Zhurong rover's Martian observations has advanced the understanding of lunar thermal evolution and Martian paleoenvironment. He has authored/edited 6 academic books and textbooks including Mars on Earth: A study of the Qaidam Basin and Planetary Geology, and his popular science work Unveiling Mars won the Outstanding Popular Science Work Award of Hubei Province. As a PI, he leads key national research projects, and his team has identified critical planetary analog sites (e.g., Qaidam Basin) and provided scientific support for lunar and Martian landing site selection. His research bridges planetary geoscience research and deep space exploration engineering, making pivotal contributions to China's planetary science development and national deep space exploration strategy.

Jessica Flahaut

Jessica Flahaut is a researcher at the French National Center for Scientific Research (CNRS), currently affiliated with the Center for Research In Petrography and Geochemistry (CRPG) in Nancy, France. Trained as a planetary geologist, her expertise lies in remote sensing data analyses, VNIR spectroscopy and landing site selection and characterization studies. She participated in numerous orbiter and rover missions (Mars Express, MSL, ELM, Bepi-Colombo, MMX...) and her work on magmatic and sedimentary processes on Mars and the Moon is featured in >50 peer-reviewed publications, white papers, and space agencies reports. Her previous experience working with remote sensing data is supplemented by laboratory work on terrestrial analogs and meteorite samples in the CRPG hyperspectral remote sensing laboratory that she is managing. Dr. Flahaut is also deeply involved with teaching and science outreach, and her dedication has been recognized through several national and international awards.

remote sensing

MDPI

Technical Note

Chronology of the Basalt Units Surrounding Chang'e-4 Landing Area

Zongyu Yue [1,2], Ke Shi [1,3], Gregory Michael [4], Kaichang Di [1,2], Sheng Gou [1,5,*], Jianzhong Liu [2,6] and Shengli Niu [5]

1 State Key Laboratory of Remote Sensing Science, Aerospace Information Research Institute, Chinese Academy of Sciences, Beijing 100101, China; yuezy@radi.ac.cn (Z.Y.); shike191@mails.ucas.ac.cn (K.S.); dikc@radi.ac.cn (K.D.)
2 Center for Excellence in Comparative Planetology, Chinese Academy of Sciences, Hefei 230026, China; liujianzhong@mail.gyig.ac.cn
3 University of Chinese Academy of Sciences, Beijing 100049, China
4 Institute of Geological Sciences, Freie Unversitaet Berlin, Malteser Strasse 74-100, Haus D, 12249 Berlin, Germany; gregory.michael@fu-berlin.de
5 State Key Laboratory of Lunar and Planetary Sciences, Macau University of Science and Technology, Macau 999078, China; 1909853ese30001@student.must.edu.mo
6 Institute of Geochemistry, Chinese Academy of Sciences, Guiyang 550081, China
* Correspondence: gousheng@aricas.ac.cn; Tel.: +86-10-64807987

Abstract: The Chang'e-4 (CE-4) lunar probe, the first soft landing spacecraft on the far side of the Moon, successfully landed in the Von Kármán crater on 3 January 2019. Geological studies of the landing area have been conducted and more intensive studies will be carried out with the in situ measured data. The chronological study of the maria basalt surrounding the CE-4 landing area is significant to the related studies. Currently, the crater size-frequency distribution (CSFD) technique is the most popular method to derive absolute model ages (AMAs) of geological units where no returned sample is available, and it has been widely used in dating maria basalt on the lunar surface. In this research, we first make a mosaic with multi-orbital Chang'e-2 (CE-2) images as a base map. Coupled with the elevation data and FeO content, nine representative areas of basalt units surrounding the CE-4 landing area are outlined and their AMAs are derived. The dating results of the nine basalt units indicate that the basalts erupted from 3.42 to 2.28 Ga ago in this area, a period much longer than derived by previous studies. The derived chronology of the above basalt units establishes a foundation for geological analysis of the returned CE-4 data.

Keywords: lunar chronology; basalt units; CE-4 landing area; absolute model age; crater size-frequency distribution (CSFD)

1. Introduction

Chang'e-4 (CE-4) was the second soft landing mission in the China's Lunar Exploration Program, and the probe successfully landed on the far side of the Moon in the Von Kármán crater inside the South Pole-Aitken (SPA) basin on 3 January 2019 [1,2]. The Yutu-2 rover was then successfully released from the lander on the same day, and it is still working as of now. The SPA basin is an elliptical impact basin with axes measuring 2400 × 2050 km [3], and it is considered as one of the three major lunar terranes according to its unique geochemistry and petrologic history [4]. Using three-dimensional numerical impact simulations, ref. [5] demonstrates that the SPA must have sampled the Moon's upper mantle to depths of 100 km. Therefore, a number of fundamental lunar science questions, such as the composition of the lunar upper mantle [6] and the reason that far side maria differ from their near side counterparts [1], are expected to be addressed with the returned CE-4 data. As a result, many studies related to the CE-4 landing area have been recently carried out (e.g., [7–13]), and more works are in progress with the returned CE-4 data. To support reliable analysis and interpretation of the scientific data from the CE-4 mission, it is necessary to create a reliable chronology of the geological units surrounding the CE-4 landing area.

Chronology here refers to the occurrence order of a series of geological events, and the occurrence of the geological events can be expressed with the absolute age or the relative age. Figure 1 shows the context of the CE-4 landing area. The local relative chronology surrounding the CE-4 landing area may be determined according to the superpositional relationships of different geological units. However, determining their absolute model ages (AMAs) is always preferable and particularly important when it is difficult to derive the relative ages between two non-adjacent geological units [14]. The principle of determining the AMA of lunar geological units was elaborated in [15] and was the subject of many papers (e.g., [14–16]). The rationale of the approach is to fit the observed crater size-frequency distribution (CSFD) of a counting area that represents the geological unit to a known crater production function (PF) (e.g., [17]), which is further used to derive the AMA along with a chronology function (CF) calibrated to radiometric dating from lunar samples [16]. Currently, the method has been widely used in dating lunar mare basalts, e.g., [17–19]. With this method, the AMA of the mare unit on the floor of Von Kármán crater has been derived as 3.35 Ga [20], 3.6 Ga [7], and 3.15 Ga [21]. However, the mare surface on the floor of Von Kármán crater was interpreted to be emplaced by one episode of basaltic volcanism in these studies, which can only give a rough estimation of the average AMA of the maria basalt. As to the mare deposits in Von Kármán M crater [20], an AMA of 3.38 Ga was derived, while [21] mapped two mare basalt units in Von Kármán M crater and derived AMAs of 3.38 Ga and 3.52 Ga, respectively. In addition, the AMA of the basalt unit in Leibnitz crater was derived as 3.44 Ga [20] and 3.37 Ga [21]. The mare basalt in Leibnitz X crater (the crater diameter is about 19.71 km), which is inside Leibnitz crater, was dated as 3.68 Ga [21]. The different dating results of the above studies to the same maria unit may arise from the different data sources and counting areas. Therefore, a strict basalt unit mapping and AMA dating is urgent for the analysis of data acquired by the CE-4 rover. In addition, because there are many ejecta in this area, CE-4 rover would probably encounter the fragments from the maria basalts nearby. Thus, the AMAs of them are also necessary to understand the whole history of the basalt formation in this area.

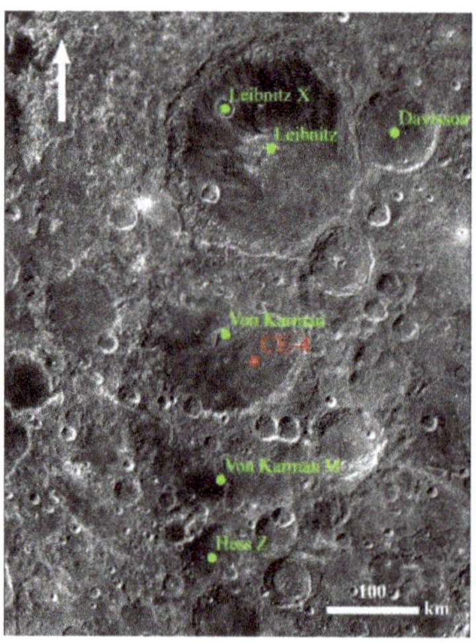

Figure 1. Geographical context of the CE4 landing area. The central location is (177.07° E, 43.19° S). The basemap is Chang'e-2 (CE-2) Digital Orthophoto Map (DOM) mosaic. The crater names are from the International Astronomical Union data base (https://planetarynames.wr.usgs.gov/, accessed on 15 November 2021). North is up.

In this research, nine basalt geological units were mapped in the CE-4 landing area according to the difference of brightness, FeO content, and the topography. Superposed craters in the counting areas used for deriving the AMAs of the above geological units were mapped in mosaics generated from the Chang'e-2 (CE-2) digital orthophoto map (DOM) [22]. The AMAs of all the nine measurement areas were obtained with the CSFD method.

2. Materials and Methods

2.1. Data

Craters for dating AMAs were mapped in a DOM mosaic generated from CE-2 orbital images. CE-2 mission was China's second lunar spacecraft and it was launched on 1 October 2010. The onboard stereo camera obtained high-resolution images, and the returned data were processed with radiometric, geometric, and photometric calibration [22]. A DOM was finally generated with a resolution of ~7 m/pixel and divided into multiple regions. The data are now publicly available through the Data Release and Information Service System of China's Lunar Exploration Program (http://moon.bao.ac.cn, accessed on 15 November 2021). Twelve DOM files were downloaded and merged to produce a DOM mosaic in ArcGIS®.

The merged digital elevation model from the Lunar Orbiter Laser Altimeter and SELENE Terrain Camera (designated as "SLDEM") [23] was used for topographic analysis of the geological context surrounding the CE-4 landing site. The product has a typical vertical accuracy of ~3–4 m and a resolution of ~42 m/pixel around the CE-4 landing area [23]. To reveal the topographic detail of the region, the SLDEM was color coded and was registered.

The FeO content product derived from the Kaguya Multiband Imager data [24] is also used to evaluate the boundaries of maria basalt units in this research. The product is released with a spatial resolution of ~60 m/pixel between 50° N and 50° S and nine maria basalts units studied in this research are included.

2.2. Methods

The process of determining the AMA of the geological unit includes mapping the surface unit and identifying the superposed craters. The AMAs can be derived by comparing the CSFD of the counted craters with the CF [15]. Here, we present the principles of mapping the measurement areas and the crater counting method. We used Craterstats (http://www.geoinf.fu-berlin.de/mex4/software/craterstats2/, accessed on 15 November 2021) to derive the AMA of each geological unit [25].

2.2.1. Maria Basalt Measurement Areas Mapping

It only makes sense to assess the AMA of a surface with the same geological history, i.e., a geologically homogeneous region that had undergone the same geological processes [25]. Significant interruptions to the homogeneity should be considered as another geological unit. For the maria basalt in this study, this means only one episode of basaltic volcanism should be emplaced in each geological unit.

In maria units mapping, the brightness differences, spectral ratios, FeO and TiO_2 contents, and morphology have usually been used to ensure homogeneity of the investigated basalts, e.g., [17–19,26–28]. In this study, we combined the derived FeO content, brightness difference from CE-2 DOM mosaic, and topographical difference from SLDEM to map the maria basalt units. The mapped measurement areas are the intersection of these above products if they are inconsistent. In addition, areas with identifiable secondary craters should be excluded in the geological mapping [25]. Finally, nine maria basalt units were outlined see below.

2.2.2. Crater Counting

The craters superimposed on the measurement areas have been mapped on the CE-2 DOM mosaics with the ArcGIS add-in "CraterTools", which measures crater diameter correctly regardless of map projection [29]. CraterTools can work well with three reference points on the crater rims, even for craters where a part of the rim is obscured or even absent. When manually digitizing a crater, the three points on the crater rim should be as evenly distributed as possible, which can give the minimum error [30]. The secondary craters should be excluded in the dating process [31,32], and they are usually recognized in chains or clusters with irregular shape. This judgement is inevitably subjective, and we have conducted an analysis of spatial randomness and clustering of the mapped craters superimposed on each geological unit following the method of mean closest two neighbors' distance (MC2ND) [32].

3. Results

3.1. Maria Basalt Units in Von Kármán Crater

The floor of Von Kármán crater (Figure 2) was mapped as mare basalts by [33], and the crater floor has significantly higher FeO contents than the surrounding terrain (Figure 2C). However, the crater floor is extensively overprinted by ejecta from nearby craters [34], making the FeO content much lower in the northern and eastern part of the crater floor. In addition, the fresh Zhinyu crater obviously excavated mare basalt deposits from the subsurface resulted in much higher FeO content (Figure 2C). Thus, we selected a small area labeled as 'a' in Figure 2 to date the maria basalt in the Von Kármán crater floor (Figure 2A,D). Figure 2B shows two local upland structures in the SLDEM, and they are more clearly shown in Figure 2E,F, respectively. In this research, they are mapped as units 'b' and 'c' for subsequent surface dating.

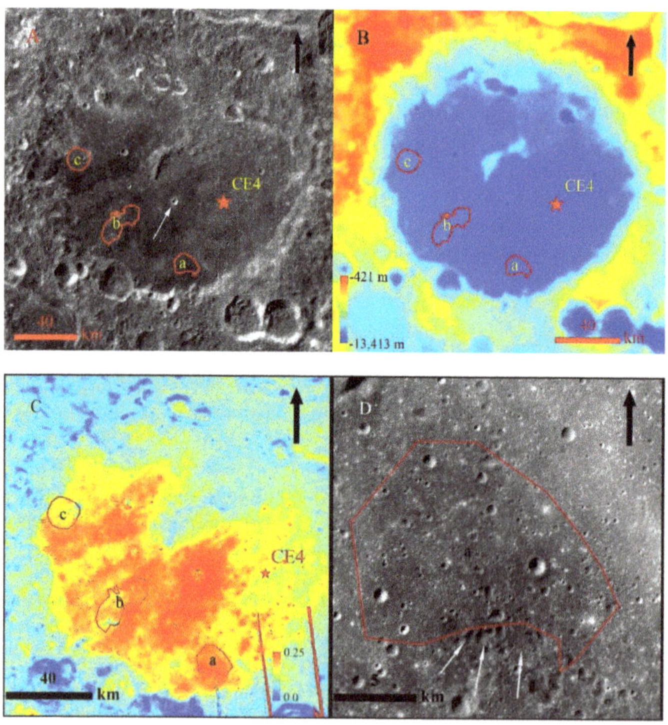

Figure 2. *Cont.*

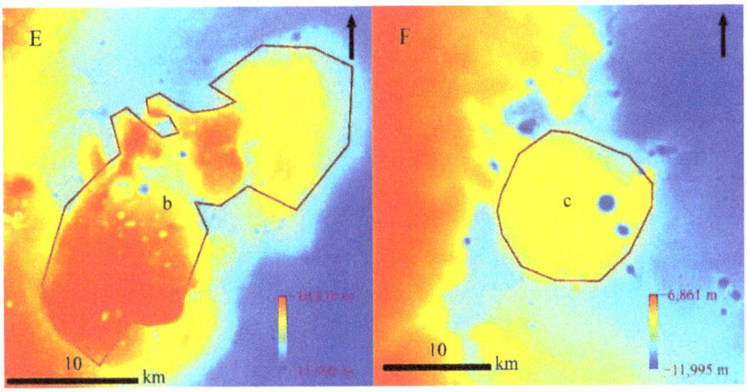

Figure 2. Three basalt measurement areas mapped by red polygons in Von Kármán crater. (**A**) CE-2 DOM mosaic (central location: 175.6° E, 44.64° S) of the Von Kármán crater with CE-4 landing site labeled. The fresh crater pointed by the white arrow is Zhinyu crater. (**B**) Local topography shown in SLDEM; the elevation ranges from −13,413 m to −421 m as coded from blue to red in this figure. (**C**) FeO content map [24] in Von Kármán crater; the values range from 0 to 24.73% as coded from blue to red in this figure. (**D**) Basalt unit 'a' shown in CE-2 DOM mosaic, with the white arrows indicating the chained secondaries. The area is 138 km^2. (**E**) Basalt unit 'b' is located in a topographical elevated terrain, and the elevation ranges from −11,990 m to −10,839 m as coded from blue to red in this figure. The area is 267 km^2. (**F**) Basalt unit 'c' is located in a topographical elevated terrain, and the elevation ranges from −11,995 m to −6861 m as coded from blue to red in this figure. The red polygon in each sub-figure is the outlined representative crater count area. The area is 151 km^2. North is up in all sub-figures.

Our CSFD measurements of the mare units inside Von Kármán crater show AMAs of 3.42 Ga, 3.34 Ga, and 3.34 Ga for the basalt units 'a', 'b', and 'c', respectively (Figure 3). The analysis results of spatial randomness and clustering of the mapped craters are on the top of each group of CSFD fitting curve. It is clear that, for the crater diameters used for surface dating, almost each of the MC2ND of the mapped craters falls within $\pm 3\sigma$ of the expectation value derived from Monte Carlo simulation, indicating that the spatial distribution of the mapped craters is random. Figure 3 also includes the areas of the counting regions, with the values ranging from 138 km^2 to 267 km^2.

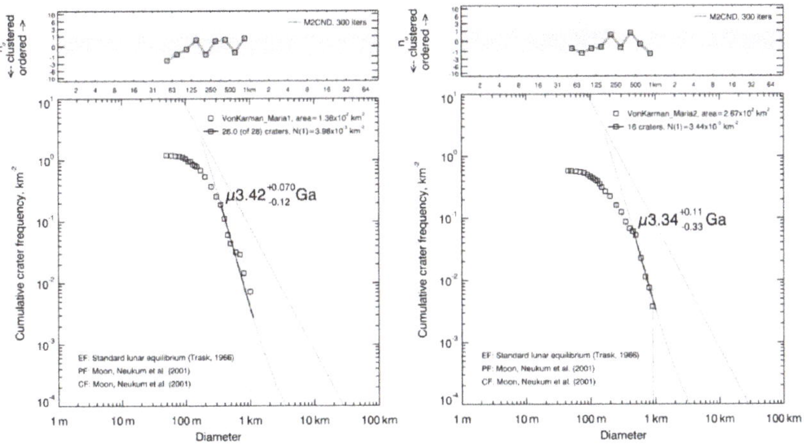

Figure 3. *Cont.*

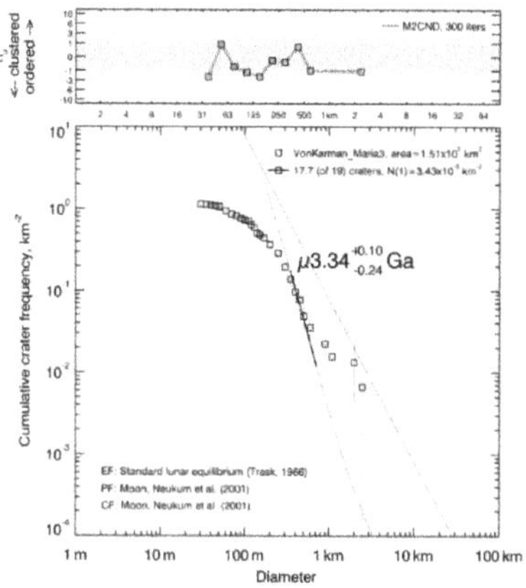

Figure 3. Dating results for basalt units in Von Kármán crater. VonKarman_Maria1, VonKarman_Maria2, and VonKarman_Maria3 correspond to the basalt units 'a', 'b', and 'c' in Figure 2, respectively. The equilibrium function is the standard lunar equilibrium of [35], while both the production and chronology functions are from [36].

3.2. Maria Basalt Units in Von Kármán M Crater

Most of the floor of Von Kármán M crater is superposed by Von Kármán crater, and the exposed southern half of the crater is filled with mare basalts [33]. The eastern part of the floor is covered by the ejecta from Alder crater (Figure 1), and there are several massifs protruding from the basalt units (Figure 4A). Similar to [21], two basalt units are mapped in this study (Figure 4), in which the raised massifs are excluded. The basalt units are relatively flat, and the topography gradually rises up from southeast to northwest (Figure 4B). The FeO contents of the basalt units are obviously much higher than surrounding terrains (Figure 4C).

The CSFD measurements of the mare units in Von Kármán M crater show AMAs of 2.36 Ga and 2.28 Ga for the basalt units 'd' and 'e', respectively (Figure 5). For the basalt unit 'd' (counting area is 2320 km^2), the minimum result of the random analysis is about -6σ of the expectation value derived from Monte Carlo simulation at the diameter of ~144 m, and the random analysis results for D >= ~200 m are within $\pm3\sigma$ of the expectation value. The minimum diameter for fitting the isochron line is at ~400 m in this study, which thus does not affect the reliability of the dating results. For the basalt unit 'e' (counting areas is 59.3 km^2), the analysis results of spatial randomness and clustering of the mapped craters fall within $\pm3\sigma$ of the expectation value derived from Monte Carlo simulation for all the diameters. The AMA difference (0.08 Ga) between the two basalt units is small, indicating the basalt in these two areas are probably from one episode.

3.3. Maria Basalt Unit in Hess Z Crater

Figure 6 shows the mapped basalt unit (counting area is 530 km^2) and the derived AMA in Hess Z crater. The southeastern part of the Hess Z crater rim was destroyed by the ejecta from the Hess crater nearby, which was superposed by the basalt herein. There are several clustered secondaries in the eastern floor of the Hess Z crater, which are deliberately avoided when map the countering area. The mapped basalt unit is relatively flat, although the topography is gradually lowered from east to west.

The CSFD measurement of the mare unit in Davisson crater shows an AMA of 3.27 Ga (Figure 8). For the craters used for dating, the MC2ND of the mapped craters falls within $\pm3\sigma$ of the expectation value derived from Monte Carlo simulation, indicating mapped craters in the counting area are random in spatial distribution.

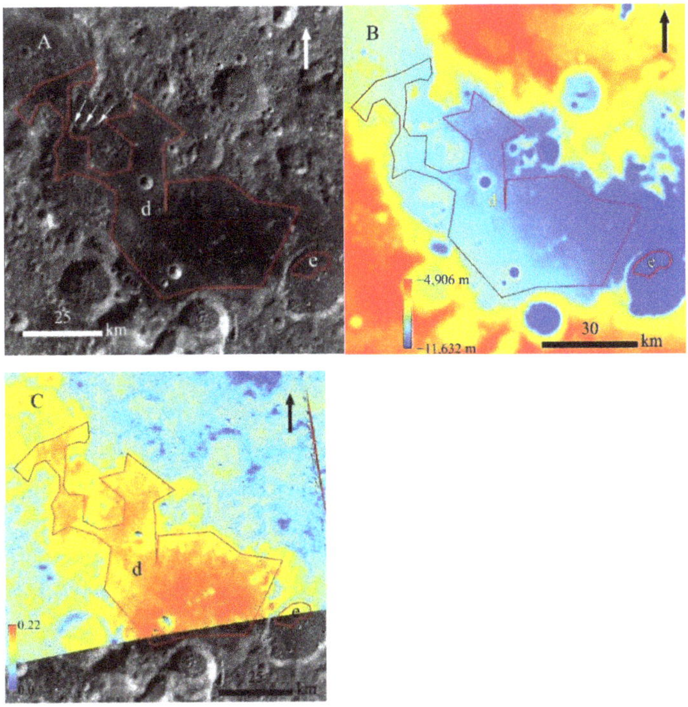

Figure 4. Two basalt measurement areas d (2320 km^2) and e (59 km^2) mapped by red polygons in Von Kármán M crater. (**A**) CE-2 DOM mosaic (central location: 173.6° E, 48.76° S) of the Von Kármán M crater. The raised massifs are indicated by the white arrows. (**B**) Local topography shown in SLDEM, and the elevation ranges from −11,632 m to −4906 m as coded from blue to red in this figure. (**C**) FeO content map [24] in Von Kármán M crater, and the maximum value is up to 22.28% as coded from blue to red in this figure. The red polygon in each sub-figure is the outlined representative crater count area. North is up in all sub-figures.

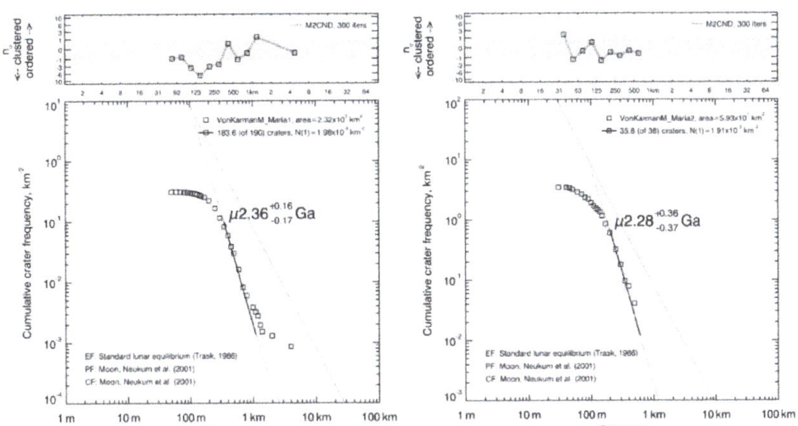

Figure 5. Dating results for basalt units in Von Kármán M crater. VonKarmanM_Maria1 and VonKarmanM_Maria2 correspond to the basalt units 'd' and 'e' in Figure 4. The equilibrium function is the standard lunar equilibrium of [35], while both the production and chronology functions are from [36].

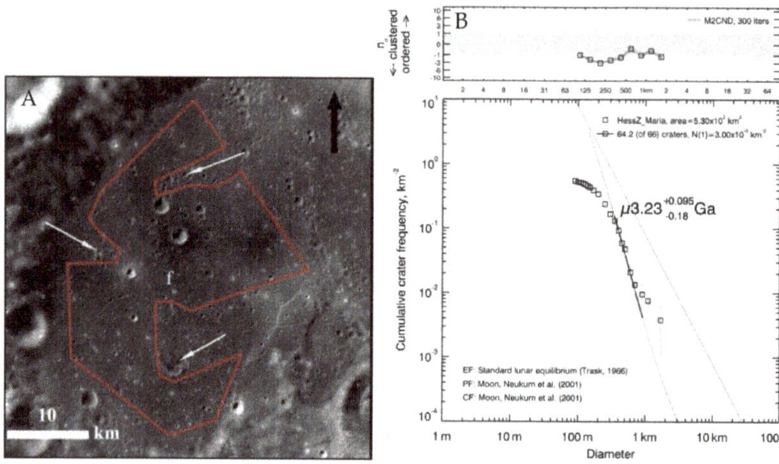

Figure 6. (**A**) CE-2 DOM mosaic of the mapped basalt unit in Hess Z crater (530 km^2) (central location: 173.49° E, 51.94° S); North is up. (**B**) derived AMA. The equilibrium function is the standard lunar equilibrium of [35], while both the production and chronology functions are from [36]. The red polygon in the left sub-figure is the outlined representative crater count area, and the arrows indicate the excluded secondary craters.

The CSFD measurement of the mare unit in Hess Z crater shows an AMA of 3.23 Ga (Figure 6). The mapped craters are randomly distributed in the counting area, as indicated by the spatial randomness and clustering analysis result (Figure 6).

3.4. Maria Basalt Unit in Davisson Crater

Most of the floor of Davisson crater is covered by basalt, as indicated by the high FeO content (Figure 7C). A central peak protrudes above the floor and several chained ejecta are distributed in the southeastern of the crater floor. They are deliberately excluded when mapping the counting area (Figure 7A). The mapped basalt unit is relatively flat, and it is lowest in the southern part of the floor (Figure 7B).

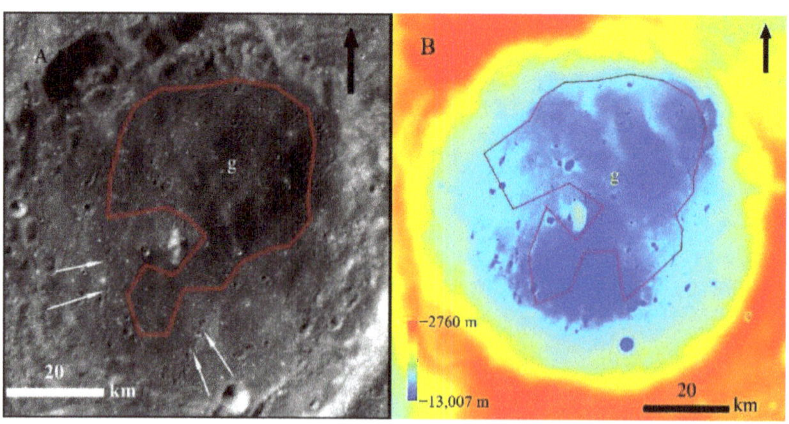

Figure 7. *Cont.*

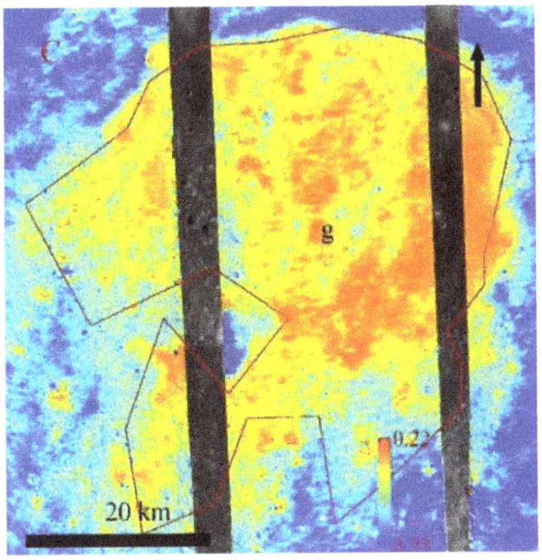

Figure 7. (**A**) CE-2 DOM mosaic (central location: 174.88° W, 37.52° S) of the mapped basalt unit (1280 km^2) in Davisson crater. The central peak and clustered secondaries (arrows) are excluded when map the counting area. (**B**) Local topography shown in SLDEM, and the elevation ranges from −13,007 m to −2760 m as coded from blue to red in this figure. (**C**) FeO content map [24] of the counting area, and the values range from 1.94% to 21.94% as coded from blue to red in this figure. The red polygon in each sub-figure is the outlined representative crater count area. North is up in all sub-figures.

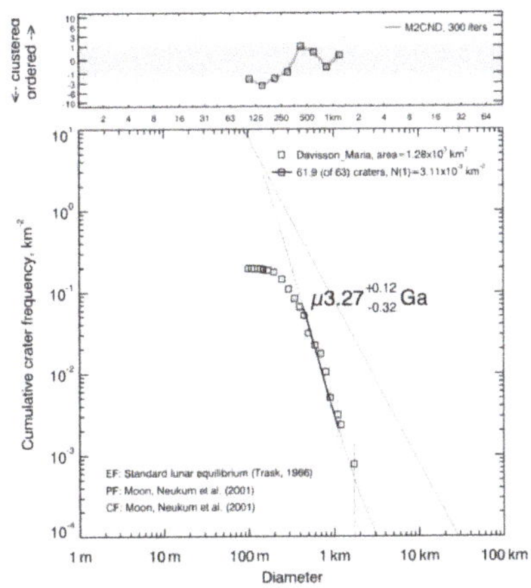

Figure 8. Dating results for basalt unit mapped in Figure 7. The equilibrium function is the standard lunar equilibrium of [35], while both the production and chronology functions are from [36].

3.5. Maria Basalt Units in Leibnitz and Leibnitz X Crater

The southeast part of the Leibnitz crater is superposed by the ejecta from Finsen crater (Figure 1), and a large amount of linear ejecta are distributed in the floor of Leibnitz crater (Figure 9A). Thus, we carefully selected a small area in the north part of the crater floor as the counting area, which is covered by basalts as proven by the much higher FeO content than surrounding ejecta (Figure 9C). It is interesting to note that the north part is higher than the south part in the floor of Leibnitz crater (Figure 9B), although the ejecta is from south.

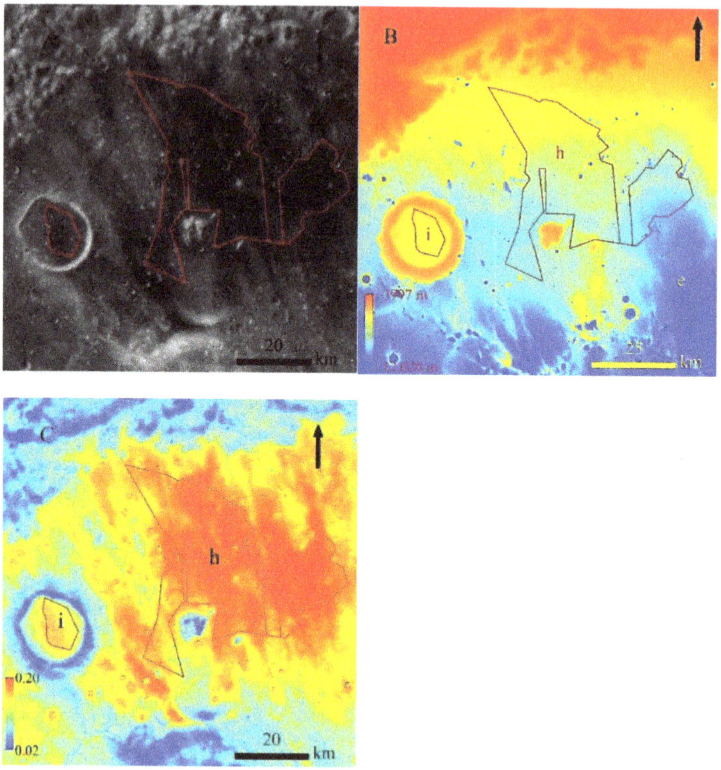

Figure 9. (**A**) CE-2 DOM mosaic (central location: 178.91° E, 36.23° S) of the mapped basalt units 'h' (1450 km^2) and 'i' (127 km^2) in Leibnitz and Leibnitz X craters. The central peak and clustered secondaries are excluded in the counting area. (**B**) Local topography shown in SLDEM, and the elevation ranges from −12,020 m to −3997 m as coded from blue to red in this figure. (**C**) FeO content map [24] of the counting areas, and the values range from 2.27% to 19.69% as coded from blue to red in this figure. The red polygons in each sub-figure are the outlined representative crater count areas. North is up in all sub-figures.

Leibnitz X crater is in the northwest of Leibnitz crater floor, and the floor of Leibnitz X crater can be divided into two parts according to the brightness difference (Figure 9A). We selected the east part of the floor as the counting area, which also has a higher FeO content than the west part (Figure 9C). The selected counting area seems unaffected by the linear ejecta from the Finsen crater. It is more interesting to note that the whole Leibnitz X crater, including its floor, is much higher than the surrounding terrains (Figure 9B). In addition, the FeO content of the maria in the counting area for Leibnitz X crater (Figure 9C) is lower than that for Leibnitz crater, indicating they are probably from different sources.

The CSFD measurements of the mare units in Leibnitz and Leibnitz X craters show AMAs of 2.61 Ga and 3.05 Ga (Figure 10). For the basalt unit in Leibnitz crater (i.e., unit 'h' in Figure 9), the random analysis is about -6σ of the expectation value at the diameter of 144 m, and the random analysis results for D >= ~400 m are within $\pm 3\sigma$ of the expectation value. Therefore, this again does not affect the reliability of our dating results, as the minimum diameter for fitting the isochron line is at ~400 m. For all the counted craters in the mare unit in Leibnitz X crater (i.e., unit 'i' in Figure 9), the spatial randomness and clustering analysis result of the mapped craters falls within $\pm 3\sigma$ of the expectation value derived from Monte Carlo simulation. It is obvious that the basalt unit in Leibnitz X crater is older than that in Leibnitz crater. Given this, coupled with the difference in FeO content, it is reasonable to conclude that they formed in two episodes of basalt formation.

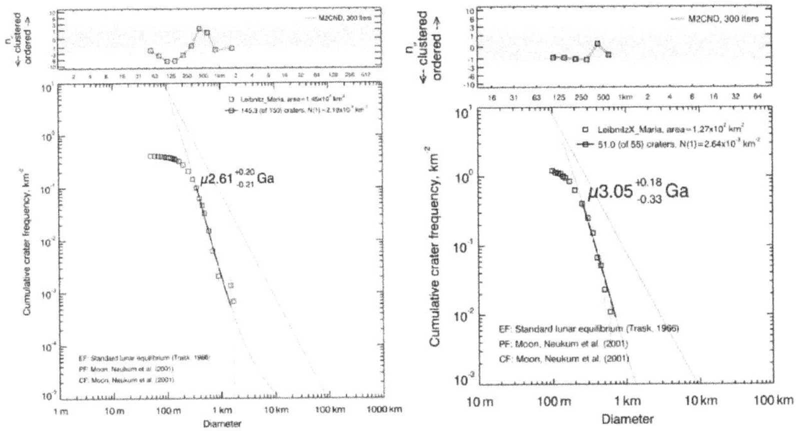

Figure 10. Dating results for the basalt units mapped in Figure 9. Leibnitz_Maria and LeibnitzX_Maria correspond to the basalt units 'h' and 'i' in Figure 9. The equilibrium function is the standard lunar equilibrium of [35], while both the production and chronology functions are from [36].

3.6. *Overview for AMAs of Basalt Units Surrounding CE-4 Landing Site*

Table 1 lists all the derived AMAs for the basalt units mapped in this study, and the N(1), i.e., the number of craters with a diameter larger than or equal to 1 km per km^2, derived by fitting a polynomial proposed in [36] to observed CSFDs, is also included for convenient comparison with other studies. The youngest basalt unit (i.e., unit 'e') is in the Von Kármán M crater, and unit 'd' in Von Kármán M crater is the second youngest among the studied basalt units. This perhaps certificates that the basalt inside Von Kármán M crater formed later than that in other places in the study region. Basalt unit 'h' in Leibnitz crater floor is the third youngest among the basalt units, and they are the only three basalt units younger than 3.0 Ga in this study. The oldest basalt unit 'a' is in the Von Kármán crater floor, which has an AMA of 3.42 Ga, and the three basalt units in the Von Kármán crater floor are the three oldest in this study. The similarity of the derived AMAs of the three basalt units inside Von Kármán crater indicates that the basalt should be first emplaced in the Von Kármán crater floor in this area. The second oldest basalt unit is in Davisson crater floor, and the above four basalt units belong to the Late Imbrian while others belong to Eratosthenian, according to [26].

Table 1. Results of AMAs for the groups of basalt units in this study.

Basalt Unit	Location	N(1) ($\times 10^{-3}$ km^{-2})	AMAs (Ga)	
			This Study	Previous Studies [1]
a	Von Kármán crater floor	3.98	3.42	3.35 *; E–I †; 3.6 ‡; 3.15 #;
b		3.44	3.34	
c		3.43	3.34	
d	Von Kármán M crater floor	1.98	2.36	3.38 ##; E–I †; 3.38 **
e		1.91	2.28	3.52 ##; E–I †; 3.38 **
f	Hess Z crater floor	3.00	3.23	E–I †; 3.51 ##

Table 1. *Cont.*

Basalt Unit	Location	N(1) ($\times 10^{-3}$ km^{-2})	AMAs (Ga)	
			This Study	Previous Studies [1]
g	Davisson crater floor	3.11	3.27	-
h	Leibnitz crater floor	2.19	2.61	3.37 [###];
i	Leibnitz X crater floor [1]	2.64	3.05	3.68 [##];

[1] Ref. [20] used the crater chronology and production functions proposed by [15], ref. [7,21] used the crater chronology and production functions proposed by [36]; * ref. [20] in which the maria in Von Kármán crater floor is measured together; ** ref. [20] in which the maria in Von Kármán M crater floor is measured together; † ref. [26] E–I refers to near the Imbrian-Eratosthenian boundary, i.e., 3.26 Ga; ‡ ref. [7] in which the maria in Von Kármán crater floor is measured by combing three subareas, while subarea 2 is approximately the same to the basalt unit 'a' in this research; # ref. [21] in which the maria in Von Kármán crater floor is measured by combing two subareas, which contain the counting areas of basalt units 'a', 'b', and 'c' in this research; ## ref. [21] in which the counting areas are roughly equivalent to that in this research; ### ref. [21] in which the counting areas are much larger than that in our research.

4. Discussion

4.1. Comparison with Previous Studies

Table 1 also lists the dating results in previous studies. For the maria in Von Kármán crater, the dated AMAs range from 3.34 to 3.42 Ga for the three counting areas mapped in our research. The results are similar to the previous results (3.26 Ga for [26]; 3.35 Ga for [20]; 3.6 Ga for [7]; 3.15 Ga for [21]). Thus, we have much confidence to analyze the data returned by CE-4 rover, which is now traversing to this area. For the results of maria in Hess Z crater, there is also not too much difference (0.28 Ga) [21,26]. However, there is slight difference of 0.63 Ga in the case of Leibnitz X crater, and large differences of ~1.0 Ga [20,21,26] in the cases of Von Kármán M crater and Leibnitz crater. For the AMA difference between this research and that of [21], given similar counting areas are mapped, the possible reason might be different imagery data used. In our study, the CE-2 DOM mosaic at a resolution of 7 m/pixel was used, while the wide angle camera (WAC) images (~100 m/pixel) from the Lunar Reconnaissance Orbiter (LRO) were used in [21]. In fact, ref. [21] also found that their dating AMAs based on the WAC images are indeed much older than those based on narrow angle camera (NAC) images (~1 m/pixel) from the LRO for the maria basalts in Antoniadi crater. For the difference of the derived AMAs between this research and that of [20], given similar image resolution used in the two studies, the most probable reason is the different locations and sizes of the counting areas. We cannot find the counting area of [20]. However, ref. [20] usually mapped the basalt in the floor of Von Kármán M crater and Leibnitz crater as one unit. If the basalt was formed through different episodes, this is unsuitable. In addition, we obtained the AMA of the maria in the floor of Davisson crater for the first time.

In early geological mapping [26], most mare deposits within SPA were mapped as late Imbrian. Ref. [20] first showed that some mare deposits in the lunar farside may have much younger ages up to Eratosthenian. Ref. [37] also proposed that the lunar farside was volcanically active for nearly as long as the lunar nearside. Our dating results surrounding the CE-4 landing area support this conclusion and indicate that small-scale lunar farside volcanism might have continued longer than previously thought.

4.2. Implications for CE-4 Mission

The dating results for the maria basalts inside Von Kármán crater are significant to the interpretation of CE-4 data. It is interesting to note that AMAs of the three basalt units are very similar, indicating that the basalts in Von Kármán crater were formed within a short period. CE-4 landed on linear impact ejecta (Figures 1 and 2), and Finsen crater was the major source [7,9–11,34]. The thickness of the ejecta might be several tens of meters in the landing area [10,11,34,38]. However, the ejecta here appear linear and are not continuous [34], and CE-4 rover may probably happen to encounter local mare basalt when it traverses westward, especially around new craters (e.g., Zhinyu crater) which can excavate subsurface material [39]. If CE-4 rover will traverse the exposed mare basalt in future, the dating results of the mare basalt inside the Von Kármán crater can provide information for analyzing the lunar regolith evolution and crater degradation, combing the spectral data and lunar ground penetrating radar data. For example, according to the data collected by the ground-penetrating radar onboard the CE-4 rover, several mare basalt layers were interpreted by [10,11]. The ejecta from Zhinyu crater also reveal at least three episodes of the basalt eruption [39], and the dating results can help to provide chronological information.

The dating results for other mare basalt is also useful in relevant studies. For example, through spatial density investigation and randomness analysis, ref. [40] pointed out that secondary craters cover at least one third of the 2.4° × 1° area surrounding CE-4 landing site. Therefore, it is very likely that CE-4 rover would encounter the fragments from the mare basalts inside Von Kármán crater dated in the manuscript, and the results in the research can provide geological context to these fragments. For example, the compositions of the fragments can be analyzed by the CE-4 rover, and their sources can thus be determined by comparing the compositions with the dated mare basalts, whose formation times have been constrained in this study.

5. Conclusions

In this research, we dated the AMAs of nine basalt measurement areas surrounding the CE-4 landing area using the crater counting method, including the mare deposits in Von Kármán, Von Kármán M, Hess Z, Davisson, Leibnitz, and Leibnitz X craters. Our results indicate that the youngest mare deposit is in the floor of Von Kármán M crater with AMAs of 2.28 Ga and 2.36 Ga. This research provides both systematic information on the history of the basalt in the region and constraints for the interpretation of the in situ measured data from the CE-4 rover. The results point to a much longer-lived volcanism in the region than was previously thought (e.g., ref. [26] proposed that most mare deposits in the SPA basin were formed during the late Imbrian epoch, i.e., 3.85–3.8 Ga.), which is also beneficial to the understanding of the evolution of lunar volcanism globally. We suggest that CE-4 rover may return data on local genetic basalts while it keeps moving westward, which will contribute to our understanding of basalt formation in the lunar farside.

Author Contributions: Conceptualization, Z.Y., K.D. and S.G.; methodology, Z.Y. and K.S.; software, Z.Y. and G.M.; investigation, Z.Y., K.S., S.G. and J.L.; data curation, S.N. and S.G.; writing—original draft preparation, Z.Y. and S.G.; writing—review and editing, Z.Y., G.M. and K.D.; funding acquisition, Z.Y., K.D., J.L. and S.G. All authors have read and agreed to the published version of the manuscript.

Funding: This research was supported by the Strategic Priority Research Program of Chinese Academy of Sciences (grant No. XDB41000000) and National Natural Science Foundation of China (grant Nos. 41972321 and 41941003), Macao Young Scholars Program (grant No. AM201902).

Institutional Review Board Statement: Not applicable.

Informed Consent Statement: Not applicable.

Data Availability Statement: All the mapped counting areas and craters are available as shapefiles in the Supporting Online Material.

Acknowledgments: The authors gratefully acknowledge China's Lunar Exploration Program at National Astronomical Observatories of Chinese Academy of Sciences for providing CE-2 DOM data (http://moon.bao.ac.cn, accessed on 15 November 2021).

Conflicts of Interest: The authors declare no conflict of interest.

References

1. Wu, W.; Li, C.; Zuo, W.; Zhang, H.; Liu, J.; Wen, W.; Su, Y.; Ren, X.; Yan, J.; Yu, D.; et al. Lunar farside to be explored by Chang'e-4. *Nat. Geosci.* **2019**, *12*, 222–223. [CrossRef]
2. Di, K.; Liu, Z.; Liu, B.; Wan, W.; Peng, M.; Wang, Y.; Gou, S.; Yue, Z.; Xin, X.; Jia, M.; et al. Chang'e-4 lander localization based on multi-source data. *J. Remote Sens.* **2019**, *23*, 177–184. [CrossRef]
3. Garrick-Bethell, I.; Zuber, M.T. Elliptical structure of the lunar South Pole-Aitken basin. *Icarus* **2009**, *204*, 399–408. [CrossRef]
4. Jolliff, B.L.; Gillis, J.J.; Haskin, L.A.; Korotev, R.L.; Wieczorek, M.A. Major lunar crustal terranes: Surface expressions and crust-mantle origins. *J. Geophys. Res. Planets* **2000**, *105*, 4197–4216. [CrossRef]
5. Melosh, H.J.; Kendall, J.; Horgan, B.; Johnson, B.C.; Bowling, T.; Lucey, P.G.; Taylor, G.J. South Pole–Aitken basin ejecta reveal the Moon's upper mantle. *Geology* **2017**, *45*, 1063–1066. [CrossRef]
6. Moriarty, D.P., III; Pieters, C.M. The Character of South Pole-Aitken Basin: Patterns of Surface and Subsurface Composition. *J. Geophys. Res. Planets* **2018**, *123*, 729–747. [CrossRef]
7. Huang, J.; Xiao, Z.; Flahaut, J.; Martinot, M.; Head, J.; Xiao, X.; Xie, M.; Xiao, L. Geological Characteristics of Von Kármán Crater, Northwestern South Pole-Aitken Basin: Chang'E-4 Landing Site Region. *J. Geophys. Res. Planets* **2018**, *123*, 1684–1700. [CrossRef]
8. Li, C.; Liu, D.; Liu, B.; Ren, X.; Liu, J.; He, Z.; Zuo, W.; Zeng, X.; Xu, R.; Tan, X.; et al. Chang'E-4 initial spectroscopic identification of lunar far-side mantle-derived materials. *Nature* **2019**, *569*, 378–382. [CrossRef]
9. Gou, S.; Di, K.; Yue, Z.; Liu, Z.; He, Z.; Xu, R.; Lin, H.; Liu, B.; Peng, M.; Wan, W.; et al. Lunar deep materials observed by Chang'e-4 rover. *Earth Planet. Sci. Lett.* **2019**, *528*, 115829. [CrossRef]
10. Zhang, J.; Zhou, B.; Lin, Y.; Zhu, M.-H.; Song, H.; Dong, Z.; Gao, Y.; Di, K.; Yang, W.; Lin, H.; et al. Lunar regolith and substructure at Chang'E-4 landing site in South Pole–Aitken basin. *Nat. Astron.* **2021**, *5*, 25–30. [CrossRef]
11. Lai, J.; Xu, Y.; Bugiolacchi, R.; Meng, X.; Xiao, L.; Xie, M.; Liu, B.; Di, K.; Zhang, X.; Zhou, B.; et al. First look by the Yutu-2 rover at the deep subsurface structure at the lunar farside. *Nat. Commun.* **2020**, *11*, 3426. [CrossRef]
12. Zhang, L.; Li, J.; Zeng, Z.; Xu, Y.; Liu, C.; Chen, S. Stratigraphy of the Von Kármán Crater Based on Chang'E-4 Lunar Penetrating Radar Data. *Geophys. Res. Lett.* **2020**, *47*, e2020GL088680. [CrossRef]
13. Lin, H.; Xu, R.; Yang, W.; Lin, Y.; Wei, Y.; Hu, S.; He, Z.; Qiao, L.; Wan, W. In Situ Photometric Experiment of Lunar Regolith With Visible and Near-Infrared Imaging Spectrometer On Board the Yutu-2 Lunar Rover. *J. Geophys. Res. Planets* **2020**, *125*, e2019JE006076. [CrossRef]
14. Stöffler, D.; Ryder, G.; Ivanov, B.A.; Artemieva, N.A.; Cintala, M.J.; Grieve, R.A.F. Cratering History and Lunar Chronology. *Rev. Mineral. Geochem.* **2006**, *60*, 519–596. [CrossRef]
15. Neukum, G. Meteoriten Bombardement und Datierung Planetarer Oberflächen. Habilitation Dissertation, University of Munich, Munich, Germany, 1983.
16. Stöffler, D.; Ryder, G. Stratigraphy and Isotope Ages of Lunar Geologic Units: Chronological Standard for the Inner Solar System. *Space Sci. Rev.* **2001**, *96*, 9–54. [CrossRef]
17. Hiesinger, H.; Jaumann, R.; Neukum, G.; Head, J.W., III. Ages of mare basalts on the lunar nearside. *J. Geophys. Res. Planets* **2000**, *105*, 29239–29275. [CrossRef]
18. Hiesinger, H.; Head, J.; Wolf, U.; Jaumann, R.; Neukum, G. Ages and stratigraphy of mare basalts in Oceanus Procellarum, Mare Nubium, Mare Cognitum, and Mare Insularum. *J. Geophys. Res. Planets* **2003**, *108*, 5065. [CrossRef]
19. Hiesinger, H.; Head, J.W., III; Wolf, U.; Jaumann, R.; Neukum, G. Ages and stratigraphy of lunar mare basalts in Mare Frigoris and other nearside maria based on crater size-frequency distribution measurements. *J. Geophys. Res. Planets* **2010**, *115*, E03003. [CrossRef]
20. Haruyama, J.; Ohtake, M.; Matsunaga, T.; Morota, T.; Honda, C.; Yokota, Y.; Abe, M.; Ogawa, Y.; Miyamoto, H.; Iwasaki, A.; et al. Long-Lived Volcanism on the Lunar Farside Revealed by SELENE Terrain Camera. *Science* **2009**, *323*, 905–908. [CrossRef]
21. Pasckert, J.H.; Hiesinger, H.; van der Bogert, C.H. Lunar farside volcanism in and around the South Pole–Aitken basin. *Icarus* **2018**, *299*, 538–562. [CrossRef]
22. Zuo, W.; Li, C.; Zhang, Z. Scientific data and their release of Chang'E-1 and Chang'E-2. *Chin. J. Geochem.* **2014**, *33*, 24–44. [CrossRef]
23. Barker, M.K.; Mazarico, E.; Neumann, G.A.; Zuber, M.T.; Haruyama, J.; Smith, D.E. A new lunar digital elevation model from the Lunar Orbiter Laser Altimeter and SELENE Terrain Camera. *Icarus* **2016**, *273*, 346–355. [CrossRef]
24. Lemelin, M.; Lucey, P.G.; Gaddis, L.R.; Hare, T.; Ohtake, M. Global map products from the Kaguya Multiband Imager at 512 ppd: Minerals, FeO and OMAT. In Proceedings of the 47th Lunar and Planetary Science Conference, The Woodlands, TX, USA, 21–25 March 2016; p. 2994.
25. Michael, G.G.; Neukum, G. Planetary surface dating from crater size–frequency distribution measurements: Partial resurfacing events and statistical age uncertainty. *Earth Planet. Sci. Lett.* **2010**, *294*, 223–229. [CrossRef]
26. Wilhelms, D.E.; McCauley, J.F.; Trask, N.J. *The Geologic History of the Moon*; U.S. Geological Survey Professional Paper 1348; U.S. Geological Survey: Denver, CO, USA, 1987. [CrossRef]
27. Nelson, D.; Koeber, S.; Daud, K.; Robinson, M.; Watters, T.; Banks, M.; Williams, N. Mapping Lunar Maria Extents and Lobate Scarps Using LROC Image Products. In Proceedings of the 45th Lunar and Planetary Science Conference, The Woodlands, TX, USA, 17–21 March 2014; p. 2861.

28. Ivanov, M.A.; Hiesinger, H.; van der Bogert, C.H.; Orgel, C.; Pasckert, J.H.; Head, J.W. Geologic History of the Northern Portion of the South Pole-Aitken Basin on the Moon. *J. Geophys. Res. Planets* **2018**, *123*, 2585–2612. [CrossRef]
29. Kneissl, T.; van Gasselt, S.; Neukum, G. Map-projection-independent crater size-frequency determination in GIS environments—New software tool for ArcGIS. *Planet. Space Sci.* **2011**, *59*, 1243–1254. [CrossRef]
30. Yue, Z.; Di, K.; Liu, Z.; Michael, G.; Jia, M.; Xin, X.; Liu, B.; Peng, M.; Liu, J. Lunar regolith thickness deduced from concentric craters in the CE-5 landing area. *Icarus* **2019**, *329*, 46–54. [CrossRef]
31. McEwen, A.S.; Bierhaus, E.B. The Importance of Secondary Cratering to Age Constraints on Planetary Surfaces. *Annu. Rev. Earth Planet. Sci.* **2006**, *34*, 535–567. [CrossRef]
32. Michael, G.G.; Platz, T.; Kneissl, T.; Schmedemann, N. Planetary surface dating from crater size–frequency distribution measurements: Spatial randomness and clustering. *Icarus* **2012**, *218*, 169–177. [CrossRef]
33. Stuart-Alexander, D.E. *Geologic Map of the Central Far Side of the Moon*; U.S. Geological Survey IMAP 1047; U.S. Geological Survey: Denver, CO, USA, 1978. [CrossRef]
34. Di, K.; Zhu, M.-H.; Yue, Z.; Lin, Y.; Wan, W.; Liu, Z.; Gou, S.; Liu, B.; Peng, M.; Wang, Y.; et al. Topographic evolution of Von Kármán crater revealed by the lunar rover Yutu-2. *Geophys. Res. Lett.* **2019**, *46*, 12764–12770. [CrossRef]
35. Trask, N.J. Size and spatial distribution of craters estimated from the ranger photographs. In *Ranger VIII and IX, Part II-Experimenters' Analyses and Interpretations*; Technical Report No. 32-800; Jet Propulsion Laboratory, California Institute of Technology: Pasadena, CA, USA, 1966; pp. 252–263.
36. Neukum, G.; Ivanov, B.A.; Hartmann, W.K. Cratering records in the inner solar system in relation to the lunar reference system. *Space Sci. Rev.* **2001**, *96*, 55–86. [CrossRef]
37. Pasckert, J.H.; Hiesinger, H.; van der Bogert, C.H. Small-scale lunar farside volcanism. *Icarus* **2015**, *257*, 336–354. [CrossRef]
38. Fu, X.-H.; Qiao, L.; Zhang, J.; Ling, Z.-C.; Li, B. The subsurface structure and stratigraphy of the Chang'E-4 landing site: Orbital evidence from small craters on the Von Kármán crater floor. *Res. Astron. Astrophys.* **2020**, *20*, 8. [CrossRef]
39. Gou, S.; Yue, Z.; Di, K.; Bugiolacchi, R.; Zhu, M.-H.; Pinet, P.; Cai, Z. Mare basalt flooding events surrounding Chang'e-4 landing site as revealed by Zhinyu crater ejecta. *Icarus* **2021**, *360*, 114370. [CrossRef]
40. Jia, M.; Di, K.; Yue, Z.; Liu, B.; Wan, W.; Niu, S.; Liu, J.; Cheng, W.; Lin, Y. Multi-scale morphologic investigation of craters in the Chang'e-4 landing area. *Icarus* **2021**, *355*, 114164. [CrossRef]

remote sensing

MDPI

Technical Note

Basalt Chronology of the Orientale Basin Based on CE-2 CCD Imaging and Implications for Lunar Basin Volcanism

Jingwen Liu [1,2], Jianzhong Liu [1,3,*], Juntao Wang [1,3], Kai Zhu [1,3] and Li Zhang [1]

1 Center for Lunar and Planetary Sciences, Institute of Geochemistry, Chinese Academy of Sciences, Guiyang 550081, China; liujingwen@mail.gyig.ac.cn (J.L.); wangjuntao@mail.gyig.ac.cn (J.W.); zhukai@mail.gyig.ac.cn (K.Z.); zhangli@mail.gyig.ac.cn (L.Z.)
2 University of Chinese Academy of Sciences, Beijing 100049, China
3 Center for Excellence in Comparative Planetology, Chinese Academy of Sciences, Hefei 230026, China
* Correspondence: liujianzhong@mail.gyig.ac.cn; Tel.: +86-137-1861-1998

Abstract: The specific duration between the impact event and subsequent volcanic flows is highly variable based on previous works. The method of crater size-frequency distribution (CSFD) has been previously used to date the basalt in Orientale Basin, which yielded inconsistent resultant Absolute Model Age (AMA) ranges. The inconsistency may be attributed to the choice of counting area and identified superposed craters. In this study, we integrated the Chang'E-2 (CE-2) imaging data (7 m/pix) and the IIM and 20 m CE-2 DTMS data, re-divided Mare Orientale, and re-estimated the age of the basalts there. The ages revealed that (1) the central basalts had multiphase eruptions, beginning at 3.77 Ga (30 My after the impact event) with the longest duration of 1.51 Gy; (2) the edge basalts have a similar features as the central basalts, beginning at 3.75–3.50 Ga (50–300 My after the impact) with the longest duration of 0.67 Gy. Compared with the basalts along the basinal margin, the central basalts have higher Ti but lower Mg# contents, consistent with the basaltic magma fractionation trend. Spatial distribution characteristics indicate that the basalt eruption occurred in the impact direction upstream and in the center, but almost absent in the impact direction downstream. Accordingly, we speculate that the longevity of the lunar mare basaltic volcanism was affected by gravity changes, material balance, and other post-impact processes.

Keywords: CE-2 CCD image; Orientale Basin; Absolute Model Age (AMA); volcanic formation

1. Introduction

Lunar volcanic history is a key to understanding lunar thermal evolution [1]. After the formation of the Moon at ~4.5 Ga [2], melt in the lunar magma ocean resided in the lunar mantle and affected the duration of lunar volcanism, according to the lunar thermal evolution model [3]. The Mg-rich cumulates at the bottom of the lunar mantle (formed in the early magma ocean period) are less dense than the Ti-rich cumulates at the top (formed in the late magma ocean period) and the original KREEP (potassium (K), rare earth elements (REE), and phosphorus (P)) layer [1]. With later geological influence, the gravitational imbalance between the Mg-/Ti-rich cumulates may have caused the primitive lunar mantle to overturn. The accompanied decompression melting may have formed the lunar mare basaltic volcanism [4,5], which constituted the main endogenic dynamics of the Moon. Lunar mare basalts, although accounting for only a very minor proportion of the lunar crust (1%), their petrogenesis and temporal-spatial distribution have recorded the geochemistry and evolution of the lunar mantle reservoir.

Because most lunar mare basalts are still unsampled, absolute radiometric age data for the majority of basalts are still lacking. Fortunately, by making use of remote sensing techniques, we can derive relative and absolute model ages for these unsampled regions [6–9]. Hiesinger et al. [6–9] performed Crater Size-Frequency Distribution (CSFD) measurement on many Procellarum KREEP Terrane (PKT) maria using Lunar Orbiter images (50–150 m/pix), with the counting crater diameter ranging between 1 and 10 km. Morota et al. [10], Lucey, P.G. et al. [11] used the SELENE (Kaguya) (10 m/pixel) to estimate the age of the main stratigraphic units in the lunar PKT area, using two CSFD statistical dating methods and craters of 250 m diameter. Pasckert [12] utilized the Lunar Reconnaissance Orbiter Camera (LROC) Wide-Angle Camera images (WAC) (108 m/pix) to divide the basalts in/around the South Pole-Aitken basin and the CSFD to estimate the age, using craters of 1 km diameter. These CSFD results and limited basalt samples collected by the Apollo, Luna, and CE-5 missions indicate that lunar volcanism may have been active for almost 3 Gy, starting at 3.9–4.0 Ga and ceasing at 1.2 Ga ago, with most basalts erupting in the late Imbrian period (3.6–3.8 Ga) [6,13]. The prolonged eruption and compositional heterogeneity of the lunar basalts imply compositional heterogeneity of the lunar interior. The lunar basalts can thus reveal the lunar interior composition, source region, and thermal evolution.

A total of 17% of the lunar surface is covered by 23 lunar maria [14], lying within or adjacent to large impact basins. Most of the lunar maria (20 out of 23) are located in the PKT, and the remaining three (i.e., Mare Orientale, Mare Moscoviense, and Mare Ingenii) are located in the highlands [15]. Almost all maria belong to impact basins, generally distributed in the center of the basins, which is a large basalt area in the center of a basin. However, not all basins have basalt, and not all basalt areas in the center of the basin can be defined as the mare, only the deposits of basaltic composition (i.e., mafic impact melts) in the center of basins can be defined as mare materials.

As shown in Figure 1: (1) In and near the PKT, basalt filled more than two annular structural regions or even all basins, the diameter of basins ranging from 100+ km (e.g., Iridum Basin) to 1000 km (e.g., Serenitatis Basin); (2) In the transition zones between the PKT and highland, and South Pole-Aitken Terrian (SPAT), some other basins only have basalts in the center and margin, distributed, with a diameter spanning from 100+ km (e.g., Schrödinger Basin) to 1000 km (e.g., Orientale Basin); (3) Some basins are exceptions, such as Austral Basin, but it is all filled with basalt.

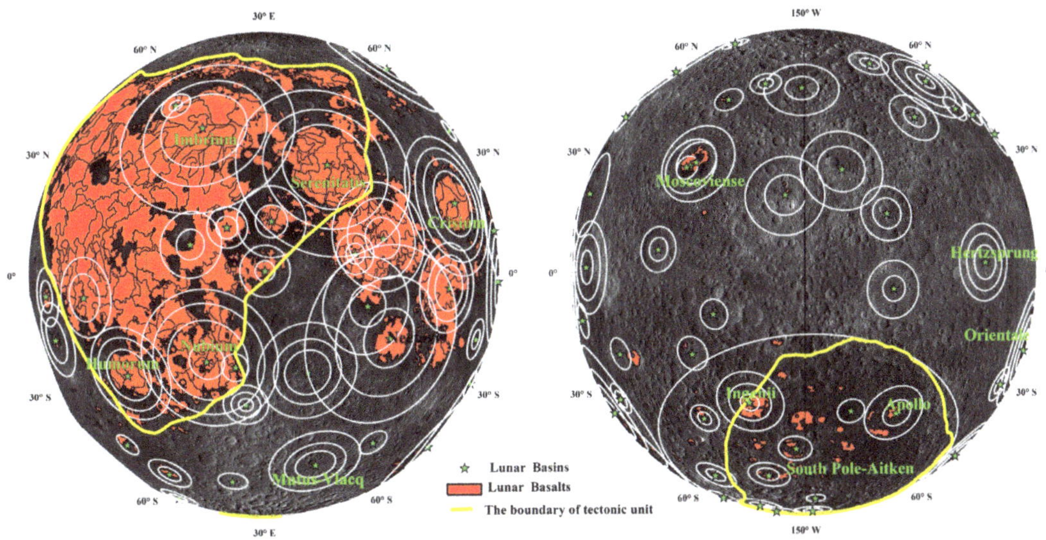

Figure 1. Distribution of the lunar basalts and basins in WAC image (108 m/pix). The boundary of the technical unit based on the work of Jolliff et al., [16], the rings the basins based on the work of Liu et al. [17]; the boundary line of basalt based on the work of Hiesinger et al. [6–9], Morota et al. [10], and Pasckert et al. [12].

Basin impact events are the major exogenic geological process on the Moon during 3.8–4.3 Ga. Exploring the interrelationship between the lunar basalt and basins is the key to uncovering the connection between lunar exogenic and endogenic geological processes. Lunar volcanism did not begin immediately after the basin-forming impacts [18], which implies that the impacts did not directly trigger the volcanism [19], yet their close spatial link implies some sort of genetic connection. To determine any potential interrelationship between them, the following questions need to be answered: (1) the timing of mare volcanism in/around the basin, and any migration of volcanism with time; (2) the possible cause of age difference (if exists) between the basalts and the impact event(s); (3) any spatial distribution trend (and the cause of it) of basalts in the basin.

The multi-ring Orientale Basin, which is locally filled with mare basalts, offers an opportunity to answer these questions. By studying the age and geological characteristics of the Mare Orientale basalts, the relationship between mare basaltic volcanism and the basin formation can be readily established [20]. Although the region has not been sampled, the method of crater size-frequency distribution (CSFD) has been used to date the eruption age of basalts in previous studies. Unfortunately, the resultant Absolute Model Age (AMA) ranges inconsistency, which hinders understanding of the relationship between basins and subsequent volcanic flows after impact. The inconsistency may be attributed to the choice of counting area and identified superposed craters. In this study, we investigate the described discrepancy in AMAs between mare basalts in/around the Orientale basin Chang'E series data. In addition, the element compositions and impact direction are combined to reveal the relationship between the basin volcanism and impact events, which provides new ideas for explaining the age differences.

2. Orientale Basin

The Orientale Basin is the youngest and best-preserved multi-ring lunar impact basin and has only undergone mild modification by subsequent minor impacts and volcanism [21]. The basin is scientifically interesting since its interior remains largely unobscured by mare deposits (unlike most other prominent nearside basins [22]) and allows investigation of an early-filling mare stage.

The circular Orientale Basin was formed in the Imbrian period and has four ring structures, with diameter of approximately 313, 520, 615, and 919 km based on the Lunar Orbiter Laser Altimeter (LOLA) data (Figure 2a). The AMA of the Orientale Basin was estimated to be 3.8 Ga [23], with 38 craters mapped (a few 20 km in diameter) on the Orientale Basin ejecta [24]. Similar work by Baldwin [25] yielded an AMA of ~3.79 Ga. Recently, Yue et al. [21] reported a AMA of ~3.80 Ga for the Orientale Basin, using craters of 0.7 km diameter in the entire ejecta around the basin.

After the basin formation, lava erupted and deposited locally in the basin, forming basalt units in its center and rim: (1) Mare Orientale, the basalt areas in the center of Orientale Basin; (2) Lacus Veris, the basalt areas between the Inner Rook Ring and the Outer Rook Ring; (3) Lacus Autumni, the basalt areas between the Outer Rook Ring and Cordillera Ring [26] (Figure 2b). However, there are certain deviations in the division of geological units and age statistics of basalts in this area, as shown in Table 1 and Figure S1.

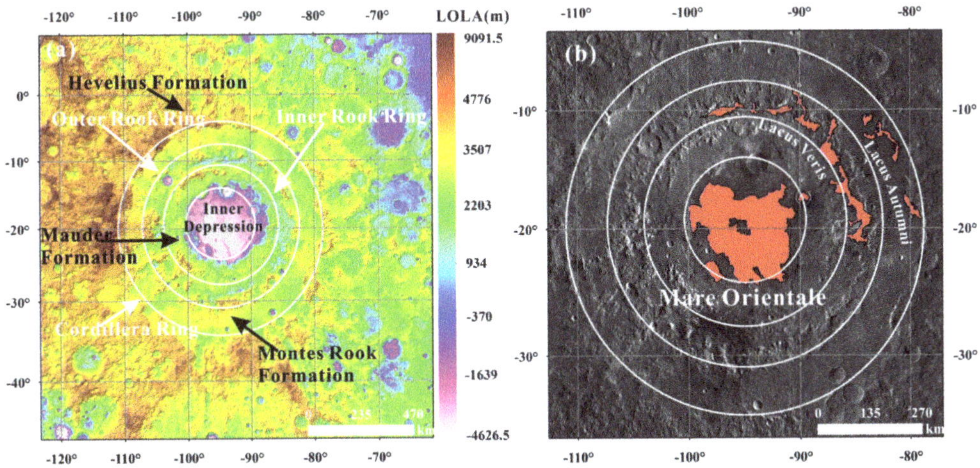

Figure 2. (**a**) Location map of Orientale Basin with four main rings (the rings the basins based on the work of Liu et al. [17]) and deposits labeled (the deposits based on the work of Whitten et al. [20]) in LOLA data; (**b**) The lunar basalts (the basalts based on the work of Morota et al. [10]) of the Orientale Basin in WAC image (108 m/pix).

Table 1. The AMAs ages of basalts in the Orientale Basin.

Basalts Name	Reference	Subunit Name	The surface Age of Basalts (Ga)
Mare Orientale	**Greely et al., 1993 [27]**	Mare Orientale West (1)	3.45
		Mare Orientale Southeast (2)	3.45
		Mare Orientale South-Central (3)	3.70
	Cho et al., 2011 [28]	Southwest	3.77
		SW Polygon	3.77
		east	2.91
	Whitten et al., 2011 [20]	Mare sheet	3.64
		Mare Orientale 7	3.57
		Kopff	3.36
Lacus Veris	**Greely et al., 1993 [27]**	Lacus Veris	2.85
	Cho et al., 2011 [28]	Noth	2.16, 3.36, 3.56
		South	2.20
	Whitten et al., 2011 [20]	9	3.44
		11	3.22
		12	3.36
		14	3.20
Lacus Autumni	**Greely et al., 1993 [27]**	Lacus Autumni	3.85
	Cho et al., 2011 [28]	North	2.03
		Middle	1.75
		South	2.16
	Whitten et al., 2011 [20]	21	3.47
		22	1.66
		23	2.38

3. Data and Methods

3.1. Crater Size-Frequency Distribution (CSFD) Model

CSFD measurement is a powerful remote sensing technique to obtain relative age and AMA for unsampled planetary surfaces, such as the Orientale basalts. The surface area of the unit and the primary impact crater within the unit were mapped with the ArcGIS add-in "CraterTools", which measures the unit area and the crater diameter [29]. Subsequently, the crater distributions measured on geologic units of different ages can be aligned along a complex continuous curve in CraterStats2 [30].

It is reasonable to assume that the physical properties of the target (e.g., density) can influence the crater size. However, the errors introduced by different crater sizes (due to differences in target material) are irrelevant because only the Orientale mare basalts were dated here. Secondary craters formed from the ejected material of a primary crater are thus not correlated to the impact rate, yet most secondary craters can be readily identified and excluded from the crater counts by their cluster/chain occurrence, elongated/irregular outline, and the herringbone pattern of their ejecta [31–33].

Moreover, the resurfacing of a magma eruption would preferentially cover smaller craters in the older unit, while larger craters would remain detectable. Continued impact cratering would form a new crater population in the younger unit. The selective masking of smaller craters would cause a deflection in the distribution between two segments on the age curve in the CSFD, which have different tangential isochrones [10]. The lower isochron can be used to approximate the end time of the resurfacing event (i.e., the age of surface basalt), whereas the upper isochron can be used to approximate the underlying basalt age [10]. The smaller impact craters would be destroyed by later impact craters and basalt eruption, whereas the larger impact craters would not be completely filled with basalt due to their greater depth. As a result, two or three isochrons would appear in the CSFD.

3.2. Data

The CSFD measurement method is generally dependent on the data image quality, i.e., the spatial resolution and the illumination conditions, on which the crater counts are performed [9]. The datasets used here include those from CE-2 (http://moon.bao.ac.cn/, accessed on 15 November 2021): 20 m-resolution CE-2 DTMs (Digital Terrain Models) (ID of the data used: H005, H006, I005, I006) and 7 m-resolution CE-2 CCD images (charge-coupled-device) images (ID of the data used: H113, H114, H115, H214, H213, H212, H211, I113, I211, I212, I213, I111, I112).

Although the CE-2 CCD images (7 m/pixel, incidence angle: 5°) [34] have a higher sun-angle than the WAC (108 m/pixel, incidence angle: 60°) (Figure 3a,d), an order of magnitude better than WAC. As a result, taking three different types of the crater as an example, we can issue these three craters measurable in the CE-2 CCD images (Figure 3c) are either barely visible or saturated in the WAC image (Figure 3e), which would yield better correct crater statistics and reliable ages using CE-2 CCD images. (1) There is a crater in area "I" (diameter: 415 m). Its display is dimmer in the original 7 m-resolution CE-2 CCD image (Figure 3b), but clearly displayed in the 7 m resolution CE-2 CCD effect after partial stretching (Figure 3c). The display is the clearest when the 7 m resolution CE-2 CCD is superimposed on the 20 m resolution CE-2 DTMs (Figure 3f), but the crater is still the least visible in the WAC image even after partial stretching (Figure 3e); (2) In area "II", a degraded crater is identified in the 7 m resolution CE-2 CCD data superimposed on the 20 m-resolution CE-2 DTMs as shown in Figure 3b,c,e,f; (3) In "III" area, small-scale impact craters, such as a 62-m impact crater, are identified on the CE-2 data (Figure 3c,f), which diameter has exceeded the size of a WAC pixel (108 m/pix).

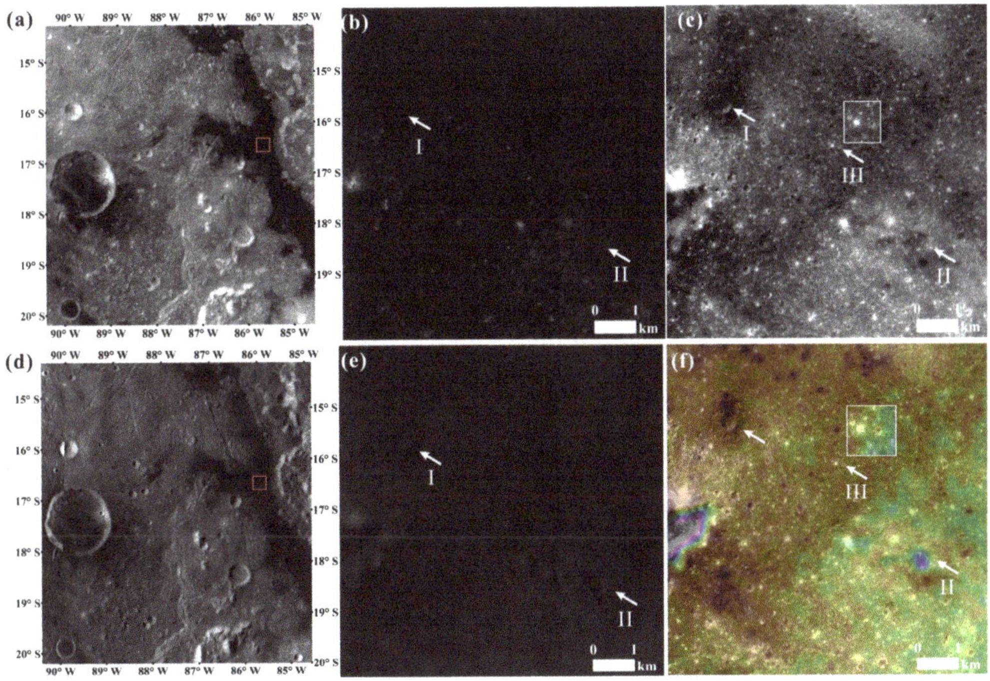

Figure 3. Comparison between CE-2 data with WAC data (108 m/pix): (**a**) 7 m resolution CE-2 CCD images of part of Orientale Basin: CE2_GRAS_DOM_07m_I112_17S088W_A.tif, the red box indicates the location of (**b**,**c**,**e**,**f**); (**b**) original 7 m resolution CE-2 CCD image of the red box area in a and d; (**c**) the 7 m resolution CE-2 CCD image of the red box area in a and d after partial stretching, the white box area is the enlarged image of the area indicated by the white arrow III; (**d**) WAC (108 m/pix) of part of Orientale Basin, the red box indicate the location of (**b**,**c**,**e**,**f**); (**e**) WAC image of the red box area in a and d after partial stretching; (**f**) the 7 m resolution CE-2 CCD data superimposed on 20 m-resolution CE-2 DTMs of the red box area in a and d, the white box area is the enlarged image of the area indicated by the white arrow III.

In summary, despite the higher sun-angle of CE-2 CCD data, the impact features with small diameters have a higher definition in CE-2 CCD images after local stretching, which is more conducive to identifying these small craters. After superimposing CE-2 DTMs on CCD data, some severely degraded craters become clearer. Therefore, these advantages of CE-2 data all would help to facilitate the crater counting and yield a more reliable AMAS in the CSFD using CE-2 data.

The TiO_2 and FeO content product derived from the Chang'E-1 (CE-1) Interference Imaging Spectrometer (IIM) data with the algorithms described by Wu et al. [35] is used to evaluate the boundaries of maria basalt units in this research. The global product is released with a spatial resolution of ~200 m/pixel. The $Mg^{\#}$ are calculated by the MgO and FeO abundance from Lunar Prospector (LP) [36]

4. Results

4.1. Basaltic Area Identification

Hiesinger et al. [6] assume that the spectrally-homogenous units were formed within a relatively short period, and have first-order similarity in mineralogy, representing a single eruptive phase. However, from the TiO_2 and FeO contents calculated from the Chang'E-1 Interference Imaging Spectrometer (IIM) data with the algorithms described by Wu et al. [35] (Figure 4a,b), the basalts in the center of the Orientale Basin are not a single geological unit. Greeley et al. [27] used the Galileo SSI albedo (0.56 um) image superimposed on a shaded airbrush map and divided this basalt field into four sub-regions. Nonetheless, due to the low data spatial resolution, this approach was not adopted by some subsequent studies [28,37,38]. Consequently, Mare Orientale is still regarded as a single geological unit for age calculation and other related research. Combined with the work of Greeley et al. [27], the estimated TiO_2 and FeO contents, and CE-2 CCD data, we have divided the sub-geological units of Mare Orientale in greater detail (Figure 4c). Among them, (1) the positions of c1 and c9 are separated from other subunits; (2) c2 has lower FeO content and TiO_2 content; (3) the FeO content and TiO_2 content of c3 and c5 are both higher, but there is a transition zone between them, which is more likely to belong to two eruptions; (4) c4 has highest FeO content and TiO2 content; (5) both c6 and c8 have lower TiO_2 content, but c6 has higher FeO content; (6) c7 has higher FeO content and lower TiO_2 content. The boundary (white line in Figure 4c, 4d) between them can be shown in combination with CE-2 CCD images.

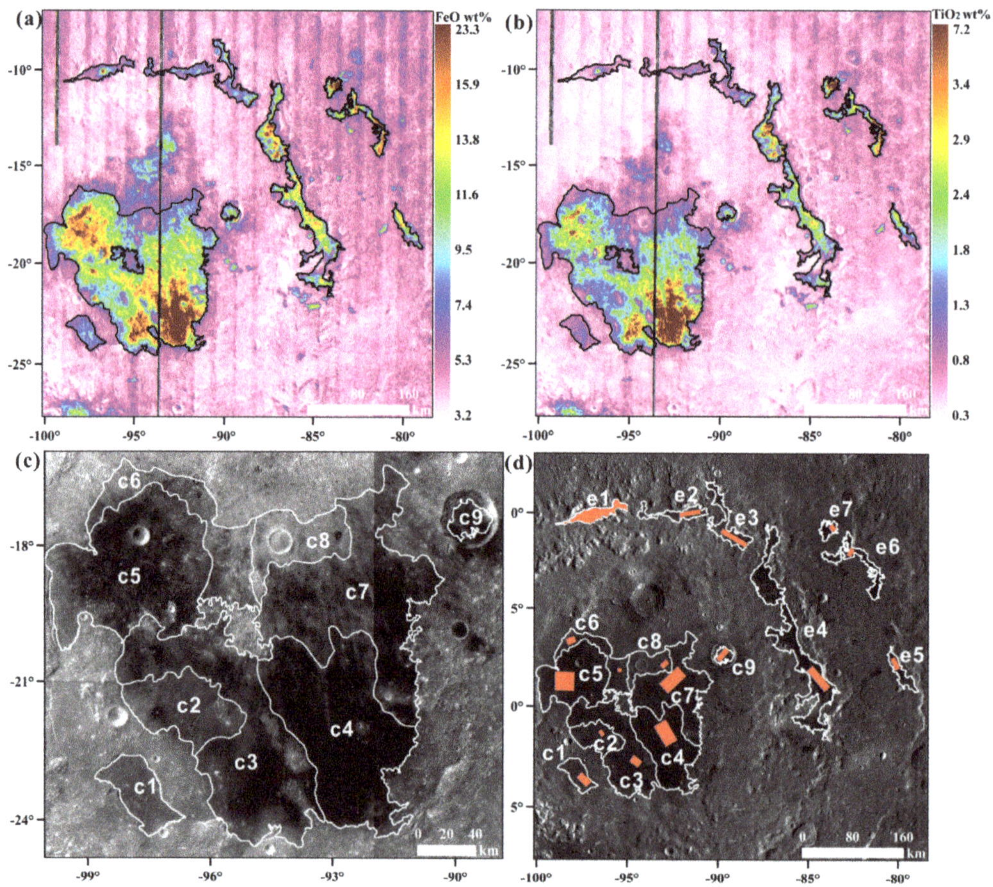

Figure 4. (**a**) IIM FeO abundance (wt%) map of the study area; (**b**) IIM TiO_2 abundance (wt%) map of the study area; (**c**) Newly defined geologic (basalt) units of the Orientale Basin. White lines denote the central basalts. As the background image is spliced from 7 CE-2 CCD images, there is a slight banding; (**d**) Basalts and crater counting areas of this study in WAC data. Red boxes denote the crater-counting areas.

Based on the basalt distribution, Lacus Veris and Lacus Autumni are divided into four and three sub-geological units, respectively.

When calculating the age, the counting areas are located within spectrally homogenous areas of a single mapped unit. For reliable CSFD measurements, we paid particular attention to excluding areas affected by other geological events, endogenic and secondary craters (Figure 4d).

4.2. Crater Retention Ages

In total, 16,823 craters (0.06 km diameter) were mapped in the Orientale Basin basalts. The mapped crater dataset and their information are given in the supplementary material (the craters counted in the different areas.scc). In contrast, the smaller craters (0.2 km diameter) are only partially recognized, with the diameter-number curve flattening toward smaller diameters, inconsistent with the CSFD rule. As a result, 3296 craters (0.2 km diameter) were used here to estimate the AMAs of the Orientale Basin basalts.

As shown in Figures 5–7, the ages obtained by CSFD of subunits c2, c5, c7, c9, e3, e4, and e7, as described in Section 3.1, appear segmented. The appearance of this phenomenon indicates that there may be multiple eruptions in the basalt units of the Orientale Basin. The younger age represents the surface age of the basalts, whilst the older age represents the underlying age of basalt. Meanwhile, combined with the impact age of the Orientale Basin, the oldest of the segmented ages obtained may be the age of the first large-scale basalt eruption after the basin impact in the subunits. The results are described in detail as follows.

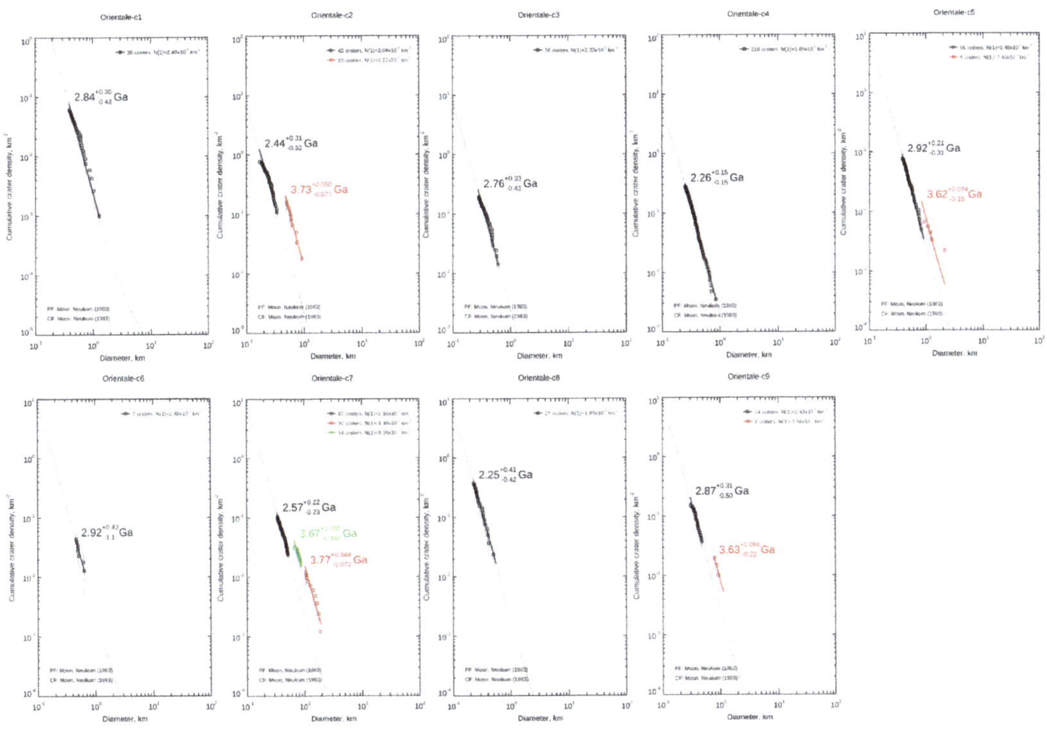

Figure 5. CSFD and mean AMAs of sub-units c1–c9 in Mare Orientale. Black characters denote the surface age of basalts, the red and green colors represent the older age of the underlying basalt. The AMAs were estimated with the Craterstats2 Tool with the production and chronology functions of [37].

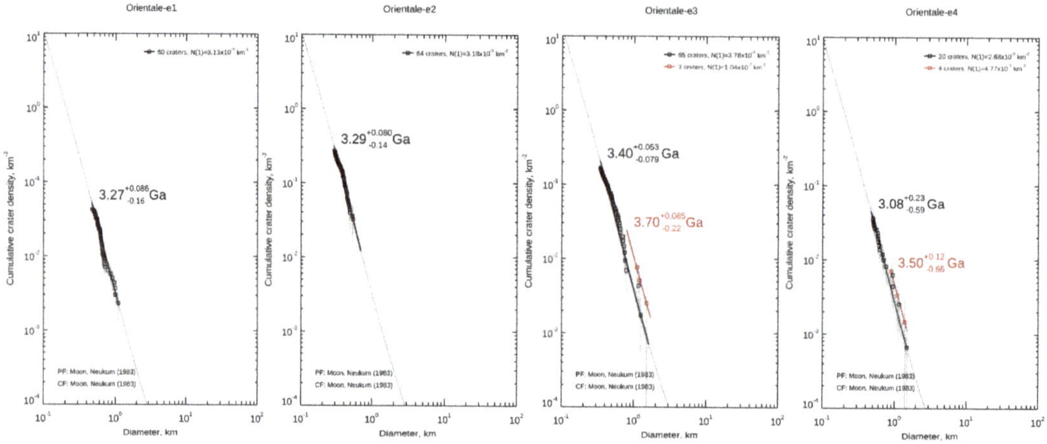

Figure 6. CSFD and mean AMAs of sub-units e1–e4 in Lacus Veris. Black characters denote the surface age of basalts, the red and green colors represent the older age of the underlying basalt. The AMAs were estimated with the Craterstats2 Tool with the production and chronology functions of [37].

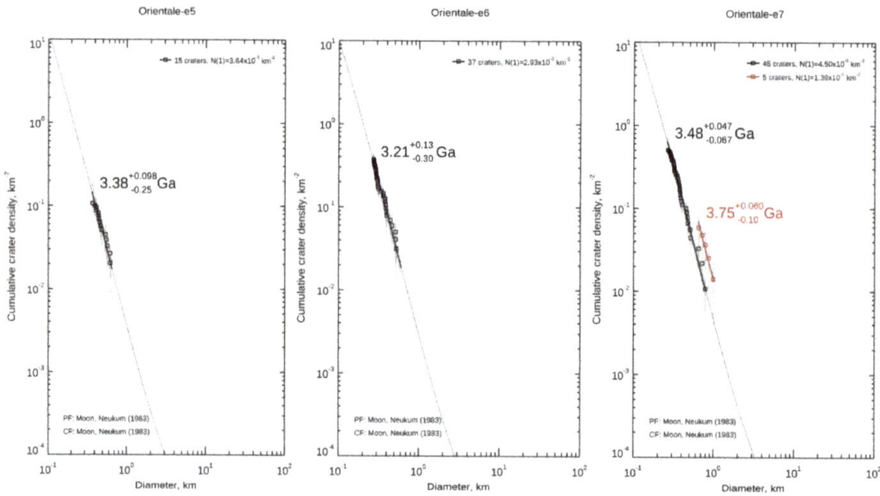

Figure 7. CSFD and mean AMAs of sub-units e5–e7 in Lacus Autumni. Black characters denote the surface age of basalts, the red and green represent the older age of the underlying basalt. The AMAs were estimated with the Craterstats2 Tool with the production and chronology functions of [37].

In Mare Orientale, the estimated surface ages of the nine spectrally-homogeneous central Orientale basalt units range from 2.26 to 2.92 Ga (Figure 5), which are much younger than those estimated by Greely et al. [27] and Whitten et al. [20], and slightly younger than those estimated by Cho et al. [28] (Table 2). The elimination of secondary craters made possible by the high-resolution observations may account for the younger age we obtained. Moreover, the larger craters (D 1 km) of the distribution in c2, c5, c7, and c9 lie on the older isochrons, which is simply interpreted as the AMA of the pre-existing basalt surface. These older ages are similar to the age of the basin impact event and likely represent the age of the first basalt eruption after the impact event.

Table 2. Model ages of the basalts in Orientale Basin.

Unite Name	Neukum et al., (1983) (Ga)		Neukum et al., (2001) (Ga)		Age of Similar Geological Units in Previous Study (Ga)
	sAMA [1] (Ga)	eAMA [2] (Ga)	sAMA (Ga)	eAMA (Ga)	
Mare Orientale					
c1	2.84 (+0.30; −0.43)	\	2.82 (+0.30; −0.43)	\	N [3], 3.57 [4], 3.77 [5]
c2	2.44 (+0.31; −0.32)	3.73 (+0.050; −0.077)	2.33 (+0.30; −0.31)	3.73 (+0.050; −0.077)	N [3], 3.64 [4], 3.77 [5]
c3	2.76 (+0.33; −0.43)	\	2.68 (+0.35; −0.42)	\	3.70 [3], 3.64 [4], 3.77 [5]
c4	2.26 (+0.15; −0.15)	\	2.18 (+0.15; −0.15)	\	3.45 [3], 3.64 [4], 2.91 [5]
c5	2.92 (+0.21; −0.31)	3.62 (+0.074; −0.15)	2.88 (+0.22; −0.31)	3.62 (+0.075; −0.16)	3.45 [4], 3.64 [4], 2.91 [5]
c6	2.92 (+0.42; −1.3)	\	2.88 (+0.45; −1.1)	\	3.45 [a], 3.64 [4], 2.91 [5]
c7	2.57 (+0.22; −0.23)	3.77 (+0.048; −0.032)	2.49 (+0.22; −0.22)	3.76 (+0.048; −0.072)	N [3], 3.64 [4], 2.91 [5]
c8	2.25 (+0.41; −0.42)	\	2.15 (+0.40; −0.40)	\	N [3], 3.64 [4], 2.91 [5]
c9	2.87 (+0.31; −0.50)	3.63 (+0.005; −0.22)	2.78 (+0.35; −0.49)	3.63 (+0.085; −0.22)	N [3], 3.36 [4], N [5]
Lacus Veris					
e1	3.27 (+0.086; −0.16)	\	3.27 (+0.086; −0.16)	\	N [3], 3.44 [4], N [5]
e2	3.29 (+0.080; −0.14)	\)	3.25 (+0.090; −0.17)	\	N [3], 3.22 [4], N [5]
e3	3.40 (+0.053; −0.079)	3.70 (+0.085; −0.22)	3.39 (+0.055; −0.083)	3.70 (+0.085; −0.22)	N [3], 3.36 [4], N [5]
e4	3.08 (+0.23; −0.59)	3.50 (+0.12; −0.65)	3.08 (+0.23; −0.59)	3.51 (+0.11; −0.60)	2.85 [3], 3.20 [4], 2.16, 3.36, 3.56, 2.20 [5]
Lacus Autumni					
e5	3.38 (+0.098; −0.25)	\	3.37 (+0.10; −0.27)	\	N [3], 3.47 [4], 2.03 [5]
e6	3.21 (+0.13; −0.30)	\	3.16 (+0.15; −0.35)	\	3.85 [3], 1.66 [4], 1.75 [5]
e7	3.48 (+0.0.047; −0.067)	3.75 (+0.060; −0.10)	3.46 (+0.0.049; −0.071)	3.76 (+0.060; −0.10)	3.85 [3], 2.38 [4], 2.16 [5]

[1] sAMA is the surface age of basalts; [2] eAMA is the oldest age of the underlying basalt. [3] Reference, Greely et al., (1993); [4] Reference Whitten et al., (2011); [5] Reference Cho et al., (2011).

Meanwhile, the ages calculated for the four main basalt ponds in Lacus Veris range from 3.08 to 3.40 Ga (Figure 6), consistent with previous estimates [20,27,28] (Table 2). Moreover, the larger craters (D 1 km) of the distribution in e3 and e4 lie on the older isochrons, which may represent the timing of the basalt eruption of the first phase of Lacus Veris. For Lacus Autumni, the estimated ages of the three spectrally homogeneous rim basalt units range from 3.21 to 3.48 Ga (Figure 7), which are much older than previous estimates [20,27,28] (Table 2). Partially degraded craters are identified in high-resolution images, causing older ages in these small dating sites. The larger craters (D 1 km) of the distribution in e9 lie on the older isochron, which may represent the age of the first basalt eruption at Lacus Autumni.

5. Discussion

5.1. Implications to the Volcanic History in the Orientale Region

The AMAs of basalt subunits (Figures 5–7) provide information of the volcanic history in the Orientale Region. Based on the estimated ages of mare deposits, we reconstruct the following volcanic history for the Orientale region: (1) 3.80 Ga impact occurred in the Orientale Basin [21]; (2) Basalt eruptions in Mare Orientale started soon after the impact (3.77 Ga); (3) Large part of basalts in the basin center was last resurfaced at 2.26 Ga, with an age difference of 1.5 Gy; (4) Lacus Veris may have erupted during 3.70–3.50 Ga, but the latest eruption occurred at 3.08 Ga after an interval of 0.62 Gy; (5) Lacus Autumni erupted at 3.75 Ga, and ceased to erupt at 3.21 to 3.48 Ga (~0.4 Ga after the eruption).

We suggest that the mare volcanism was first focused in the basin center, and then occurred in the basin margin. However, due to some unknown mechanisms, the volcanism in the center lasted longer, resulting in a younger surface age.

5.2. Implications for the Basin Volcanism

5.2.1. TiO_2 and $Mg^{\#}$ Contents of the Orientale Basalts

Although global remote sensing data indicate no simple relationship between the Ti content and the mare basalt age, and that basalts with varying Ti content could have erupted coevally [39–41]. Therefore, Ti content cannot be used as a research indicator when studying the eruption characteristics of global lunar basalts. However, remote sensing data also revealed that the basin-scale mare volcanism evolved independently of neighboring regions [42], a possible relationship between Ti content and volcanic age has been reported in every separate basin [42]. The wide Ti range of mare basalts suggests diverse mantle compositions and/or complex mantle dynamics. Taylor et al. suggested that TiO_2 can be used to reflect the magma source region, with higher TiO_2 content implying more primitive lunar mantle-derived magma (and vice versa for lunar crustal-derived magma) [43]. Mg-number ($Mg^{\#}$ = atomic Mg/(Mg + Fe)) is an important petrologic discriminator for both terrestrial and lunar rocks [44] and can reflect the degree of basalt fractional crystallization.

In this study, we have calculated the average TiO_2 and $Mg^{\#}$ contents of the basalts from the margin of seven basins and the center of nine basins (Table S1, Figure 8). The basalts in the basin center and rim are named the Central Basin Basalt (CBB) and Rim Basin Basalt (RBB) in this paper, respectively (Figure 4d). The nine geological units of Mare Orientale belong to the CBB, while the seven geological units of Lacus Veris and Lacus Autumni belong to the RBB. Accordingly, the Mare Orientale basalts have higher TiO_2 but lower $Mg^{\#}$ contents than the Lacus Veris and Lacus Autumni basalts. The low-Ti magma was generated in the earlier stage, whilst the high-Ti magma was formed in the later stage [42].

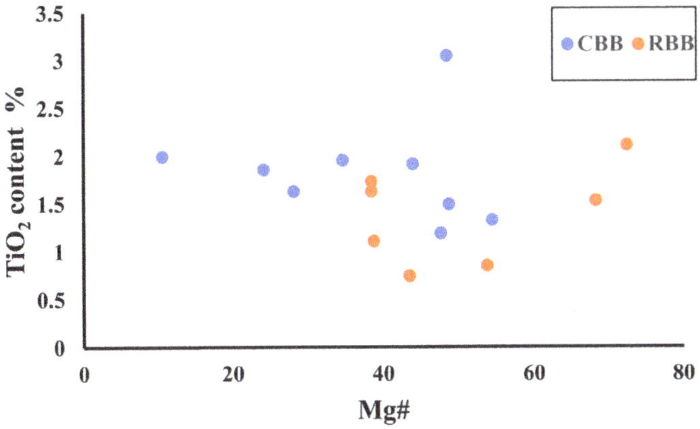

Figure 8. TiO_2 and $Mg^{\#}$ content of the Orientale basalts.

This also shows that basalts in different regions may come from different sources: (1) if the basalts were sourced from other locations, the basalt composition on the basin margin should be similar to that in the center; (2) the basalt in the basin was likely formed from different stages and different source regions: basalt in the center is more fractionated and has a deeper source region, while that on the margin is less fractionated and has a shallower source region.

5.2.2. Relationship between Impact Direction and Basalt Distribution

The overall pattern of ejecta and offset mascon [45] indicates that the Orientale Basin was formed from an oblique impact. The focused stream spatial data analyzed using the LOLA DEM data can identify the ejecta patterns generated by oblique impacts [46] and the impact direction along a northeast-southwest line. Detailed ejecta facies of the Orientale Basin indicates the same northeast-southwest impact direction [46,47]. As shown in Figure 9, the line of bilateral symmetry is represented by the red line and the impact direction is indicated by the arrow at the southwestern end.

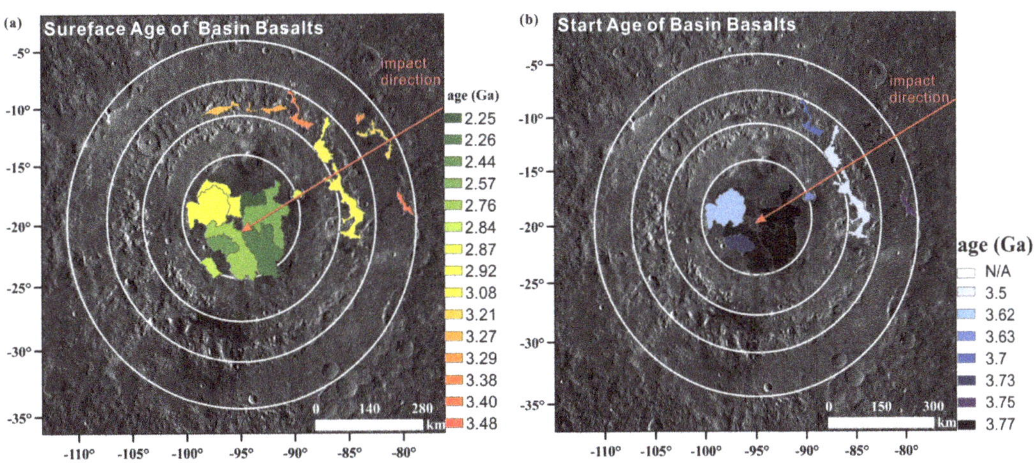

Figure 9. Relationship between impact direction and basalt distribution.

We found that there are much more basalts in the upstream than downstream of the Orientale impact direction, indicating that the basalt eruption occurred more likely toward the uprange. Combining the numerical simulation by Stickle et al. [48], the gravity gradient estimated by Andrews-Hanna et al. [49], and the findings on the Malt Wilson impact crater [50], we suggest that the impact occurred in the uprange and the center impact basin with more deep fractures, which led to the appearance of more basalt in the corresponding area.

5.2.3. Model for the Basin Volcanic Formation

Integrating our results and published works, we suggest that at 3.80 Ga, an impact body (50–80 km diameter) [51] may have struck the lunar surface obliquely forming the Orientale Basin [46,47]. The basin excavation process depressurized the underlying material and in-situ decompression melting occurred. Besides, more impact fractures were formed at the center and the upstream of the impact, forming a basin mason and alternating gravity gradients within the basin [49]. The mascon associated with the lunar maria may have created stress difference (~100 bar) within the upper 300 km of the lunar crust and mantle [52], and impact melting of the lunar mantle gradually formed a magma chamber. Approximately 30 My after the impact, the basalt may have erupted to the surface in the center. Due to the deeper faults in the upper reaches of the basin, some basalts may have erupted at the negative gravity anomalies upstream. Subsequently, the thermal energy caused by the impact continued to weaken, and the basalt eruption in the basin margin ceased. The thermal energy in the center and the gravity anomalies [52] may have continued the eruption for 1.51 Gy.

6. Conclusions

Based on the derived model ages, we conclude the following: (1) Both the central basalts and the margin basalts are characterized by multiple eruptions. The longest volcanism of the central basalt and margin basalt lasted for 1.51 and 0.67 Gy, respectively. The results provide a foundation to estimate the lunar basalt eruption rate and duration [53]; (2) spatial distribution of the basalts suggest that the basalt eruption occurred in the upstream and center of the Orienlate impact tract and that there was almost no basalt eruption in the downstream of the tract; (3) the higher TiO_2 and lower $Mg^{\#}$ characteristics in the center basalt suggest that the continuous basalt eruption was sustained by the post-impact internal thermal evolution and gravity gradient changes. For future studies, it is necessary to verify whether such eruption pattern exists also in the volcanic formation of other basins. Meanwhile, numerical simulation of the volcanic formation can help to better understand it.

Supplementary Materials: The following supporting information can be downloaded at: https://www.mdpi.com/article/10.3390/rs14061426/s1, The craters counted in the different areas.scc; Figure S1: The division of geological units in previous papers; Table S1: TiO_2 content and Mg# of the Orientale basalts. Refs [11,35,36,44] are cited in the supplementary materials.

Author Contributions: Conceptualization, J.L. (Jingwen Liu) and J.L. (Jianzhong Liu); methodology, J.L. (Jingwen Liu).; software, J.L. (Jingwen Liu) and J.W.; validation, J.L. (Jingwen Liu) and J.W.; formal analysis, J.L. (Jingwen Liu); investigation, J.L. (Jingwen Liu); resources, J.L. (Jingwen Liu); data curation, J.L. (Jingwen Liu) and L.Z.; writing—original draft preparation, J.L. (Jingwen Liu); writing—review and editing, J.L. (Jingwen Liu) and J.L. (Jianzhong Liu); visualization, J.L. (Jingwen Liu) and K.Z.; supervision, J.L. (Jianzhong Liu); project administration, J.L. (Jianzhong Liu); funding acquisition, J.L. (Jianzhong Liu). All authors have read and agreed to the published version of the manuscript.

Funding: This work was supported by the Strategic Priority Research Program of the Chinese Academy of Sciences (Grant No. XDB 41000000), the Pre-research project on Civil Aerospace Technologies by CNSA (Grant No. D020201), the National Science and Technology Infrastructure Work Projects (Grant No. 2015FY210500), Beijing Municipal Science and Technology Commission (Grant No. Z181100002918003), the Key Program of Frontier Science of Chinese Academy of Sciences (Grant

No. QYZDY-SSW-DQC028), and the National Natural Science Foundation of China (Grant No. 41773065 and No. 41941003).

Institutional Review Board Statement: Not applicable.

Informed Consent Statement: Not applicable.

Data Availability Statement: The Chang'E series data (http://moon.bao.ac.cn/, accessed on 1 September 2019). The SLDEM2015 is available from LOLA Planetary Data System (PDS) Data Note (http://imbrium.mit.edu/EXTRAS/, accessed on 30 August 2020 SLDEM2015/). The LRO WAC images are available from the LRO data collections. (https://astrogeology.usgs.gov/search/map/Moon/LRO/LROC_WAC/Lunar_LRO_LROC-WAC_Mosaic_global_100m_June2013, accessed on 3 February 2020). The thorium abundance data are available from the PDS Geoscience Node (https://pds-geosciences.wustl.edu/missions/lunarp/reduced_special.html, accessed on 30 August 2020).

Acknowledgments: We gratefully acknowledge all those who worked on the Chang'E series data archive to make 7 m resolution CE-2 CCD s, the 20 m resolution CE-2 DTMs, and IIM data publicly available. We also thank the Planetary Data System archive for providing the LRO WAC imagery and SLDEM2015. We are grateful to the anonymous reviewers and editors for their constructive feedback, which improved the quality of this paper.

Conflicts of Interest: The authors declare no conflict of interest.

References

1. Moriarty, D.P., 3rd; Dygert, N.; Valencia, S.N.; Watkins, R.N.; Petro, N.E. The search for lunar mantle rocks exposed on the surface of the Moon. *Nat. Commun.* **2021**, *12*, 4659. [CrossRef] [PubMed]
2. Kruijer, T.S.; Kleine, T. Tungsten isotopes and the origin of the Moon. *Earth Planet. Sci. Lett.* **2017**, *475*, 15–24. [CrossRef]
3. Ziethe, R.; Seiferlin, K.; Hiesinger, H. Duration and extent of lunar volcanism: Comparison of 3D convection models to mare basalt ages. *Planet. Space Sci.* **2009**, *57*, 784–796. [CrossRef]
4. Dygert, N.; Hirth, G.; Liang, Y. A flow law for ilmenite in dislocation creep: Implications for lunar cumulate mantle overturn. *Geophys. Res. Lett.* **2016**, *43*, 532–540. [CrossRef]
5. Hess, P.C.; Parmentier, E. A model for the thermal and chemical evolution of the Moon's interior: Implications for the onset of mare volcanism. *Earth Planet. Sci. Lett.* **1995**, *134*, 501–514. [CrossRef]
6. Hiesinger, H.; Jaumann, R.; Neukum, G.; Head, J.W. Ages of mare basalts on the lunar nearside. *J. Geophys. Res. Planets* **2000**, *105*, 29239–29275. [CrossRef]
7. Hiesinger, H.; Head, J.W.; Wolf, U.; Jaumann, R.; Neukum, G. Ages and stratigraphy of lunar mare basalts in Mare Frigoris and other nearside maria based on crater size-frequency distribution measurements. *J. Geophys. Res. Planets* **2010**, *115*, 1–22. [CrossRef]
8. Hiesinger, H.; Head, J.W.; Wolf, U.; Jaumann, R.; Neukum, G. Ages and stratigraphy of mare basalts in Oceanus Procellarum, Mare Nubium, Mare Cognitum, and Mare Insularum. *J. Geophys. Res. Planets* **2003**, *108*, 1–27. [CrossRef]
9. Hiesinger, H.; Head, J.; Wolf, U.; Jaumann, R.; Neukum, G.; Advances, J.R. Ages and stratigraphy of lunar mare basalts: A synthesis. *Lunar Stratigr.* **2011**, *477*, 1–51. [CrossRef]
10. Morota, T.; Haruyama, J.; Ohtake, M.; Matsunaga, T.; Honda, C.; Yokota, Y.; Kimura, J.; Ogawa, Y.; Hirata, N.; Demura, H.; et al. Timing and characteristics of the latest mare eruption on the Moon. *Earth Planet. Sci. Lett.* **2011**, *302*, 255–266. [CrossRef]
11. Lucey, P.G.; Blewett, D.T.; Hawke, B.R. Mapping the FeO and TiO_2 content of the lunar surface multispectral imagery. *J. Geophys. Res. Planets* **1998**, *103*, 3679–3699. [CrossRef]
12. Pasckert, J.H.; Hiesinger, H.; van der Bogert, C.H. Lunar farside volcanism in and around the South Pole–Aitken basin. *Icarus* **2018**, *299*, 538–562. [CrossRef]
13. Head, J.W.; Wilson, L. Lunar Mare Volcanism: Stratigraphy, Eruption Conditions, and the Evolution of Secondary Crusts. *Geochim. Cosmochim. Acta* **1992**, *56*, 2155–2175. [CrossRef]
14. Head, J.W.; Hawke, B.R. Geology of the Apollo 14 region/Fra Mauro/-Stratigraphic history and sample provenance. In Proceedings of the 6th Lunar Science Conference, Houston, TX, USA, 17–21 March 1975; Volume 6, pp. 2483–2501.
15. Arivazhagan, S. Quantitative Characterization of lunar Mare Orientale basalts detected by moon mineralogical mapper on Chandrayaan-1. In *Planetary Exploration and Science: Recent Results and Advances*; Springer: Berlin/Heidelberg, Germany, 2015; pp. 21–43.
16. Jolliff, B.L.; Gillis, J.J.; Haskin, L.A.; Korotev, R.L.; Wieczorek, M.A. Major lunar crustal terranes: Surface expressions and crust-mantle origins. *J. Geophys. Res. Planets* **2000**, *105*, 4197–4216. [CrossRef]
17. Liu, J.; Liu, J.; Yue, Z.; Zhang, L.; Wang, J.; Zhu, K. Characterization and interpretation of the global lunar impact basins based on remote sensing. *Icarus* **2022**, *378*, 114952. [CrossRef]
18. Baldwin, R.B. Summary of arguments for a hot moon. *Science* **1970**, *170*, 1297–1300. [CrossRef] [PubMed]
19. Arkani-Hamed, J. On the thermal history of the moon. *Moon* **1973**, *6*, 380–383. [CrossRef]

20. Whitten, J.; Head, J.W.; Staid, M.; Pieters, C.M.; Mustard, J.; Clark, R.; Nettles, J.; Klima, R.L.; Taylor, L. Lunar mare deposits associated with the Orientale impact basin: New insights into mineralogy, history, mode of emplacement, and relation to Orientale Basin evolution from Moon Mineralogy Mapper (M3) data from Chandrayaan-1. *J. Geophys. Res. Planets* **2011**, *116*, E00G09. [CrossRef]
21. Yue, Z.; Yang, M.; Jia, M.; Michael, G.; Di, K.; Gou, S.; Liu, J. Refined model age for Orientale Basin derived from zonal crater dating of its ejecta. *Icarus* **2020**, *346*, 113804. [CrossRef]
22. Nahm, A.L.; Ohman, T.; Kring, D.A. Normal faulting origin for the Cordillera and Outer Rook Rings of Orientale Basin, the Moon. *J. Geophys. Res. Planets* **2013**, *118*, 190–205. [CrossRef]
23. Wilhelms, D.E.; John, F.; Trask, N.J. *The Geologic History of the Moon*; United States Government Publishing Office: Washington, DC, USA, 1987; pp. 2330–7102. [CrossRef]
24. Neukum, G.; König, B.; Arkani-Hamed, J. A study of lunar impact crater size-distributions. *Moon* **1975**, *12*, 201–229. [CrossRef]
25. Baldwin, R.B. On the Relative and Absolute Ages of 7 Lunar Front Face Basins: II. From Crater Counts. *Icarus* **1987**, *71*, 19–29. [CrossRef]
26. Greeley, R. Modes of emplacement of basalt terrains and an analysis of mare volcanism in the Orientale Basin. In Proceedings of the Lunar and Planetary Science Conference Proceedings, Houston, TX, USA, 15–19 March 1976; pp. 2747–2759.
27. Greeley, R.; Kadel, S.D.; Williams, D.A.; Gaddis, L.R.; Head, J.W.; McEwen, A.S.; Murchie, S.L.; Nagel, E.; Neukum, G.; Pieters, C.M.; et al. Galileo imaging observations of lunar maria and related deposits. *J. Geophys. Res. Planets* **1993**, *98*, 17183–17205. [CrossRef]
28. Cho, Y.; Morota, T.; Haruyama, J.; Yasui, M.; Hirata, N.; Sugita, S. Young mare volcanism in the Orientale region contemporary with the Procellarum KREEP Terrane (PKT) volcanism peak period ~2 billion years ago. *Geophys. Res. Lett.* **2012**, *39*, 1–5. [CrossRef]
29. Kneissl, T.; van Gasselt, S.; Neukum, G. Map-projection-independent crater size-frequency determination in GIS environments—New software tool for ArcGIS. *Planet. Space Sci.* **2011**, *59*, 1243–1254. [CrossRef]
30. Michael, G.; Neukum, G. Planetary surface dating from crater size–frequency distribution measurements: Partial resurfacing events and statistical age uncertainty. *Earth Planet. Sci. Lett.* **2010**, *294*, 223–229. [CrossRef]
31. Pike, R.J. *Geometric Interpretation of Lunar Craters*; U.S. Government Printing Office: Washington, DC, USA, 1980.
32. Oberbeck, V.R.; Morrison, R.H. Laboratory simulation of the herringbone pattern associated with lunar secondary crater chains. *Moon* **1974**, *9*, 415–455. [CrossRef]
33. Melosh, H.J. *Impact Cratering: A Geologic Process*; Oxford University Press: New York, NY, USA, 1989.
34. Zuo, W.; Li, C.; Zhang, Z. Scientific data and their release of Chang'E-1 and Chang'E-2. *Chin. J. Geochem.* **2014**, *33*, 24–44. [CrossRef]
35. Wu, Y.; Xue, B.; Zhao, B.; Lucey, P.; Chen, J.; Xu, X.; Li, C.; Ouyang, Z. Global estimates of lunar iron and titanium contents from the Chang'E-1 IIM data. *J. Geophys. Res. Planets* **2012**, *117*, 1–23. [CrossRef]
36. Prettyman, T.H.; Hagerty, J.J.; Elphic, R.C.; Feldman, W.C.; Lawrence, D.J.; McKinney, G.W.; Vaniman, D.T. Elemental composition of the lunar surface: Analysis of gamma ray spectroscopy data from Lunar Prospector. *J. Geophys. Res. Planets* **2006**, *111*, 1–41. [CrossRef]
37. Morota, J.; Ohtake, M.; Matsunaga, T.; Yokota, Y.; Honda, C.; Sugihara, T.; Kimura, J.; Ishihara, Y. Mare Volcanism on the Farside and in the Orientale Region of the Moon. In Proceedings of the 41st Lunar and Planetary Science Conference, The Woodlands, TX, USA, 1–5 March 2010; p. 1309.
38. Spudis, P.D.; Martin, D.J.; Kramer, G. Geology and composition of the Orientale basin impact melt sheet. *J. Geophys. Res. Planets* **2014**, *119*, 19–29. [CrossRef]
39. Kodama, S.; Yamaguchi, Y. Lunar mare volcanism in the eastern nearside region derived from Clementine UV/VIS data. *Meteorit. Planet. Sci.* **2003**, *38*, 1461–1484. [CrossRef]
40. Hiesinger, H.; Jaumann, R.; Neukum, G.; Head, J. On the relation of age and titanium content of lunar mare basalts. In Proceedings of the 29th Annual Lunar and Planetary Science Conference, Houston, TX, USA, 16–20 March 1998; p. 1243.
41. Hiesinger, H.; Head, J.; Wolf, U.; Neukum, G. Lunar mare basalts: Mineralogical variations with time. In Proceedings of the 32nd Lunar and Planetary Science Conference, Houston, TX, USA, 12–16 March 2001; p. 1826.
42. Kodama, S.; Yamaguchi, Y. Mare volcanism on the Moon inferred from Clementine UVVIS data. In Proceedings of the 36th Annual Lunar and Planetary Science Conference, League City, TX, USA, 14–18 March 2005; p. 1641.
43. Taylor, G.J.; Warren, P.; Ryder, G.; Delano, J.; Pieters, C.; Lofgren, G. Lunar rocks. In *Lunar Sourcebook*; Cambridge University Press: Cambridge, UK, 1991; pp. 183–284.
44. Cahill, J.; Lucey, P.; Gillis, J.; Steutel, D. Global Mapping of Mg-Number Derived from Clementine Data. In *AGU Fall Meeting Abstracts*; American Geophysical Union: San Francisco, CA, USA, 17 December 2004; p. P23A-0232.
45. Schultz, P. Making the man in the moon: Origin of the Imbrium Basin. In Proceedings of the Twenty-sixth Lunar and Planetary Science Conference, Houston, TX, USA, 13–17 March 1995; pp. 1251–1252.
46. Liu, J.W.; Liu, J.Z.; Guo, D.J.; Ji, J.Z.; Wang, Q.L.; Li, S.J. Comprehensive analysis of the lunar Orientale Basin and research of the initial impact condition. *Acta Petrol. Sin.* **2016**, *32*, 135–143.
47. Morse, Z.R.; Osinski, G.R.; Tornabene, L.L. New morphological mapping and interpretation of ejecta deposits from Orientale Basin on the Moon. *Icarus* **2018**, *299*, 253–271. [CrossRef]

48. Stickle, A.; Schultz, P. Investigating pressure magnitudes at depth for oblique impacts into layered targets: Applications to terrestrial impacts in sedimentary targets. *Meteorit. Planet. Sci.* **2013**, *48*, 1638–1650. [CrossRef]
49. Andrews-Hanna, J.C.; Head, J.W.; Johnson, B.; Keane, J.T.; Kiefer, W.S.; McGovern, P.J.; Neumann, G.A.; Wieczorek, M.A.; Zuber, M.T. Ring faults and ring dikes around the Orientale basin on the Moon. *Icarus* **2018**, *310*, 1–20. [CrossRef]
50. Kenkmann, T.; Poelchau, M.H. Low-angle collision with Earth: The elliptical impact crater Matt Wilson, Northern Territory, Australia. *Geology* **2009**, *37*, 459–462. [CrossRef]
51. Potter, R.W.; Kring, D.A.; Collins, G.S.; Kiefer, W.S.; McGovern, P.J. Numerical modeling of the formation and structure of the Orientale impact basin. *J. Geophys. Res. Planets* **2013**, *118*, 963–979. [CrossRef]
52. Arkani-Hamed, J. Effect of a giant impact on the thermal evolution of the Moon. *Moon* **1974**, *9*, 183–209. [CrossRef]
53. Head, J.W. Lava Flooding of Ancient Planetary Crusts—Geometry, Thickness, and Volumes of Flooded Lunar Impact Basins. *Moon Planets* **1982**, *26*, 61–88. [CrossRef]

remote sensing

MDPI

Review

Lunar Mare Fecunditatis: A Science-Rich Region and a Concept Mission for Long-Distance Exploration

Siyuan Zhao [1], Yuqi Qian [1], Long Xiao [1,*], Jiannan Zhao [1], Qi He [1], Jun Huang [1], Jiang Wang [1], Hui Chen [2] and Weiyang Xu [2]

[1] State Key Laboratory of Geological Processes and Mineral Resources, Planetary Science Institute, School of Earth Sciences, China University of Geosciences, Wuhan 430074, China; siyuanzhao@cug.edu.cn (S.Z.); yuqi_qian@cug.edu.cn (Y.Q.); jnzhao@cug.edu.cn (J.Z.); he_qi@cug.edu.cn (Q.H.); junhuang@cug.edu.cn (J.H.); j.wang@cug.edu.cn (J.W.)

[2] Aerospace System Engineering Shanghai, Shanghai 201109, China; 13761181357@163.com (H.C.); weiyang.xu@outlook.com (W.X.)

* Correspondence: longxiao@cug.edu.cn

Abstract: Mare Fecunditatis is a ~310,000 km^2 flat basalt plain located in the low-latitude area of the Moon. Plenty of volcanic features (multiple episodes of mare basalts, sinuous rilles, lava tubes, pyroclastic deposits, domes, irregular mare patches (IMP), ring-moat dome structures (RMDS), floor-fractured craters), tectonic features (grabens and wrinkle ridges), impact-related features, and other features (swirls, pit craters) are identified in Mare Fecunditatis. An in-situ mission to Mare Fecunditatis is scientifically significant to better understand the lunar thermal histories and other questions. All previous in-situ and human missions (Apollo, Luna, Chang'E) were limited to small areas, and no traverse longer than 40 km has been made yet. With the development of technology, long-distance movement will be possible in the future on the lunar surface, providing opportunities to explore multiple sites at one mission with complete documentation of the regional geology. Eight high-value targets (pit crater, IMPs, RMDSs, young basalts, high-Al basalts, pyroclastic deposits, swirls, and fresh craters) were found in Mare Fecunditatis, and a ~1400 km-traverse in 5 years is proposed to explore them to solve the most fundamental lunar questions.

Keywords: Moon; Mare Fecunditatis; long-distance exploration

1. Introduction

The Soviet Union, United States, Europe, China, Japan, and India have successively conducted their lunar robotic or crewed missions since the 1960s. Extraordinary scientific discoveries have been made, but even more questions have been proposed, including (1) bombardment history, (2) lunar interior, (3) crust and regolith, (4) polar region and volatiles, (5) volcanism and thermal evolution, (6) impact process, and (7) space weathering [1], and more scientific landing sites across the Moon have been proposed (e.g., [2–7]). However, different high-value targets are distributed all over the Moon and are normally far away from each other (Figure 1). The short detection distance (less than 39 km, made by Lunokhod 2 in 1973) of previous missions limited the complete scientific documentation of the investigation area, and exploring multiple sites in large distance at one mission was impossible in the era. With the development of technologies, multi-target long-distance missions [8,9] in the near future may become reality. The Intrepid mission [8] concept is one of the examples. The designed rover plans to travel ~1800 km in the Oceanus Procellarum in a 4-year period, which will explore Reiner Gamma, Marius Hills, young mare basalts, Aristarchus Plateau, Aristarchus Crater, and irregular mare patches (IMPs). Both 4 and 1.2 billion years of volcanic activity on the Moon could be revealed by Intrepid. The long travel distance would make the scientific outcomes several times larger than a normal rover mission.

Mare Fecunditatis (centered at 7.8° S, 53.7° E) is a low-latitude area with flat terrain on the nearside of the Moon. The basin formed in the pre-Nectarian period (>3.92 Ga), while the mare basalts were mainly filled in the Imbrian period (3.2–3.85 Ga) and continued until the Eratosthenian period (1.1–3.2 Ga) [10–12]. There is a diversity of volcanic features (e.g., mare deposits, sinuous rilles, lava tubes, pyroclastic deposits, domes, irregular mare patches, ring-moat dome structures), tectonic features (e.g., graben and wrinkled ridge), impact-related features, and other features (e.g., swirls, pit craters) in Mare Fecunditatis. A composite of in-situ investigations of these features will provide a window to understand the external and internal processes. In this study, the scientific value of eight targets (pit crater, IMPs, RMDSs, young basalts, high-Al basalts, pyroclastic deposits, swirls, and fresh craters) was listed, and a ~1400 km traverse routine through Mare Fecunditatis in 5 years was proposed, which would help the planning of long-distance missions in the future.

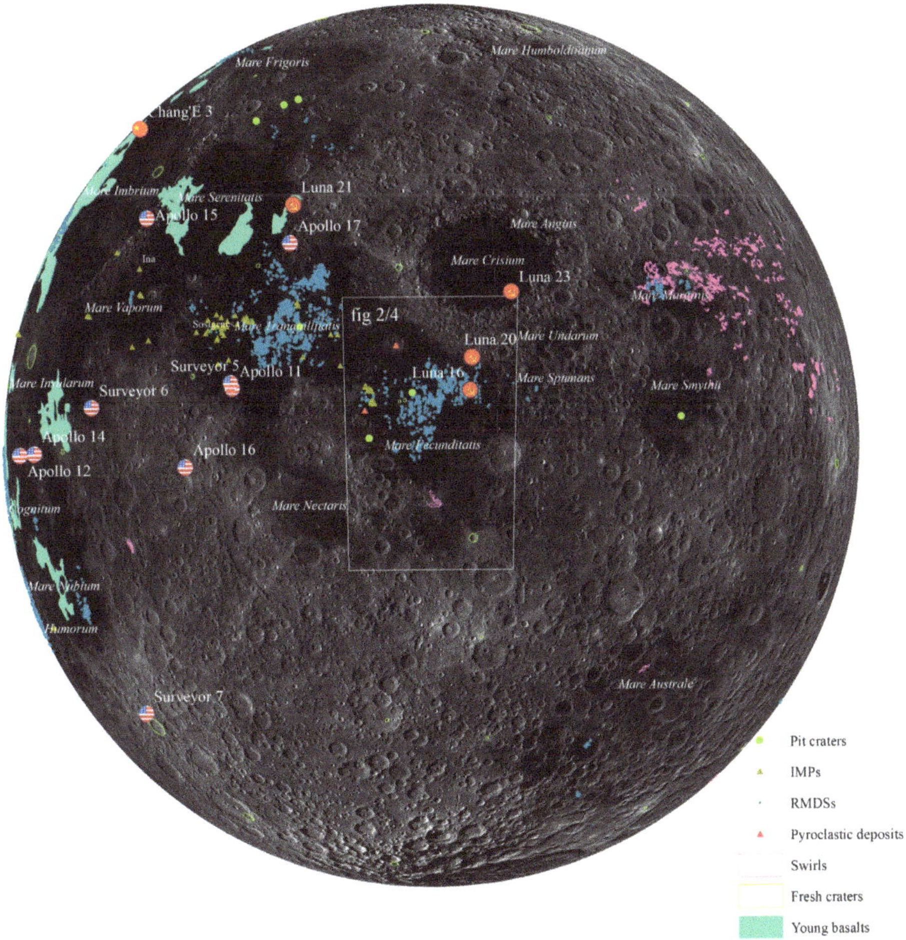

Figure 1. Orthographic map of Wide Angle Camera (WAC, [13]) image with projection center in Mare Fecunditatis (7.8° S, 53.7° E). The flag icons stand for the previous soft-landing missions of the United States, Soviet Union, and China, respectively. The green points, dark yellow triangles, blue points, red triangles, pink outlines, green outlines, and jadeite areas represent identified pit craters [14], IMPs [15], RMDSs [16], pyroclastic deposits [17,18], swirls [19], fresh craters [20], and young basalts (≤3.0 Ga, [21–25]), respectively. The white box represents the locations of other figures in this paper.

2. Topography of Mare Fecunditatis

Mare Fecunditatis is located in the low-latitude area on the eastern nearside of the Moon, covering an area of ~310,000 km^2 (Figures 1 and 2a). SELENE and LRO DEM 2015 (SLDEM2015, 60 m/pixel, [26]) data show that Mare Fecunditatis has a higher elevation in the southwest and a lower elevation in the northeast, and the elevation of 99% of Mare Fecunditatis varies from −0.3 km to −2.7 km (Figure 2b). The peak ring of the Mare Fecunditatis basin is strongly degraded, and the outline is difficult to distinguish [27]. The slope in the basin interior is relatively gentle (average slope of 3.56°); more than 99% of the area has a slope of less than 20°, and 84% of the area has a slope of less than 5° (Figure 2c). Roughness [28] based on the median slope of the 57 m-baseline Lunar Orbiter Laser Altimeter (LOLA, [29,30]) data shows an average value is 1.7°, while the roughness of more than 90% of the area is less than 2.6° (Figure 2d). Mare Fecunditatis is relatively flat, and only crater rims appear with large roughness.

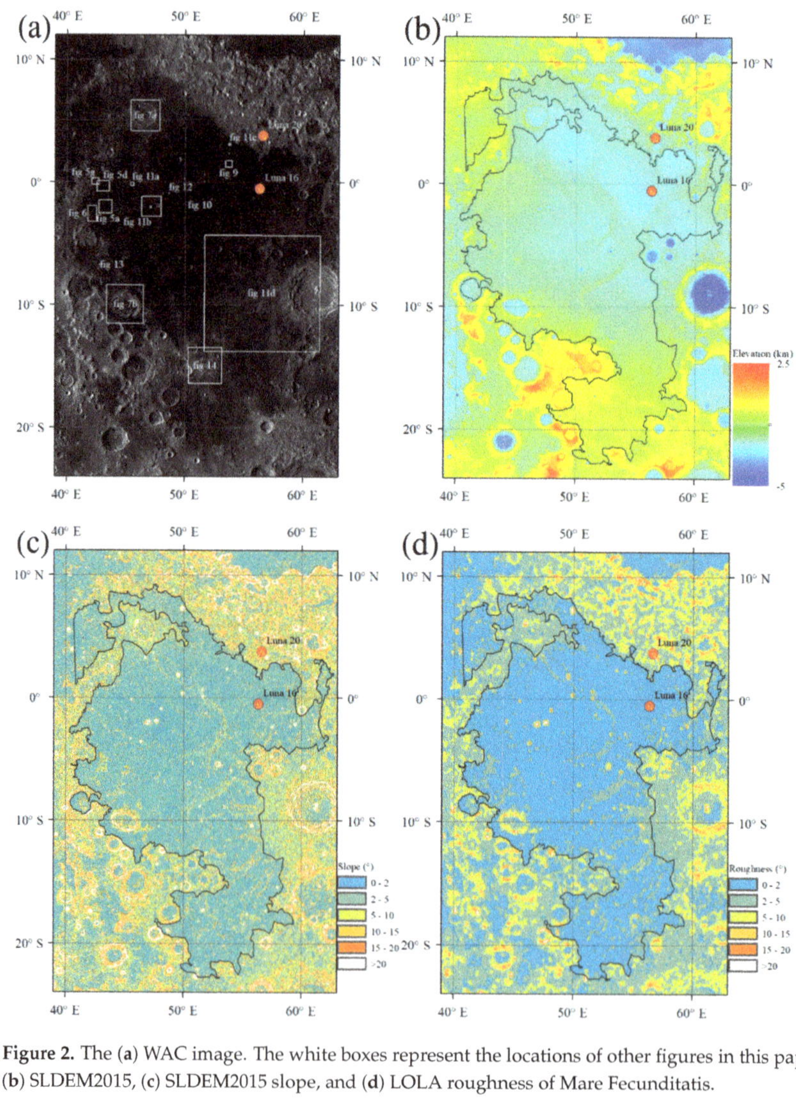

Figure 2. The (**a**) WAC image. The white boxes represent the locations of other figures in this paper (**b**) SLDEM2015, (**c**) SLDEM2015 slope, and (**d**) LOLA roughness of Mare Fecunditatis.

3. Geology and Geomorphology of Mare Fecunditatis

Mare Fecunditatis basin was formed in the pre-Nectarian period, followed by the mare basalts eruption in the Imbrian period, and the volcanic activity continued until the early Eratosthenian period [10–12]. There is no mass concentration in the center of Mare Fecunditatis [31], while there are positive Bouguer anomalies on the east and west sides [32] of the basin. A diversity of geological features (Figure 3) is found in Mare Fecunditatis.

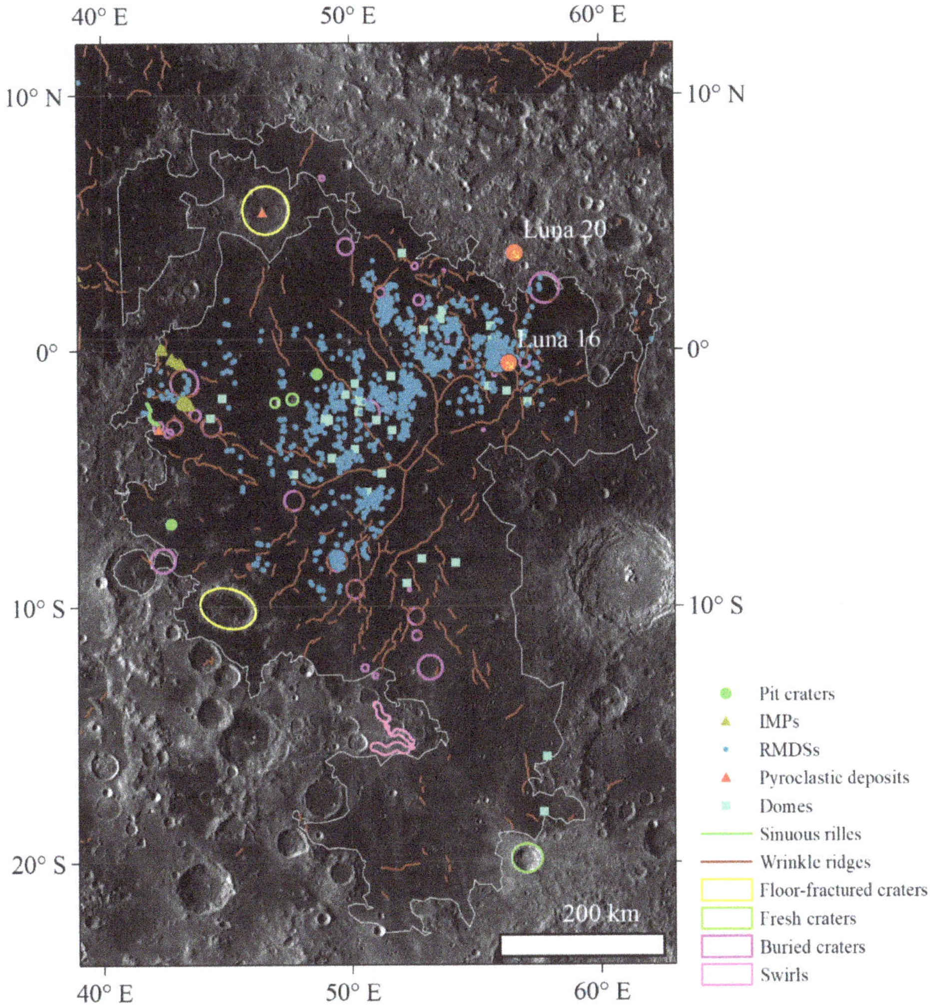

Figure 3. Geological features in Mare Fecunditatis, shown with a WAC image. The green points, dark yellow triangles, blue points, red triangles, bluish green boxes, green lines, brown lines, yellow outlines, green outlines, purple outlines, and pink outlines represent identified pit craters [14], IMPs (updated from [15]), RMDSs [16], pyroclastic deposits [17,18], domes, sinuous rilles [33], wrinkle ridges [34], floor-fractured craters [35], fresh craters [20], buried craters, and swirls [19], respectively.

3.1. Volcanic Features

3.1.1. Mare Basalts

Lunar mare basalt is the main volcanic product of lunar mare volcanism, originated from the partial melting of the lunar mantle and filling some of the impact basins after eruption, including Fecunditatis basin, forming Mare Fecunditatis.

The basalt in Mare Fecunditatis has an average thickness of ~500 m [36], with a thickness of up to 1500 m in the center [37]. The Multiband Imager (MI, [38]) FeO map shows that Mare Fecunditatis is 17–20 wt% in the center and 14–17 wt% in the south (Figure 2a, [39,40]). The WAC TiO_2 map [41] shows high-Ti (TiO_2 > 6 wt%) and low-Ti (TiO_2: 2–6 wt%) basalts are mainly distributed in the center of the basin, while very low-Ti (TiO_2 < 2 wt%) basalts are mainly distributed in the and southern region (Figure 2b). Interference Imaging Spectrometer (IIM, [42,43]) Al maps [44] show high-Al (Al_2O_3 > 11 wt%) basalts are widely distributed in the Mare Fecunditatis, except for the northeastern part; the results are similar with the compositional constraints based on FeO, TiO_2, and Th maps [45,46].

Luna 16 landed in the northeastern part of Mare Fecunditatis (0.7° S, 56.3° E) in 1970. All returned basalt fragments have high-Al (Al_2O_3 > 11 wt%) characteristics [47], with a TiO_2 content of 1–5 wt%, Al_2O_3 content of 11–20 wt%, and Mg# (Mg/(Mg+Fe) in mole percent) of 0.3–0.4 [47,48]. The isotopic ages of basalt samples show three-episode magmatic activities between ~3.14 and 3.42 Ga, but concentrate in ~3.4 Ga [49–52]. According to the mare unit map [12] (Figure 4d), the age of the unit where Luna 16 landed is 3.36 Ga, which is in good agreement with the 3.4 Ga isotopic age. The age of the low-Ti, high-Al basalt in the central and southern parts is 3.5–3.7 Ga, while the high-Ti basalt in the northeast is relatively young, at 3.34–3.36 Ga, and the magmatic activity of some basalts in the southwest continues to 3.14 Ga. In addition, we find a small number of areas with lower albedo, fewer craters, and fewer ejecta materials. Among them, three areas show relatively younger ages (2.2–2.9 Ga; Figure 5).

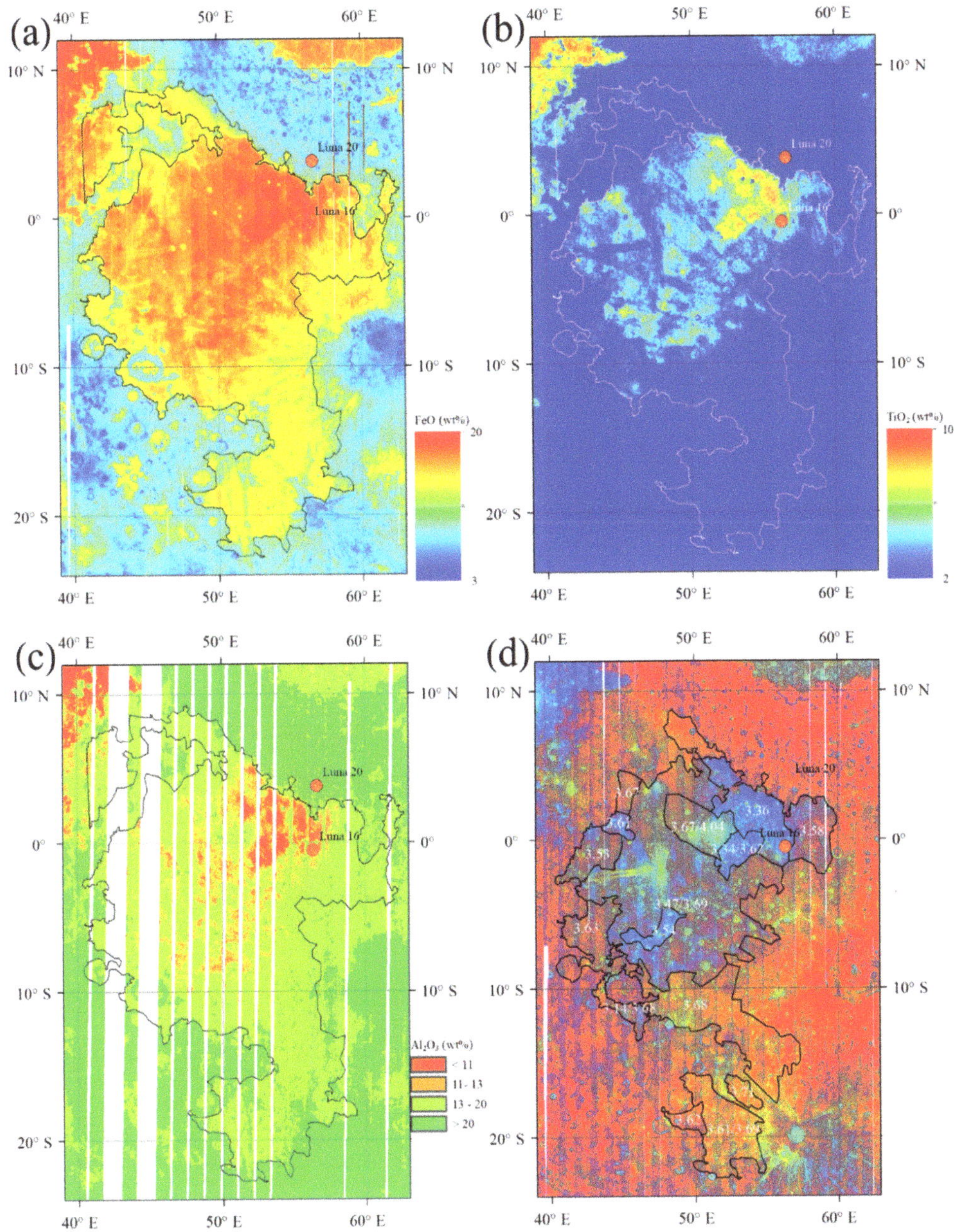

Figure 4. The (**a**) MI FeO map, (**b**) WAC TiO_2 map, (**c**) IIM Al_2O_3 map, and (**d**) MI color ratio map and the age of different geological units (adapted from [12]) of Mare Fecunditatis.

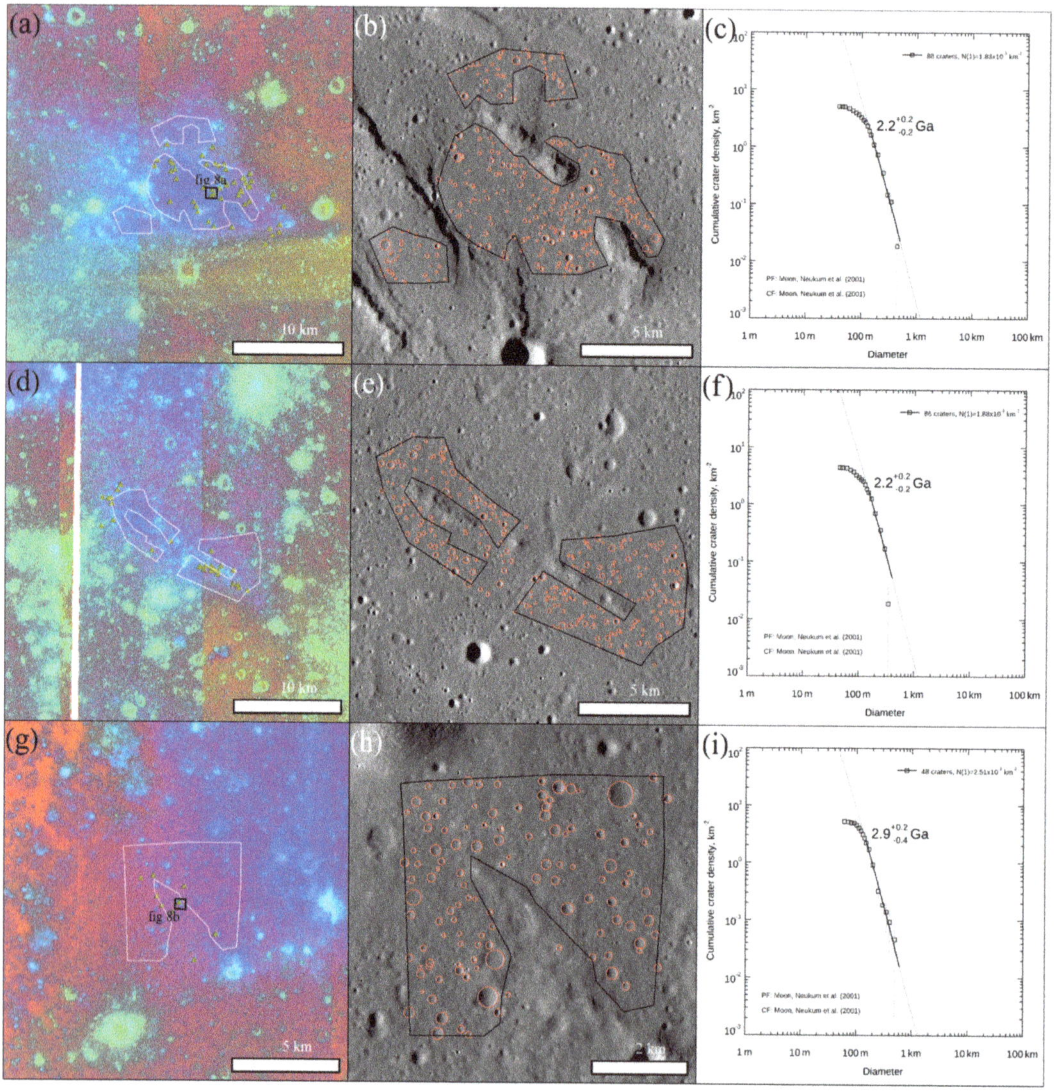

Figure 5. Crater size–frequency distribution measurements of three Eratosthenian-aged basalt areas with IMPs (dark yellow triangles); craters with diameter > 150 m are used for dating. (**a**) MI color ratio map of counting area 1 (2.0° S, 43.2° E); (**b**) crater counting of area 1, shown with a Terrain Camera (TC, [53]) image; (**c**) absolute model age of area 1; (**d**) MI color ratio map of counting area 2 (0.4° S, 43.0° E); (**e**) crater counting of area 2, shown with a TC image; (**f**) absolute model age of area 2; (**g**) MI color ratio map of counting area 3 (0.1° N, 42.4° E); (**h**) crater counting of area 3, shown with a TC image; (**i**) absolute model age of area 3. The black boxes represent the locations of other figures in this paper.

3.1.2. Sinuous Rilles

Sinuous rilles are formed by the erosion of the lunar surface or the continuous collapse of lava tubes during the flow of high-temperature lava, which usually originates from a depression, extending from high terrain to low terrain, and gradually disappears on the smooth lunar mare [54]. A short small-scale sinuous rille (Figures 3 and 6) is found in the western part of Mare Fecunditatis, adjacent to pyroclastic deposits. This sinuous rille is 26.8 km in length, 1000 m wide, and 69 m deep. A possible 17.5 km long collapsed lava tube with 0.69 km in width and 65 m in depth is also discovered next to the sinuous rille.

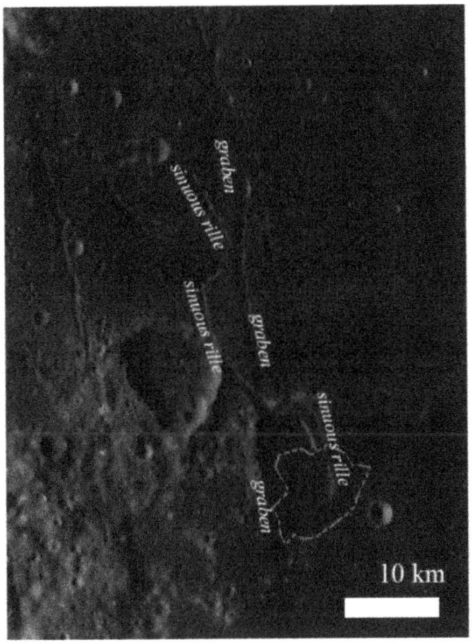

Figure 6. Sinuous rille, graben, and pyroclastic deposits (red dotted outline) in western Mare Fecunditatis (2.5° S, 42.0° E), shown with a WAC image.

3.1.3. Floor-Fractured Craters

The Taruntius crater (5.5° N, 46.5° E, Figure 7a) and Goclenius crater (10.0° S, 45.0° E, Figure 7b) are found in the northern and western part of the Mare Fecunditatis, respectively. Their floors are cut by concentric or polygonal fractures, known as floor-fractured craters (FFCs) [35,55,56]. There are two formation hypotheses for FFCs: (1) viscous relaxation, wherein the crater floor rebounds to fill the crater at a rate controlled by the subsurface viscosity structure, resulting in an overall amplitude shallowing of long-wavelength crater topography [57]; and (2) magma intrusions and sills form beneath the crater, and the magma lifting produces the laccolith and fractures the overlying the crater floor [35,55,56,58]. Recent numerical simulations do not support the viscous relaxation hypothesis [59], but high-resolution topographic and gravity data advocate the magma intrusion hypothesis [35,55,56].

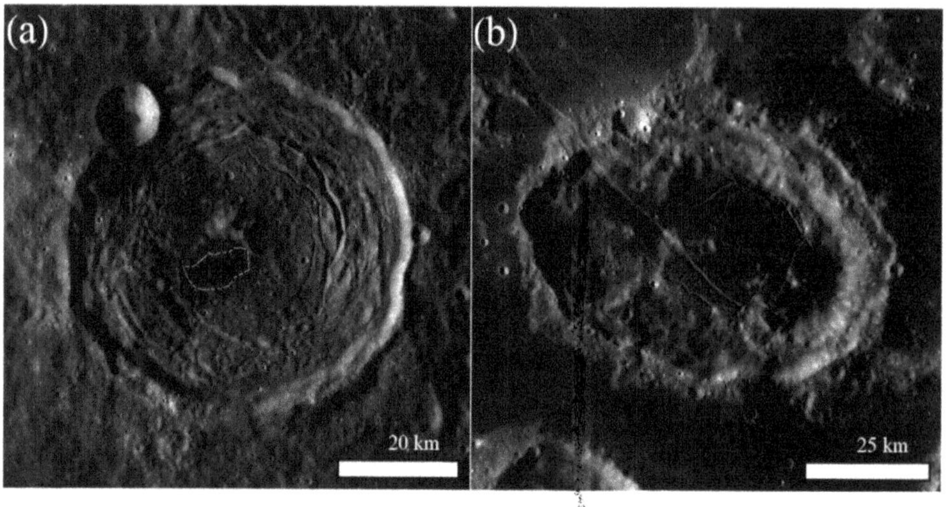

Figure 7. Floor-fractured craters, shown with WAC images. (**a**) Crater Taruntius (5.5° N, 46.5° E) with dark mantle pyroclastic deposits in the center (red dotted line); (**b**) Crater Goclenius (10.0° S, 45.0° E).

3.1.4. Pyroclastic Deposits

Pyroclastic deposits are the products of explosive volcanism, and the characteristic richness in glass and titanium make them darker than overflow basalts [60]. They have a smooth surface and usually occur close to sinuous rilles, irregular depressions, and the boundary between the maria and highlands [61]. Two areas in Mare Fecunditatis have pyroclastic deposits: in the Taruntius crater [62] on the northern side of the center peak (Figure 7a, 5.4° N, 46.5° E), and the area connecting to the sinuous rille (Figure 6, 3.0° S, 42.3° E).

3.1.5. Irregular Mare Patches

Irregular Mare Patches (IMPs) are enigmatic features occurring in the lunar mare. Typical IMPs (such as Ina (18.65° N, 5.30° E) and Sosigenes (8.335° N, 19.071° E), Figure 1) are one to several kilometers in size and composed of positive-relief mounds surrounding low rough hummocky and/or blocky floor units. The IMPs in Mare Fecunditatis are relatively small (tens to hundreds of meters in size) and only develop irregular, rough, bright, and pit features. They may be formed by (1) sublimation [63]; (2) small magma intrusion on the top of the dome [64–66]; (3) clearing of the overlying lunar regolith by outgassing activity within 10 Ma [67]; (4) lava flow inflation [68]; (5) basalt eruption within 100 Ma [15]; (6) pyroclastic eruption [69]; and (7) lava lake process and foamy magma extrusion [70–73]. The most fundamental scientific question of IMP is whether they are young or not. Some authors [15] suggest that IMPs have ages younger than 100 Ma based on impact crater chronologies. Others proposed IMP formed contemporarily with the Imbrian-aged host basalts [68,70].

In the western part of Mare Fecunditatis, dozens of IMPs (Figure 3) are identified within a length of tens to hundreds of meters. They are distributed on three small Eratosthenian-aged areas (Figure 5), where TiO_2 content is >4 wt%. Most of the IMPs are located next to the rim of craters (Figure 8a), and a few IMPs are not connected to the crater (Figure 8b).

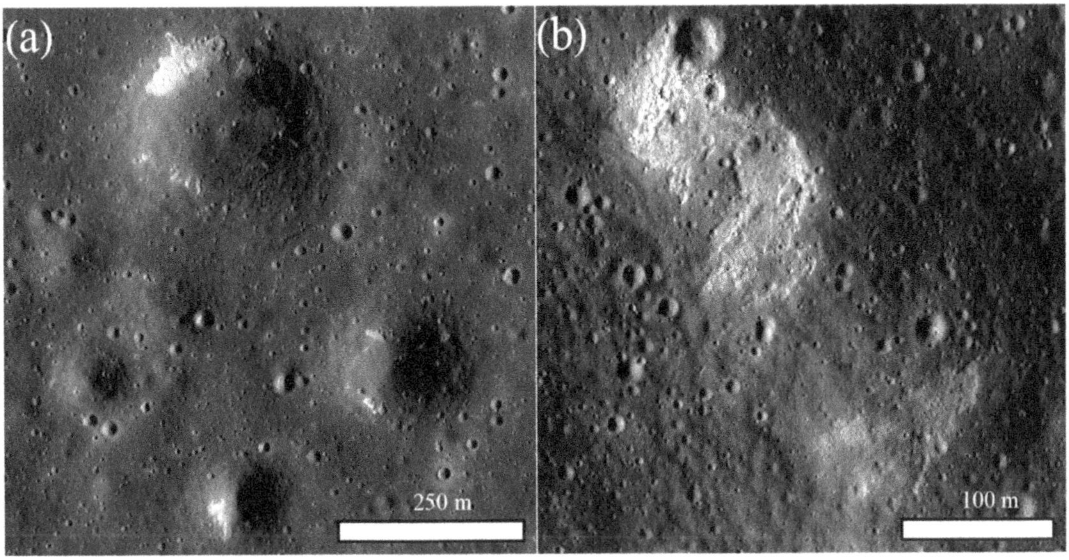

Figure 8. Two sites of IMP in Mare Fecunditatis, shown with Narrow Angle Camera (NAC, [13]) images. (**a**) IMPs next to the crater rim (2.02° S, 43.31° E); (**b**) IMPs on the mare (0.09° N, 42.41° E). Data source: NAC M1230441915RE and NAC M150239673LE.

3.1.6. Domes

The lunar volcanic domes are formed by (1) cooling limited lava flows [74]; (2) subsurface intrusion [75]; and combination of (2) and (1) [76]. There are at least 38 domes (e.g., Figure 9) with a diameter of more than 500 m in the Mare Fecunditatis, mainly distributed in the center of Fecunditatis where the content of TiO_2 generally exceeds 3 wt%.

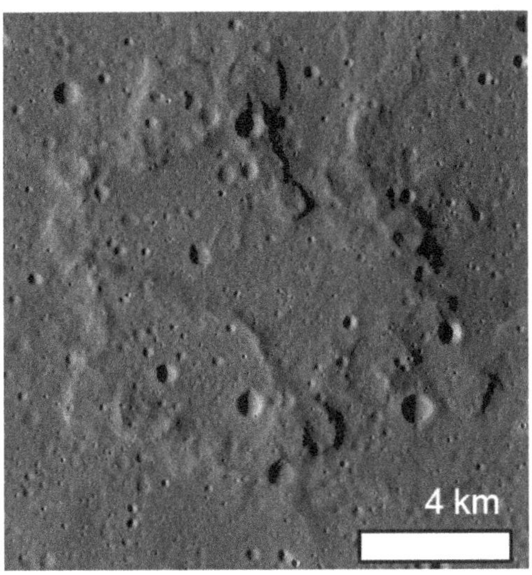

Figure 9. The largest dome (1.5° N, 53.7° E) in Mare Fecunditatis, shown with a TC image.

3.1.7. Ring-Moat Dome Structures

RMDS is a newly discovered lunar volcanic feature [77]. RMDSs are small circular mounds hundreds of meters in diameter and ~3–4 m in height, surrounded by narrow, shallow moats (Figure 10). Four hypotheses were proposed for the formation of RMDSs [77]: (1) high-viscous lava eruption soon after the mare basalt; (2) geologically very recent eruption; (3) small-scale squeezing features formed at the time of the mare basalt; and (4) volatility-rich magmatic foam extrusion. More than 1,600 RMDSs have been identified in the Mare Fecunditatis [16] and they are concentrated in the northern sector, where TiO_2 contents of the RMDS-bearing basalt are larger than 3 wt% and absolute model ages range from 3.36 to 3.67 Ga [12]. The height of RMDSs is very low (usually a few meters), which should be flattened now if RMDSs have similar ages with the surrounding mare basalts. Some RMDSs were found in the craters with a low degree of degradation (130–1500 Ma); thus, RMDSs are also potentially very young features [78,79].

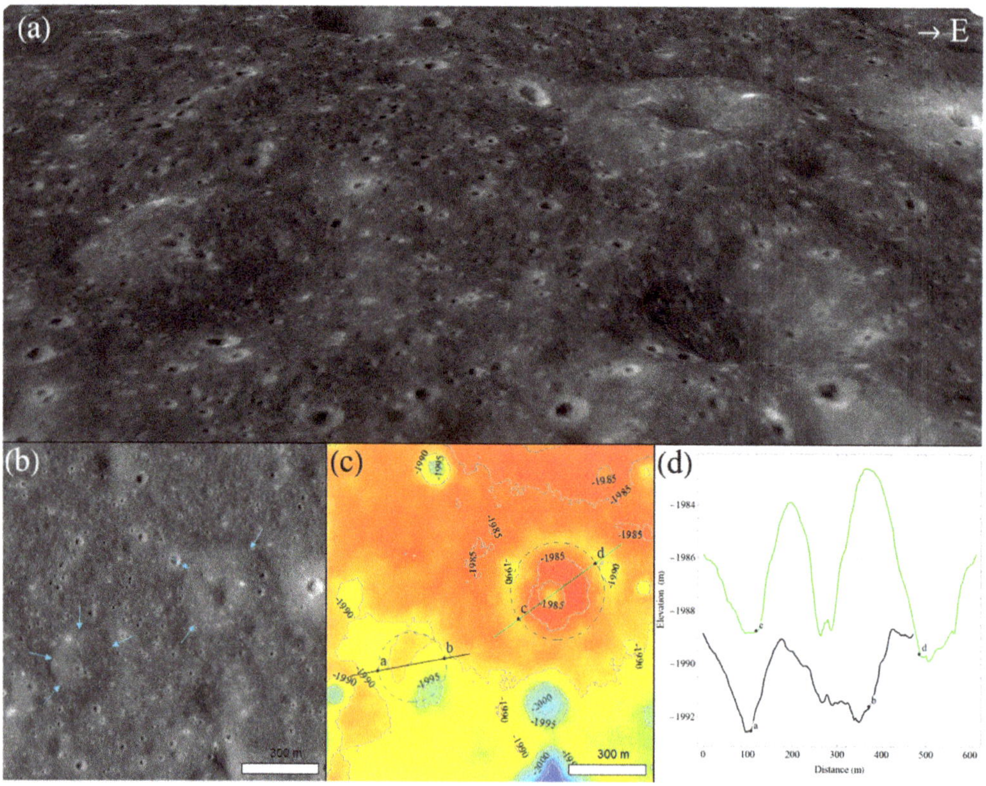

Figure 10. Two RMDSs (1.38° S, 52.27° E and 1.37° S, 52.29° E) in Mare Fecunditatis. (**a**) 3D image of these two RMDSs (five times vertical exaggeration); (**b**) NAC orthoimage; (**c**) NAC DTM; (**d**) elevation of the profiles of (**c**). Data source: NAC M1126787189LE and NAC M106533362LE.

3.2. Tectonic Features

3.2.1. Wrinkle Ridges

A wrinkle ridge is a linear positive relief landform on the lunar surface, which extends up to hundreds of kilometers, mainly distributed in the lunar maria. There are three main hypotheses on the formation of wrinkle ridges: (1) tectonic origins [80–82]; (2) magmatic origins [83–85]; and (3) tectonic and magmatic origins [85–88]. There are more than two hundred wrinkled ridges (Figure 3) in Mare Fecunditatis [89]. Their lengths vary from 1 km to 250 km; most of them are less than 50 km long, mainly with an NS trend, consistent with the global trend [34].

3.2.2. Arcuate Rilles and Grabens

In addition to the sinuous rilles and floor fractures, the tectonic movement also forms long and narrow grooves, including arcuate rilles and grabens (straight rilles) (Figure 6). Arcuate rilles are usually on the edge of the lunar mare with smooth curves [90]. The graben (straight rilles) forms under the extensional stresses, and a block of the crust cracks and drops down to create the valley floor [91–93].

3.3. Impact Craters

Impact craters are the most common geomorphologic features on the lunar surface. According to their size and shape, craters are divided into simple craters (<20 km, bowl-shaped, Figure 11a), complex craters (tens to hundreds of kilometers, with terraced walls, central peaks, and flat floor), and multi-ringed basins (>290 km, with multiple peak rings) [94]. Impact craters in Mare Fecunditatis are mainly simple impact craters, with several complex impact craters on the edges. There are also some atypical craters such as elliptical craters (Figure 11b), buried craters (Figure 11c), and crater chains (Figure 11d) in Mare Fecunditatis.

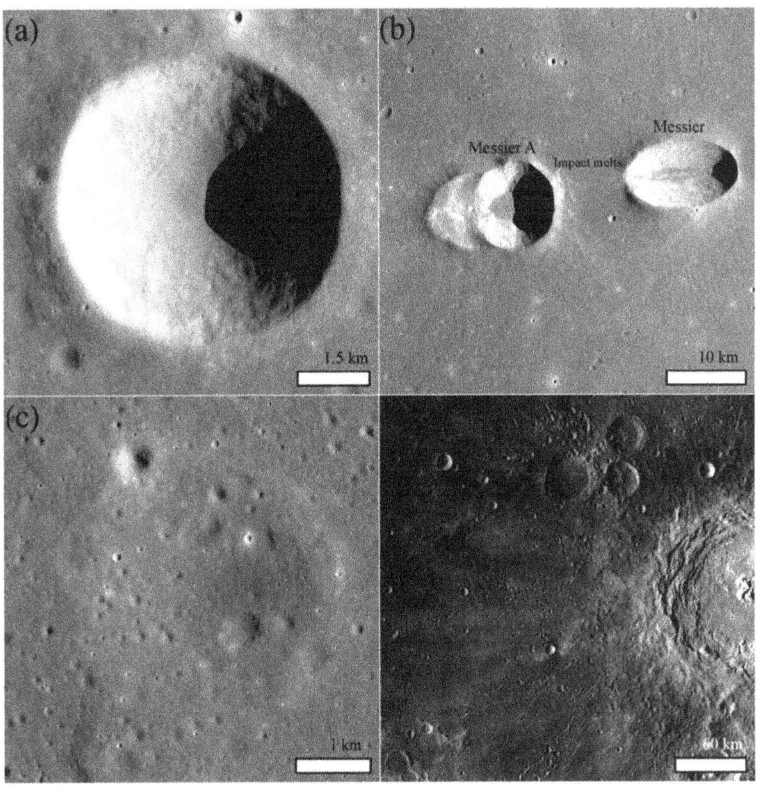

Figure 11. Different types of craters in Mare Fecunditatis. (**a**) A simple crater (0.2° S, 45.5° E), shown with a TC image; (**b**) Messier (1.9° S, 47.7° E), Messier A (1.97° S, 46.95° E), and smooth impact melts, shown with a TC image; (**c**) a buried crater (3.10° N, 53.78° E), shown with a TC image; (**d**) crater chains on the south of Langrenus (8.8° S, 61.2° E), shown with a WAC image.

Messier Crater (1.9° S, 47.67° E) is an elliptical crater with a major axis ~14.3 km and a minor axis of ~8.3 km, while the Messier A crater (1.97° S, 46.95° E) has a major axis ~15.8 km and a minor axis ~11 km (Figure 11b) [95]. Messier was likely formed by a low-angle westward impact, and Messier A formed following a rebound by the impacting body [96]. Major vertical rays extend over 100 km north and south from Messier and horizontal rays extend over 100 km west from Messier A. The results of Monte Carlo simulations [97] show that the simulating ray patterns by oblique impact are very similar to reality. In addition, there are at least 29 buried impact craters in the Mare Fecunditatis. The crater chains refer to a row of craters distributed linearly. Crater chains formed by secondary craters and collapse of the lava tube are identified in Mare Fecunditatis.

3.4. Other Features

3.4.1. Pit Craters

Lunar pit craters formed by the collapse of underground space. A total of 15 pit craters have been discovered on the maria and 5 on the highlands [14,98–100]. The underground space may be formed by the lava tube or the stoping of magma [14,100]. Recently, the underground space of the caves in Marius Hills has been detected through radar and gravity analyses [101,102]. Pit craters stay stable for at least tens of years. Some pits (such as King Crater Bridge) found at the Apollo period still show no change with the NAC data [14]. There is a pit crater (0.917° S, 48.660° E, Figure 12) in the center of Fecunditatis, and a highland-type pit crater (6.752° S, 42.759° E, Figure 13) at the southwest.

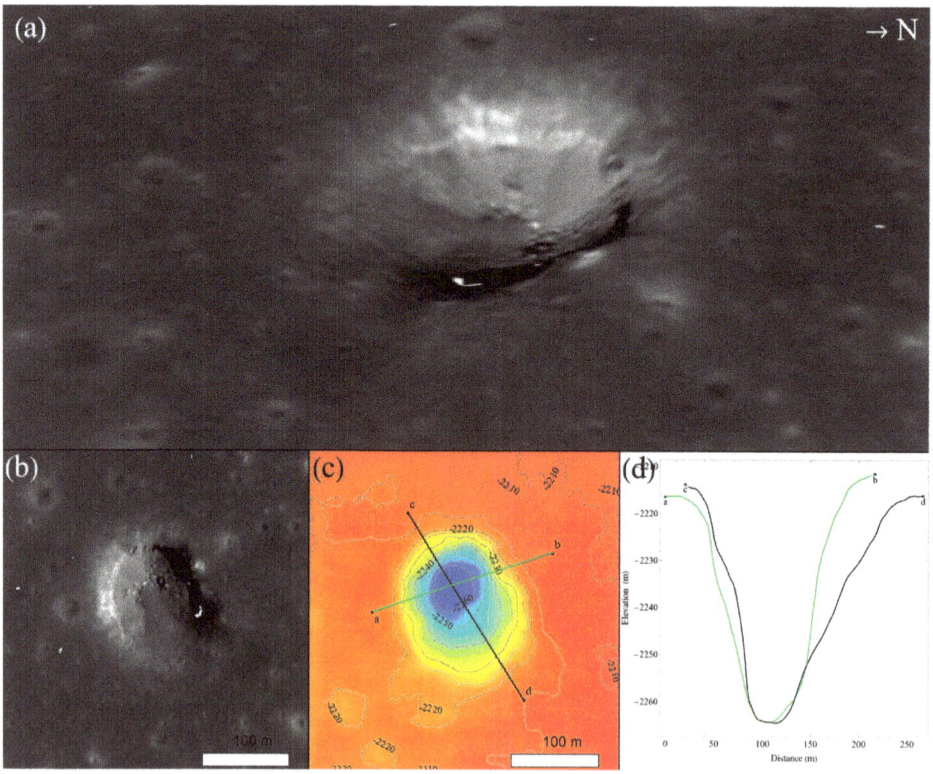

Figure 12. Pit crater (0.917° S, 48.660° E) in the central Fecunditatis. (**a**) 3D image (real elevation values are used); (**b**) NAC orthoimage; (**c**) NAC DTM; (**d**) elevation of the profiles of (**c**). Data source: NAC M1105602888RE and NAC M1197457075RE.

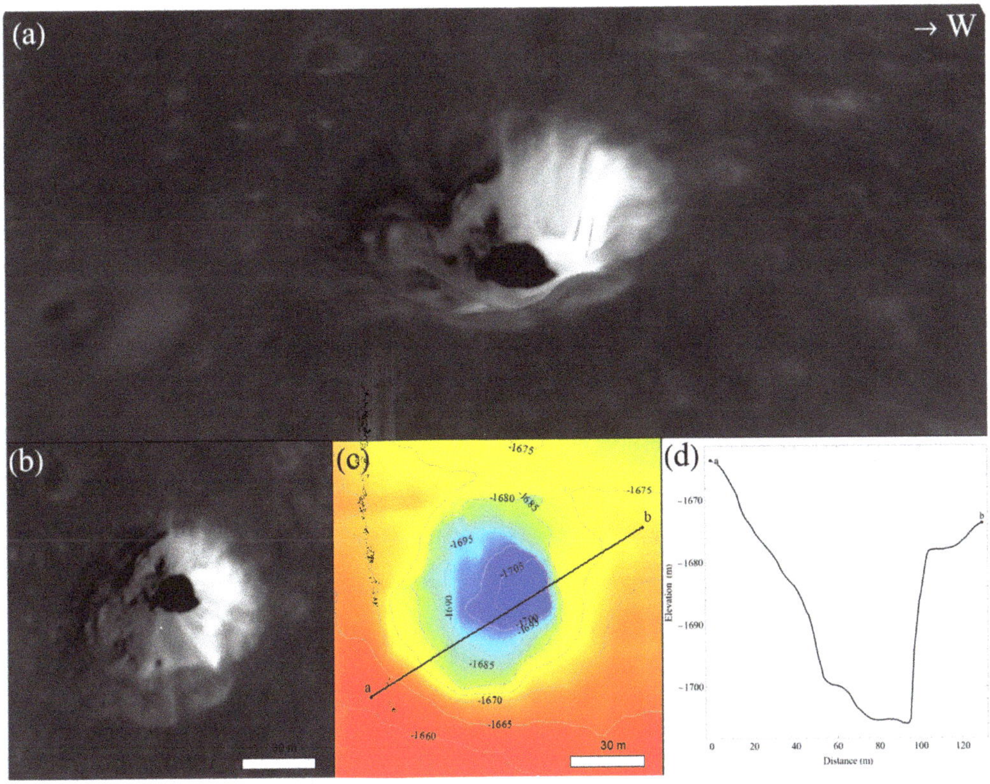

Figure 13. Pit crater (6.752° S, 42.759° E) on the west of Fecunditatis. (**a**) 3D image (real elevation values are used); (**b**) NAC orthoimage; (**c**) NAC DTM; (**d**) elevation of the profiles of (**c**). Data source: NAC M1100931431LE and NAC M1120953701RE.

This paper employs Integrated Software for Imagers and Spectrometers (ISIS) to make the high-resolution NAC DTM (Figures 12c and 13c) of two pit craters. The entrance to the central Mare Fecunditatis pit crater (Figure 12) is about 125 × 100 m in size and 35 m in depth. Deposits lie on the southeast and northwest wall. The deposits on the northwest wall extend ~30 m laterally, with a slope of 20–65°. The southeast wall has collapsed severely, with a relatively gentle slope (10–35°). Most of the deposits are finer than the resolution of the NAC image (~1.1 m), with only a few meter-level rocks at the bottom of the pit crater. The bottom of the pit is relatively flat (<10°), with an area of 27 × 23 m^2 lower than 15°. The highland pit crater on the southwest side of Mare Fecunditatis (Figure 13) is about 70 m in diameter, and the walls are very steep in all directions. The gentlest southwest wall is 30–60°, and the maximum slope of the northeast wall exceeds 80°. The bottom of the pit is relatively flat (<10°) with an area of 23 × 10 m^2.

3.4.2. Swirls

Lunar swirls are high-albedo loops and ribbons occurring in both maria and highlands associated with strong crustal magnetic fields. The formation of the swirls may be related to (1) the magnetic anomalies blocking the solar wind ion bombardment, which reduces the degree of space weathering (darkening with time) [103–106]; (2) comet impacts or micrometeoroid swarms scouring the top-most surface regolith, which exposes fresh material and imparts a remnant magnetization [107–109]; (3) weak electric fields attract the high-albedo, fine-grained, feldspar-rich dust [110]. The interaction between the solar wind and silicate regolith produces water (e.g., [111,112]). Moon Mineralogy Mapper (M^3, [113]) spectral data show that the regolith on the swirl has lower water than the surrounding area [106], which supports the hypothesis (1).

There are three swirls (Figure 14, identified by [19]) on the highland at the southwest side of Mare Fecunditatis with a low optical maturity (OMAT, Figure 14b, [40]), and the surface vector mapping (SVM, [114]) of magnetic field (B-flied) shows this area has a magnitude of up to 100 nT (Figure 14d). The MI mineral data (Figure 14c, [40]) of the swirl show consistent plagioclase with the surrounding area, which contradicts hypothesis (3).

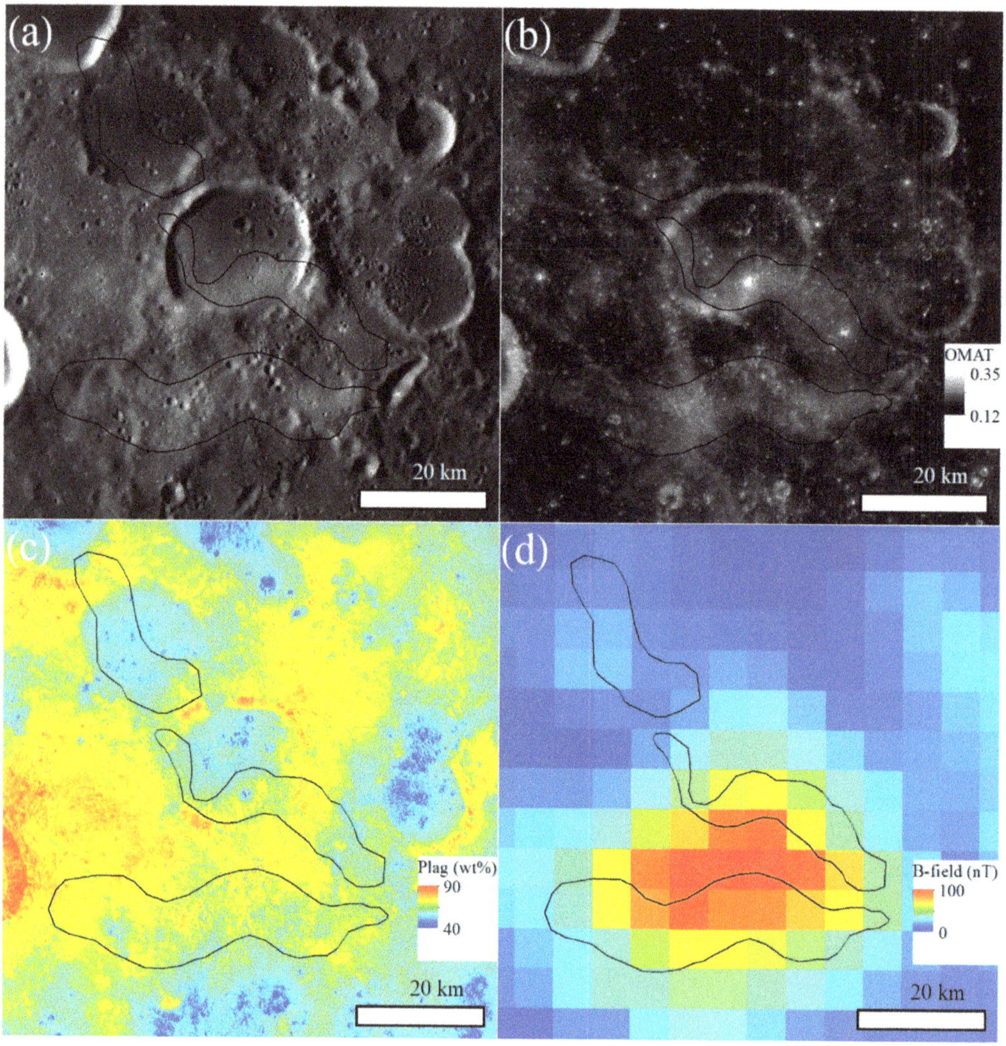

Figure 14. Swirls (14.9° S, 51.8° E) on the west of Mare Fecunditatis. (**a**) WAC image; (**b**) MI OMAT image (brighter tones indicate lower maturity); (**c**) MI Plagioclase abundance; (**d**) SVM magnetic field magnitude map.

4. Discussion

After analyzing the geological features in Mare Fecunditatis, eight targets are selected with high scientific values. First, we summarize their scientific significance and then design a traverse to detect all of these targets for a composite long-distance mission.

4.1. High-Value Detection Targets

4.1.1. Pit Craters

Pit craters are the natural cross-section to the subsurface of mare basalts, which could be used to study multiple eruptions of lunar volcanism and ancient regolith. Pit craters also connect to underground space, which has excellent conditions for the construction of a long-term lunar research station [115,116], such as (1) relatively constant temperature [117,118], (2) low cosmic ray flux [119,120], (3) little lunar dust [98], and (4) shielding from micrometeorite impact [120]. However, the entrance of the pit crater is usually very steep. The pit crater in the middle of Mare Fecunditatis has a collapsed wall (10–35°) comprised of deposits on the southeastern side, which makes it easy to enter the underground space technically. A robot with inflatable composite structures [121] with laser scanners [122] and other payloads may be considered to enter the cave. If successful, this project be the first to enter the moon and investigate in-situ lava outcrops.

4.1.2. Young Volcanism

Crater size–frequency distribution (CSFD) measurements are widely used to date the planetary surface. The CSFD measurements rely heavily on the lunar crater chronology function (e.g., [123–127]), which is based on the returned lunar samples with known sources. However, due to the lack of young samples (<3 Ga) at the age range of 1–3 Ga, the chronology function maybe not be constrained well [128]. The lunar chronology function is gentle, between 1 and 3 Ga, plus there is a small crater population in the young region, which will bring additional errors of dating a young lunar surface [129]. Although Chang'E-5 brought back some young lunar basalts, with the youngest dating age being ~2.0 Ga [130,131], more samples are needed to better constrain the lunar chronology function within the age between 1.0 and 3.0 Ga. Basalts in Mare Fecunditatis with ages between ~2.2 and ~2.9 Ga are ideal places to constrain the chronology function.

The timing of the latest lunar volcanic activity is extremely important for understanding the thermal evolution history of the Moon. The Moon is expected to cool rapidly and freeze the molten mantle layer quickly after its formation due to its small size [132]. Some authors simulated that the heat is enough to maintain partial melting from the mantle and eruption of mare basalts, until 2.2–3.4 Ga [133]. The youngest mare basalts are mainly distributed in Oceanus Procellarum (e.g., [21,25,134]) with the concentration of heat-producing elements (e.g., [135–137]). However, the 2.0 Ga Chang'E-5 basalts are neither admixing KREEP (enriched in potassium, rare-earth elements, and phosphorus) [130,138,139] nor enriching volatiles [140], raising the question of what derives the late-stage basalts. The young mare basalts within Mare Fecunditatis are far away from the Lunar Procellarum KREEP Terrane (PKT) and may have a distinct origin with Chang'E-5 samples. All of the IMPs are emplaced in the Eratosthenian basalts, and RMDSs are distributed in the Imbrian basalts with TiO_2 > 3 wt% in Mare Fecunditatis; their spatial and time relationship may provide additional clues of the young lunar volcanism. Plenty of in-situ chemical detections, sample returns, and rock dating are required, and it is noteworthy that there may be some sub-meter rocks in IMPs.

4.1.3. High-Al Basalts

Lunar basalt samples have significantly enhanced our understanding of the lunar mantle composition and structures. High-Al basalts are widely distributed across the Moon [45,46]. However, high-Al basalt samples are relatively few (Apollo 12, Luna 16, and one sample each from Apollo 14 and Apollo 16). Moreover, the age of these samples is bimodal, with few 3.4–4.0 Ga-old samples and no 3.5–3.8 Ga-old samples. Remote sensing data show that the southern Mare Fecunditatis is filled with low-Ti (<2 wt%) high-Al basalts of 3.6–3.7 Ga [60], which is distinct from the collected samples. Thus, the high-Al basalts in the southern Mare Fecunditatis could provide samples to study this type of special lunar volcanism.

4.1.4. Pyroclastic Deposits

Pyroclastic deposits result from explosive volcanic eruptions, originating from the deep mantle (~300 to ~400 km) [141,142]. The pyroclastic magma rises faster and may carry some lunar mantle xenoliths. Sample analysis (e.g., [143–146]) and infrared remote sensing data [147,148] showed over a hundred μg/g water in the pyroclastic glasses, indicating the existence of volatile reservoirs in the lunar mantle. More pyroclastic glasses are required to understand the water content and distribution in the lunar mantle. On the other hand, the water in the pyroclastic deposits may be an important in-situ resource.

There are two sites of pyroclastic deposits in the Mare Fecunditatis: (1) one in Crater Taruntius (5.4° N, 46.5° E), whose wall has a relatively high slope (15–30°), but is difficult to access; (2) another in the western region of the Mare Fecunditatis (3.0° S, 42.3° E), which is flat and easy to access.

4.1.5. Swirls

Lunar swirls are essential to understand (1) the history of the lunar core dynamo; (2) the magnetic effects of large-scale impacts; (3) the formation of the swirls; and (4) the specific space weathering effects (including the production of nanophase iron and water) to the regolith with weak solar wind ion bombardment. Hundreds of swirls have been identified on the lunar mare and highlands. However, none of the swirls have been detected by the in-situ missions. Lunar swirls on the west side of Mare Fecunditatis are across both highland and mare and most areas have a slope less than 15°, which is possible to detect. Swirls near Mare Fecunditatis are across the highlands and maria, which require detailed route planning with high-resolution images (e.g., NAC) and a rover with excellent climbing ability.

4.1.6. Fresh Craters

The nature and extent of lateral and vertical mixing of local and ejecta material are poorly constrained [1]. The eject rays from the fresh craters are less affected by space weathering, and it is easy to separate the young ejecta from local materials if they have distinct compositions. Petavius B (19.9°S, 57.0°E, 32.0 km in diameter), located in the southeastern Mare Fecunditatis, is the ideal target to study the mixing of local and ejected materials. It has a high albedo butterfly-shaped ray pattern stretching ~200 km with low-Fe and low-Ti contents.

In addition, sample returned from impact melts of the fresh crater, such as impact melts (tens of km^2) between Messier and Messier A, would provide additional calibration points at the younger range of the lunar chronology function, as a supplementary of the young impact melt samples of Copernicus, Tycho, North Ray, and Cone craters collected in the Apollo missions.

4.2. *Traverse Design for Long-Distance Detection*

Here, we envision a long-range exploration mission, referring to previous rovers and the Intrepid [8,149] (Table 1). The concept mission temporarily named Fengfu is under the brainstorming stage, and some content may be unrealistic right now. Most of the high-value detection targets selected in this study are distributed in the center and southern Mare Fecunditatis. A 1408 km traverse (Figure 15; Table 2) is designed with reference to the 60 m spatial resolution SLDEM2015 slope map and 10 m-spatial-resolution TC images [150]. The detection traverse starts from an RMDS (S1-1, 1.31° S, 52.50° E), passing through several RMDSs (S1), domes (S1-4 and S3), pit crater (S2), Messier Crater and its impact melt and ray system (S4), young mare basalt (S5), IMPs (S5), pyroclastic deposits (S6), sinuous rilles

(S6), high-Al basalt, swirls (S7), and impact blanket of the Petavius B. The average slope of the traverse is 2.58°, the slope of more than 94% in the traverse is less than 5°, and almost 100% has a slope of less than 10°. After entering the swirl on the highland, the slope is steeper, but still <20°.

Table 1. Travel distance, time, and speed of previous lunar rovers, the Intrepid, and the Fengfu.

Rover	Landing Time	Travel Distance	Travel Time	Average Speed	Top Speed
Lunokhod 1	1970	10.54 km	321 days	0.97 km/lunar day	0.1 km/h
Lunar Roving Vehicle (Apollo 15)	1971	27.76 km	3 h 02 min	9.15 km/h	13 km/h
Lunar Roving Vehicle (Apollo 16)	1972	26.55 km	3 h 26 min	7.73 km/h	13 km/h
Lunar Roving Vehicle (Apollo 17)	1972	35.89 km	4 h 26 min	8.10 km/h	13 km/h
Lunokhod 2	1973	39 km	136 days	8.46 km/lunar day	2 km/h
Yutu	2013	0.1148 km	42 days	80.63 m/lunar day	0.2 km/h
Yutu-2	2018	1.0039 km (10 January 2022)	1103 days	26.85 m/lunar day	0.2 km/h
Intrepid	-	1800 km	4 years	37.5 km/lunar day	1 km/h
Fengfu (this study)	-	1400 km	5 years	22.58 km/lunar day	1 km/h

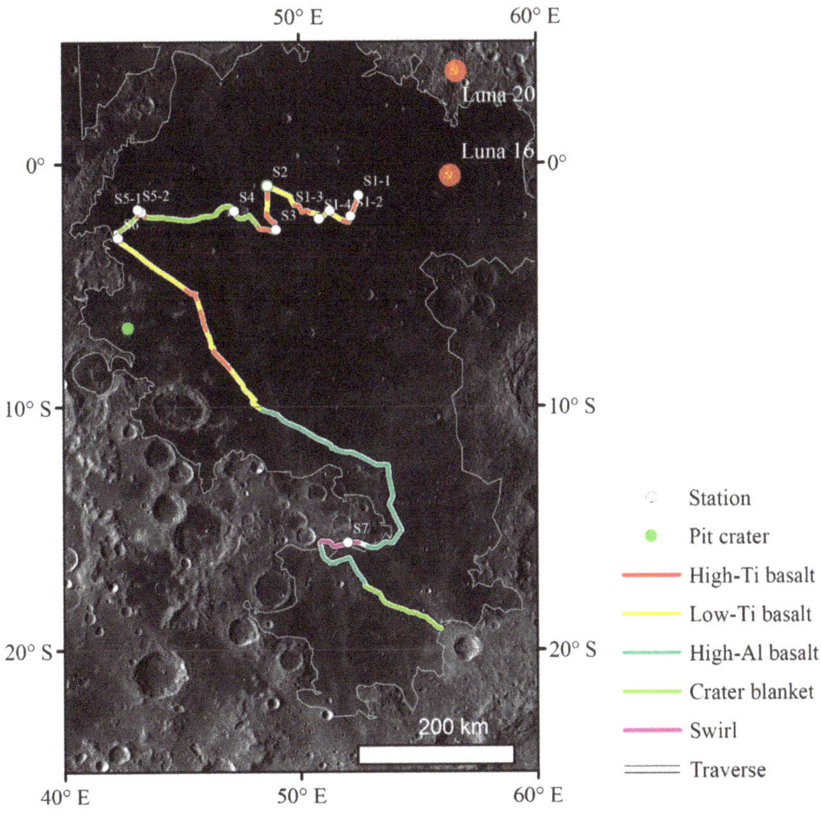

Figure 15. Long-distance detection traverse in Mare Fecunditatis, shown with a WAC image. White points represent the stations in the mission, green points represent the pit craters, and lines with different colors represent segments of the traverse. Considering the error of remote sensing data, in this figure, the high-Ti basalt-traverses are defined as TiO_2 > 5 wt% on the WAC TiO_2 map, and high-Al basalt-traverses are defined as Al_2O_3 > 13 wt% on the IIM chemical map.

Table 2. Stations of the long-distance detection traverse in Mare Fecunditatis.

Station	Latitude	Longitude	Exploration Time (Lunar Day)	Description
S1-1	1.31471° S	52.5026° E	1	RMDS (High-Ti)
S1-2	2.16267° S	52.1579° E	1	RMDS (Low-Ti)
S1-3	1.95172° S	51.265° E	1	RMDS (Low-Ti)
S1-4	2.26975° S	50.8258° E	1	RMDS and Dome
S2	0.917° S	48.66° E	5	Pit crater
S3	2.71559° S	49.0189° E	1	Dome
S4	1.94542° S	47.2645° E	1	Impact melt
S5-1	1.883° S	43.176° E	1	IMP in a crater
S5-2	2.16852° S	43.4695° E	1	IMP on the mare
S6	3.04264° S	42.348° E	1	Pyroclastic deposits and Sinuous rilles
S7	15.607° S	51.9968° E	1	Swirl
Exploration traverse: High-Ti basalt, Low-Ti basalt, High-Al basalt, Impact blanket, and Swirl				

The traverse needs an unmanned rover with scientific payloads and some small robots for sample collections. A small rover is greatly affected by the ambient temperature; for example, Yutu-2 (135 kg) has to finish working at noon due to the high temperature. Although a large rover is likely to have a good temperature-control system, we assume that the Fengfu rover only has a third of the time (236 h) to work per lunar day and has two working modes: (1) target-exploration mode, focusing on exploring targets and hardly moving; and (2) navigation mode, traveling and detecting passing regolith and rocks. Fifteen lunar days are allotted to the target–exploration mode (Table 2), including five lunar days for the pit crater, and one lunar day for another target. Complex targets (such as pit craters) would be explored in conjunction with astronauts or dedicated robots at the same time—the samples would be returned, and the rover would be repaired. Although the Fengfu could be designed to move very quickly, the rapid movement of the lunar rover faces the risk of dust accumulation [151,152]. If the average navigation speed of the Fengfu is 0.5 km/h (the same as the Intrepid), and (semi-automatically) travels 30 km (60 h) every lunar day, there are another 176 h to explore passing regolith and rocks (roughly traveling for 1 h and exploring for 3 h). It will take ~47 lunar days to finish the navigation. At the end of this mission, the rest of samples would be packeted and returned by another mission. A total of 62 lunar days (~5 years) are required to complete the mission.

Compared with detection targets in the Intrepid mission [8,149], both missions contain IMPs (potential very young volcanism), young basalts (modification the crater chronology curve and understanding the lunar thermal history), pyroclastic deposits (understanding the deep mantle), and swirls (understanding lunar core dynamo and space weathering effects). The targets in our work also include the pit crater (to detect underground space), RMDSs (potential very young volcanism), high-Al basalt (filling the gap between Apollo 14 and Luna 16), and large fresh craters with melt sheet (modification of the crater chronology curve) or distinct composition (understanding the mixture of local and ejecta material processes). The exploration of Mare Fecunditatis is significant to fully understand the history of lunar evolution.

5. Conclusions

A long-distance lunar traverse could solve more fundamental scientific questions compared with a nominal short-distance mission. Mare Fecunditatis has a multitude of high-value features and a flat terrain, which is very suitable for long-distance detection. In total, several outstanding detection targets are selected: (1) pit crater may connect vast underground void space, which is an ideal place to build a research station; (2) young basalts (2.2–2.9 Ga) and the impact melts of Messier Crater may modify the chronology functions; (3) young basalts and potential very young volcanism (IMP and RMDS) are key to the thermal evolution history of the Moon; (4) basalts in the southern Fecunditatis may be a type of high-Al basalt that differs from previous samples; (5) pyroclastic deposits are windows to understanding the composition of the deep mantle; (6) swirls and magnetic anomalies are important to the history of lunar core dynamo and space weathering effects; and (7)the ray system of the fresh crater may disclose the mixing of local and ejecta material processes. Based on this, a ~1408 km traverse over 5 years has been proposed to explore Mare Fecunditatis, which may help future planning of this class of long-distance missions on the lunar surface.

Author Contributions: Conceptualization, L.X.; methodology, S.Z.; investigation, S.Z.; writing—original draft preparation, S.Z.; writing—review and editing, S.Z., Y.Q., L.X., J.Z., Q.H., J.H., J.W., H.C. and W.X.; visualization, S.Z.; funding acquisition, L.X. All authors have read and agreed to the published version of the manuscript.

Funding: This research was funded by the National Key R & D Program of China (2020YFE0202100) MOST Special Fund from the State Key Laboratory of Geological Processes and Mineral Resources, China University of Geosciences (MSFGPMR05), and the 111 project of China (BP0820004).

Data Availability Statement: The WAC images, NAC images, SLDEM2015, Clementine color ratio map, WAC TiO_2 map, and MI mineral and composition maps are available at PDS Geosciences Node (https://pds-geosciences.wustl.edu/, accessed on 7 January 2022). The TC images and TCDTM are available at SELENE Data Archive (https://darts.isas.jaxa.jp/planet/pdap/selene/index.html.en, accessed on 7 January 2022). The IIM chemical maps are available at https://mega.nz/#F!6ThgVK6a!_x1Pq1p5UW4SVvbK640wow, accessed on 7 January 2022. The SVM magnetic field maps are available at http://www.geo.titech.ac.jp/lab/tsunakawa/Kaguya_LMAG, accessed on 7 January 2022.

Acknowledgments: The authors would like to thank three anonymous reviewers for their insightful and constructive comments and suggestions.

Conflicts of Interest: The authors declare no conflict of interest.

References

1. National Research Council. *The Scientific Context for Exploration of the Moon*; The National Academies Press: Washington, DC, USA, 2007; ISBN 978-0-309-10919-2.
2. Flahaut, J.; Blanchette-Guertin, J.-F.; Jilly, C.; Sharma, P.; Souchon, A.; van Westrenen, W.; Kring, D.A. Identification and characterization of science-rich landing sites for lunar lander missions using integrated remote sensing observations. *Adv. Space Res.* **2012**, *50*, 1647–1665. [CrossRef]
3. Jawin, E.R.; Valencia, S.N.; Watkins, R.N.; Crowell, J.M.; Neal, C.R.; Schmidt, G. Lunar Science for Landed Missions Workshop Findings Report. *Earth Space Sci.* **2019**, *6*, 2–40. [CrossRef]
4. Kring, D.A.; Durda, D.D. (Eds.) *A Global Lunar Landing Site Study to Provide the Scientific Context for Exploration of the Moon*; National Academies Press: Washington, DC, USA, 2012.
5. Hiesinger, H.; van der Bogert, C.H.; Wedler, A.; Jaumann, R.; Mall, U.; Head, J.W.; Anand, M.; Jolliff, B.; Pinet, P.; Xiao, L.; et al. The Rima Bode Region—Candidate for a Future Lunar Landing Site. In Proceedings of the 52nd Lunar and Planetary Science Conference, Virtual Event, 15–19 March 2021; p. 1485.
6. Flahaut, J.; Carpenter, J.; Williams, J.-P.; Anand, M.; Crawford, I.A.; van Westrenen, W.; Füri, E.; Xiao, L.; Zhao, S. Regions of interest (ROI) for future exploration missions to the lunar South Pole. *Planet. Space Sci.* **2019**, *180*, 104750. [CrossRef]
7. Lemelin, M.; Blair, D.M.; Roberts, C.E.; Runyon, K.D.; Nowka, D.; Kring, D.A. High-priority lunar landing sites for in situ and sample return studies of polar volatiles. *Planet. Space Sci.* **2014**, *101*, 149–161. [CrossRef]
8. Robinson, M.S.; Blewett, D.T.; Frank, E.; Illsley, P.; Lawrence, S.; Mahanti, P.; Rampe, E.B.; Speyerer, E.J.; Stopar, J.; Tikoo, S.; et al. Intrepid: Lunar Roving Prospector. In Proceedings of the AGU Fall Meeting, San Francisco, CA, USA, 9–13 December 2019; American Geophysical Union: Washington, DC, USA, 2019; p. P33D-03.
9. Zhang, C.; Xu, W.; Wang, Y. Development Ideas of Manned Lunar Surface Exploration. *Aerosp. Shanghai* **2021**, *38*, 109–118. [CrossRef]
10. Wilhelms, D.E.; McCauley, J.F. *Geologic Map of the Near Side of the MOON*; IMAP 1971; USGS: Reston, VI, USA, 1971. [CrossRef]
11. Wilhelms, D.E.; McCauley, J.F.; Trask, N.J. *The Geologic History of the Moon*; Professional Paper; USGS: Reston, VI, USA, 1987. [CrossRef]
12. Hiesinger, H.; Head, J.W.; Wolf, U.; Jaumann, R.; Neukum, G. New Ages for Basalts in Mare Fecunditatis Based on Crater Size-Frequency Measurements. In Proceedings of the 37th Lunar and Planetary Science Conference, The Woodlands, Houston, TX, USA, 13–27 March 2006; p. 1151.
13. Robinson, M.S.; Brylow, S.M.; Tschimmel, M.; Humm, D.; Lawrence, S.J.; Thomas, P.C.; Denevi, B.W.; Bowman-Cisneros, E.; Zerr, J.; Ravine, M.A.; et al. Lunar Reconnaissance Orbiter Camera (LROC) Instrument Overview. *Space Sci. Rev.* **2010**, *150*, 81–124. [CrossRef]

14. Wagner, R.V.; Robinson, M.S. Occurrence and Origin of Lunar Pits: Observations from a New Catalog. In Proceedings of the 52nd Lunar and Planetary Science Conference, Virtual Event, 15–19 March 2021.
15. Braden, S.E.; Stopar, J.D.; Robinson, M.S.; Lawrence, S.J.; van der Bogert, C.H.; Hiesinger, H. Evidence for basaltic volcanism on the Moon within the past 100 million years. *Nat. Geosci.* **2014**, *7*, 787–791. [CrossRef]
16. Zhang, F.; Head, J.W.; Wöhler, C.; Bugiolacchi, R.; Wilson, L.; Basilevsky, A.T.; Grumpe, A.; Zou, Y.L. Ring-Moat Dome Structures (RMDSs) in the Lunar Maria: Statistical, Compositional, and Morphological Characterization and Assessment of Theories of Origin. *J. Geophys. Res.* **2020**, *125*, e2019JE005967. [CrossRef]
17. Gaddis, L.R.; Staid, M.I.; Tyburczy, J.A.; Hawke, B.; Petro, N.E. Compositional analyses of lunar pyroclastic deposits. *Icarus* **2003**, *161*, 262–280. [CrossRef]
18. Gustafson, J.O.; Bell, J.F.; Gaddis, L.R.; Hawke, B.R.; Giguere, T.A. Characterization of previously unidentified lunar pyroclastic deposits using Lunar Reconnaissance Orbiter Camera data. *J. Geophys. Res.* **2012**, *117*, e12. [CrossRef]
19. Denevi, B.W.; Robinson, M.S.; Boyd, A.K.; Blewett, D.T.; Klima, R.L. The distribution and extent of lunar swirls. *Icarus* **2016**, *273*, 53–67. [CrossRef]
20. Ravi, S.; Meyer, H.M.; Mahanti, P.; Robinson, M.S. On the usefulness of optical maturity for relative age classification of fresh craters. In Proceedings of the AGU Fall Meeting, San Francisco, CA, USA, 12–16 December 2016; p. P53A-2166.
21. Hiesinger, H.; Head, J.W.; Wolf, U.; Jaumann, R.; Neukum, G. Ages and stratigraphy of lunar mare basalts: A synthesis. In *Geological Society of America Special Papers: Recent Advances and Current Research Issues in Lunar Stratigraphy*; Geological Society of America: Boulder, CO, USA, 2011; p. 477. [CrossRef]
22. Hiesinger, H.; Head, J.W. Ages and stratigraphy of mare basalts in Oceanus Procellarum, Mare Nubium, Mare Cognitum, and Mare Insularum. *J. Geophys. Res.* **2003**, *108*, e7. [CrossRef]
23. Hiesinger, H.; Head, J.W.; Wolf, U.; Jaumann, R.; Neukum, G. Lunar mare basalt flow units: Thicknesses determined from crater size-frequency distributions. *Geophys. Res. Lett.* **2002**, *29*, 89-1–89-4. [CrossRef]
24. Hiesinger, H.; Head, J.W.; Wolf, U.; Jaumann, R.; Neukum, G. Ages and stratigraphy of lunar mare basalts in Mare Frigoris and other nearside maria based on crater size-frequency distribution measurements. *J. Geophys. Res.* **2010**, *115*, e3. [CrossRef]
25. Hiesinger, H.; Jaumann, R.; Neukum, G.; Head, J.W. Ages of mare basalts on the lunar nearside. *J. Geophys. Res.* **2000**, *105*, 29239–29275. [CrossRef]
26. Barker, M.K.; Mazarico, E.; Neumann, G.A.; Zuber, M.T.; Haruyama, J.; Smith, D.E. A new lunar digital elevation model from the Lunar Orbiter Laser Altimeter and SELENE Terrain Camera. *Icarus* **2016**, *273*, 346–355. [CrossRef]
27. McCauley, J.F.; Scott, D.H. The geologic setting of the Luna 16 landing site. *Earth Planet. Sci. Lett.* **1972**, *13*, 225–232. [CrossRef]
28. Rosenburg, M.A.; Aharonson, O.; Head, J.W.; Kreslavsky, M.A.; Mazarico, E.; Neumann, G.A.; Smith, D.E.; Torrence, M.H.; Zuber, M.T. Global surface slopes and roughness of the Moon from the Lunar Orbiter Laser Altimeter. *J. Geophys. Res.* **2011**, *116*, E11007. [CrossRef]
29. Zuber, M.T.; Smith, D.E.; Zellar, R.S.; Neumann, G.A.; Sun, X.; Katz, R.B.; Kleyner, I.; Matuszeski, A.; McGarry, J.F.; Ott, M.N.; et al. The Lunar Reconnaissance Orbiter Laser Ranging Investigation. *Space Sci. Rev.* **2010**, *150*, 63–80. [CrossRef]
30. Smith, D.E.; Zuber, M.T.; Jackson, G.B.; Cavanaugh, J.F.; Neumann, G.A.; Riris, H.; Sun, X.; Zellar, R.S.; Coltharp, C.; Connelly, J.; et al. The Lunar Orbiter Laser Altimeter Investigation on the Lunar Reconnaissance Orbiter Mission. *Space Sci. Rev.* **2010**, *150*, 209–241. [CrossRef]
31. Muller, P.M.; Sjogren, W.L. Mascons: Lunar Mass Concentrations. *Science* **1968**, *161*, 680–684. [CrossRef]
32. Neumann, G.A.; Zuber, M.T.; Wieczorek, M.A.; Head, J.W.; Baker, D.M.H.; Solomon, S.C.; Smith, D.E.; Lemoine, F.G.; Mazarico, E.; Sabaka, T.J.; et al. Lunar impact basins revealed by Gravity Recovery and Interior Laboratory measurements. *Sci. Adv.* **2015**, *1*, e1500852. [CrossRef]
33. Hurwitz, D.M.; Head, J.W.; Hiesinger, H. Lunar sinuous rilles: Distribution, characteristics, and implications for their origin. *Planet. Space Sci.* **2013**, *79–80*, 1–38. [CrossRef]
34. Yue, Z.; Li, W.; Di, K.; Liu, Z.; Liu, J. Global mapping and analysis of lunar wrinkle ridges. *J. Geophys. Res.* **2015**, *120*, 978–994. [CrossRef]
35. Jozwiak, L.M.; Head, J.W.; Zuber, M.T.; Smith, D.E.; Neumann, G.A. Lunar floor-fractured craters: Classification, distribution, origin and implications for magmatism and shallow crustal structure. *J. Geophys. Res.* **2012**, *117*, 6. [CrossRef]
36. Rajmon, D.; Spudis, P. Distribution and stratigraphy of basaltic units in Maria Tranquillitatis and Fecunditatis: A Clementine perspective. *Meteorit. Planet. Sci.* **2004**, *39*, 1699–1720. [CrossRef]
37. Dehon, R.A. Mare Fecunditatis: Basin Configuration. In Proceedings of the Conference on Origins of Mare Basalts and Their Implications for Lunar Evolution, The Woodlands, Houston, TX, USA, 17–19 November 1975; Lunar Science Institute: Houston, TX, USA, 1975; p. 32.
38. Ohtake, M.; Haruyama, J.; Matsunaga, T.; Yokota, Y.; Morota, T.; Honda, C. Performance and scientific objectives of the SELENE (KAGUYA) Multiband Imager. *Earth Planets Space* **2008**, *60*, 257–264. [CrossRef]
39. Otake, H.; Ohtake, M.; Hirata, N. Lunar Iron and Titanium Abundance Algorithms Based on SELENE (Kaguya) Multiband Imager Data. In Proceedings of the 43rd Lunar and Planetary Science Conference, The Woodlands, Houston, TX, USA, 19–23 March 2012; p. 1905.

40. Lemelin, M.; Lucey, P.G.; Gaddis, L.R.; Hare, T.; Ohtake, M. Global Map Products from the Kaguya Multiband Imager at 512 ppd: Minerals, FeO, and OMAT. In Proceedings of the 47th Lunar and Planetary Science Conference, The Woodlands, Houston, TX, USA, 21–25 March 2016; p. 2994.
41. Sato, H.; Robinson, M.S.; Lawrence, S.J.; Denevi, B.W.; Hapke, B.; Jolliff, B.L.; Hiesinger, H. Lunar mare TiO2 abundances estimated from UV/Vis reflectance. *Icarus* **2017**, *296*, 216–238. [CrossRef]
42. Wu, Y.; Besse, S.; Li, J.-Y.; Combe, J.-P.; Wang, Z.; Zhou, X.; Wang, C. Photometric correction and in-flight calibration of Chang' E-1 Interference Imaging Spectrometer (IIM) data. *Icarus* **2013**, *222*, 283–295. [CrossRef]
43. Ouyang, Z.; Li, C.; Zou, Y.; Zhang, H.; Lü, C.; Liu, J.; Liu, J.; Zuo, W.; Su, Y.; Wen, W.; et al. Primary scientific results of Chang'E-1 lunar mission. *Sci. China Earth Sci.* **2010**, *53*, 1565–1581. [CrossRef]
44. Xia, W.; Wang, X.; Zhao, S.; Jin, H.; Chen, X.; Yang, M.; Wu, X.; Hu, C.; Zhang, Y.; Shi, Y.; et al. New maps of lunar surface chemistry. *Icarus* **2019**, *321*, 200–215. [CrossRef]
45. Kramer, G.Y.; Jolliff, B.L.; Neal, C.R. Distinguishing high-alumina mare basalts using Clementine UVVIS and Lunar Prospector GRS data: Mare Moscoviense and Mare Nectaris. *J. Geophys. Res.* **2008**, *113*, 10. [CrossRef]
46. Kramer, G.Y.; Jolliff, B.L.; Neal, C.R. Searching for high alumina mare basalts using Clementine UVVIS and Lunar Prospector GRS data: Mare Fecunditatis and Mare Imbrium. *Icarus* **2008**, *198*, 7–18. [CrossRef]
47. Kurat, G.; Kracher, A.; Keil, K.; Warner, R.; Prinz, M. Composition and origin of Luna 16 aluminous mare basalts. In *Lunar and Planetary Science Conference Proceedings*; Pergamon Press: New York, NY, USA, 1976; Volume 2, pp. 1301–1321.
48. Ma, M.-S.; Schmitt, R.A.; Nielsen, R.L.; Taylor, G.J.; Warner, R.D.; Keil, K. Petrogenesis of Luna 16 aluminous Mare basalts. *Geophys. Res. Lett.* **1979**, *6*, 909–912. [CrossRef]
49. Huneke, J.C.; Podosek, F.A.; Wasserburg, G.J. Gas retention and cosmic-ray exposure ages of a basalt fragment from Mare Fecunditatis. *Earth Planet. Sci. Lett.* **1972**, *13*, 375–383. [CrossRef]
50. Cadogan, P.H.; Turner, G. 40Ar-39Ar dating of Luna 16 and Luna 20 samples. *Phil. Trans. R. Soc. Lond. A* **1977**, *284*, 167–177. [CrossRef]
51. Cohen, B.A.; Snyder, G.A.; Hall, C.M.; Taylor, L.A.; Nazarov, M.A. Argon-40-argon-39 chronology and petrogenesis along the eastern limb of the Moon from Luna 16, 20 and 24 samples. *Meteorit. Planet. Sci.* **2001**, *36*, 1345–1366. [CrossRef]
52. Fernades, V.A.; Burgess, R. Volcanism in Mare Fecunditatis and Mare Crisium: Ar-Ar age studies. *Geochim. Cosmochim. Acta* **2005**, *69*, 4919–4934. [CrossRef]
53. Haruyama, J.; Ohtake, M.; Matsunaga, T.; Morota, T.; Yokota, Y.; Honda, C.; Hirata, N.; Demura, H.; Iwasaki, A.; Nakamura, R.; et al. Planned radiometrically calibrated and geometrically corrected products of lunar high-resolution Terrain Camera on SELENE. *Adv. Space Res.* **2008**, *42*, 310–316. [CrossRef]
54. Greeley, R. Lunar hadley rille: Considerations of its origin. *Science* **1971**, *172*, 722–725. [CrossRef]
55. Jozwiak, L.M.; Head III, J.W.; Neumann, G.A.; Wilson, L. Observational constraints on the identification of shallow lunar magmatism: Insights from floor-fractured craters. *Icarus* **2017**, *283*, 224–231. [CrossRef]
56. Jozwiak, L.M.; Head, J.W.; Wilson, L. Lunar floor-fractured craters as magmatic intrusions: Geometry, modes of emplacement, associated tectonic and volcanic features, and implications for gravity anomalies. *Icarus* **2015**, *248*, 424–447. [CrossRef]
57. Hall, J.L.; Solomon, S.C.; Head, J.W. Lunar floor-fractured craters: Evidence for viscous relaxation of crater topography. *J. Geophys. Res. Solid Earth* **1981**, *86*, 9537–9552. [CrossRef]
58. Wichman, R.W.; Schultz, P.H. Floor-fractured craters in Mare Smythii and west of Oceanus Procellarum: Implications of crater modification by viscous relaxation and igneous intrusion models. *J. Geophys. Res.* **1995**, *100*, 21201–21218. [CrossRef]
59. Dombard, A.J.; Gillis, J.J. Testing the viability of topographic relaxation as a mechanism for the formation of lunar floor-fractured craters. *J. Geophys. Res.* **2001**, *106*, 27901–27909. [CrossRef]
60. Hiesinger, H.; Head, J.W. New Views of Lunar Geoscience: An Introduction and Overview. *Rev. Mineral. Geochem.* **2006**, *60*, 1–81. [CrossRef]
61. Gaddis, L.R. Progress toward characterization of juvenile materials in lunar pyroclastic deposits. In *Workshop on New Views of the Moon 2: Understanding the Moon Through the Integration of Diverse Datasets*; Gaddis, L., Shearer, C.K., Eds.; Lunar and Planetary Institute: Houston, TX, USA, 1999.
62. Spudis, P.D. Young dark mantle deposits on the Moon. In Proceedings of the Workshop on Lunar Volcanic Glasses: Scientific and Resource Potential, Houston, TX, USA, 10–11 October 1989; LPI Technical Report 90-02. Delano, J.W., Heiken, G.H., Eds.; Lunar and Planetary Institute: Houston, TX, USA, 1990; p. 60.
63. Whitaker, E.A. An Unusual Mare Feature. In *Apollo 15: Preliminary Science Report*; Swann, G.A., Bailey, N.G., Batson, R.M., Freeman, V.L., Hait, M.H., Head, J.W., Holt, H.E., Howard, K.A., Irwin, J.B., Larson, K.B., et al., Eds.; NASA: Washington, DC, USA, 1972; p. 84.
64. El-Baz, F. New geological findings in Apollo 15 lunar orbital photography. In Proceedings of the Third Lunar Science Con-ference, Houston, TX, USA, 10 January 1972.
65. El-Baz, F. "D-Caldera": New Photographs of a Unique Feature. In *Apollo 17: Preliminary Science Report*; Renner, K., Ed.; NASA: Washington, DC, USA, 1973.
66. Strain, P.L.; El-Baz, F. The geology and morphology of Ina. In Proceedings of the 11th Lunar and Planetary Science Conference, The Woodlands, Houston, TX, USA, 17–21 March 1980; Pergamon Press: New York, NY, USA, 1980; pp. 2437–2446.
67. Schultz, P.H.; Staid, M.I.; Pieters, C.M. Lunar activity from recent gas release. *Nature* **2006**, *444*, 184–186. [CrossRef]

68. Garry, W.B.; Robinson, M.S.; Zimbelman, J.R.; Bleacher, J.E.; Hawke, B.R.; Crumpler, L.S.; Braden, S.E.; Sato, H. The origin of Ina: Evidence for inflated lava flows on the Moon. *J. Geophys. Res.* **2012**, *117*, e12. [CrossRef]
69. Carter, L.M.; Hawke, B.R.; Garry, W.B.; Campbell, B.A.; Giguere, T.A.; Bussey, D.B.J. Radar Observations of Lunar Hollow Terrain. In Proceedings of the 44th Lunar and Planetary Science Conference, The Woodlands, Houston, TX, USA, 18–22 March 2013; p. 2146.
70. Qiao, l.; Head, J.; Wilson, L.; Xiao, L.; Kreslavsky, M.; Dufek, J. Ina pit crater on the Moon: Extrusion of waning-stage lava lake magmatic foam results in extremely young crater retention ages. *Geology* **2017**, *45*, 455–458. [CrossRef]
71. Wilson, L.; Head, J.W. Eruption of magmatic foams on the Moon: Formation in the waning stages of dike emplacement events as an explanation of "irregular mare patches". *J. Volcanol. Geotherm. Res.* **2017**, *335*, 113–127. [CrossRef]
72. Qiao, l.; Head, J.W.; Xiao, L.; Wilson, L.; Dufek, J.D. The role of substrate characteristics in producing anomalously young crater retention ages in volcanic deposits on the Moon: Morphology, topography, subresolution roughness, and mode of emplacement of the Sosigenes lunar irregular mare patch. *Meteorit. Planet. Sci.* **2018**, *53*, 778–812. [CrossRef]
73. Qiao, l.; Head, J.W.; Ling, Z.; Wilson, L.; Xiao, L.; Dufek, J.D.; Yan, J. Geological Characterization of the Ina Shield Volcano Summit Pit Crater on the Moon: Evidence for Extrusion of Waning-Stage Lava Lake Magmatic Foams and Anomalously Young Crater Retention Ages. *J. Geophys. Res.* **2019**, *124*, 1100–1140. [CrossRef]
74. Head, J.W.; Wilson, L. Generation, ascent and eruption of magma on the Moon: New insights into source depths, magma supply, intrusions and effusive/explosive eruptions (Part 2: Predicted emplacement processes and observations). *Icarus* **2017**, *283*, 176–223. [CrossRef]
75. Wöhler, C.; Lena, R.; Lazzarotti, P.; Phillips, J.; Wirths, M.; Pujic, Z. A combined spectrophotometric and morphometric study of the lunar mare dome fields near Cauchy, Arago, Hortensius, and Milichius. *Icarus* **2006**, *183*, 237–264. [CrossRef]
76. Huang, Q.; Zhao, J.; Wang, X.; Wang, T.; Zhang, F.; Qiao, L.; Chen, Y.; Qiu, D.; Yang, Y.; Xiao, L. A large long-lived central-vent volcano in the Gardner region: Implications for the volcanic history of the nearside of the Moon. *Earth Planet. Sci. Lett.* **2020**, *542*, 116301. [CrossRef]
77. Zhang, F.; Head, J.W.; Basilevsky, A.T.; Bugiolacchi, R.; Komatsu, G.; Wilson, L.; Fa, W.; Zhu, M.-H. Newly Discovered Ring-Moat Dome Structures in the Lunar Maria: Possible Origins and Implications. *Geophys. Res. Lett.* **2017**, *44*, 9216–9224. [CrossRef]
78. Basilevsky, A.T.; Zhang, F.; Wöhler, C.; Bugiolacchi, R.; Head, J.W.; Wilson, L. Lunar Ring-Moat Dome Structures and Their Relationships with Small Impact Craters. In Proceedings of the 50th Lunar and Planetary Science Conference, The Woodlands, Houston, TX, USA; 2019; p. 1507.
79. Zhang, F.; Head, J.W.; Wöhler, C.; Basilevsky, A.T.; Wilson, L.; Xie, M.; Bugiolacchi, R.; Wilhelm, T.; Althoff, S.; Zou, Y.L. The Lunar Mare Ring-Moat Dome Structure (RMDS) Age Conundrum: Contemporaneous with Imbrian-Aged Host Lava Flows or Emplaced in the Copernican? *J. Geophys. Res. Planets* **2021**, *126*, e2021JE006880. [CrossRef]
80. Bryan, W.B. Wrinkle-ridges as deformed surface crust on ponded mare lava. In Proceedings of the Fourth Lunar Science Conference, The Woodlands, Houston, TX, USA, 17 March 1973; Pergamon Press: New York, NY, USA, 1973.
81. Howard, K.A.; Muehlberger, W.R. Mare ridges and related studies: Part C: Lunar thrust faults in the Taurus-Littrow region. In *Apollo 17: Preliminary Science Report*; Renner, K., Ed.; NASA: Washington, DC, USA, 1973.
82. Lucchitta, B.K. Mare ridges and related highland scarps-Result of vertical tectonism. In Proceedings of the Seventh Lunar Science Conference, Houston, TX, USA, 15–19 March 1976.
83. Quaide, W. Rilles, ridges, and domes—Clues to maria history. *Icarus* **1965**, *4*, 374–389. [CrossRef]
84. Strom, R.G. Lunar Mare Ridges, Rings and Volcanic Ring Complexes. In *The Moon, Proceedings from IAU Symposium No. 47, the University of Newcastle-Upon-Tyne England, 22–26 March 1971*; Runcorn, S.K., Urey, H.C., Eds.; Springer International Publishing: Cham, Switzerland, 1972; pp. 187–215. ISBN 978-94-010-2861-5.
85. Scott, D.H. Mare ridges and related studies: Part D: Small structures of the Taurus-Littrow region. In *Apollo 17: Preliminary Science Report*; Renner, K., Ed.; NASA: Washington, DC, USA, 1973.
86. Young, R.A.; Brennan, W.J.; Wolfe, R.W.; Nichols, D.J. Mare ridges and related studies: Part A: Volcanism in the lunar maria. In *Apollo 17: Preliminary Science Report*; Renner, K., Ed.; NASA: Washington, DC, USA, 1973.
87. Scott, D.H.; Diaz, J.M.; Watkins, J.A. The geologic evaluation and regional synthesis of metric and panoramic photographs. In Proceedings of the Sixth Lunar Science Conference, Houston, TX, USA, 17–21 March 1975; pp. 2531–2540.
88. Yue, Z.; Michael, G.G.; Di, K.; Liu, J. Global survey of lunar wrinkle ridge formation times. *Earth Planet. Sci. Lett.* **2017**, *477*, 14–20. [CrossRef]
89. Thompson, T.J.; Robinson, M.S.; Watters, T.R.; Johnson, M.B. Global Lunar Wrinkle Ridge Identification and Analysis. In Proceedings of the 48th Lunar and Planetary Science Conference, The Woodlands, Houston, TX, USA, 20–24 March 2017; p. 2665.
90. Maxwell, T.A.; El-Baz, F.; Ward, S.H. Distribution, Morphology, and Origin of Ridges and Arches in Mare Serenitatis. *Geol. Soc. Am. Bull.* **1975**, *86*, 1273. [CrossRef]
91. Fielder, G. Topography and tectonics of the lunar straight wall. *Planet. Space Sci.* **1963**, *11*, 23–30. [CrossRef]
92. Middlehurst, B.M. Lunar Tidal Phenomena and the Lunar Rille System. In *The Moon, Proceedings from IAU Symposium No. 47, the University of Newcastle-Upon-Tyne, UK, 22–26 March 1971*; Runcorn, S.K., Urey, H.C., Eds.; Springer International Publishing: Cham, Switzerland, 1972; pp. 450–457. ISBN 978-94-010-2861-5.

93. Solomon, S.C.; Head, J.W. Vertical movement in mare basins: Relation to mare emplacement, basin tectonics, and lunar thermal history. *J. Geophys. Res. Solid Earth* **1979**, *84*, 1667–1682. [CrossRef]
94. Melosh, H.J. *Impact Cratering: A Geologic Process*; Oxford University Press: New York, NY, USA, 1989; ISBN 9780195042849.
95. Herrick, R.R.; Forsberg-Taylor, N.K. The shape and appearance of craters formed by oblique impact on the Moon and Venus. *Meteorit. Planet. Sci.* **2003**, *38*, 1551–1578. [CrossRef]
96. Gault, D.E.; Wedekind, J.A. Experimental studies of oblique impact. In Proceedings of the 9th Lunar and Planetary Science Conference, The Woodlands, Houston, TX, USA, 17 March 1978; Pergamon Press: New York, NY, 1978.
97. Zhang, X.; Xia, X.; Wang, Y. The origin of Messier crater rays and the model of oblique impact. *Prog. Geophys.* **2015**, *30*, 488–492. [CrossRef]
98. Haruyama, J.; Hioki, K.; Shirao, M.; Morota, T.; Hiesinger, H.; van der Bogert, C.H.; Miyamoto, H.; Iwasaki, A.; Yokota, Y.; Ohtake, M.; et al. Possible lunar lava tube skylight observed by SELENE cameras. *Geophys. Res. Lett.* **2009**, *36*. [CrossRef]
99. Wagner, R.V.; Robinson, M.S. Distribution, formation mechanisms, and significance of lunar pits. *Icarus* **2014**, *237*, 52–60. [CrossRef]
100. Robinson, M.S.; Ashley, J.W.; Boyd, A.K.; Wagner, R.V.; Speyerer, E.J.; Ray Hawke, B.; Hiesinger, H.; van der Bogert, C.H. Confirmation of sublunarean voids and thin layering in mare deposits. *Planet. Space Sci.* **2012**, *69*, 18–27. [CrossRef]
101. Kaku, T.; Haruyama, J.; Miyake, W.; Kumamoto, A.; Ishiyama, K.; Nishibori, T.; Yamamoto, K.; Crites, S.T.; Michikami, T.; Yokota, Y.; et al. Detection of Intact Lava Tubes at Marius Hills on the Moon by SELENE (Kaguya) Lunar Radar Sounder. *Geophys. Res. Lett.* **2017**, *44*, 10155–10161. [CrossRef]
102. Chappaz, L.; Sood, R.; Melosh, H.J.; Howell, K.C.; Blair, D.M.; Milbury, C.; Zuber, M.T. Evidence of large empty lava tubes on the Moon using GRAIL gravity. *Geophys. Res. Lett.* **2017**, *44*, 105–112. [CrossRef]
103. Hood, L.L.; Schubert, G. Lunar Magnetic Anomalies and Surface Optical Properties. *Science* **1980**, *208*, 49–51. [CrossRef]
104. Neish, C.D.; Blewett, D.T.; Bussey, D.; Lawrence, S.J.; Mechtley, M.; Thomson, B.J. The surficial nature of lunar swirls as revealed by the Mini-RF instrument. *Icarus* **2011**, *215*, 186–196. [CrossRef]
105. Glotch, T.D.; Bandfield, J.L.; Lucey, P.G.; Hayne, P.O.; Greenhagen, B.T.; Arnold, J.A.; Ghent, R.R.; Paige, D.A. Formation of lunar swirls by magnetic field standoff of the solar wind. *Nat. Commun.* **2015**, *6*, 6189. [CrossRef] [PubMed]
106. Li, S.; Garrick-Bethell, I. Surface Water at Lunar Magnetic Anomalies. *Geophys. Res. Lett.* **2019**, *46*, 14318–14327. [CrossRef]
107. Schultz, P.H.; Srnka, L.J. Cometary collisions on the Moon and Mercury. *Nature* **1980**, *284*, 22–26. [CrossRef]
108. Pinet, P.C.; Shevchenko, V.V.; Chevrel, S.D.; Daydou, Y.; Rosemberg, C. Local and regional lunar regolith characteristics at Reiner Gamma Formation: Optical and spectroscopic properties from Clementine and Earth-based data. *J. Geophys. Res.* **2000**, *105*, 9457–9475. [CrossRef]
109. Syal, M.B.; Schultz, P.H. Cometary impact effects at the Moon: Implications for lunar swirl formation. *Icarus* **2015**, *257*, 194–206. [CrossRef]
110. Garrick-Bethell, I.; Head, J.W.; Pieters, C.M. Spectral properties, magnetic fields, and dust transport at lunar swirls. *Icarus* **2011**, *212*, 480–492. [CrossRef]
111. Managadze, G.G.; Cherepin, V.T.; Shkuratov, Y.G.; Kolesnik, V.N.; Chumikov, A.E. Simulating OH/H2O formation by solar wind at the lunar surface. *Icarus* **2011**, *215*, 449–451. [CrossRef]
112. Chaussidon, M. Lunar water from the solar wind. *Nat. Geosci.* **2012**, *5*, 766–767. [CrossRef]
113. Green, R.O.; Pieters, C.; Mouroulis, P.; Eastwood, M.; Boardman, J.; Glavich, T.; Isaacson, P.; Annadurai, M.; Besse, S.; Barr, D.; et al. The Moon Mineralogy Mapper (M3) imaging spectrometer for lunar science: Instrument description, calibration, on-orbit measurements, science data calibration and on-orbit validation. *J. Geophys. Res.* **2011**, *116*, 367. [CrossRef]
114. Tsunakawa, H.; Takahashi, F.; Shimizu, H.; Shibuya, H.; Matsushima, M. Surface vector mapping of magnetic anomalies over the Moon using Kaguya and Lunar Prospector observations. *J. Geophys. Res.* **2015**, *120*, 1160–1185. [CrossRef]
115. Benaroya, H.; Bernold, L.; Chua, K.M. Engineering, Design and Construction of Lunar Bases. *J. Aerosp. Eng.* **2002**, *15*, 33–45. [CrossRef]
116. Daga, A.W.; Daga, M.A.; Wendel, W.R. A Preliminary Assessment of the Potential of Lava Tube-Situated Lunar Base Architecture. In *Engineering, Construction, and Operations in Space II*; ASCE: Reston, VI, USA, 1990; pp. 568–577.
117. Horvath, T.; Hayne, P.O.; Paige, D.A. Thermal Environments and Illumination in Lunar Pits and Lava Tubes. In *AGU Fall Meeting Abstracts 2019*; AGU: Washington, DC, USA, 2019; p. P31C-3454.
118. Haruyama, J.; Sawai, S.; Mizuno, T.; Yoshimitsu, T.; Fukuda, S.; Nakatani, I. Exploration of Lunar Holes, Possible Skylights of Underlying Lava Tubes, by Smart Lander for Investigating Moon (SLIM). *Trans. JSASS Aerosp. Tech. Jpn.* **2012**, *10*, Pk_7–Pk_10. [CrossRef]
119. De Angelis, G.; Wilson, J.W.; Clowdsley, M.S.; Nealy, J.E.; Humes, D.H.; Clem, J.M. Lunar lava tube radiation safety analysis. *J. Radiat. Res.* **2002**, *43*, S41–S45. [CrossRef]
120. Haruyama, J.; Morota, T.; Kobayashi, S.; Sawai, S.; Lucey, P.G.; Shirao, M.; Nishino, M.N. Lunar Holes and Lava Tubes as Resources for Lunar Science and Exploration. In *Moon: Prospective Energy and Material Resources*; Badescu, V., Ed.; Springer: Berlin/Heidelberg, Germany, 2012; pp. 139–163. ISBN 978-3-642-27968-3.
121. Cadogan, D.; Stein, J.; Grahne, M. Inflatable composite habitat structures for lunar and mars exploration. *Acta Astronaut.* **1999**, *44*, 399–406. [CrossRef]

122. Katzer, J.; Suchocki, C.; Błaszczak-Bąk, W.; Zarzycki, P.K.; Damięcka-Suchocka, M. Reliability and effectiveness of laser scanners in future construction efforts on the Moon and Mars. *Autom. Constr.* **2021**, *132*, 103979. [CrossRef]
123. Draper, D.S.; Lawrence, S.J.; Klima, R.S.; Denevi, B.W.; van der Bogert, C.H.; Elardo, S.M.; Hiesinger, H.H. The Inner Solar System Chronology (ISOCHRON) Lunar Sample Return Mission Concept: Revealing Two Billion Years of History. *Planet. Sci. J.* **2021**, *2*, 79. [CrossRef]
124. Robbins, S.J. New crater calibrations for the lunar crater-age chronology. *Earth Planet. Sci. Lett.* **2014**, *403*, 188–198. [CrossRef]
125. Neukum, G.; Ivanov, B.A.; Hartmann, W.K. Cratering Records in the Inner Solar System in Relation to the Lunar Reference System. In *Chronology and Evolution of Mars*; Kallenbach, R., Geiss, J., Hartmann, W.K., Eds.; Springer: Dordrecht, The Netherlands, 2001; pp. 55–86. ISBN 978-94-017-1035-0.
126. Marchi, S.; Mottola, S.; Cremonese, G.; Massironi, M.; Martellato, E. A new chronology for the Moon and Mercury. *Astron. J.* **2009**, *137*, 4936–4948. [CrossRef]
127. Hartmann, W.K.; Quantin, C.; Mangold, N. Possible long-term decline in impact rates. *Icarus* **2007**, *186*, 11–23. [CrossRef]
128. van der Bogert, C.H.; Hiesinger, H. Which Samples are Needed for Improved Calibration of the Lunar Cratering Chronology? In Proceedings of the 51st Lunar and Planetary Science Conference, Virtual Event, 15–19 March 2021; p. 2088.
129. Williams, J.-P.; van der Bogert, C.H.; Pathare, A.V.; Michael, G.G.; Kirchoff, M.R.; Hiesinger, H. Dating very young planetary surfaces from crater statistics: A review of issues and challenges. *Meteorit. Planet. Sci.* **2018**, *53*, 554–582. [CrossRef]
130. Che, X.; Nemchin, A.; Liu, D.; Long, T.; Wang, C.; Norman, M.D.; Joy, K.H.; Tartese, R.; Head, J.; Jolliff, B.; et al. Age and composition of young basalts on the Moon, measured from samples returned by Chang'e-5. *Science* **2021**, *374*, 887–890. [CrossRef] [PubMed]
131. Li, Q.-L.; Zhou, Q.; Liu, Y.; Xiao, Z.; Lin, Y.; Li, J.-H.; Ma, H.-X.; Tang, G.-Q.; Guo, S.; Tang, X.; et al. Two billion-year-old volcanism on the Moon from Chang'E-5 basalts. *Nature* **2021**, *600*, 54–58. [CrossRef]
132. Stevenson, D.J. Origin of the Moon-The Collision Hypothesis. *Annu. Rev. Earth Planet. Sci.* **1987**, *15*, 271–315. [CrossRef]
133. Spohn, T. The Longevity of Lunar Volcanism: Implications of Thermal Evolution Calculations with 2D and 3D Mantle Convection Models. *Icarus* **2001**, *149*, 54–65. [CrossRef]
134. Schultz, P.H.; Spudis, P.D. Beginning and end of lunar mare volcanism. *Nature* **1983**, *302*, 233–236. [CrossRef]
135. Haskin, L.A.; Gillis, J.J.; Korotev, R.L.; Jolliff, B.L. The materials of the lunar Procellarum KREEP Terrane: A synthesis of data from geomorphological mapping, remote sensing, and sample analyses. *J. Geophys. Res.* **2000**, *105*, 20403–20415. [CrossRef]
136. Jolliff, B.L.; Gillis, J.J.; Haskin, L.A.; Korotev, R.L.; Wieczorek, M.A. Major lunar crustal terranes: Surface expressions and crust-mantle origins. *J. Geophys. Res.* **2000**, *105*, 4197–4216. [CrossRef]
137. Wieczorek, M.A.; Phillips, R.J. The "Procellarum KREEP Terrane": Implications for mare volcanism and lunar evolution. *J. Geophys. Res.* **2000**, *105*, 20417–20430. [CrossRef]
138. Tian, H.-C.; Wang, H.; Chen, Y.; Yang, W.; Zhou, Q.; Zhang, C.; Lin, H.-L.; Huang, C.; Wu, S.-T.; Jia, L.-H.; et al. Non-KREEP origin for Chang'E-5 basalts in the Procellarum KREEP Terrane. *Nature* **2021**, *600*, 59–63. [CrossRef]
139. Li, C.; Hu, H.; Yang, M.-F.; Pei, Z.-Y.; Zhou, Q.; Ren, X.; Liu, B.; Liu, D.; Zeng, X.; Zhang, G.; et al. Characteristics of the lunar samples returned by Chang'E-5 mission. *Natl. Sci. Rev.* **2021**, *9*, nwab188. [CrossRef]
140. Hu, S.; He, H.; Ji, J.; Lin, Y.; Hui, H.; Anand, M.; Tartèse, R.; Yan, Y.; Hao, J.; Li, R.; et al. A dry lunar mantle reservoir for young mare basalts of Chang'E-5. *Nature* **2021**, *600*, 49–53. [CrossRef] [PubMed]
141. Delano, J.W. Pristine lunar glasses: Criteria, data, and implications. *J. Geophys. Res.* **1986**, *91*, 201–213. [CrossRef]
142. Shearer, C.; Papike, J. Basaltic magmatism on the Moon: A perspective from volcanic picritic glass beads. *Geochim. Cosmochim. Acta* **1993**, *57*, 4785–4812. [CrossRef]
143. Saal, A.E.; Hauri, E.H.; Cascio, M.L.; van Orman, J.A.; Rutherford, M.C.; Cooper, R.F. Volatile content of lunar volcanic glasses and the presence of water in the Moon's interior. *Nature* **2008**, *454*, 192–195. [CrossRef]
144. Saal, A.E.; Hauri, E.H.; van Orman, J.A.; Rutherford, M.J. Hydrogen Isotopes in Lunar Volcanic Glasses and Melt Inclusions Reveal a Carbonaceous Chondrite Heritage. *Science* **2013**, *340*, 1317–1320. [CrossRef]
145. Hauri, E.H.; Weinreich, T.; Saal, A.E.; Rutherford, M.C.; van Orman, J.A. High Pre-Eruptive Water Contents Preserved in Lunar Melt Inclusions. *Science* **2011**, *333*, 213–215. [CrossRef]
146. Hauri, E.H.; Saal, A.E.; Rutherford, M.J.; van Orman, J.A. Water in the Moon's interior: Truth and consequences. *Earth Planet. Sci. Lett.* **2015**, *409*, 252–264. [CrossRef]
147. Milliken, R.E.; Li, S. Remote detection of widespread indigenous water in lunar pyroclastic deposits. *Nat. Geosci.* **2017**, *10*, 561–565. [CrossRef]
148. Trang, D.; Tonkham, T.; Filiberto, J.; Li, S.; Lemelin, M.; Elder, C.M. Eruption characteristics of lunar localized pyroclastic deposits as evidenced by remotely sensed water, mineralogy, and regolith. *Icarus* **2022**, *375*, 114837. [CrossRef]
149. Robinson, M.; Elliott, J. Intrepid Planetary Mission Concept Study Report. 2020. Available online: https://science.nasa.gov/files/science-red/s3fs-public/atoms/files/Lunar%20INTREPID.pdf (accessed on 22 September 2020).
150. Haruyama, J.; Matsunaga, T.; Ohtake, M.; Morota, T.; Honda, C.; Yokota, Y.; Torii, M.; Ogawa, Y. Global lunar-surface mapping experiment using the Lunar Imager/Spectrometer on SELENE. *Earth Planets Space* **2008**, *60*, 243–255. [CrossRef]

151. Li, M.; Tian, D.-B.; Bai, Y.; Wang, Z.-H.; Yu, Q.; Li, Y. Simulation test on lunar dust raising by lunar rover wheel. In *Design, Manufacturing and Mechatronics, Proceedings of the 2015 International Conference on Design, Manufacturing, and Mechatronics (ICDMM2015), Wuhan, China, 17–18 April 2015*; Shahhosseini, A.M., Ed.; World Scientific: Hackensack, NJ, USA, 2016; pp. 35–40, ISBN 978-981-4730-50-1.
152. Li, M.; Wang, H.; Tian, D.; Tong, J.; Lv, Z.; Li, Y. Discrete Element Simulation of lunar dust suspension caused by lunar rover wheel. In Proceedings of the International Conference on Transportation, Mechanical, and Electrical Engineering (TMEE), Changchun, China, 16–18 December 2011; IEEE: New York, NY, USA, 2011; pp. 316–319, ISBN 978-1-4577-1701-7.

remote sensing

MDPI

Article

Robotic Mapping Approach under Illumination-Variant Environments at Planetary Construction Sites

Sungchul Hong [1,*], Pranjay Shyam [2], Antyanta Bangunharcana [2] and Hyuseoung Shin [3]

1 Department of Geoinformatic Engineering, Inha University, Incheon 22212, Korea
2 Department of Mechanical Engineering, Korea Advanced Institute of Science and Technology, College of Engineering, Daejeon 34141, Korea; pranjayshyam@kaist.ac.kr (P.S.); antabangun@kaist.ac.kr (A.B.)
3 Department of Future Technology and Convergence Research, Korea Institute of Civil Engineering and Building Technology, Goyang-si 10223, Korea; hyushin@kict.re.kr
* Correspondence: schong@inha.ac.kr

Abstract: In planetary construction, the semiautonomous teleoperation of robots is expected to perform complex tasks for site preparation and infrastructure emplacement. A highly detailed 3D map is essential for construction planning and management. However, the planetary surface imposes mapping restrictions due to rugged and homogeneous terrains. Additionally, changes in illumination conditions cause the mapping result (or 3D point-cloud map) to have inconsistent color properties that hamper the understanding of the topographic properties of a worksite. Therefore, this paper proposes a robotic construction mapping approach robust to illumination-variant environments. The proposed approach leverages a deep learning-based low-light image enhancement (LLIE) method to improve the mapping capabilities of the visual simultaneous localization and mapping (SLAM)-based robotic mapping method. In the experiment, the robotic mapping system in the emulated planetary worksite collected terrain images during the daytime from noon to late afternoon. Two sets of point-cloud maps, which were created from original and enhanced terrain images, were examined for comparison purposes. The experiment results showed that the LLIE method in the robotic mapping method significantly enhanced the brightness, preserving the inherent colors of the original terrain images. The visibility and the overall accuracy of the point-cloud map were consequently increased.

Keywords: planetary construction; robotic mapping; SLAM; low-light enhancement; 3D point-cloud map; deep learning

1. Introduction

The international space community has been increasingly committed to returning humans to the Moon, leading to the establishment of a permanent base and, eventually, the Mars base (Figure 1) [1]. In situ resource utilization (ISRU) technology, which aims to produce consumables from raw materials on the surfaces of the Moon and Mars, is one of the key enablers of sustainable human exploration and habitation [2–5]. For instance, water ice, recently discovered on the Moon and Mars [6–9], is used to generate hydrogen, oxygen, water, fuel, and propellants for life support [10,11]. Regolith is another available resource to produce construction materials such as bricks and blocks [12,13]. Three-dimensional (3D) construction printing technology has been proposed as a robust and efficient means to construct planetary infrastructure under extreme environments [14–17].

Planetary construction necessitates collaborative efforts from a variety of fields, one of which is 3D terrain mapping for on-site construction planning and management. Different from Earth construction, various types of teleoperated robots are expected to build a planetary base and infrastructure [14,18,19]. A highly detailed and accurate 3D terrain map is essential for planning robotic construction operations such as site preparation and infrastructure emplacement. Although planetary remote sensing data have been used to create global terrain maps [20,21], their spatial resolutions are insufficient to support robotic construction. Rover-based 3D mapping is compulsory to understand the topographic aspect of construction candidate sites. However, due to the homogeneous and rugged planetary terrains in a global navigation satellite system (GNSS)-denied environment, the rover's mapping capabilities are limited. Thus, recent studies have presented numerous 3D simultaneous location and mapping (SLAM) techniques that can dynamically estimate the current position and automatically build a 3D point-cloud map of surrounding environments.

Figure 1. Planetary base and ISRU facilities for human presence.

Sensors mainly used in SLAM techniques include light detection and ranging (LiDAR) and cameras. LiDAR SLAM, which allows high-resolution and long-range observations, can improve the autonomous navigation of planetary rovers. Tong et al. [22] and Merali et al. [23] presented a 3D SLAM framework using rover-mounted LiDAR and demonstrated globally consistent mapping results over test areas. Shaukat et al. [24] presented camera–LiDAR fusion SLAM, in which LiDAR overcomes the camera's field of view and point density restrictions and allows the camera to obtain a semantic interpretation of a scene. However, planetary rovers typically suffer from limited capabilities in terms of power, computational resources, and data storage [25]. Furthermore, no LiDAR exists that can be applied to contemporary planetary rovers. In this regard, visual SLAM, which mainly uses a camera system, has been considered to develop a 3D robotic mapping method for planetary surfaces. In comparison to LiDAR, the advantage of a camera system is that it is relatively lighter and consumes less power per unit. Additionally, as point clouds with color information can be generated, static and kinematic objects can be easily detected and identified in the dynamic nature of worksites.

In this paper, two types of visual SLAM-based mapping approaches, monocular SLAM and stereo SLAM, are considered. Both approaches rely on the quantity and quality of feature points on image frames from a camera system. Monocular SLAM searches for correspondences using a feature descriptor for each feature point between consecutive image frames from a single camera. The corresponding feature points are then used to estimate the camera pose, as well as to create point clouds. Thus, it is crucial not only to choose suitable feature points as a landmark but also to use a robust feature-matching method. For example, Bajpai et al. [26] proposed a biologically inspired visual saliency model that semantically detects feature points. Tseng et al. [27] improved oriented FAST and rotated BRIEF (ORB) to enhance feature matching in a single image sequence. However, a single camera is not able to handle the sudden movement of a mobile platform. Additional inertial and range sensors are required to avoid scale ambiguity and measurement drift on rough terrains. In comparison to monocular SLAM, stereo SLAM matches corresponding feature points on a pair of two image frames, recovering relatively accurate pose estimations and 3D measurements [28]. Hidalgo-Carrió et al. [29] developed a sensor fusion framework using a stereo camera, IMU, and wheel odometer. A Gaussian-based odometry error model was designed to predict nonsystematic errors to increase the localization and mapping accuracy. Additionally, for longer and faster traversal in future planetary missions, a field-programmable gate array (FPGA) was implemented in the rover as a computing device for terrain mapping in combination with the IMU and wheel encoder [25].

However, there are limitations pertaining to the photogrammetric mapping principle, such as the sparse nature of point clouds due to uniform or repetitive texture of surfaces, inconsistent color of images due to daylight fluctuations, and even infeasibility of image acquisition in dark illumination conditions. Therefore, with the recognized technical limitations, the primary objective of this paper is to develop a novel approach employing modern visual SLAM and low-light image enhancement (LLIE) methods robust to variant illumination conditions at unstructured worksites. The remainder of this paper is organized as follows. Section 2 briefly describes the LLIE methods. Section 3 describes the proposed pipeline of the 3D robotic mapping approach. Section 4 presents the experimental setup to validate the proposed approach, followed by visibility and accuracy assessments of the mapping results. Finally, conclusions are discussed in Section 5.

2. Low-Light Image Enhancement Method

Generally, LLIE methods can be categorized into histogram equalization (HE), retinex theory, and deep-learning-based methods. HE increases the global contrast of a given image by mapping the pixel intensity distribution to a uniform frequency distribution and thereby spreading out the intensity values, resulting in the enhancement of areas with low contrast. However, relying on global contrast to improve the visibility of a diverse range of objects in the foreground and background may result in an inferior image due to the overenhancement of certain regions within the foreground. To avoid this limitation, adaptive histogram equalization (AHE) was proposed, where the complete image is divided into smaller patches, and histogram equalization is subsequently applied to each patch [30,31]. The contrast-limited adaptive histogram (CLAHE) method was proposed to reduce noise amplification by clipping the histogram at a predefined threshold depending on the normalization of the histogram and, in turn, the size of the patch [32]. This clipped region, when redistributed among neighboring histograms, tends to improve overall image quality. Other methods extend this principle by dividing the image into sub-histograms based on the mean pixel value (bi-histogram equalization [33]), median pixel value (dualistic subimage histogram equalization [34], recursive subimage histogram equalization [35]), and minimum mean brightness error bi-histogram equalization [36]. These parts are subsequently processed individually before merging to obtain an enhanced image. While these algorithms provide robust performance, they cannot be directly applied to individual color channels in RGB images, as they dramatically change the color balance.

The retinex theory is modeled on the human visual system and decomposes an image into an illumination map and a reflectance map in the scene [37]. Classical retinex-based algorithms such as SSR [38] and MSR [39] estimate and remove the illumination map to obtain enhanced images. However, since the illumination map represents scene lighting, removing it results in enhanced images having artifacts near the edges. To overcome this limitation, naturalness preserved enhancement (NPE) was proposed to obtain illumination and reflectance maps using a bright-pass filter, thus preserving the naturalness of the image [40]. Another approach formulates the decomposition process as an optimization problem, wherein Fu et al. [41] proposed a weighted variational model for simultaneously estimating reflectance and illumination maps. However, this method is sensitive to the presence of noise in low-contrast regions, resulting in halo artifacts in the enhanced images. LIME relies on illumination map estimation with the structure known prior to enhancing low-light images [42]. Subsequently, BM3D is used for denoising with postprocessing. However, due to less reliance on the reflectance map, amplified noise results in low-contrast regions of the image.

Deep-learning-based models rely on the feature extraction and representation capabilities of convolutional neural networks (CNNs) to enhance low-light images. LLNet was proposed as a stacked autoencoder framework for jointly performing the task of low-light image enhancement and denoising of grayscale images [43]. Based on the retinex model, LightenNet relies on enhancing the illumination map to obtain a natural-looking enhanced image with better detail [44]. RetinexNet estimates both reflectance and illumination maps using two CNNs that decompose an image into reflectance and illumination maps, followed by enhancement of the illumination map while denoising the reflectance map [45]. MBLLEN utilizes multiple branches across different feature scales, subsequently enhancing and fusing these features to generate enhanced images [46]. DeepFuse images captured at multiple exposure levels are fused to generate well-lit images [47]. While deep-learning-based methods represent the current state of the art, paired data sets representing low illumination and well-lit conditions for a particular scene are required for training. Additionally, a collection of such data sets is logistically expensive. Thus, to overcome the reliance on paired training samples, EnlightenGAN [48] and Low-LightGAN [49] utilize generative adversarial networks (GANs) to generate paired data sets without explicitly constructing such data sets. A more recent model, DALE, relies on a visual attention module to identify dark regions and uses these regions to estimate visual attention [50].

As illumination variations affect the performance of perception tasks in visual SLAM, contrast enhancement was employed to improve the detection and matching of feature points [51]. The color space conversion and guided filter-based method was introduced to extract illumination-insensitive feature points [52]. However, these mathematically predefined methods are not scalable for diverse illumination environments. Deep-learning-based LLIE methods, which have been shown to outperform traditional methods [53,54], have been increasingly adopted to visual SLAM [55–57]. Therefore, in this paper, modern deep-learning-based LLIE methods are experimentally investigated to enhance the visual SLAM-based construction mapping approach, concerning the illumination-variant conditions at planetary worksites.

3. Robotic Construction Mapping Approach

The 3D robotic mapping approach is designed to build a highly detailed 3D point-cloud map for construction purposes. Different from planetary exploration into unknown environments during a mission, planetary construction requires a rover to traverse within a limited extent of a worksite. In planetary construction, rovers are required to revisit the same site multiple times. However, uniform illumination conditions are not guaranteed because viewpoints and solar altitudes are different for each visit to the worksite. Thus, the visual SLAM technique with the LLIE method is designed to create a 3D dense point-cloud map, ensuring a perceptually color-consistent representation of the test site.

A 3D robotic mapping approach, an extension of the visual stereo SLAM-based robotic mapping method [58], was designed in response to construction mapping requirements. The proposed method adopts stereo parallel tracking and mapping (S-PTAM) [59] as the base SLAM framework, to which the dense mapping and LLIE methods are combined to enhance construction mapping capabilities. The proposed method consists of three threads: preprocessing, mapping, and localization (Figure 2). In the preprocessing thread, the deep-learning-based LLIE method is selectively applied to stereo images from the robotic mapping system. The images obtained under varying illumination conditions are enhanced without oversaturating, contrasting, or stylizing the image. The LLIE method leverages feature extraction and matching correspondences between stereo images. Because the training data set for planetary terrain is not typically available, the deep learning model is trained based on a subset of planetary terrain images. Furthermore, the enhanced image set is used to train the disparity estimation deep network [60], and the disparity map is then created for localization and dense mapping threads.

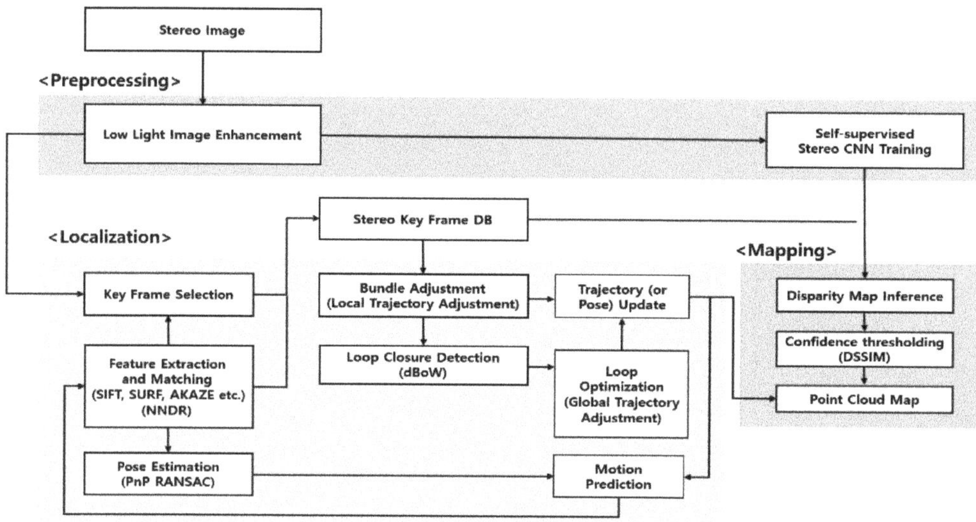

Figure 2. Conceptual process of the robotic construction mapping method.

In the localization thread, the camera trajectory relies on the quality and quantity of feature matches between stereo pairs. In the feature-matching procedure, the disparity map, along with the nearest neighbor distance ratio (NNDR) constraint, is used as an additional constraint to increase feature matching concerning homogeneous planetary terrains. In the first sequence, the camera pose, which is estimated by matching correspondences between the terrain features and identical features, is set up as an origin. The sequent stereo pairs are selected as a keyframe if the feature-matching number is less than 90% of that in the previous keyframe. When the threshold is satisfied, the camera pose is sequentially updated through triangulation between the feature matches of neighboring keyframes. These keyframes are stored in the stereo keyframe database. For local adjustment of the camera trajectory from the beginning, the bundle adjustment is repeatedly used to refine the estimated camera pose. However, the locational error is inevitably accumulated and propagated into the camera trajectory. It is necessary to recognize the already-visited place for the global adjustment of the camera trajectory. Loop closure detection with 3D points from the disparity map is used for global adjustment of the camera trajectory using the graph optimization process.

In the mapping thread, the disparity map is the basis for creating a dense 3D point cloud. However, an error in disparity values, which is caused by homogeneous (or textureless) terrains, is another concern for creating accurate point clouds. The structural dissimilarity (DSSIM) threshold [61] is used to examine a pixel-wise correspondence between stereo pairs. The estimated disparity with DSSIM below the predefined threshold is mapped to the 3D point cloud. In addition, the 3D point cloud from each image frame is referenced in the local coordinate system. As the accuracy of stereo depth estimation reduces at a longer distance, only points with a distance of less than 20 m are reprojected and mapped.

All point clouds from every key frame are reprojected at a reference coordinate system from the first frame in the image sequence. Additionally, a 3D grid with a predefined resolution is used to register and manage 3D points. To manage a 3D point-cloud map effectively, the 3D space is divided into a set of 3D grids with a predefined resolution. Specifically, each of the image points reprojected onto the 3D space, carrying the RGB pixel color information, are registered into the nearest grid in the 3D coordinate. Multiple points originating from images at various timesteps can land on the same grid. Grids with the number of points less than the predefined threshold are considered noisy measurements to be filtered out and recorded as unoccupied.

In addition, the number of reprojected points from each image frame can reach the order of 100,000. The 3D grid helps in limiting memory usage. In the implementation, the grid size was $2 \times 2 \times 2$ cm^3, and the minimal number of points in each grid was 5. The RGB color associated with the grid was computed simply as the mean RGB values of the contained points. Then, for each 3D grid, the associated occupancy and color information were updated at every timestep. Application of the LLIE method at the initial stage helps in providing consistent RGB color information across multiple frames over a long period of mapping.

4. Experiments and Results

4.1. Overview

The proposed approach in Section 3 is applied to the emulated planetary worksite (test site, hereinafter). To simulate the robotic mapping process, the robotic mapping system collects terrain images to build a point-cloud map depicting the entire test site in Section 4.2. Although all terrain images were taken during the day, changes in lighting conditions caused the point clouds to have inconsistent color properties. The topographic aspects of the test site are not clearly distinguished. Therefore, in the experiments, the deep-learning-based LLIE methods are examined and selected to enhance the terrain images in Section 4.3. The proposed mapping method combined with the selected LLIE method is used to create the 3D point-cloud map for construction purposes. Additionally, for comparison purposes, the visibility and the accuracy of the point-cloud maps based on original and enhanced images (hereinafter referred to as original and enhanced point-cloud maps, respectively) are assessed in Section 4.4.

4.2. Robotic Mapping System and Test Site

The robotic mapping system employed in this research is a four-wheeled mobile platform with payloads consisting of a stereo camera system and a Wi-Fi router (Figure 3). The stereo camera system has a 20 cm baseline, from which 15 stereo image frames with a 484×366 resolution are collected per second while the mapping system is moving. A camera mast with manual pan and tilt capabilities is ideal for effectively collecting terrain images, minimizing the rover's motions. The pan axis can rotate $\pm 180^\circ$ from the center, and the tilt axis range is $\pm 45^\circ$. Additionally, with the Wi-Fi-enabled router, the rover and its camera system can be remotely operated to simulate the planetary construction mapping process.

Figure 3. Robotic mapping system.

In planetary construction, flat areas with sparse rock distribution are likely to be construction candidate sites due to their topographic and geotechnical feasibility to strengthen the foundation and stabilize the surface [62,63]. The field test site, which is located at the Korea Research Institute of Civil Engineering and Building Technology (KICT) (Figure 4), is designed for robotic mapping systems to simulate a construction mapping process. The dimensions are 40 m × 50 m, and rounded-pile and hollowed-out areas of different sizes are distributed over a flat ground with few rocks. In the experiment, the robotic mapping system slowly traversed the entire test site from noon to late afternoon, occasionally stopping to recognize the current location and find a path to the next terrain feature. In the robotic operation, major terrain features (e.g., rock, crater, and mound) were repeatedly visited from different angles and positions. In the following sections, the terrain image set from the mapping system was used to experimentally select the LLIE method and to create the 3D point-cloud map of the test site.

Figure 4. Emulated planetary worksite (test site).

4.3. Image Enhancement under Varying Illumination Conditions

In the experiment, six different LLIE methods (RetinexNet, DALE, DLN, DSLR, GLAD, KinD) were initially selected and trained on the LOL data set [45]. The LLIE methods were sequentially fine-tuned using 440 image pairs from an emulated terrain in the indoor laboratory, where diverse illumination conditions from a bright level to a dark level could be manually adjusted [64]. Another 60 images were used to validate the LLIE methods. The proposed approach was implemented using Python due to the easy integration of deep learning models. The computation time per image frame was approximately 0.2 s on the graphics processing unit (RTX 3070). Specifically, 0.1 s was used to perform the image enhancement procedure on both stereo image inputs and another 0.1 s for visual tracking with AKAZE [65].

To determine the robust LLIE method, four terrain images under different illumination conditions were obtained from the test site. As cloud coverage and solar altitude continuously change, the terrain images in Figure 5 have a different magnitude of brightness and color property. In Figure 5a, the terrain image was taken under clear sky at noon, in which the rock has a bright color with a shadow. However, the image quality depends on the cloud coverage that reflects the sunlight. In Figure 5b,c, the terrain images become incrementally darker as the cloud coverage increases. In addition, the low solar altitude in the late afternoon caused the onboard camera to become underexposed. The terrain image in Figure 5d consequently has the lowest brightness compared to the other images, with the entire visual content degraded and the mound in the background hardly discernible.

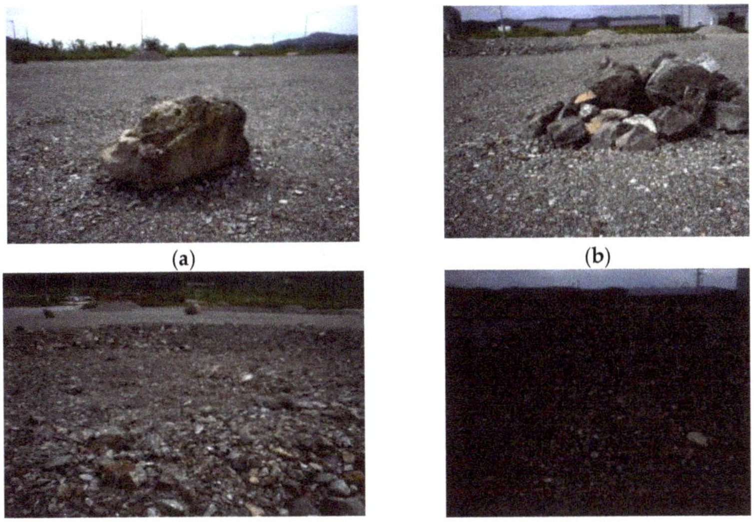

(a) (b) (c) (d)

Figure 5. Terrain images with decreasing illumination conditions (**a**–**d**): (**a**) bright condition with clear sky at noon; (**b**) overcast condition, partly cloudy in the afternoon; (**c**) overcast condition with full cloud coverage in the afternoon; (**d**) darkest condition with low solar altitude in the late afternoon.

To ensure the image enhancement results from the darkest condition in the image set, the LLIE methods were applied to the terrain image in Figure 5d. The image enhancement results were then evaluated quantitatively and qualitatively, as shown in Table 1 and Figure 6 respectively. In Table 1, Naturalness Image Quality Evaluator (NIQE) [66] and Blind/Referenceless Image Spatial Quality Evaluator (BRISQUE) [67], which are no-reference image evaluation metrics, are used to quantify the enhanced image qualities. NIQE measures the distance between the multivariate Gaussian model of the natural image and the test image. Additionally, BRIQUE evaluates the test image based on natural scene statistics. A smaller score from both NIQE and BRISQUE indicates better visually perceptual quality. Table 1 shows that GLAD is the best method from both evaluation metrics. However, in some cases, different evaluation metrics assign a different score to the same LLIE method in that each metric evaluates different aspects of the enhanced image. For example, NIQE indicates that DLN and Kind achieve the second and third best scores, respectively. However, the second and third best scores in BRISQUE are achieved by DSLR and DLN, respectively.

Table 1. Performance of different models on dark images.

LLIE methods	NIQE	BRISQUE
RetinexNet	5.48	67.40
DALE	8.40	65.56
DLN	4.27	55.74
DSLR	5.07	52.38
GLAD	3.82	45.33
KinD	4.67	60.60

Figure 6. Darkest image enhancement results in the test image in Figure 5d: (**a**) RetinexNet; (**b**) DALE; (**c**) DLN; (**d**) DSLR; (**e**) GLAD; (**f**) KinD.

These evaluation metric scores are not completely consistent with subjective human perception; thus, the image enhancement results in Figure 6 is visually inspected. In Figure 6a, RetinexNet severely stylizes and distorts the colors of the terrain features in the image. DSLR in Figure 6 has the second highest score in BRISQUE, but random and gridding artifacts are introduced within the image. Additionally, the entire image is oversmoothed, which causes a loss of detail, especially for the gravels on the mound. In Figure 6d, DALE reveals chromatic distortion surrounding the different artifacts. KinD in Figure 6f improved the contrast, but visible color distortion is shown in the lower part of the mound. Overall, GLAD (Figure 6e) and DLN (Figure 6c) improved the brightness the best, maintaining clarity and color in the darkest conditions in Figure 5.

A second experiment was conducted to ensure consistent color restorations over the terrain images with a different degree of brightness (Figure 5). GLAD and DLN were selected from the first experiment due to their quantitatively and qualitatively stable performances. Additionally, the image brightness is increased without severely degrading the image quality. Figure 7 shows the image enhancement results from the four terrain images in Figure 5. In Figure 7a, DLN generates slightly overexposed images but blurs the extent of terrain features such as gravel, rock, and mound. Additionally, the different brightness is obviously observed when the first and last images are compared. However, in Figure 7b, GLAD tends to demonstrate consistent color enhancement over all images, maintaining the image quality. Additionally, the features in each image can be clearly identified in comparison to those in the DLN-enhanced images. Therefore, GLAD is eventually selected for enhancing the entire terrain image frames from the robotic mapping system.

(a)

(b)

Figure 7. Image enhancement results over the test images in Figure 5: (**a**) DLN; (**b**) GLAD.

4.4. Robotic Mapping Results

Similar to construction on Earth, the first phase in planetary construction is site preparation. The teleoperated construction robots are then employed to clear and grade the ground before emplacing infrastructure such as roads, landing pads, and ISRU utilities.

Therefore, to understand the terrain morphological characteristics of the worksite, on-site robotic mapping is essential to create a highly detailed and accurate point-cloud map.

The proposed mapping method sequentially involved enhanced image pairs from GLAD. Three-dimensional (3D) point clouds were then generated along the rover trajectory, by which the test site in Figure 8 was reconstructed as a mapping result. In Figure 9, the original and enhanced point-cloud maps are shown for comparison purposes. Multiple circles in both point-cloud maps are blind spots that were inevitably made when the rover-mounted camera was fully rotated. Additionally, to build the point-cloud map, the rover slowly traversed around the entire test site, repeatedly visiting the same locations. Terrain images were continuously influenced by changes in cloud coverage and solar altitude from noon to late afternoon. Therefore, in Figure 9a, the original point-cloud map consequently has the mixed nature of colors from beige to dark gray. However, in Figure 9), the LLIE method visually improved the original terrain images to have relatively brighter and more consistent colors irrespective of the illumination condition changes. Shapes and distributions of terrain features (e.g., mound, crater) and obstacles (e.g., rock) are easily identified in the enhanced point-cloud map.

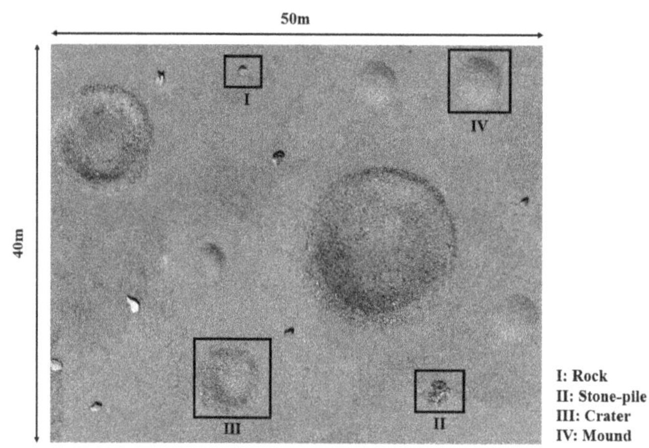

Figure 8. Aerial photo of the test site.

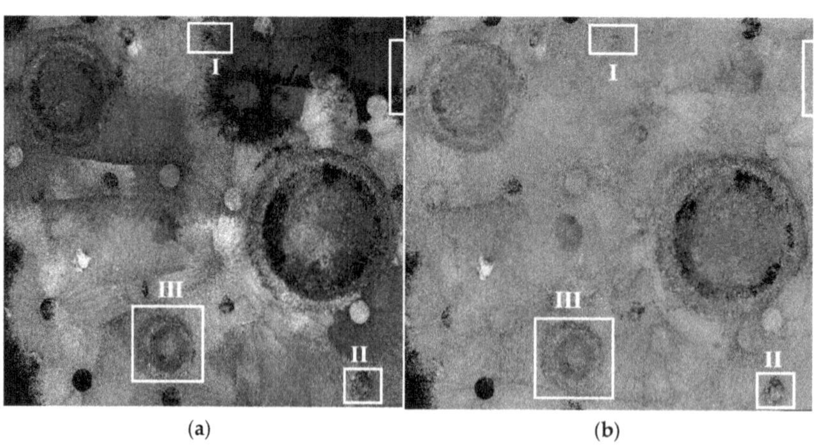

Figure 9. Robotic mapping results to terrain images from the robotic mapping system: (**a**) original point-cloud map; (**b**) enhanced point-cloud map.

In Figures 10 and 11, point clouds of the rock, stone pile, crater, and mound in both point-cloud maps are selected for visual inspection, each of which is indexed in Figures 8 and 9. In the robotic mapping process, images of the same feature were taken from different angles and distances. Point clouds are repeatedly created, blurring the outlines of the terrain features. Therefore, even though images were obtained in better lighting conditions, small features (e.g., pebble and stone) in all point clouds are hardly discernible in Figures 10 and 11. However, in Figure 10, the perceptual appearance of terrain features is degraded by the dark colors of the point clouds. Images of the stone pile were taken under partial or full cloud coverage, and therefore, its appearance has distorted color properties with indistinct outlines. Furthermore, the mound has an empty space on the top due to the lower height of the rover-mounted camera system. Although prior knowledge is given from the terrain image, the defects in point clouds make it more difficult to identify the mound from the background. In contrast, Figure 11 shows that the overall qualities of the point clouds are significantly improved, especially for the stone pile and the mound. The rock, stone pile, and crater still have blurred outlines, but the enhanced image set creates point clouds closer to their natural appearances, preserving inherent colors and details. Additionally, for the mound, the brightness of the point clouds is enhanced while restoring the colors from dark gray to beige. The visual perception is much improved when the mounds in Figures 10 and 11 are compared to each other.

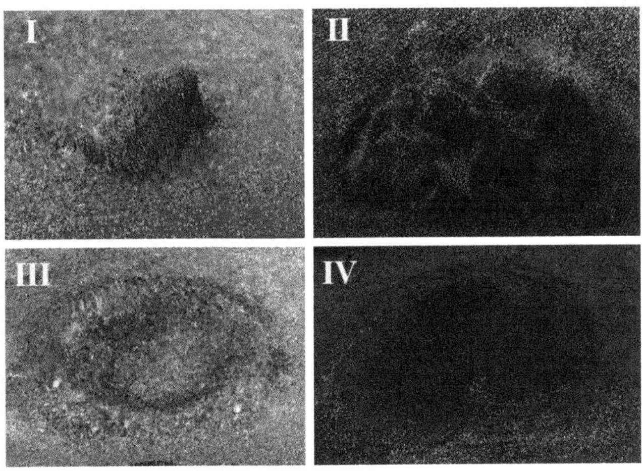

Figure 10. Terrain features selected from original point cloud in Figure 9.

Accuracy assessment was conducted to investigate the influence of the enhanced images on the quality of the point-cloud map. Terrestrial LiDAR (Trimble X7) was employed to obtain the point cloud as a reference, to which the iterative closest point (ICP) was used to optimally align the point-cloud maps in Figure 9. The root mean square error (RMSE) measurements for the original and enhanced point-cloud maps are 0.29 m and 0.25 m, respectively. As the test site has a flat and smooth surface, the enhanced image set does not lead to a significant reduction in RMSE measurements. To rigorously analyze the positional errors, the positional error distribution is geographically represented along with the positional error histogram in Figure 12. The color index, ranging from blue to yellow-green, represents the magnitude of positional errors along the x-axis of the histogram. Generally, positional errors in both point-cloud maps are not uniformly distributed over the test site. As the robotic mapping system mostly moved around the center of the test site, positional errors in the marginal areas are often larger than the central area. However, when the positional error distributions and histograms in Figure 12a,b are compared, the enhanced point-cloud map has a larger blue extent, and its histogram has a narrower width.

Accuracy assessment results indicate that the enhanced image set increases the overall accuracy of the original point clouds.

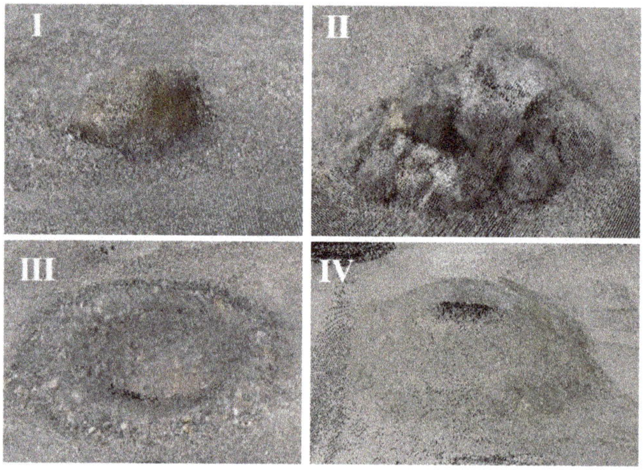

Figure 11. Terrain features selected from the enhanced point-cloud maps in Figure 9.

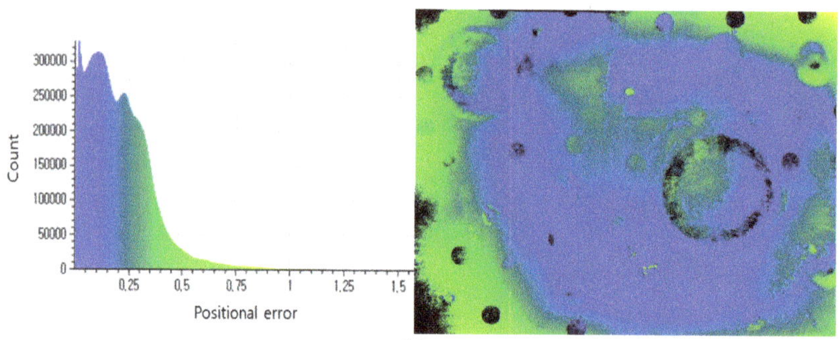

(**a**)

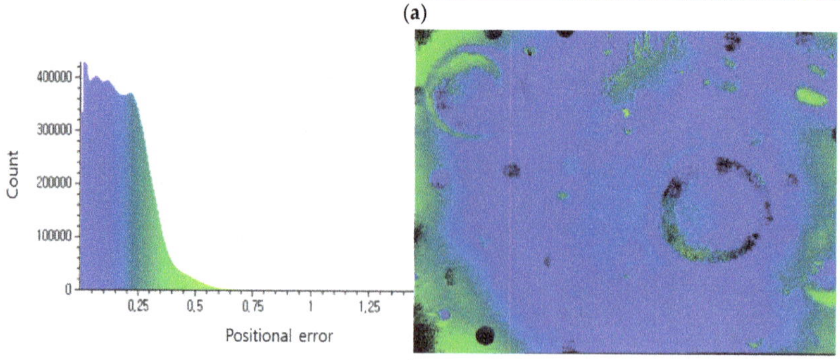

(**b**)

Figure 12. Positional error histogram (left) and positional error distribution over the test site (right). Identical color index is used to represent a magnitude of positional errors in the original and enhanced point-cloud maps: (**a**) positional errors in the original point-cloud map; (**b**) positional errors in the enhanced point-cloud map.

5. Discussion and Conclusions

Many space agencies worldwide have renewed their planetary exploration roadmap to return humans to the surfaces of the Moon and Mars. Numerous ideas and proposals have been presented to build a planetary base for sustained human exploration and habitation. Planetary construction implies that various types of construction robots will repeatedly visit the same worksite to perform complex tasks such as excavation, grading, leveling, and additive construction. These robotic construction operations solely rely on images from the onboard camera system. However, illumination conditions may vary for each visit due to illumination-variant conditions such as solar altitudes and different weather, especially for Mars. The inconsistent color property of images hampers the ability to recognize terrain features, even at the returned location. Additionally, in the unknown environments of worksites, prior knowledge of detailed topographic information is essential for construction planning and management. It is critical to ensure that images from robots are maintained in bright and consistent colors despite continually changing illumination conditions.

The proposed robotic mapping approach for planetary construction aims to build a 3D dense point-cloud map that depicts the worksite with perceptually color-consistent properties. In the proposed approach, the stereo SLAM-based robotic mapping method, which is based on S-PTAM, is combined with the LLIE method to enhance mapping capabilities. The robotic mapping system, which was deployed at the emulated planetary worksite, collected terrain images in a topographic surveying manner. All terrain images were collected during the daytime from noon to late afternoon. However, changes in weather and solar altitude resulted in the terrain images having inconsistent colors and consequently affected the point-cloud map. Dark point clouds hid many terrain features of interest, making it difficult to understand the topographic properties of the test site. Therefore, the deep-learning-based LLIE method was experimentally selected, enabling the original terrain images to have structural consistency with improved brightness and color. The enhanced point-cloud map was then created and compared with the original point-cloud map. The experiment results indicate that the robotic mapping method leverages the LLIE method to improve the visibility and the overall accuracy of the original point-cloud map. The brightness of the original point-cloud map was significantly improved, preserving inherent colors and details. Additionally, the point-cloud map became closer to the natural appearance of the test site.

The proposed approach shows promising results for planetary construction mapping. Additionally, color restoration and consistent imagery have potential for use in scientific investigations, such as geologic mapping. However, path planning, along with global localization, is another concern to minimize robot motions for efficient mapping and to historically align updated mapping results for construction monitoring. In addition, technical constraints still remain for implementation. In the planetary construction process, humans and robots are required to repetitively visit the worksite. The terrain image set is then accumulated to fine-tune the LLIE methods for future robots. However, the LLIE method in the proposed approach requires robots to use a large image set for training purposes. Additionally, during mapping, a disparity map is computed at a high frame rate. Heavy computation and a large data set pose great challenges to its application. The proposed approach should be adjusted to be more lightweight and optimized with respect to technical progress on the mechanical and hardware components of future planetary construction robots.

Author Contributions: Conceptualization, S.H.; methodology, P.S., S.H., A.B. and H.S.; validation, S.H.; formal analysis, S.H. and P.S.; resources, H.S.; data curation, P.S. and A.B.; writing—original draft preparation, S.H. and P.S.; writing—review and editing, S.H., P.S. and A.B; supervision, S.H.; project administration, H.S.; funding acquisition, H.S. All authors have read and agreed to the published version of the manuscript.

Funding: This research was supported by the research project "Development of environmental simulator and advanced construction technologies over TRL6 in extreme conditions" funded by Korea Institute of Civil Engineering and Building Technology.

Data Availability Statement: Not applicable.

Acknowledgments: This research was supported by the research project "Development of environmental simulator and advanced construction technologies over TRL6 in extreme conditions" funded by Korea Institute of Civil Engineering and Building Technology and an Inha University research grant.

Conflicts of Interest: The authors declare no conflict of interest.

References

1. ISECG. Global Exploration Roadmap Supplement; Lunar Surface Exploration Scenario Update. Available online: https://www.globalspaceexploration.org/?p=1049 (accessed on 10 January 2022).
2. Arya, A.S.; Rajasekhar, R.P.; Thangjam, G.A.; Kumar, A.K. Detection of potential site for future human habitability on the Moon using Chandrayaan-1 data. *Curr. Sci.* **2011**, *100*, 524–529.
3. Spudis, P.; Bussey, D.; Baloga, S.; Cahill, J.; Glaze, L.; Patterson, G.; Raney, R.; Thompson, T.; Thomson, B.; Ustinov, E. Evidence for water ice on the Moon: Results for anomalous polar craters from the LRO Mini-RF imaging radar. *J. Geophys. Res. Planets* **2013**, *118*, 2016–2029. [CrossRef]
4. Ralphs, M.; Franz, B.; Baker, T.; Howe, S. Water extraction on Mars for an expanding human colony. *Life Sci. Space Res.* **2015**, *7*, 57–60. [CrossRef]
5. Thangavelautham, J.; Robinson, M.S.; Taits, A.; McKinney, T.; Amidan, S.; Polak, A. Flying, hopping Pit-Bots for cave and lava tube exploration on the Moon and Mars. *arXiv Preprint* **2017**, arXiv:1701.07799.
6. Li, S.; Lucey, P.G.; Milliken, R.E.; Hayne, P.O.; Fisher, E.; Williams, J.-P.; Hurley, D.M.; Elphic, R.C. Direct evidence of surface exposed water ice in the lunar polar regions. *Proc. Natl. Acad. Sci. USA* **2018**, *115*, 8907–8912. [CrossRef]
7. Deutsch, A.N.; Head, J.W., III; Neumann, G.A. Analyzing the ages of south polar craters on the Moon: Implications for the sources and evolution of surface water ice. *Icarus* **2020**, *336*, 113455. [CrossRef]
8. Piqueux, S.; Buz, J.; Edwards, C.S.; Bandfield, J.L.; Kleinböhl, A.; Kass, D.M.; Hayne, P.O.; MCS; THEMIS Teams. Widespread shallow water ice on Mars at high latitudesand midlatitudes. *Geophys. Res. Lett.* **2019**, *46*, 14290–14298. [CrossRef]
9. Titus, T.N.; Kieffer, H.H.; Christensen, P.R. Exposed water ice discovered near the south pole of Mars. *Science* **2003**, *299*, 1048–1051. [CrossRef]
10. Anand, M.; Crawford, I.A.; Balat-Pichelin, M.; Abanades, S.; Van Westrenen, W.; Péraudeau, G.; Jaumann, R.; Seboldt, W. A brief review of chemical and mineralogical resources on the Moon and likely initial In Situ Resource Utilization (ISRU) applications. *Planet. Space Sci.* **2012**, *74*, 42–48. [CrossRef]
11. Arney, D.C.; Jones, C.A.; Klovstad, J.; Komar, D.; Earle, K.; Moses, R.; Bushnell, D.; Shyface, H. Sustaining Human Presence on Mars Using ISRU and a Reusable Lander. In Proceedings of the AIAA Space 2015 Conference and Exposition, Pasadena, CA, 31 August–2 September 2015; p. 4479.
12. Naser, M. Extraterrestrial construction materials. *Prog. Mater. Sci.* **2019**, *105*, 100577. [CrossRef]
13. Lee, W.-B.; Ju, G.-H.; Choi, G.-H.; Sim, E.-S. Development Trends of Space Robots. *Curr. Ind. Technol. Trends Aerosp.* **2011**, *9*, 158–175.
14. Khoshnevis, B. Automated construction by contour crafting—Related robotics and information technologies. *Autom. Constr.* **2004**, *13*, 5–19. [CrossRef]
15. Khoshnevis, B.; Yuan, X.; Zahiri, B.; Zhang, J.; Xia, B. Construction by Contour Crafting using sulfur concrete with planetary applications. *Rapid Prototyp. J.* **2016**, *22*, 848–856. [CrossRef]
16. Cesaretti, G.; Dini, E.; De Kestelier, X.; Colla, V.; Pambaguian, L. Building components for an outpost on the Lunar soil by means of a novel 3D printing technology. *Acta Astronaut.* **2014**, *93*, 430–450. [CrossRef]
17. Roman, M.; Yashar, M.; Fiske, M.; Nazarian, S.; Adams, A.; Boyd, P.; Bentley, M.; Ballard, J. 3D-printing lunar and martian habitats and the potential applications for additive construction. In Proceedings of the International Conference on Environmental Systems, Lisbon, Portugal, 31 July 2020.
18. Kelso, R.; Romo, R.; Andersen, C.; Mueller, R.; Lippitt, T.; Gelino, N.; Smith, J.; Townsend, I.I.; Schuler, J.; Nugent, M. Planetary Basalt Field Project: Construction of a Lunar Launch/Landing Pad, PISCES and NASA Kennedy space center project update. In *Earth and Space 2016: Engineering for Extreme Environments*; American Society of Civil Engineers: Reston, VA, USA, 2016; pp. 653–667.
19. Mueller, R.; Fikes, J.C.; Case, M.P.; Khoshnevis, B.; Fiske, M.R.; Edmunson, J.; Kelso, R.; Romo, R.; Andersen, C. Additive construction with mobile emplacement (ACME). In Proceedings of the 68th International Astronautical Congress (IAC), Adelaide, Australia, 25–29 September 2017; pp. 25–29.
20. Jawin, E.R.; Valencia, S.N.; Watkins, R.N.; Crowell, J.M.; Neal, C.R.; Schmidt, G. Lunar science for landed missions workshop findings report. *Earth Space Sci.* **2019**, *6*, 2–40. [CrossRef]

21. Pajola, M.; Rossato, S.; Baratti, E.; Kling, A. Planetary mapping for landing sites selection: The Mars Case Study. In *Planetary Cartography and GIS*; Springer: Cham, Switzerland, 2019; pp. 175–190.
22. Tong, C.H.; Barfoot, T.D.; Dupuis, É. Three-dimensional SLAM for mapping planetary work site environments. *J. Field Robot.* **2012**, *29*, 381–412. [CrossRef]
23. Merali, R.S.; Tong, C.; Gammell, J.; Bakambu, J.; Dupuis, E.; Barfoot, T.D. 3D surface mapping using a semiautonomous rover: A planetary analog field experiment. In Proceedings of the 2012 International Symposium on Artificial Intelligence, Robotics and Automation in Space (i-SAIRAS), Fort Lauderdale, Florida, USA, 9–11 January 2012.
24. Shaukat, A.; Blacker, P.C.; Spiteri, C.; Gao, Y. Towards camera-LIDAR fusion-based terrain modelling for planetary surfaces: Review and analysis. *Sensors* **2016**, *16*, 1952. [CrossRef]
25. Schuster, M.J.; Brunner, S.G.; Bussmann, K.; Büttner, S.; Dömel, A.; Hellerer, M.; Lehner, H.; Lehner, P.; Porges, O.; Reill, J. Towards autonomous planetary exploration. *J. Intell. Robot. Syst.* **2019**, *93*, 461–494. [CrossRef]
26. Bajpai, A.; Burroughes, G.; Shaukat, A.; Gao, Y. Planetary monocular simultaneous localization and mapping. *J. Field Robot.* **2016**, *33*, 229–242. [CrossRef]
27. Tseng, K.-K.; Li, J.; Chang, Y.; Yung, K.; Chan, C.; Hsu, C.-Y. A new architecture for simultaneous localization and mapping: An application of a planetary rover. *Enterp. Inf. Syst.* **2021**, *15*, 1162–1178. [CrossRef]
28. Lemaire, T.; Berger, C.; Jung, I.-K.; Lacroix, S. Vision-based slam: Stereo and monocular approaches. *Int. J. Comput. Vis.* **2007**, *74*, 343–364. [CrossRef]
29. Hidalgo-Carrió, J.; Poulakis, P.; Kirchner, F. Adaptive localization and mapping with application to planetary rovers. *J. Field Robot.* **2018**, *35*, 961–987. [CrossRef]
30. Hummel, R. Image enhancement by histogram transformation Comp. *Graph. Image Process* **1975**, *6*, pp. 184–195. [CrossRef]
31. Ketcham, D.J. Real-Time Image Enhancement Techniques. In Proceedings of the Image Processing, International Society for Optics and Photonics. Pacific Grove, CA, USA, 9 July 1976; pp. 120–125.
32. Zuiderveld, K. Contrast limited adaptive histogram equalization. In *Graphics gems IV*; Academic Press: Cambridge, MA, USA, 1994; pp. 474–485.
33. Kim, Y.-T. Contrast enhancement using brightness preserving bi-histogram equalization. *IEEE Trans. Consum. Electron.* **1997**, *43*, 1–8.
34. Wang, Y.; Chen, Q.; Zhang, B. Image enhancement based on equal area dualistic sub-image histogram equalization method. *IEEE Trans. Consum. Electron.* **1999**, *45*, 68–75. [CrossRef]
35. Sim, K.; Tso, C.; Tan, Y. Recursive sub-image histogram equalization applied to gray scale images. *Pattern Recognit. Lett.* **2007**, *28*, 1209–1221. [CrossRef]
36. Chen, S.-D.; Ramli, A.R. Preserving brightness in histogram equalization based contrast enhancement techniques. *Digit. Signal Process.* **2004**, *14*, 413–428. [CrossRef]
37. Land, E.H. The retinex theory of color vision. *Sci. Am.* **1977**, *237*, 108–129. [CrossRef]
38. Jobson, D.J.; Rahman, Z.-u.; Woodell, G.A. Properties and performance of a center/surround retinex. *IEEE Trans. Image Process.* **1997**, *6*, 451–462. [CrossRef]
39. Jobson, D.J.; Rahman, Z.-u.; Woodell, G.A. A multiscale retinex for bridging the gap between color images and the human observation of scenes. *IEEE Trans. Image Process.* **1997**, *6*, 965–976. [CrossRef]
40. Wang, S.; Zheng, J.; Hu, H.-M.; Li, B. Naturalness preserved enhancement algorithm for non-uniform illumination images. *IEEE Trans. Image Process.* **2013**, *22*, 3538–3548. [CrossRef] [PubMed]
41. Fu, X.; Zeng, D.; Huang, Y.; Zhang, X.-P.; Ding, X. A weighted variational model for simultaneous reflectance and illumination estimation. In Proceedings of the IEEE Conference on Computer Vision and Pattern Recognition, Las Vegas, NV, USA, 27–30 June 2016; pp. 2782–2790.
42. Guo, X.; Li, Y.; Ling, H. LIME: Low-light image enhancement via illumination map estimation. *IEEE Trans. Image Process.* **2016**, *26*, 982–993. [CrossRef] [PubMed]
43. Lore, K.G.; Akintayo, A.; Sarkar, S. LLNet: A deep autoencoder approach to natural low-light image enhancement. *Pattern Recognit.* **2017**, *61*, 650–662. [CrossRef]
44. Li, C.; Guo, J.; Porikli, F.; Pang, Y. LightenNet: A convolutional neural network for weakly illuminated image enhancement. *Pattern Recognit. Lett.* **2018**, *104*, 15–22. [CrossRef]
45. Wei, C.; Wang, W.; Yang, W.; Liu, J. Deep retinex decomposition for low-light enhancement. *arXiv Preprint* **2018**, arXiv:1808.04560.
46. Lv, F.; Lu, F.; Wu, J.; Lim, C. MBLLEN: Low-light image/video enhancement using CNNs. In Proceedings of the BMVC, Newcastle, UK, 3–6 September 2018; p. 220.
47. Prabhakar, K.R.; Srikar, V.S.; Babu, R.V. DeepFuse: A deep unsupervised approach for exposure fusion with extreme exposure image Pairs. In Proceedings of the ICCV, Venice, Italy, 22–29 October 2017; p. 3.
48. Jiang, Y.; Gong, X.; Liu, D.; Cheng, Y.; Fang, C.; Shen, X.; Yang, J.; Zhou, P.; Wang, Z. Enlightengan: Deep light enhancement without paired supervision. *arXiv Preprint* **2019**, arXiv:1906.06972. [CrossRef]
49. Kim, G.; Kwon, D.; Kwon, J. Low-lightgan: Low-light enhancement via advanced generative adversarial network with task-driven training. In Proceedings of the 2019 IEEE International Conference on Image Processing (ICIP), Taipei, Taiwan, 22–25 September 2019; pp. 2811–2815.
50. Kwon, D.; Kim, G.; Kwon, J. DALE: Dark region-aware low-light image enhancement. *arXiv Preprint* **2008**, arXiv:12493 2020.

51. Wang, X.; Christie, M.; Marchand, E. Optimized contrast enhancements to improve robustness of visual tracking in a SLAM relocalisation context. In Proceedings of the 2018 IEEE/RSJ International Conference on Intelligent Robots and Systems (IROS), Madrid, Spain, 1–5 October 2018; pp. 103–108.
52. Sun, P.; Lau, H.Y. RGB-Channel-based Illumination Robust SLAM Method. *J. Control Autom. Eng.* **2019**, *7*, 61–69. [CrossRef]
53. Lee, H.S.; Park, J.M.; Hong, S.; Shin, H.-S. Low light image enhancement to construct terrain information from permanently shadowed region on the moon. *J. Korean Soc. Geospat. Inf. Syst.* **2020**, *28*, 41–48.
54. Fan, W.; Huo, Y.; Li, X. Degraded image enhancement using dual-domain-adaptive wavelet and improved fuzzy transform. *Math. Probl. Eng.* **2021**, *2021*, 1–12. [CrossRef]
55. Gomez-Ojeda, R.; Zhang, Z.; Gonzalez-Jimenez, J.; Scaramuzza, D. Learning-based image enhancement for visual odometry in challenging HDR environments. In Proceedings of the 2018 IEEE International Conference on Robotics and Automation (ICRA), Brisbane, QLD, Australia, 21–25 May 2018; pp. 805–811.
56. Jung, E.; Yang, N.; Cremers, D. Multi-frame GAN: Image enhancement for stereo visual odometry in low light. In Proceedings of the Conference on Robot Learning, 16–18 November 2020; pp. 651–660.
57. Hu, J.; Guo, X.; Chen, J.; Liang, G.; Deng, F.; Lam, T.L. A two-stage unsupervised approach for low light image enhancement. *IEEE Robot. Autom. Lett.* **2021**, *6*, 8363–8370. [CrossRef]
58. Hong, S.; Bangunharcana, A.; Park, J.-M.; Choi, M.; Shin, H.-S. Visual SLAM-based robotic mapping method for planetary construction. *Sensors* **2021**, *21*, 7715. [CrossRef] [PubMed]
59. Pire, T.; Fischer, T.; Castro, G.; De Cristóforis, P.; Civera, J.; Berlles, J.J. S-PTAM: Stereo parallel tracking and mapping. *Robot. Auton. Syst.* **2017**, *93*, 27–42. [CrossRef]
60. Zhong, Y.; Dai, Y.; Li, H. Self-supervised learning for stereo matching with self-improving ability. *arXiv Preprint* **2017**, arXiv:1709.00930.
61. Loza, A.; Mihaylova, L.; Canagarajah, N.; Bull, D. Structural similarity-based object tracking in video sequences. In Proceedings of the 2006 9th International Conference on Information Fusion, Florence, Italy, 10–13 July 2006; pp. 1–6.
62. Bussey, B.; Hoffman, S.J. Human Mars landing site and impacts on Mars surface operations. In Proceedings of the 2016 IEEE Aerospace Conference, Big Sky, MT, USA, 5–12 March 2016; pp. 1–21.
63. Green, R.D.; Kleinhenz, J.E. In-situ resource utilization (ISRU) Living off the Land on the Moon and Mars. In Proceedings of the American Chemical Society National Meeting & Exposition, Oraldo, FL, USA, 31 March–4 April 2019; pp. 1–4.
64. Hong, S.; Shin, H.-S. Construction of indoor and outdoor testing environments to develop rover-based lunar geo-spatial information technology. In Proceedings of the the KSCE Conference, Jeju, Korea, 10–23 October 2020; pp. 455–456.
65. Alcantarilla, P.F.; Solutions, T. Fast explicit diffusion for accelerated features in nonlinear scale spaces. *IEEE Trans. Pattern Anal. Mach. Intell.* **2011**, *34*, 1281–1298.
66. Mittal, A.; Soundararajan, R.; Bovik, A.C. Making a "completely blind" image quality analyzer. *IEEE Signal Process. Lett.* **2012**, *20*, 209–212. [CrossRef]
67. Mittal, A.; Moorthy, A.K.; Bovik, A.C. No-reference image quality assessment in the spatial domain. *IEEE Trans. Image Process.* **2012**, *21*, 4695–4708. [CrossRef]

remote sensing

MDPI

Article

High-Precision Registration of Lunar Global Mapping Products Based on Spherical Triangular Mesh

Zheng Bo [1,2], Kaichang Di [1,3,*], Bin Liu [1], Jia Wang [4], Zhaoqin Liu [1], Xin Xin [1,5], Ziqing Cheng [4] and Jinkuan Yin [6]

1 State Key Laboratory of Remote Sensing Science, Aerospace Information Research Institute, Chinese Academy of Sciences, Beijing 100101, China; bozheng@aircas.ac.cn (Z.B.); liubin201656@aircas.ac.cn (B.L.); liuzq@aircas.ac.cn (Z.L.); xinxin@tcu.edu.cn (X.X.)
2 University of Chinese Academy of Sciences, Beijing 100049, China
3 CAS Center for Excellence in Comparative Planetology, Hefei 230026, China
4 Beijing Aerospace Control Center, Beijing 100094, China; 15210106156@139.com (J.W.); yzlunehlzu@163.com (Z.C.)
5 School of Geology and Geomatics, Tianjin Chengjian University, Tianjin 300384, China
6 Beijing Yingce Spatial Info Tech Limited Company, Beijing 100043, China; jinkuan0528@souhu.com
* Correspondence: dikc@aircas.ac.cn; Tel.: +86-10-64868229

Abstract: Lunar global mapping products provide a solid data foundation for lunar scientific research and exploration. As the widespread geometric inconsistencies among multi-source mapping products seriously affect the synergistic use of the products, high-precision registration of multiple lunar global products is critical, and it is highly challenging due to large coverage and complex local geometric inconsistencies. In this research, we propose a spherical triangular-mesh-based method for high-precision registration of lunar global mapping products, which involves four steps: data preprocessing, feature point extraction and matching, spherical Delaunay triangulation, and geometric correction with spherical barycentric coordinates. This global registration method avoids map projection distortions by using spherical coordinates directly, and achieves high precision by confining the geometric models to spherical triangular facets. Experiments are conducted using two groups of different types of mapping products to verify the proposed method quantitatively and qualitatively. The results show that the geometric inconsistencies are reduced from hundreds of pixels to sub-pixel level globally after the registration, demonstrating the effectiveness of the proposed method.

Keywords: lunar global mapping products; registration; spherical triangular mesh; spherical barycentric coordinates

1. Introduction

Since the late 1950s, over a hundred lunar exploration missions have been carried out, from which massive data have been acquired by multiple sensors, such as cameras, spectrometers, laser altimeters, etc. The multi-source data and derived mapping products have greatly contributed to studies of the topography, geomorphology, mineralogy, and geologic evolution of the Moon. Particularly, various types of lunar global mapping products, which are essential for lunar global analyses, have been produced based on orbital data and photogrammetry and remote sensing technologies. For example, the U.S. Clementine mission acquired topographic data and multispectral imagery covering almost the entire Moon for the first time, from which lunar basemaps were produced [1–3]. These maps were later spatially registered to the Unified Lunar Control Network 2005 (ULCN2005) [4,5]. Based on the images obtained by the Wide-Angle Camera onboard the Lunar Reconnaissance Orbiter, a digital elevation model (DEM) covering the region between 79°N and 79°S and a global digital orthophoto mosaic (100 m/pixel) were produced [6–8]. Haruyama et al. [9] used data acquired from the SELENE Terrain Camera (TC) and Multi-band Imager (MI) from the JAXA SELENE/Kaguya mission to produce five lunar mosaic mapping products, namely

TCOrtho_MAP, DTM_MAP, TC_Morning_MAP, and SLDEM. In addition, 59 m/pixel of multispectral mosaic products covering the region between 50°N and 50°S were created from the topographically corrected MI reflectance data [10,11]. Li et al. used Chang'E-1 (CE-1) three line-array image data and laser altimeter data to produce a lunar global digital topographic product (CE1TMap2009), including the 120 m/pixel digital orthophoto map (DOM) CE1_TMap_GDOM_120m, 500 m/pixel DEM CE1_TMap_GDEM_500m [12], and 3000 m/pixel DEM CE1_TMap_GDEM_3000m [13]. A new global lunar topographic product CE2TMap2015 (including DOM and DEM) was produced with different resolutions of 7 m, 20 m, and 50 m based on Chang'E-2 (CE-2) stereo images [14].

Based on lunar global mapping products from multiple sources, the advantages of different mapping products can be combined and fully utilized to provide a solid data foundation for lunar scientific research and exploration. However, varying degrees of geometric inconsistencies among lunar global mapping products from multiple sources widely exist [13,15,16]. This is due to the uncertainties in positions and attitudes of the different sensors onboard different orbiters or the same sensor at different times, and the differences in data processing and map production methods [17]. Geometric inconsistency seriously restricts the comparative study and synergistic use of lunar global mapping products; therefore, it is a prerequisite to accurately register these mapping products to the same reference before further exploitation [18]. Unlike local products, global products cover a wide range, and the map projection of a global product can have different distortions in different regions, which brings difficulties for global registration. Moreover, transformation models of traditional registration methods usually fail because global mapping products typically exhibit varying degrees of geometric inconsistencies in different local regions. Hence, it is challenging to register lunar global mapping products with high precision.

Kirk et al. [18] elucidated the significance of unifying the current lunar data, and presented the processing principles of how to register datasets to a common reference frame or to each other. Many studies have been performed for lunar local data registration. For example, Wu et al. [19] proposed a multi-feature-based surface matching method for the co-registration of multi-source lunar DEMs to produce high-precision lunar topographic models; For the co-registration between lunar orbital imagery and reference DEM, a pixel-based matching method considering both geometric and photometric information was developed by Xin et al. [20]. However, no result about lunar global mapping product registration has been reported as far as we know.

In the field of remote sensing, image registration has been extensively studied over the past few decades. Registration methods can be broadly divided into area-based and feature-based methods [21]. Compared with area-based algorithms, feature-based methods have the advantage of reducing computational cost, especially in the case of complex deformation [22]. Therefore, in this paper, we adopt feature-based methods to solve the lunar global mapping product registration problem. Feature detection and matching, transformation model establishment, image resampling, and registration accuracy assessment are the basic steps of feature-based methods. The first two steps play the most important role in the accuracy of image registration. During the past decades, most of the research has focused on feature detection and matching. The traditional local invariant features, such as scale-invariant feature transform (SIFT) [23], affine-SIFT (ASIFT) [24], speeded-up robust feature (SURF) [25], and so on have been widely used for the matching of remote sensing images. For multimodal remote sensing image registration, the most direct strategy achieves feature detection and description by directly modifying off-the-shelf methods, such as uniform robust SIFT (UR-SIFT) [26], scale restrict (SR-SIFT) [27], uniform nonlinear diffusion-based Harris (UND-Harris) [28], and so on. A growing number of studies are focusing on designing valid descriptors to construct more reliable feature matches so that transformation parameters can be estimated correctly, such as multiscale self-similarities (MSS) [29], distinctive order-based self-similarity descriptor (DOBSS) [30], rank-based local self-similarity (RLSS) [31], histogram of orientated phase congruency (HOPC) [32], channel feature of oriented gradient (CFOG) [33], guided local outlier factor (GLOF) [34], and advanced neighborhood topology consensus (ANTC) [35]. Deep-learning-based feature extraction methods have been developed recently, e.g., Li et al. [36] introduced a multi-task feature extraction network and self-supervised feature points. With feature points obtained, a transformation model such as affine [37], polynomial [38,39], projective [40], and so on can be estimated. The transformation model used in image registration can either be a global transformation or a local transformation. Most of the existing image registration research adopts the global transformation model [41,42], which usually ignores the local geometric deformation between images. Especially for images with wide coverage, large terrain variation, or complicated geometrical inconsistency, local geometric deformation cannot be effectively corrected by a global transformation model. In this case, researchers have studied the problem of complex local deformation and proposed some local transformation models to improve the accuracy of registration, e.g., piecewise linear model (PWLM) [43–46], local weighted mean method [47], thin plate spline [48–50], multiquadric functions [51], and as-projective-as-possible algorithm [52]. A few works on complex local registration in the remote sensing field have been reported, among which a triangular irregular network (TIN)-based method is most commonly used [53–59]. It first triangulates the image into 2D triangular facets using the Delaunay triangulation method [60] based on a control point set (the control points and their correspondences are given), and then the local transformation model within every triangular facet is estimated and applied for precise registration. The local transformation model adopted in this method is usually a 2D affine transformation model.

In this work, we propose a spherical triangular-mesh-based registration method as a solution to high-precision registration of lunar global mapping products. By combining the dense spherical triangular mesh and the accurate local transformation model of spherical barycentric coordinates, the proposed method overcomes the difficulties of accurate registration for lunar global products caused by the complex projection transformation and local geometric inconsistency. The remainder of this paper is organized as follows. Section 2 introduces the spherical triangular-mesh-based method in detail. Experimental data and results are presented in Section 3. The conclusions and discussions are given in Section 4.

2. Methodology

Figure 1 shows the flowchart of the proposed spherical triangular-mesh-based method, which consists of four major steps; namely, data preprocessing, feature point extraction and matching, spherical Delaunay triangulation, and geometric correction with spherical barycentric coordinates. The technical details are elucidated in the following sub-sections.

2.1. Data Preprocessing

The Moon 2000 planetocentric coordinate system was selected as the spatial reference system in our research as it is most widely used and is convenient for lunar global analysis. If the lunar global mapping products to be processed are not in the Moon 2000 coordinate system, they should first be converted to the Moon 2000 system with longitude and latitude as the unit. After unifying the coordinate system, we subdivide each product into regular blocks in longitude and latitude for efficient handling of large data. In order to avoid resolution loss, there is usually a large amount of redundancy (repeated pixels) in high latitudes in the image files stored in the spherical coordinate system, which will cause difficulties in feature point extraction. Thus, blocks at high latitudes need to be converted to the projected coordinate system (i.e., polar stereographic projection). This map projection process is only used to facilitate feature point extraction and matching. The subsequent Delaunay triangulation and geometric correction are all realized in the spherical coordinate system.

Additionally, when registering two global DEMs or a DEM to lunar imagery, we use the hillshading technique to generate simulated imagery of DEM data [16,20]. The simulated image is a gray-scale representation of the terrain surface under a certain illumination condition (i.e., azimuth and elevation of the Sun), and is similar to a real lunar image. The 3D-3D or 3D-2D matching is then transferred into the 2D-2D matching based on the simulated imagery. We have developed a software tool to create simulated images of DEM data for batch processing based on the hillshading technique and GDAL (Geospatial Data Abstraction Library) open-source library. The solar azimuth angle and solar altitude angle of reference and source DEM data need to be set to be the same, and the default values are 315° and 45°, respectively.

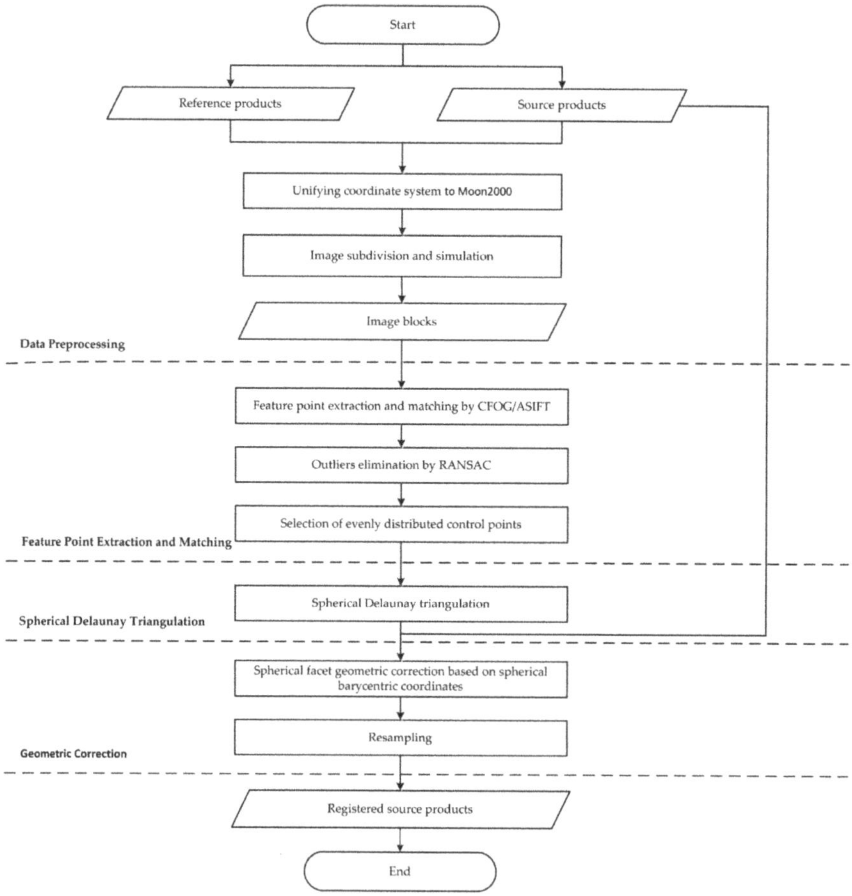

Figure 1. Overall flowchart of the proposed method.

2.2. Feature Point Extraction and Matching

We chose the ASIFT algorithm [24] to extract and match the feature points of DEM simulated images under the same illumination condition. As a variant of SIFT, ASIFT is not only invariant to scale, rotation, and translation, but also invariant to affine transformation. A comparative study has shown that ASIFT outperforms SIFT and SURF in terms of more keypoints and fewer mismatches [61]. A two-resolution scheme further reduces the ASIFT complexity to about twice that of SIFT [24]. In this study, an open-source web code [62] developed by Yu and Morel [63] was used to extract and match the feature points on the simulated images in corresponding blocks.

A fast and robust matching for multimodal remote sensing images proposed by Ye et al. [33] was also adopted in this research, and the software package can be downloaded from the GitHub website [64]. This method first extracts the channel feature of orientated gradient (CFOG) of each pixel to generate the feature representation for each pixel. A similarity measure based on the feature representation is then established in the frequency domain to accelerate the image matching. Subsequently, corresponding feature points between images are detected in a template matching manner.

The random sample consensus (RANSAC) [65] algorithm is used to eliminate mismatches in both of the two methods. Moreover, for lunar global mapping product registration with complex geometric inconsistency, even distribution of control points covering every possible local deformation region is critical. In order to obtain evenly distributed tie points to ensure the uniformity of the spherical triangular mesh, the redundant points are removed from the set of tie points. For this purpose, each block is divided into dense grids, where the point closest to the center of the grid is retained as the control point to construct the spherical triangulation if multiple points exist in a grid.

The settings in the feature point extraction and matching step include search radius (e.g., 100 pixels), template radius (e.g., 100 pixels), grid number (e.g., 225), error elimination threshold (e.g., 2.5 pixels), and so on, which can be set and modified flexibly according to the reference and source data.

2.3. Spherical Delaunay Triangulation

The construction of a spherical Delaunay triangulation network is an extension of the traditional 2D Delaunay triangulation network. The rules for constructing the 2D Delaunay triangulation are (1) the 2D Delaunay triangulation is unique, and the circumcircle of any triangle does not contain any fourth point; (2) among all possible triangulations, the smallest angle of the Delaunay triangle is the largest.

Specifically, the rules mean that after the diagonals of the convex quadrilateral formed by two adjacent triangles are exchanged with each other, the minimum angle of the six new internal angles no longer increases. By extending the criteria of 2D Delaunay triangulation to the sphere, the following rules [66] can be obtained: (1) the inferior arc surface stretched by the circumscribed circle of any spherical triangle contains no other point; that is, the center of the sphere and all other points are located on the same side of the plane of the triangle; (2) the minimum of the six new interior angles of two adjacent spherical triangles does not increase after their diagonals are swapped.

In this paper, the incremental algorithm is used to generate the spherical Delaunay triangulation. The algorithm was first proposed by Lawson [67] and improved by Bowyer [68]. Renka [69] then proposed a spherical Delaunay triangulation algorithm based on the incremental algorithm. The open-source code for this incremental algorithm is available on the GitHub website [70]. The basic steps of the algorithm are as follows.

(1) The convex hull H of the spherical scatter point set V is established to enclose all the data points. A convex hull is a minimum convex polyhedron containing a series of known vertices in spatial geometry. In this algorithm, the convex hull consists of eight triangular faces determined by six vertices.
(2) Find the spherical triangle T that contains the insertion point P in the constructed convex hull H or triangulation network. T is then triangulated by connecting the insertion point P to the three vertices of T to form three new spherical triangles.
(3) A local optimization process (LOP) is used to optimize the triangle triangulation; that is, by exchanging diagonals to ensure that the triangle formed is a Delaunay triangle.
(4) Repeat (2) and (3) until the last point has been inserted.

2.4. Spherical Geometric Correction Based on Spherical Barycentric Coordinates

A barycentric coordinate is a special local coordinate, which takes the vertices of a polygon as the reference points to locate an arbitrary point inside the polygon. Barycentric coordinates are widely used in computer graphics and other fields for their excellent characteristics such as normality, linear reconstruction, and affine invariance. In this research, we apply the barycentric coordinates to the deformation of spherical triangular facets.

Barycentric coordinates were first introduced by Möbius [71]. Let T be a triangle with vertices $(v_i)_{i=1,2,3}$. The position of any point v inside the triangle T on the plane

$\mathbb{E}^2$ can be obtained by taking an adequately weighted sum of the three vertices. These three weight coefficients $\lambda_i(v,T)_{i=1,2,3}$ are the barycentric coordinates of the point v in the reference triangle, and the barycentric coordinates are continuous functions that satisfy the following properties.

$$\forall i \quad \lambda_i(v,T) > 0; \tag{1}$$

$$\sum_i \lambda_i(v,T) = 1; \tag{2}$$

$$\sum_i \lambda_i(v,T)v_i = v. \tag{3}$$

Möbius [72] first extended the concept of barycentric coordinates from a plane to a sphere. Similar to the barycentric coordinates of a planar triangle, the spherical barycentric coordinates represent the position of arbitrary point v on a sphere relative to the three vertices of a reference spherical triangle P. Previous studies [73,74] have shown that, unlike planar triangles, it is impossible for the barycentric coordinates of any point on a spherical triangle to satisfy properties (1), (2), and (3) at the same time. Specifically, properties (2) and (3) are contradictory to each other on spheres. Property (2) satisfies the request that any point v represented by the barycentric coordinates be located in a plane, whereas the property (3) demands that the point v be located on a spherical surface rather than a plane. Consequently, to use the barycentric approach as a solution to the spherical problem, it is essential to relax the above properties to obtain the spherical barycentric coordinates. Alfeld et al. [75] and Langer et al. [74] strictly retained property (3), but relaxed property (2) to

$$\sum_i \lambda_i(v,P) \geq 1; \tag{4}$$

This is because their main goal was to define curves and surfaces over the sphere. Lei et al. [76] proposed an opposite strategy, which abandons the preservation of property (3) but strictly preserves property (2), dedicated to constructing spherical grid systems. For the problem to be solved in this paper, specifically referring to determining the new position of the sphere point on the transformed sphere surface by the unique spherical barycentric coordinates, it is essential to retain property (3). Hence, we follow the first strategy, retaining property (3) and replacing property (2) with (4).

This paper follows the formula of spherical barycentric coordinates given by Alfeld et al. [75]. Once more, it is assumed that the vertices of a spherical triangle P are the points in the set $V = \{v_1, v_2, v_3\}$; let n_i denote the unit normal vectors to the planes $P_i = span(V \backslash \{v_i\})$, $i = 1,2,3$. For a point $v \in$ sphere S, let the angles α_i, β_i be defined by the dot products

$$sin\alpha_i := v \cdot n_i, \quad sin\beta_i := v_i \cdot n_i, \quad i = 1,2,3. \tag{5}$$

where α_i represent oriented angles between the vector v and the planes P_i, while β_i are the analogues angles between v_i and P_i (see Figure 2). For nontrivial spherical triangles, $sin\beta_i \neq 0, i = 1,2,3$. The spherical barycentric coordinates of the point $v \in S$ with respect to the triangle P are given by

$$\lambda_i(v) = \frac{sin\alpha_i}{sin\beta_i}, \quad i = 1,2,3. \tag{6}$$

The main idea of geometric correction based on spherical barycentric coordinates is to establish the transformation relationship between the source spherical triangle and the reference spherical triangle. The vertex of a spherical triangle P is denoted as $V = \{v_1, v_2, v_3\}$, and the positions of each pixel v inside the spherical triangle are the spherical barycentric coordinates $\lambda_i(v,P)$. P is deformed to form a reference triangle P' by moving the position of the vertex, the source vertex V corresponds to the new reference vertex $V' = \{v_1', v_2', v_3'\}$, and the internal pixels also change accordingly. The position of each point v' inside the P' can be calculated using the spherical barycentric coordinates as follows.

$$v' = \sum_i \lambda_i(v,P) \cdot v_i', \quad i = 1,2,3. \tag{7}$$

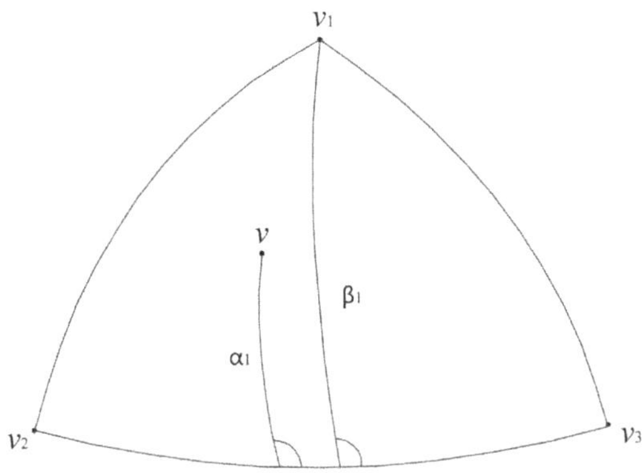

Figure 2. Computing spherical barycentric coordinates by Equation (5).

The geometric correction is realized by calculating the transformed position of each pixel of the source data using the above equation for each facet. The final product is generated after resampling.

3. Experimental Results

We performed registration experiments to verify the performance of the proposed high-precision registration method using four different lunar global mapping products; namely, Moon Clementine UVVIS Warped Color Ratio Mosaic 200 m, CE1_TMap_GDOM_120m, SLDEM2015, and CE2TMap2015_DEM_20m. Two separate experiments were conducted to evaluate registration of imagery products and registration of DEMs. In the first experiment, Moon Clementine UVVIS Warped Color Ratio Mosaic 200 m was taken as the reference data and the CE1_TMap_GDOM_120m was taken as source data to be registered to the reference data. The data for the second experiment are both DEM products, with SLDEM2015 and CE2TMap2015_DEM_20m as reference and source data, respectively.

3.1. Experimental Data

The Clementine Ultraviolet-Visible (UVVIS) warped color-ratio mineral map is a false color band ratio image covering the global lunar surface and published by United States Geological Survey (USGS) Astrogeology Science Center. It was generated from the Clementine UVVIS mosaics using three spectral filters (central wavelengths of 415, 750, and 1000 nm) [77,78], and multispectral mosaics with highly accurate radiometric calibrations, photometric corrections, and geodetical controls have been registered to ULCN2005 [4]. The false color band ratio image is a composite image, including the ratio of the 750/415 nm bands used for the red-channel brightness, 750/1000 nm for the green channel, and the 415/750 nm ratio for the blue channel. For detailed information on what color ratios represent, please refer to Lucey et al. [79]. This global product with the spatial resolution of 200 m/pixel is available in GeoTIFF format at the USGS website [80]. Some small gaps in the image are due to lack of coverage in one or more of the bands used in the color ratio.

CE1_TMap_GDOM_120m is a lunar global DOM made of images obtained by a stereo camera with three-line CCDs onboard the CE-1. The intersection angles of the forward and backward relative to the nadir images of the stereo camera are $\pm 16.7^{\circ}$. CE-1 has an orbital altitude of about 200 km and an orbital inclination of nearly 90°. During the flight around the Moon, the three-linear CCD stereo camera takes images in the linear array push-broom way, with a spatial resolution of about 120 m and an image width of about 60 km. The overlap rate of image data in adjacent orbits is more than 40%, and the planimetric positioning accuracy is about 100 m~1.5 km [81]. Li et al. [12] processed the image data according to the mapping specification, carried out photometric correction, improved the consistency of the images, and finally produced a lunar global orthophoto map with a simple cylindrical projection and a spatial resolution of 120 m. The data are provided by the Lunar and Planetary Data Release System [82].

SLDEM2015 is a near-global DEM product constructed by combining the altimeter data from the Lunar Reconnaissance Orbiter Laser Altimeter (LOLA) and the DEM generated by stereo images obtained by SELENE TC. Barker et al. [15] divided the SELENE DEM into $1^{\circ} \times 1^{\circ}$ tiles, taking the LOLA point cloud data as reference. Each tile was then adjusted by the tile-averaged transformation parameters (three translation parameters and two tilt parameters) solved by a bounded downhill simplex minimization algorithm. The resultant SLDEM2015 combines the advantages of LOLA's accurate reference geodetic framework with the high resolution of the SELENE DEM. The SLDEM2015 covers the region between 60°N and 60°S, with a grid spacing of 512 pixels/degree (60 m/pixel at the equator) and a typical vertical accuracy of 3 m–4 m. This DEM product is available for download in GeoTIFF format from the USGS website [83].

CE2TMap2015_DEM_20m is a lunar global DEM, which is a major part of CE2TMap2015 [14]. It was produced from stereo images obtained by the CE-2 stereo camera, using the lunar surface positions of five laser ranging retro reflectors (LRRRs) (Apollo 11, Apollo 14, Apollo 15, Lunokhod 1, and Lunokhod 2) as the absolute control points to realize a seamless mosaic and high-precision absolute positioning [14,84]. CE2TMap2015_DEM_20m has a resolution of 20 m/pixel and 100% coverage of the Moon. It is the highest-resolution lunar global DEM product ever published, and is available through the Lunar and Planetary Data Release System [85]. This DEM product is organized in 188 subdivisions; each subdivision includes a data file (GeoTIFF format), a coordinate information file (tfw), and a projection file (prj) of the same name.

3.2. Data Preprocessing, Matching, and Spherical Delaunay Triangulation Results

In the data preprocessing step, the reference and source products were first divided into regular blocks in longitude and latitude. Within the range of the latitude $\pm 60^{\circ}$, Moon Clementine UVVIS Warped Color Ratio Mosaic 200 m and CE1_TMap_GDOM_120m were divided into $30^{\circ} \times 30^{\circ}$ blocks, and SLDEM2015 and CE2TMap2015_DEM_20m were divided into $5^{\circ} \times 5^{\circ}$ blocks. For high latitudes and polar regions beyond the $\pm 60^{\circ}$ area, the products of the first experiment were converted into Moon stereoscopic north/south pole projection and divided into 18 blocks with the same size.

In experiment 1, the multimodal remote sensing image matching method proposed by Ye et al. [33] was used to match the feature points between corresponding reference and source blocks. The ASIFT algorithm was used to extract and match the feature points of DEM simulated images under the same illumination in experiment 2. The RANSAC algorithm was used to eliminate mismatches in each block. Each block of experiment 1 and experiment 2 was then divided into 50×50 and 15×15 non-overlapping grids, respectively, and the point closest to the center of each grid was reserved as the control point of the subsequent construction of the triangulated network. Finally, 131,082 and 301,022 control points for experiment 1 and 2 were obtained, respectively. Figures 3 and 4 show the matching results of one typical block in each experiment.

We used the points of the reference image obtained above to construct the spherical Delaunay triangulated network, and the results of experiment 1 are shown in Figure 5. The numbers of spherical triangular facets in experiments 1 and 2 were 262,160 and 602,022, respectively. The spherical Delaunay triangulated network constructed by the source image is the same as that constructed by the reference image, with vertices being replaced by the corresponding vertices in the source image. As a result, based on the corresponding spherical triangular facets, the global registration problem is converted into the local problem. Figure 5 takes experiment 1 as an example to show the results of spherical Delaunay triangulation.

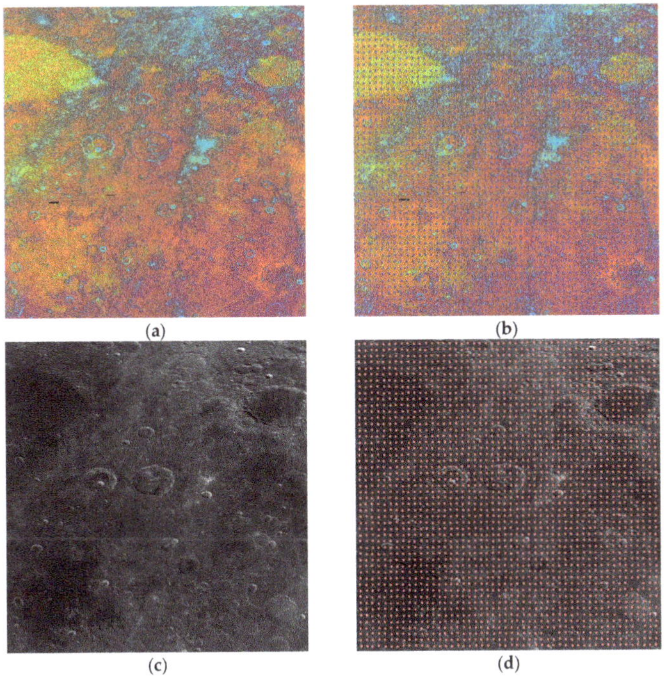

Figure 3. Examples of matching result of a 30° × 30° image block centered at (45°E, 45°N). (**a**) Clementine UVVIS Warped Color Ratio Mosaic 200 m; (**b**) final control points (Clementine UVVIS Warped Color Ratio Mosaic 200 m); (**c**) CE1_TMap_GDOM_120m; (**d**) final control points (CE1_TMap_GDOM_120m).

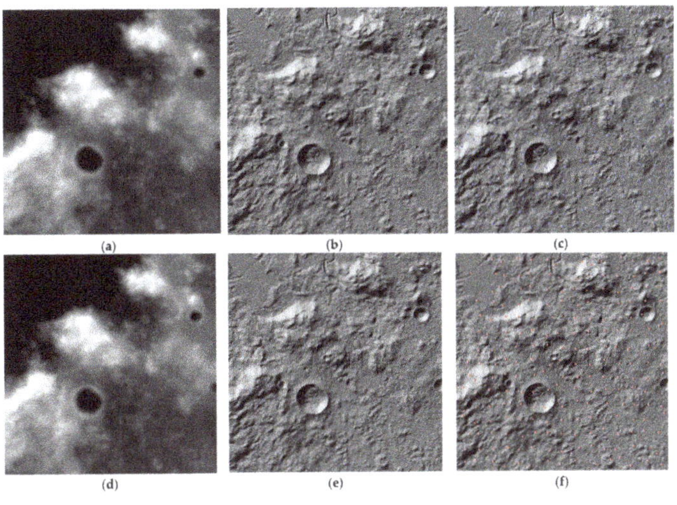

Figure 4. Examples of matching result of a 5° × 5° image block centered at (2.5°E, 22.5°N). (**a**) SLDEM2015; (**b**) simulated image (SLDEM2015); (**c**) final control points (SLDEM2015); (**d**) CE1_TMap_GDOM_120m; (**e**) simulated image (CE2TMap2015_DEM_20m); (**f**) final control points (CE2TMap2015_DEM_20m).

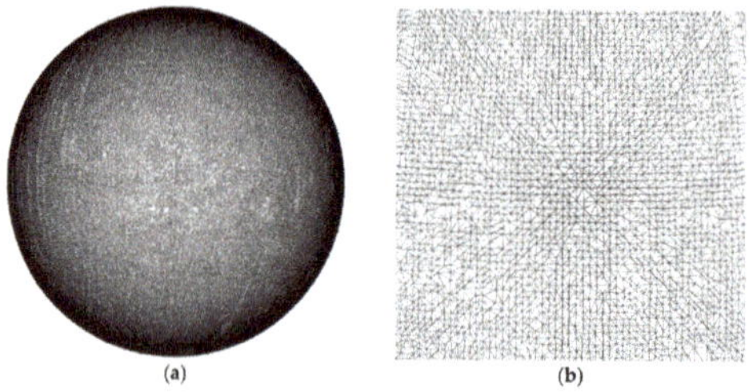

Figure 5. Spherical Delaunay triangulation result. (**a**) Global spherical Delaunay triangulation based on control points from experiment 1; (**b**) enlarged view of the spherical Delaunay triangulation of (**a**) centered at (45°E, 15°N) and a window size of 30° × 30°.

3.3. Registration Results and Analysis

3.3.1. Quantitative Analysis

In this subsection, a quantitative analysis was first conducted to evaluate the registration accuracy. A certain number of randomly distributed points were automatically selected as the checkpoints to assess the geometric displacements before and after the registration, and these checkpoints were selected from the redundant points removed from the tie points set (after outlier elimination) in the matching step of the proposed method. The number of checkpoints was 16,377 and 17,280 for the experiments 1 and 2, respectively. The distribution of the checkpoints is shown in Figure 6.

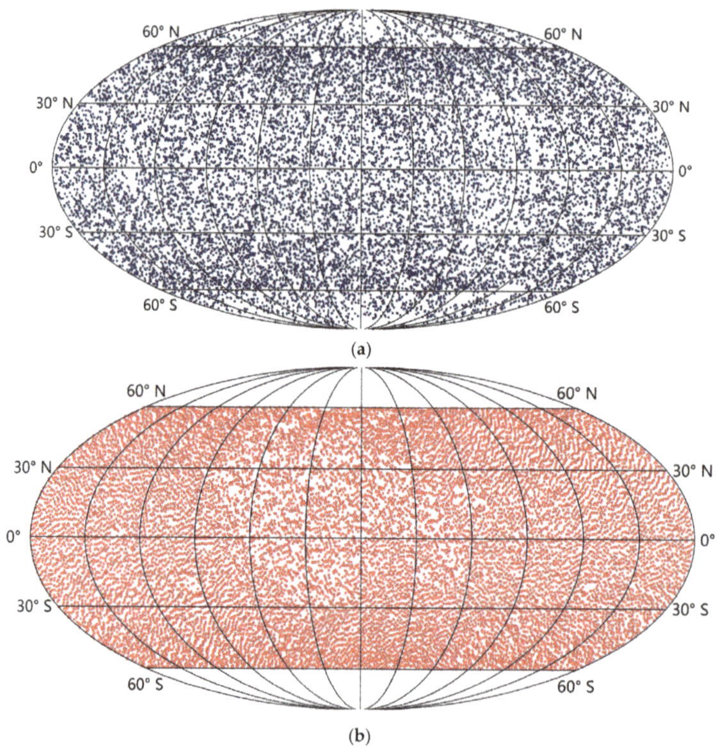

Figure 6. The distribution of the checkpoints matched from (**a**) Clementine UVVIS Warped Color Ratio Mosaic 200 m and CE1_TMap_GDOM_120m; (**b**) SLDEM2015 and CE2TMap2015_DEM_20m.

The geometric displacement, i.e., the planar residual, was evaluated based on the difference in planetocentric coordinates of each checkpoint pair. The planar residual was computed by the following formula:

$$\sigma_p = \sqrt{\Delta x^2 + \Delta y^2}, \tag{8}$$

$$\Delta x = \frac{\pi D\cos\left(lat_{ref}\right)}{360}\Delta lon, \quad \Delta y = \frac{\pi D\cos\left(lat_{ref}\right)}{360}\Delta lat,$$
$$\Delta lon = lon_{source} - lon_{ref}, \quad \Delta lat = lat_{source} - lat_{ref}.$$

where Δlon and Δlat are the longitude and latitude residuals of the checkpoints, respectively; Δx and Δy, which are converted to meters, are the residuals between the checkpoints; D is the diameter of the Moon; the coordinates of the checkpoints in source data are (lon_{source}, lat_{source}), and the coordinates of the checkpoints in reference data are (lon_{ref}, lat_{ref}).

The magnitude and distribution of planar residuals are shown by the color shading maps (Figures 7 and 8) with 1° spacing. Figure 7a shows that the differences between CE1_TMap_GDOM_120m and Clementine UVVIS Warped Color Ratio Mosaic 200 m vary widely, with smaller differences on the lunar near side compared to the far side and smaller residuals in the northern hemisphere compared to the southern. Areas with particularly high differences are clustered in the southern hemisphere (78°E–140°E, 7°N–80°S), and the highest residual is more than 5000 m, which is much larger than the differences of tens of meters in other areas. It is clear from Figure 8a that the differences between CE2TMap2015_DEM_20m and SLDEM2015 near lunar maria are smaller than those of other areas. This is attributed to the fact that five LRRRs with high positioning accuracy were used as control points in CE-2 map production and that the topography of the area is relatively flat. Figures 7b and 8b show that the planar residuals between reference data and source data after registration of both the experiments are greatly reduced, indicating that the proposed method is effective and the registration results are of high precision.

Furthermore, statistics of mean absolute error (MAE) and root-mean-square error (RMSE) of the planar residuals between the checkpoints of reference and source data have been used for quantitative evaluation. The errors are represented both in meters and pixels (with respect to the reference data). The results, given in Table 1, show that the MEANs and RMSEs are several pixels in experiment 1 and over ten pixels in experiment 2 before registration; the errors are significantly reduced to sub-pixel level after registration in both experiments. The statistical results demonstrate that our proposed method has achieved high precision over the entire lunar globe.

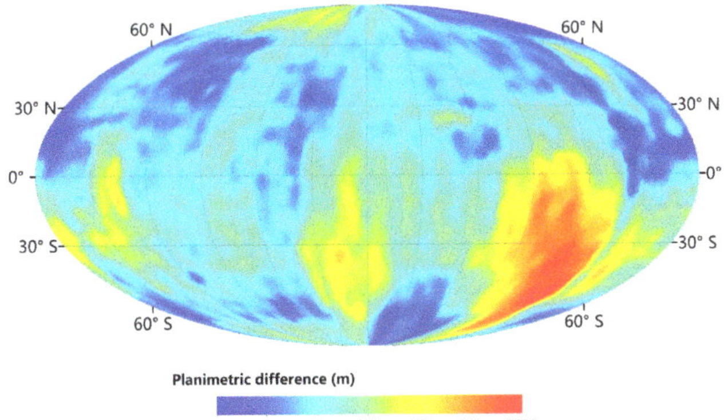

(a)

Figure 7. *Cont.*

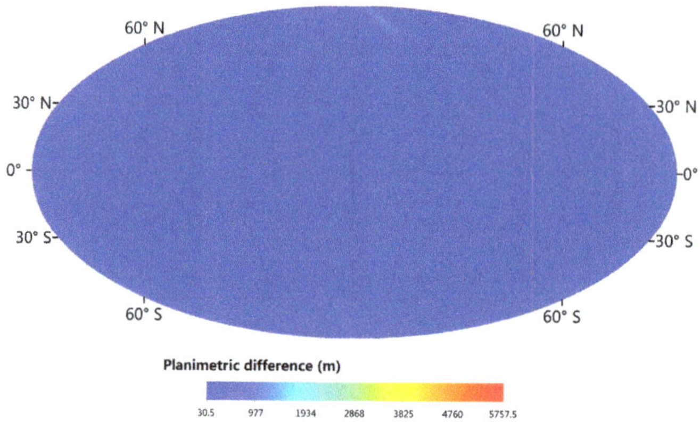

(**b**)

Figure 7. Color-coded map of planimetric difference at a 1° grid space between (**a**) Clementine UVVIS Warped Color Ratio Mosaic 200 m and CE1_TMap_GDOM_120m before registration; (**b**) Clementine UVVIS Warped Color Ratio Mosaic 200 m and CE1_TMap_GDOM_120m after registration.

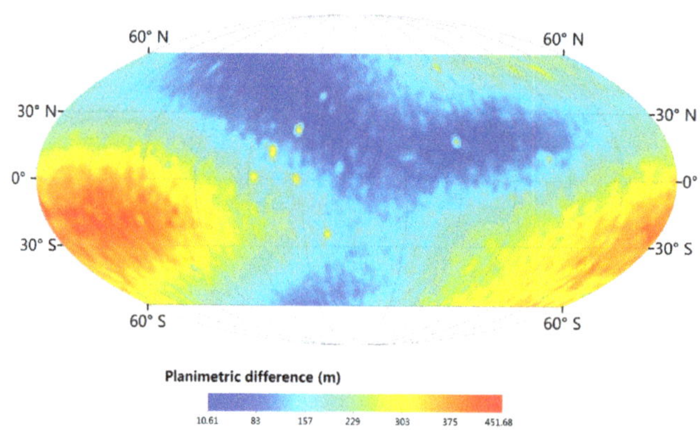

(**a**)

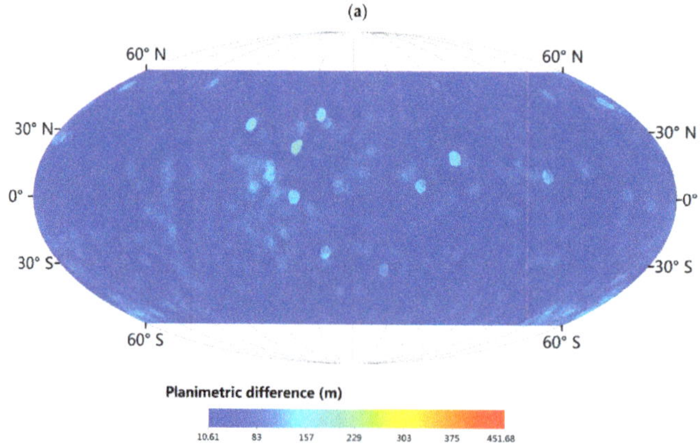

(**b**)

Figure 8. Color-coded map of planimetric difference at a 1° grid space between (**a**) SLDEM2015 and CE2TMap2015_DEM_20m before registration; (**b**) SLDEM2015 and CE2TMap2015_DEM_20m after registration.

Table 1. Quantitative statistics of the planar residuals before and after registration using our proposed method in experiments 1 and 2.

Experiment Group	Registration Status	MAE (m, pixel)	RMSE (m, pixel)
experiment 1	before	2225.0, 11.13	2442.2, 12.21
	after	135.7, 0.68	198.3, 0.99
experiment 2	before	184.5, 3.08	106.5, 1.78
	after	38.5, 0.64	43.0, 0.71

3.3.2. Qualitative Analysis

The performance of the proposed method was also evaluated qualitatively by visual inspection using checkerboard mosaic images, where neighboring squares were taken from the reference product and the source product before and after registration. The details are shown in Figures 9 and 10.

Compared with the checkerboard image before registration (Figures 9c and 10e) showing large deviations at the boundaries of adjacent squares, the regions of the checkerboard image after registration (Figures 9d and 10f) are well overlapped, and the linear features across the boundaries of adjacent squares are continuous. This visually verifies the effectiveness of the proposed registration method.

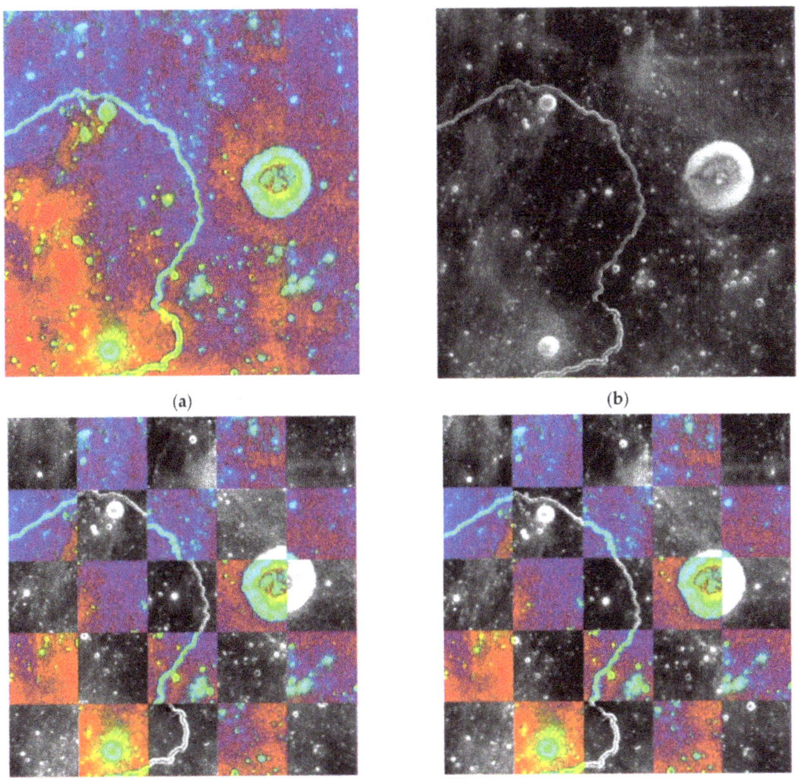

Figure 9. Examples of registration results of Moon Clementine UVVIS Warped Color Ratio Mosaic 200 m and CE1_TMap_GDOM_120m in the 2° × 2° block centered at (47.88°W, 16.27°N). (**a**) Reference image; (**b**) source image; (**c**) checkerboard composite of the reference and the pre-registered source images; (**d**) checkerboard composite of the reference and the registered source images.

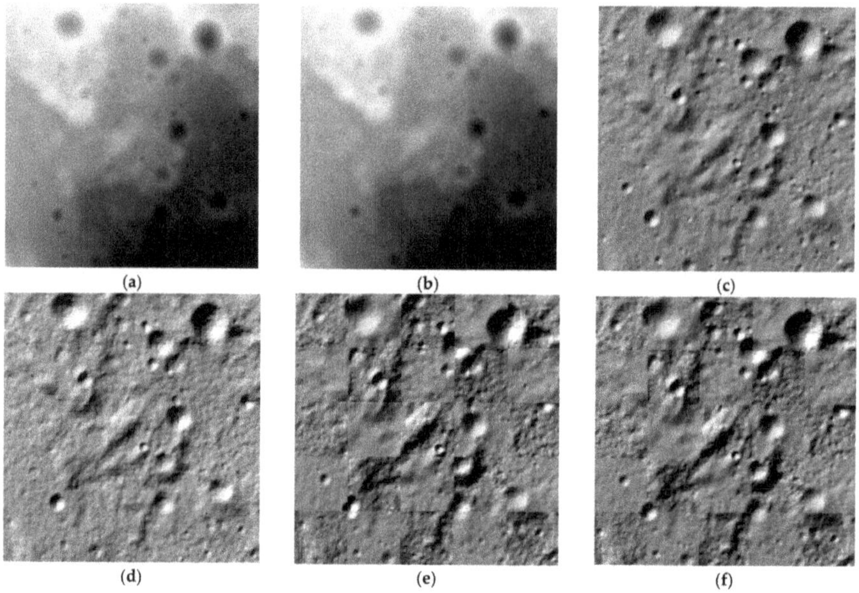

Figure 10. Examples of registration results of SLDEM2015 and CE2TMap2015_DEM_20m in the 0.5° × 0.5° block centered at (119.29°W, 3.72°N). (a) Reference DEM; (b) source DEM; (c) reference simulated image; (d) source simulated image; (e) checkerboard composite of the reference and the pre-registered source simulated images; (f) checkerboard composite of the reference and the registered source simulated images.

In principle, the proposed method needs a large number of evenly distributed control points to represent the local variation in geometric inconsistency. Therefore, the quantity and spatial distribution of the extracted control points are crucial to the registration precision of lunar global products with complex local deformation. More control points generally result in higher registration precision, but too many control points will inevitably reduce the efficiency of spherical triangulation and geometric correction. Therefore, the effect of control point density on registration precision needs to be studied in future research.

4. Conclusions

This paper proposed a spherical triangular-mesh-based method for high-precision registration of lunar global mapping products. The proposed method has three main innovations. First, the global spherical registration is not affected by the distortion caused by map projection, especially in high latitude and polar regions. Second, the spherical triangular facets avoid using a single transformation model to register the whole lunar global product, and thus resolve the problem of local geometric distortion. Finally, the local transformation model of spherical barycentric coordinates can achieve high-precision of registration and ensure that the registered product after transformation is continuous at the edges of the spherical triangular mesh.

In the experiments, the performance of the proposed method was evaluated quantitatively and qualitatively. The quantitative results show that the planar residuals between source data and reference data after registration are greatly reduced to sub-pixel level. The qualitative results conducted with the same data are consistent with the quantitative analysis. Both the quantitative and qualitative results show that our proposed method achieved high precision for the registration of lunar global products, confirming the effectiveness of this method. The method can be applied to register global mapping products of the same or different types, e.g., optical image products, spectral image products, and DEM.

Author Contributions: Conceptualization, K.D.; methodology, Z.B., K.D., B.L., J.W. and Z.L.; software, Z.B., X.X. and J.Y.; validation, Z.B., B.L. and Z.C.; investigation, K.D., J.W. and Z.L.; writing—original draft preparation, Z.B.; writing—review and editing, K.D.; project administration, K.D. All authors have read and agreed to the published version of the manuscript.

Funding: This work was supported by the Strategic Priority Research Program of the Chinese Academy of Sciences (grant no. XDB41000000) and the National Natural Science Foundation of China (grant nos. 41941003 and 41771490).

Institutional Review Board Statement: Not applicable.

Informed Consent Statement: Not applicable.

Data Availability Statement: Not applicable.

Conflicts of Interest: The authors declare no conflict of interest.

References

1. Smith, D.E.; Zuber, M.T.; Neumann, G.A.; Lemoine, F.G. Topography of the Moon from the Clementine lidar. *J. Geophys. Res.* **1997**, *102*, 1591–1611. [CrossRef]
2. Robinson, M.S.; McEwen, A.S.; Eliason, E.; Lee, E.M.; Malaret, E.; Lucey, P.G. Clementine UVVIS global mosaic: A new tool for understanding the lunar crust. In Proceedings of the Lunar and Planetary Science Conference, Houston, TX, USA, 15–29 March 1999.
3. Eliason, E.; Isbell, C.; Lee, E.; Becker, T.; Gaddis, L.; McEwen, A.; Robinson, M. *The Clementine UVVIS Global Lunar Mosaic*; Lunar and Planetary Institute: Houston, TX, USA, 1999.
4. Hare, T.M.; Archinal, B.A.; Becker, T.L.; Gaddis, L.R.; Lee, E.M.; Redding, B.L.; Rosiek, M.R. Clementine mosaics warped to ULCN2005 network. In Proceedings of the 39th Annual Lunar and Planetary Science Conference, League City, TX, USA, 10–14 March 2008; p. 2337.
5. Lee, E.M.; Gaddis, L.R.; Weller, L.; Richie, J.O.; Becker, T.; Shinaman, J.; Rosiek, M.R.; Archinal, B.A. A new Clementine basemap of the Moon. In Proceedings of the 40th Annual Lunar and Planetary Science Conference, The Woodlands, TX, USA, 23–27 March 2009; p. 2445.
6. WAC Global Morphologic Map. Available online: http://wms.lroc.asu.edu/lroc/view_rdr/WAC_GLOBAL (accessed on 10 April 2018).
7. LRO CameraTeam Release High Resolution Global Topographic Map of Moon. Available online: http://www.nasa.gov/mission_pages/LRO/news/lro-topo.html (accessed on 10 April 2018).
8. Wagner, R.V.; Speyerer, E.J.; Robinson, M.S.; LROC Team. New mosaicked data products from the LROC team. In Proceedings of the 46th Annual Lunar and Planetary Science Conference, The Woodlands, TX, USA, 16–20 March 2015; p. 1473.
9. Haruyama, J.; Ohtake, M.; Matsunaga, T.; Otake, H.; Yamamoto, S. Data Products of SELENE (Kaguya) Terrain Camera for Future Lunar Missions. In Proceedings of the 45th Annual Lunar and Planetary Science Conference, The Woodlands, TX, USA, 17–21 March 2014.
10. Taylor, L.A.; Pieters, C.; Patchen, A.; Taylor, D.S.; Morris, R.V.; Keller, L.P.; McKay, D.S. Mineralogical and chemical characterization of lunar highland soils: Insights into the space weathering of soils on airless bodies. *J. Geophys. Res.* **2010**, *115*, E02002. [CrossRef]
11. Ohtake, M.; Pieters, C.M.; Isaacson, P.; Besse, S.; Green, R.O. One Moon, many measurements 3: Spectral reflectance. *Icarus* **2013**, *226*, 364–374. [CrossRef]
12. Li, C.; Su, Y.; Wen, W.; Bian, W.; Zhao, B.; Yang, J.; Zou, X.; Wang, M.; Xu, C.; Kong, D. The global image of the Moon obtained by the chang'e-1: Data processing and lunar cartography. *Sci. China* **2010**, *53*, 1091–1102. [CrossRef]
13. Li, C.; Ren, X.; Liu, J.; Zou, X.; Mu, L.; Wang, J.; Shu, R.; Zou, Y.; Zhang, H.; Chang, L. Laser altimetry data of chang'e-1 and the global lunar dem model. *Sci. China Ser. D Earth Sci.* **2010**, *40*, 281–293. [CrossRef]
14. Li, C.; Liu, J.; Ren, X.; Yan, W.; Zuo, W.; Mou, L.; Zhang, H.; Su, Y.; Wen, W.; Tan, X.; et al. Lunar Global High-precision Terrain Reconstruction Based on Chang'e-2 Stereo Images. *Geomat. Inf. Sci. Wuhan Univ.* **2018**, *43*, 485–495.
15. Barker, M.K.; Mazarico, E.; Neumann, G.A.; Zuber, M.T.; Haruyama, J.; Smith, D.E. A new lunar digital elevation model from the Lunar Orbiter Laser Altimeter and SELENE Terrain Camera. *Icarus* **2016**, *273*, 346–355. [CrossRef]
16. Xin, X.; Liu, B.; Di, K.; Yue, Z.; Gou, S. Geometric quality assessment of chang'e-2 global DEM product. *Remote Sens.* **2020**, *12*, 526. [CrossRef]
17. Di, K.; Liu, B.; Xin, X.; Yue, Z.; Ye, L. Advances and applications of lunar photogrammetric mapping using orbital images. *Acta Geod. Et Cartogr. Sin.* **2019**, *48*, 1562.
18. Kirk, R.L.; Archinal, B.A.; Gaddis, L.R.; Rosiek, M.R. Lunar Cartography: Progress in the 2000s and prospects for the2010s. In Proceedings of the 2012 ISPRS-International Archives of the Photogrammetry Remote Sensing and Spatial Information Sciences, Melbourne, Australia, 25 August–1 September 2012; Volume XXXIX-B4, pp. 489–494.
19. Wu, B.; Guo, J.; Hu, H.; Li, Z.; Chen, Y. Co-registration of lunar topographic models derived from Chang'E-1, SELENE, and LRO laser altimeter data based on a Novel surface matching method. *Earth Planet Sci. Lett.* **2013**, *364*, 68–84. [CrossRef]

20. Xin, X.; Liu, B.; Di, K.; Jia, M.; Oberst, J. High-precision co-registration of orbiter imagery and digital elevation model constrained by both geometric and photometric information. *ISPRS J. Photogramm. Remote Sens.* **2018**, *144*, 28–37. [CrossRef]
21. Zitova, B.; Flusser, J. Image registration methods: A survey. *Image Vis. Comput.* **2003**, *21*, 977–1000. [CrossRef]
22. Li, B.; Wei, W.; Hao, Y. Automatic registration of multisensor images with affine deformation based on triangle area representation. *J. Electron. Imaging* **2013**, *22*, 3002. [CrossRef]
23. Lowe, D.G. Distinctive image features from scale-invariant keypoints. *Int. J. Comput. Vis.* **2004**, *60*, 91–110. [CrossRef]
24. Morel, J.M.; Yu, G. ASIFT: A New Framework for Fully Affine Invariant Image Comparison. *SIAM J. Imaging Sci.* **2009**, *2*, 438–469. [CrossRef]
25. Bay, H.; Tuytelaars, T.; Van Gool, L. Surf: Speeded up robust features. In Proceedings of the European Conference on Computer Vision, Graz, Austria, 7–13 May 2006; pp. 404–417.
26. Yi, Z.; Zhiguo, C.; Yang, X. Multi-spectral remote image registration based on SIFT. *Electron. Lett.* **2008**, *44*, 107–108. [CrossRef]
27. Sedaghat, A.; Mokhtarzade, M.; Ebadi, H. Uniform robust scale-invariant feature matching for optical remote sensing images. *IEEE Trans. Geosci. Remote Sens.* **2011**, *49*, 4516–4527. [CrossRef]
28. Fan, J.; Yan, W.; Ming, L.; Liang, W.; Cao, Y. Sar and optical image registration using nonlinear diffusion and phase congruency structural descriptor. *IEEE Trans. Geosci. Remote Sens.* **2018**, *56*, 5368–5379. [CrossRef]
29. Sun, H.; Lei, L.; Zou, H.; Wang, C. Multimodal remote sensing image registration using multiscale self-similarities. In Proceedings of the IEEE International Conference on Computer Vision in Remote Sensing, Xiamen, China, 16–18 December 2012; pp. 199–202.
30. Sedaghat, A.; Ebadi, H. Distinctive order based self-similarity descriptor for multi-sensor remote sensing image matching. *ISPRS J. Photogramm. Remote Sens.* **2015**, *108*, 62–71. [CrossRef]
31. Xiong, X.; Xu, Q.; Jin, G.; Zhang, H.; Gao, X. Rank-based local self-similarity descriptor for optical-to-sar image matching. *IEEE Geosci. Remote Sens. Lett.* **2019**, *17*, 1742–1746. [CrossRef]
32. Ye, Y.; Shan, J.; Bruzzone, L.; Shen, L. Robust registration of multimodal remote sensing images based on structural similarity. *IEEE Trans. Geosci. Remote Sens.* **2017**, *55*, 2941–2958. [CrossRef]
33. Ye, Y.; Bruzzone, L.; Shan, J.; Bovolo, F.; Zhu, Q. Fast and Robust Matching for Multimodal Remote Sensing Image Registration. *IEEE Trans. Geosci. Remote Sens.* **2019**, *57*, 9059–9070. [CrossRef]
34. Wang, G.; Chen, Y. Robust feature matching using guided local outlier factor. *Pattern Recognit.* **2021**, *117*, 107986. [CrossRef]
35. Liu, Y.; Li, Y.; Dai, L.; Yang, C.; Wei, L.; Lai, T.; Chen, R. Robust feature matching via advanced neighborhood topology consensus. *Neurocomputing* **2021**, *421*, 273–284. [CrossRef]
36. Li, G.; Yu, L.; Fei, S. A deep-learning real-time visual slam system based on multi-task feature extraction network and self-supervised feature points. *Measurement* **2021**, *168*, 108403. [CrossRef]
37. Song, Z.; Zhou, S.; Guan, J. A novel image registration algorithm for remote sensing under affine transformation. *IEEE Trans. Geosci. Remote Sens.* **2014**, *52*, 4895–4912. [CrossRef]
38. Dai, X.; Khorram, S. A feature-based image registration algorithm using improved chain-code representation combined with invariant moments. *IEEE Trans. Geosci. Remote Sens.* **1999**, *37*, 2351–2362.
39. Moigne, J.L. Introduction to remote sensing image registration. In Proceedings of the International Geoscience and Remote Sensing Symposium. Software Engineering Division, NASA Goddard Space Flight Center, Greenbelt, MD, USA, 23–28 July 2017; pp. 2565–2568.
40. Wong, A.; Clausi, D.A. Arrsi: Automatic registration of remote-sensing images. *IEEE Trans. Geosci. Remote Sens.* **2007**, *45*, 1483–1493. [CrossRef]
41. Gong, M.; Zhao, S.; Jiao, L.; Tian, D.; Wang, S. A novel coarse-to-fine scheme for automatic image registration based on SIFT and mutual information. *IEEE Trans. Geosci. Remote Sens.* **2014**, *52*, 4328–4338. [CrossRef]
42. Liang, J.; Liu, X.; Huang, K.; Li, X.; Wang, D.; Wang, X. Automatic registration of multisensor images using an integrated Spatial and Mutual Information (SMI) metric. *IEEE Trans. Geosci. Remote Sens.* **2014**, *52*, 603–615. [CrossRef]
43. Goshtasby, A. Piecewise linear mapping functions for image registration. *Pattern Recognit.* **1986**, *19*, 459–466. [CrossRef]
44. Goshtasby, A. Piecewise cubic mapping functions for image registration. *Pattern Recognit.* **1987**, *20*, 525–533. [CrossRef]
45. Goshtasby, A. Image registration by local approximation methods. *Image Vis. Comput.* **1988**, *6*, 255–261. [CrossRef]
46. Flusser, J. An adaptive method for image registration. *Pattern Recognit.* **1992**, *25*, 45–54. [CrossRef]
47. Goshtasby, A. Registration of images with geometric distortions. *IEEE Trans. Geosci. Remote Sens.* **1988**, *26*, 60–64. [CrossRef]
48. Bookstein, F.L. Principal warps: Thin-plate splines and the decomposition of deformations. *IEEE Trans. Pattern Anal. Mach. Intell.* **1989**, *11*, 567–585. [CrossRef]
49. Bentoutou, Y.; Taleb, N.; Kpalma, K.; Ronsin, J. An automatic image registration for applications in remote sensing. *IEEE Trans. Geosci. Remote Sens.* **2005**, *43*, 2127–2137. [CrossRef]
50. Di, K.; Jia, M.; Xin, X.; Wang, J.; Liu, B.; Li, J.; Xie, J.; Liu, Z.; Peng, M.; Yue, Z.; et al. High-resolution large-area digital orthophoto map generation using lroc nac images. *Photogramm. Eng. Remote Sens.* **2019**, *85*, 481–491. [CrossRef]
51. Ehlers, M.; Fogel, D.N. High-precision geometric correction of airborne remote sensing revisited: The multiquadric interpolation. *Proc. SPIE Int. Soc. Opt. Eng.* **1994**, *2315*, 814–824.
52. Zagorchev, L.; Goshtasby, A. A comparative study of transformation functions for nonrigid image registration. *IEEE Trans. Image Process.* **2006**, *15*, 529–538. [CrossRef]

53. Zhang, Z.; Zhang, J. Facet based differential registration of remote sensing images. In Proceedings of the International Geoscience and Remote Sensing Symposium, Sydney, NSW, Australia, 9–13 July 2001; pp. 712–714.
54. Yu, L.; Zhang, D.; Holden, E. A fast and fully automatic registration approach based on point features for multi-source remote-sensing images. *Comput. Geosci.* **2008**, *34*, 838–848. [CrossRef]
55. Arevalo, V.; Gonzalez, J. Improving piecewise linear registration of high-resolution satellite images through mesh optimization. *IEEE Trans. Geosci. Remote Sens.* **2008**, *46*, 3792–3803. [CrossRef]
56. Lin, X.; Zhang, Y.; Yang, Y. Application of Triangulation-Based Image Registration Method in the Remote Sensing Image Fusion. In Proceedings of the International Conference on Environmental Science and Information Application Technology, Wuhan, China, 4–5 July 2009; pp. 501–504.
57. Xie, J.; Li, B.; Han, W.; Bao, J.; Gu, F.; Guo, F. A tiny facet primitive remote sensing image registration method based on SIFT key points. In Proceedings of the 2012 International Symposium on Instrumentation & Measurement, Sensor Network and Automation (IMSNA), Sanya, China, 25–28 August 2012; pp. 138–141.
58. Han, Y.; Byun, Y.; Choi, J.; Han, D.; Kim, Y. Automatic registration of high-resolution images using local properties of features. *Photogramm. Eng. Remote Sens.* **2012**, *78*, 211–221. [CrossRef]
59. Han, Y.; Choi, J.; Byun, Y.; Kim, Y. Parameter optimization for the extraction of matching points between high-resolution multisensor images in urban areas. *IEEE Trans. Geosci. Remote Sens.* **2014**, *52*, 5612–5621. [CrossRef]
60. Lee, D.T.; Schachter, B.J. Two algorithms for constructing a Delaunay triangulation. *Internat. J. Comput. Inform. Sci.* **1980**, *9*, 219–242. [CrossRef]
61. Wu, J.; Cui, Z.; Sheng, V.S.; Zhao, P.; Su, D.; Gong, S. A comparative study of sift and its variants. *Meas. Sci. Rev.* **2013**, *13*, 123–131. [CrossRef]
62. ASIFT: An Algorithm for Fully Affine Invariant Comparison. Available online: http://demo.ipol.im/demo/my_affine_sift/ (accessed on 13 March 2022).
63. Yu, G.; Morel, J.M. ASIFT: An Algorithm for Fully Affine Invariant Comparison. *Image Process. Line* **2011**, *1*, 11–38. [CrossRef]
64. CFOG. Available online: https://github.com/yeyuanxin110/CFOG (accessed on 13 March 2022).
65. Fischler, M.A.; Bolles, R.C. Random Sample Consensus: A Paradigm for Model Fitting with Applications to Image Analysis and Automated Cartography. *Read. Comput. Vis.* **1987**, *24*, 726–740. [CrossRef]
66. Zhou, L.; Huang, D.; Li, C.; Zhou, D. Algorithm for gps network construction based on spherical delaunay triangulated irregular network. *Xinan Jiaotong Daxue Xuebao J. Southwest Jiaotong Univ.* **2007**, *42*, 380–383.
67. Lawson, C.L. Software for C1 Surface Interpolation. In Proceedings of the Symposium Conducted by the Mathematics Research Center, the University of Wisconsin–Madison, Madison, WI, USA, 28–30 March 1977; pp. 161–194.
68. Bowyer, A. Computing Dirichlet tessellations. *Comput. J.* **1981**, *24*, 162–166. [CrossRef]
69. Robert, J.R. Algorithm 772: STRIPACK: Delaunay triangulation and Voronoi diagram on the surface of a sphere. *ACM Trans. Math. Softw.* **1997**, *23*, 416–434.
70. DelaunayTriangulation. Available online: https://github.com/xialinbo/DelaunayTriangulation (accessed on 13 March 2022).
71. Möbius, A.F. *Der Barycentrische Calcul*; Johann Ambrosius Barth: Leipzig, Germany, 1827.
72. Möbius, A.F. *Über Eine Neue Behandlungsweise der Analytischen Sphärik*; Weidmann: Leipzig, Germany, 1846.
73. Bronw, J.; Worsey, A. Problems with defining barycentric coordinates for the sphere. *RAIRO Modélisation Mathématique Et Anal. Numérique* **1992**, *26*, 37–49.
74. Langer, T.; Belyaev, A.; Seidel, H.P. Spherical barycentric coordinates. In Proceedings of the 4th Eurographics Symposium on Geometry Processing (SGP '06), Cagliari, Italy, 26–28 June 2006; pp. 81–88.
75. Alfeld, P.; Neamtu, M.; Schumaker, L.L. Bernstein-Bézier polynomials on spheres and sphere-like surfaces. *Comput. Aided Geom. Des.* **1996**, *13*, 333–349. [CrossRef]
76. Lei, K.; Qi, D.; Tian, X. A new coordinate system for constructing spherical grid systems. *Appl. Sci.* **2020**, *10*, 655. [CrossRef]
77. Nozette, S.; Rustan, P.; Pleasance, L.P.; Kordas, J.F.; Lewis, I.T.; Park, H.S.; Priest, R.E. The Clementine Mission to the Moon: Scientific Overview. *Science* **1994**, *266*, 1835–1839. [CrossRef] [PubMed]
78. McEwen, A.S.; Robinson, M.S. Mapping of the Moon by Clementine. *Adv. Space Res.* **1997**, *19*, 1523–1533. [CrossRef]
79. Lucey, P.G.; Blewett, D.T.; Taylor, G.J.; Hawke, B.R. Imaging of lunar surface maturity. *J. Geophys. Res.* **2000**, *105*, 20377–20386. [CrossRef]
80. Moon Clementine UVVIS Warped Color Ratio Mosaic 200m v1. Available online: https://astrogeology.usgs.gov/search/map/Moon/Clementine/UVVIS/Lunar_Clementine_UVVIS_Warp_ClrRatio_Global_200m (accessed on 13 March 2022).
81. Zhao, B.; Yang, J.; Wen, D.; Gao, W.; Ruan, P.; He, Y. Design and On-orbit Measurement of Chang' E-1 Satellite CCD Stereo Camera. *Spacecr. Eng.* **2009**, *18*, 30–36.
82. CE1_TMap_GDOM_120m. Available online: https://clpds.bao.ac.cn/ce5web/searchOrder_hyperSearchData.search?pid=CE1/CCD/level/DOM-120m (accessed on 13 March 2022).
83. SLDEM2015. Available online: https://astrogeology.usgs.gov/search/map/Moon/LRO/LOLA/Lunar_LRO_LrocKaguya_DEMmerge_60N60S_512ppd (accessed on 13 March 2022).
84. Ren, X.; Liu, J.; Li, C.; Li, H.; Yan, W.; Wang, F.; Wang, W.; Zhang, X.; Gao, X.; Chen, W. A global adjustment method for photogrammetric processing of Chang'E-2 stereo images. *IEEE Trans. Geosci. Remote Sens.* **2019**, *57*, 6832–6843. [CrossRef]
85. CE2TMap2015_DEM_20m. Available online: https://clpds.bao.ac.cn/ce5web/searchOrder_hyperSearchData.search?pid=CE2/CCD/level/DEM-20m (accessed on 13 March 2022).

remote sensing

MDPI

Article

Remote Sensing Survey of Altiplano-Puna Volcanic Complex Rocks and Minerals for Planetary Analog Use

Gen Ito [1,*], Jessica Flahaut [1], Osvaldo González-Maurel [2], Benigno Godoy [3], Vincent Payet [1,4] and Marie Barthez [1]

1 Centre de Recherche Pétrographiques et Géochimiques (CRPG), Centre National de la Recherche Scientifique (CNRS)/Université de Lorraine, 15 Rue Notre Dame des Pauvres, 54500 Vandoeuvre-lès-Nancy, France; jessica.flahaut@ens-lyon.org (J.F.); vincent.payet@ipsl.fr (V.P.); marie.barthez@univ-lorraine.fr (M.B.)
2 Department of Geological Sciences, University of Cape Town, Rondebosch 7700, South Africa; osvaldo.gonzalezmaurel@uct.ac.za
3 Centro de Excelencia en Geotermia de los Andes and Departamento de Geología, Universidad de Chile, Plaza Ercilla, Santiago 803, Chile; bgodoy@uchile.cl
4 Institut Pierre Simon Laplace, Centre National de la Recherche Scientifique (CNRS), 91120 Palaiseau, France
* Correspondence: gen.ito@univ-lorraine.fr

Abstract: The Altiplano-Puna Volcanic Complex (APVC) of the Central Andes is an arid region with extensive volcanism, possessing various geological features comparable to those of other solar system objects. The unique features of the APVC, e.g., hydrothermal fields and evaporite salars, have been used as planetary analogs before, but the complexity of the APVC presents a wealth of opportunities for more analog studies that have not been exploited previously. Motivated by the potential of using the APVC as an analog of the volcanic terrains of solar system objects, we mapped the mineralogy and silica content of the APVC up to ~100,000 km^2 in northern Chile based on a combination of remote sensing data resembling those of the Moon and Mars. The band ratio indices of Landsat 8 Operational Land Imager multispectral images and mineral classifications based on spectral hourglass approach using Earth Observing-1 Hyperion hyperspectral images (both in the visible to shortwave infrared wavelengths) were used to map iron-bearing and alteration minerals. We also used Hyperion imagery to detect feldspar spectral signatures and demonstrated that feldspar minerals can be detected on non-anorthosites, which may influence interpretations of feldspar spectral signatures on Mars. From the Terra Advanced Spaceborne Thermal Emission and Reflection Radiometer Global Emissivity Dataset, we derived the silica percentage of non-evaporite rocks within errors of approximately 2–3 wt.% SiO_2 for those in the 60–70 wt.% range (about 8 wt.% errors for the 50–60 wt.% range). Based on an integrated assessment of the three datasets, we highlighted three regions of particular interest worthy of further field investigation. We also evaluated the benefits and limitations of all three remote sensing methods for mapping key minerals and capturing rock diversity, based on available samples and existing geological maps.

Keywords: spectroscopy; mineral mapping; Atacama; Altiplano-Puna; planetary analog; Landsat; ASTER; Hyperion; iron; alteration; feldspar; silica; reflectance; emissivity

1. Introduction

Planetary analog studies are comparisons of properties of areas on Earth that have similar characteristics as extraterrestrial bodies in order to gain insights about solar system objects and beyond. Analogs on Earth are often the only opportunity for researchers to conduct rigorous investigations with a suite of advanced instruments, test models, and verify remote sensing observations with ground-truth experiments; thus, planetary analog studies have been one of the crucial methods for understanding extraterrestrial bodies. As new missions to various solar system objects by multiple countries progress this decade [1–4], new questions that cannot be answered by spacecraft alone are expected to arise, and to respond to them, planetary analog studies will continue to play an important role.

Many planetary analog studies with a variety of objectives have been conducted in various parts of the world since the beginning of the modern exploration of extraterrestrial bodies. One of the most often-used terrain types for analog studies are volcanic fields, as large proportions of lunar and Martian surfaces are volcanic in nature [5–7]. In recent years, multi-institutional programs, such as NASA Desert Research and Technology Studies (Desert RATS), Solar System Exploration Research Virtual Institute (SSERVI), and Planetary Science and Technology from Analog Research (PSTAR), have made considerable progress at volcanic analog sites including the Hawaiian Islands and lava fields in the western United States (e.g., Craters of the Moon National Monument, Idaho) [8–12]. These large programs used the analog sites for scientific investigations, technology developments, and operations testing. Other analog studies focused on the geochemical aspects of volcanic fields, mainly alterations in minerals and rocks. For example, at the Atacama Desert of northern Chile, Ruff et al. [13] investigated the silica deposits of the El Tatio hydrothermal field, which closely resembled observations made by the Mars Exploration Rover. Flahaut et al. [14] and Yant et al. [15] compared the Solfatara volcanic crater in central Italy and Kilauea's December 1974 Flow in Hawaii, respectively, to fumarolic alterations on Mars. For astrobiology, alteration and other extreme environments are relevant for life-supporting conditions on extraterrestrial bodies, and altered and evaporite areas such as those in the Central Andes and Australia have been used for astrobiology analog studies [16,17]. The cold environments found in Antarctica are also studied often for gully geomorphology comparisons with their Martian counterparts [18].

The Altiplano-Puna Volcanic Complex (APVC) [19] of the Central Andes is one of the volcanic analog sites that contains a variety of evaporite deposits, alteration minerals, and a diverse range of mafic-to-silicic volcanic features suited for comparisons to the surfaces of the Moon and Mars. Comparisons of the APVC to the Moon and Mars in recent works (e.g., Ruff et al. [13], Aerts et al. [17], Flahaut et al. [20,21]) gave insights into major questions of these bodies (e.g., silica deposits and salars on Mars, dome formation on the Moon), and they have re-emphasized the strong potential of the APVC to better understand the Moon and Mars. There is an interest in the use of the variety of features available at the APVC for further analog studies; however, this region is still not fully characterized and has unexplored areas due to harsh conditions for field expeditions. For this reason, we used a set of orbital multispectral and hyperspectral images to map the minerals of the APVC to survey areas of interest for ongoing and future analog studies and field expeditions. We used Landsat 8 Operational Land Imager (OLI) visible near-infrared (VNIR; 0.5–0.9 μm) and shortwave infrared (SWIR; 1.2–2.5 μm) bands, Terra Advanced Spaceborne Thermal Emission and Reflection Radiometer (ASTER) thermal infrared (TIR; 8–15 μm) bands, and Earth Observing-1 (EO-1) Hyperion VNIR-SWIR bands to simulate a set of instruments available on lunar and Martian orbiters. This set of instruments covers wide ranges of wavelength, spectral resolution, spatial coverage, and spatial resolution, and the instruments complement each other's strengths and weaknesses.

The Altiplano-Puna Volcanic Complex

The Atacama Desert of northern Chile, the driest non-polar desert on Earth, is a natural laboratory to study the diversity of volcanic geoforms. Given the relative scarcity of vegetation and clouds and the low atmospheric water content during the four seasons of the year, it is very plausible to obtain detailed insights from orbit. The Atacama Desert has thus been proposed as an excellent analog environment for some planetary bodies due to its comparable climate, volcanic structures, and lava flow morphologies (e.g., Flahaut et al. [20]). A wide range of volcanic edifices has been found, particularly in the modern volcanic arc of the Central Andes, including stratovolcanoes, calderas, monogenetic cones, and domes, which are frequently very well-preserved due to the hyperarid climate of the region [22,23]. Remarkable compositional heterogeneity (from ca. 48 to 80 wt.% SiO_2) defines Central Andean volcanism, with andesites and dacites being the most common erupted materials at many volcanoes [24]. Basalts and basaltic andesites are rare and concentrated only in a few areas of the main arc and the Altiplano-Puna back-arc [25]. Rhyolite lavas, on the other hand, are unusual in stratovolcanoes, but ignimbrite deposits related to the APVC are commonly dacitic-to-rhyolitic in composition [26]. The APVC is one of the largest ignimbrite provinces in the world, covering a surface area greater than 70,000 km^2 [19]. Our study area covers the western boundary of the APVC between 21°30′S, 67°00′W and 24°30′S, 69°00′W (Figure A1a). Within this area, common volcanic centers include Miocene-to-Quaternary andesitic-to-dacitic stratovolcanoes (e.g., Carcanal, San Pedro, Lascar), low-silica monogenetic edifices (e.g., La Poruña scoria cone; El Negrillar), and small-volume silicic domes (e.g., Chanka) that overlie more voluminous Miocene rhyodacitic-to-rhyolitic ignimbrite sheets. This area also includes closed basins infilled by groundwater (e.g., Lake Tuyajto) and/or covered by non-consolidated-salts-to-hard-saline-crusts that may form extensive evaporite deposits (e.g., Salar de Atacama).

Hubbard and Crowly [27] demonstrated a mineral mapping methodology using a combination of EO-1 Advanced Land Imager (ALI), EO-1 Hyperion, and Terra ASTER imageries in the VNIR and SWIR wavelengths at the APVC. This study succeeded in qualitatively identifying and mapping the distribution of alteration and evaporite minerals at Azufre Volcano (21°47′S, 68°13′W) and Salar de Ascotan (21°34′S, 68°16′W), although it stopped shy of deriving quantitative abundances and analyzing outside of the two example areas.

Motivated by this work, we used similar instruments to map the abundance and distribution of minerals and bulk-rock properties of mainly volcanic surfaces useful for ongoing volcanology field campaigns [28–30], analog studies of the Moon and Mars [21,31,32], and future studies.

2. Materials and Methods

2.1. Landsat 8 Operational Land Imager (OLI)

Landsat 8 began operations in 2013 and is one of the latest satellites in the Landsat series [33]. Its OLI sensor possesses 9 spectral bands in the VNIR-SWIR (0.4–2.5 μm; Figure 1). The OLI operates in the push-broom imaging method and has a spatial resolution of 30 m/pixel. Landsat 8 satellite flies in sequence with EO-1 (prior to decommission) and Terra satellites, known as the morning constellation [34]. Further specifications of the OLI can be found in Irons et al. [33].

We downloaded three Landsat 8 OLI level 2 reflectance data from the U.S. Geological Survey (USGS) EarthExplorer (file names available in Table A1). Bands 1–7 are appropriate bands for mineralogical analysis; Bands 8 and 9 are panchromatic and cirrus (cloud) sensors, respectively, and not used in this study. Landsat 8 also possesses a thermal infrared sensor, but it was not used in this study because it only contains two bands, which were mainly intended to study surface temperatures. For analysis using TIR wavelengths, we use ASTER data.

The three selected images were combined using the mosaic tool of the ENVI 5.6 software from L3Harris Geospatial (https://www.l3harrisgeospatial.com/Software-Technology/ENVI, accessed on 24 January 2022). To remove shadows and other dark areas from analysis, pixels with reflectance less than 0.1 in all bands 1–7 were masked. Band ratio 1/6 was computed to capture the high spectral contrast of snow and water, and pixels with values greater than 1.0 were masked. The selected images were almost free of clouds; a small cloud and its shadow found at 23°67′S, 67°47′W was manually masked. Pixels with Band 2 reflectance greater than 0.19 were also masked as they corresponded to salars, which are not the focus of this study.

First, we applied the decorrelation stretch (DCS [35]) technique to obtain a general idea of the variabilities within the scene (Figure A1c). Next, band ratios based on Rockwell et al. [36] were computed to map iron-bearing and alteration mineral groups (Table 1). Ferric iron 1 index captures the stronger absorption of blue/green light with respect to orange/red light that arises from charge transfer, and it is widely used to map ferric iron minerals [37]. Ferric iron 2 index also captures blue/green absorption but also crystal-field absorptions in the near-infrared. This index is more sensitive to minerals with greater abundance of ferric iron than the ferric iron 1 index [37]. In the ferrous/coarse-ferric iron index, the broad absorption characteristic of ferrous iron is captured. Additionally, oxidizing basalts leaving behind hematite in volcanic flows and cinder cones are captured [37]. The clay-sulfate-mica index captures the diagnostic absorption bands in the SWIR of these minerals. The band ratio 5/4 is an index of green vegetation, and this is subtracted from the 6/7 band ratio in order to reduce the effect of green vegetation (although green vegetation is sparse at the APVC). Landsat 8 OLI has only one band in the 2.0–2.5 μm range; thus, the distinction among clay, sulfate, and mica was not made; this was instead performed with the more appropriate hyperspectral EO-1 Hyperion. The iron-sulfate mineral index captured the strong absorption of light in Band 1 (0.43–0.45 μm) in comparison to Band 2 (0.45–0.51 μm), which is not a characteristic of other iron-bearing minerals (oxide and hydroxide minerals [36]). As with index 4, the green vegetation band ratio was subtracted from the 2/1 band ratio to further mitigate effects of vegetation.

Table 1. Landsat 8 OLI band ratios from Rockwell et al. [36] used to characterize iron-bearing and alteration minerals.

Index Number	Index Name	Formula [1]
1	Ferric iron 1 "redness"	$4/2$
2	Ferric iron 2	$4/2 \times (4+6)/5$
3	Ferrous iron, coarse-grained ferric iron, fire ash [2]	$(3+6)/(4+5)$
4	Clay, sulfate, mica, marble [2]	$6/7 - 5/4$
5	Iron sulfate	$2/1 - 5/4$

[1] Numbers correspond to Landsat 8 OLI band numbers (see Figure 1). [2] Fire ash and marble are rarely found at the field site and, thus, omitted from interpretation.

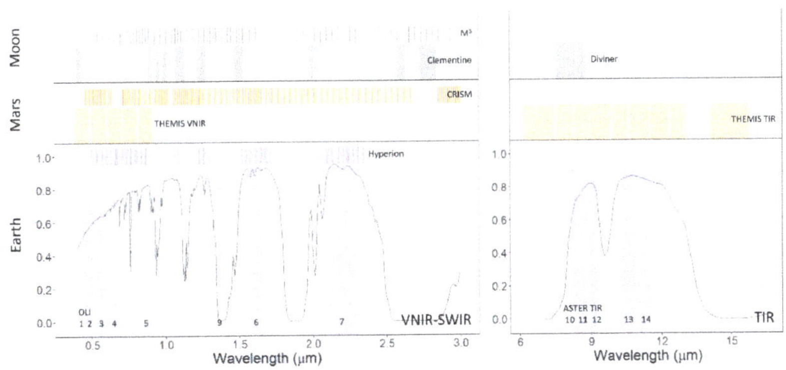

Figure 1. Overview of selected spectral bands of remote sensing instruments for the Earth, Moon, and Mars mentioned in this work in VNIR-SWIR and TIR ranges. Blue, orange, and gray colors correspond to instruments for the Earth, Mars, and the Moon, respectively. For the Earth component, the atmospheric transmittance computed with MODerate resolution atmospheric TRANsmission (MODTRAN) for one-way US standard 1976 model [38] is also shown. Band numbers for Landsat 8 OLI and Terra ASTER TIR are indicated at the bottom.

2.2. Terra Advanced Spaceborne Thermal Emission and Reflection Radiometer (ASTER)

ASTER, one of the five instruments of the satellite Terra, was built by Japan Ministry of Economy, Trade, and Industry in collaboration with NASA and launched in 1999, mainly for studying land surfaces [39]. ASTER has 15 spectral bands that span the 0.5–12 μm wavelength range, covering the VNIR, SWIR, and TIR. Bands 10–14 (8.1–11.7 μm; Figure 1) are TIR and have a spatial resolution of 90 m/pixel [39]. We used the TIR bands to retrieve a bulk-rock property, i.e., silica percentage, as a complement to mineralogy derived from

VNIR and SWIR data of Landsat 8 OLI and EO-1 Hyperion. The VNIR-SWIR bands of ASTER were not used in this work because the mineralogical information that can be derived from them is redundant to or, at least, less informative than the OLI+Hyperion combination in terms of spectral resolution, wavelength coverage, spatial resolution, and spatial coverage (one minor benefit is 15 m/pixel resolution, but only in bands 1–3; other bands have 30 m/pixel resolution).

From USGS EarthExplorer, we downloaded ASTER Global Emissivity Dataset (GED; files names available in Table A1). ASTER GED was released by NASA Jet Propulsion Laboratory and is the most extensive and accurate emissivity product of the Earth's land surface to date [40]. Emissivity is derived from the mean of cloud-free ASTER observations during the 2000–2008 period, and this dataset covers all land surfaces with spatial resolution of 100 m/pixel (coarser resolution data are also available [40]).

First, we applied the DCS technique to obtain a general idea of the variabilities. Surfaces of water and water-bearing minerals (e.g., gypsum) in salars appeared in green color with this DCS band combination (R:14 G:12 B:10; Figure A1d), and they were masked and removed from further analysis. Parts of salars that were not masked with this approach were later captured using the Spectral Angle Mapper (SAM) tool [41] in ENVI, and these pixels were also masked.

In the TIR, emissivity spectra of silicate minerals generally exhibit major absorption features in the 8–13 μm wavelength range because vibrational frequencies of Si–O bond in silicate mineral structures match the frequencies of electromagnetic energy in this wavelength range [42–44]. These absorption bands are known as reststrahlen bands, and the minima of reststrahlen bands occur at progressively longer wavelengths as the SiO_2 content of a rock decreases [42–48]. This relationship allows the derivation of silica content of rocks if the wavelength of the reststrahlen band minima can be retrieved from TIR emissivity spectra.

We used the technique of Hook et al. [45], who determined a quantitative relationship between silica percentage and minimum position of reststrahlen bands from the ASTER spectral library [49,50], to determine the silica percentage. Minima of reststrahlen bands were determined by fitting a four-term Gaussian function

$$f(\lambda) = A_0 \exp(-z^2/2) + A_3 \quad (1)$$

to emissivity spectra, where λ is the wavelength, A_0 is the amplitude, and A_3 is the offset. In the exponent, z is expressed as

$$z = (\lambda - A_1)/A_2 \quad (2)$$

where A_1 and A_2 are the wavelength position of the minimum value and the area enclosed by the function, respectively. For the ASTER dataset without ultramafics (which is appropriate for the APVC) the relationship between weight percent silica (y) and wavelength of minimum of the Gaussian function (A_1) is known as

$$y = -16.203\, A_1 + 213.97 \quad (3)$$

with r^2 of 0.7615 [45].

The computations involved in silica percentage derivation from ASTER GED were accomplished with a Python routine implemented as part of the CRPG spectral processing software, Mineral Recognizer [51]. Spectra with no distinguishable reststrahlen bands were not processed.

2.3. Earth Observing-1 (EO-1) Hyperion

Hyperion, which was decommissioned in 2017, contained 220 spectral bands in the 0.4–2.5 μm wavelength range (Figure 1), and it had a spatial resolution of 30 m/pixel [52]. Hyperion products are long, narrow strips of hyperspectral cubes with standard scene width and length of 7.7 km and 42 km, respectively (the length can extend up to 185 km) [52]. This makes Hyperion complementary to Landsat 8 OLI, which has lower spectral resolution but greater spatial coverage.

Six Hyperion radiance images available on USGS EarthExplorer were downloaded (Table A1) and converted to reflectance using the ENVI FLASSH module [53–55]. The selection of Hyperion images was first guided by areas of mineralogical interest determined from the Landsat and ASTER maps and further constrained by the availability and quality of images. The derived reflectance spectra were noisy with occasional spikes in spectra. This was expected as Hyperion data have comparatively low signal-to-noise ratio (150:1 to 60:1 for Hyperion but above 200:1 for Landsat 8 OLI [33,52,56]). We used the Mineral Recognizer denoising algorithm to eliminate spikes using a combination of percentile and z-score techniques [51].

The selected Hyperion images were processed to retrieve mineral classes using the spectral hourglass approach [57]. Data were first transformed using the Minimum Noise Fraction (MNF) technique [58]. To separate useful signals from noise, the first 18 MNF bands were kept for Inverse MNF and further processing, which is a consistent number of bands with Hubbard and Crowley [27]. Next, Pixel Purity Index [59] was computed, and pixels with potential endmember spectra were identified. Upon manual inspection, obvious spectra of non-minerals (e.g., snow, image edges) were removed. For the remaining endmember spectra, Spectral Analyst Tool based on the USGS Spectral Library version 7 Beckman group [60] was used with SAM and Spectral Feature Fitting (SFF) techniques [41,61] to identify minerals or mineral groups. The Spectral Analyst tool is sensitive to the input wavelength range; thus, we searched for consistent matches after applying it to the 0.4–2.5 μm, 0.4–1.2 μm, and 2.0–2.5 μm ranges. Next, pixels in the scene matching the selected endmember spectra were identified using the SAM technique with the default threshold angle of 0.1. Similar or redundant SAM classes were merged based on geospatial and spectral similarities and irrelevant SAM classes (e.g., shadows) were omitted upon manual inspection.

Additionally, we specifically searched for feldspar minerals, as the geological interpretation of remote sensing detection of feldspar minerals in the VNIR-SWIR on Mars is a currently debated topic [31,32,62–65]; thus, a comparable detection of feldspars at the APVC is eminently beneficial. The VNIR-SWIR reflectance spectra of feldspar minerals are mostly featureless, however, for those containing minor amounts of iron, a broad absorption band center around 1.25 μm can be detected (e.g., Adams and Goulaud [66], Cheek and Pieters [67]). Capturing this broad absorption feature requires Hyperion's spectral coverage and resolution. We computed the spectral index of feldspar minerals in a similar way as for the index BD1300 in Viviano-Beck et al. [68] as

$$B = 1 - R_c/(a \times R_s + b \times R_l) \quad (4)$$

where B is the feldspar band depth index, R_c, R_s, and R_l are the reflectance at the center, the short wavelength edge, and the long wavelength edge of the band, respectively. Wavelength parameters a and b are derived as

$$b = (\lambda_c - \lambda_s)/(\lambda_l - \lambda_s) \quad (5)$$

$$a = 1 - b \quad (6)$$

where λ_c, λ_s, and λ_l are the wavelength of the center, the short edge, and the long edge of the band, respectively. For reflectances R_c, R_s, and R_l, means were taken from reflectances at several wavelengths for each of λ_c, λ_s, and λ_l to reduce noise (exact wavelengths are indicated in Table A2). Useable Hyperion spectral bands were limited around the typical feldspar band center (around 1.25 μm); therefore, we actually computed two values of B for each spectrum using two different λ_c that together captured the feldspar band. We then computed the mean of the two preceding B values to derive the final value for B. In order to reduce false positives, first and second derivatives of the reflectances were computed to eliminate spectra without distinguished concave shapes centered around 1.25 μm. Shadows and other dark materials with low spectral reflectance (<1.5) were also removed from analysis. The computations of spectral indices were performed using Mineral Recognizer.

3. Results

The main objective of this study was to conduct a mineralogical survey of the non-evaporite surfaces of the APVC, to identify potential areas of interest to use as planetary analogs, and to support geological field expeditions in subsequent works. We determined three regions of interest that showed potential and had at least some availability of Hyperion imagery. In this process, mineral and rock characterizations capitalize on the advantage of synthesizing multiple datasets and techniques that are different and complementary to one another (Section 2). The assessment of the utilization of multiple remote sensing datasets is itself noteworthy, and it is discussed in Section 4.2.

3.1. Cerro Carcanal and Vicinity

3.1.1. Alteration Minerals

The first area of interest is Cerro Carcanal and its vicinity (Figure 2; 21°57′36″S, 68°25′48″W to 22°7′48″S, 68°17′24″W), which is a stratovolcano from the Miocene age [69]. Landsat index 4 showed elevated levels of the clay-sulfate-mica mineral group, and especially high index values were found at the summit of Carcanal volcano (Figure 3d). The Hyperion hyperspectral analysis captured the various shapes of the spectral bands in the 2.0–2.5 μm region and indicated that the summit contains argillic alteration minerals, mainly montmorillonite, alunite, and kaolinite (Figure 4). Alunite and kaolinite minerals typically indicate advanced argillic alteration states more than montmorillonite [70]; thus, pixels with spectra closely matching alunite and kaolinite in the 2.0–2.5 μm region were grouped as "High Alteration" to distinguish them from the other pixels matching montmorillonite, which were classified as "Alteration." The clay-sulfate-mica material at the summit seems to have flown down the volcano slope, and the "High Alteration" class became scarce further from the summit (Figure 3f). The flows continued beyond the coverage of the Hyperion imagery and made particularly large deposits to the south of the volcano (e.g., 68°31′12″S, 22°11′60″W; Figure 2d). Landsat index 5 showed that moderate amounts of iron-sulfate minerals were also found at the summit (Figure 3e), but the iron-sulfate index values were comparatively low in relation to those of the nearby volcanoes (Figure 2e), such as Paniri (22°3′34″S, 68°13′48″W) and Cerro del Leon (22°8′0″S, 68°6′36″W).

3.1.2. Iron-Bearing Minerals

Especially high values of the Landsat ferrous/coarse-ferric iron mineral group were found sparingly at the summit, slope, and base of Carcanal volcano (Figure 3c). The Hyperion analysis found that the spectra of these pixels highly matched hematite spectra (Figures 3f and 4), and these pixels were classified as "High-Iron-Bearing" class. One of these features, known as El Rojo III scoria cone (22°0′35″S, 68°18′46″W), has been geochemically analyzed before and was identified as a deeply weathered and oxidized basaltic-andesite deposit containing approximately 8.35 wt.% FeO [71]. This is consistent with Hyperion's spectrum-based identification as hematite and high Landsat ferrous/coarse-ferric iron mineral index. Other pixels high in Landsat ferrous/coarse-ferric iron mineral group and Hyperion "High-Iron-Bearing" class are expected to be of similar materials. In particular, outside of the Hyperion imagery coverage, Landsat index 3 highlighted a large deposit of this material to the south of Carcanal volcano (22°8′60″S, 68°24′36″W; Figure 2c). These features, however, were not highlighted by high values of ferric iron index 1 and 2 (Figure 3a,b).

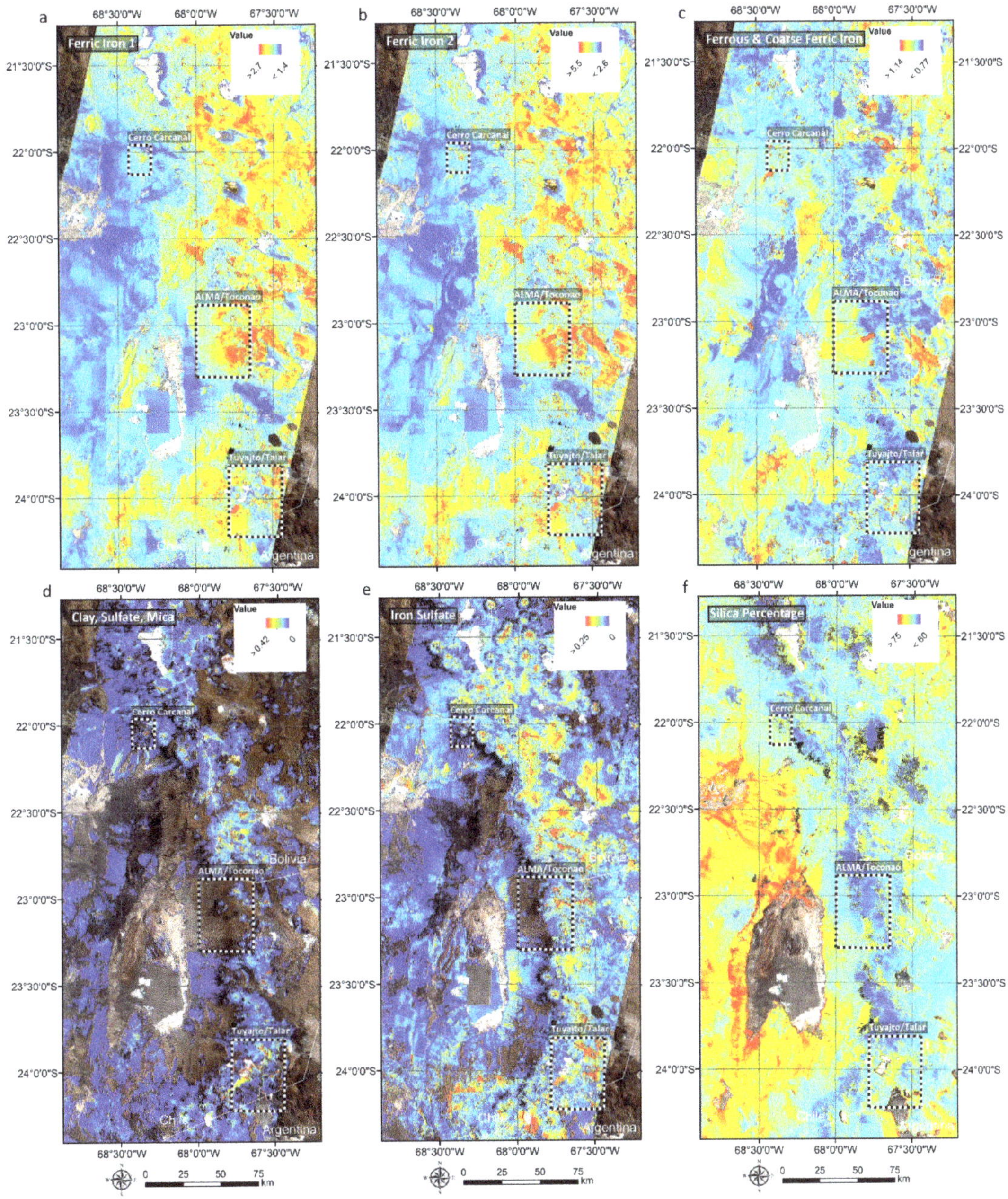

Figure 2. (**a**) Landsat index 1 "ferric iron 1 redness". (**b**) Landsat index 2 "ferric iron 2". (**c**) Landsat index 3 "ferrous and coarse ferric iron". (**d**) Landsat index 4 "clay, sulfate, and mica". (**e**) Landsat index 5 "iron sulfate". (**f**) Estimated silica percentage from ASTER imagery. Three regions selected in this study are indicated with rectangles. For Landsat index maps (**a–e**), pixels with negative values were masked. For ASTER silica percentage map (**f**), pixels corresponding to salars were masked (Section 2.2).

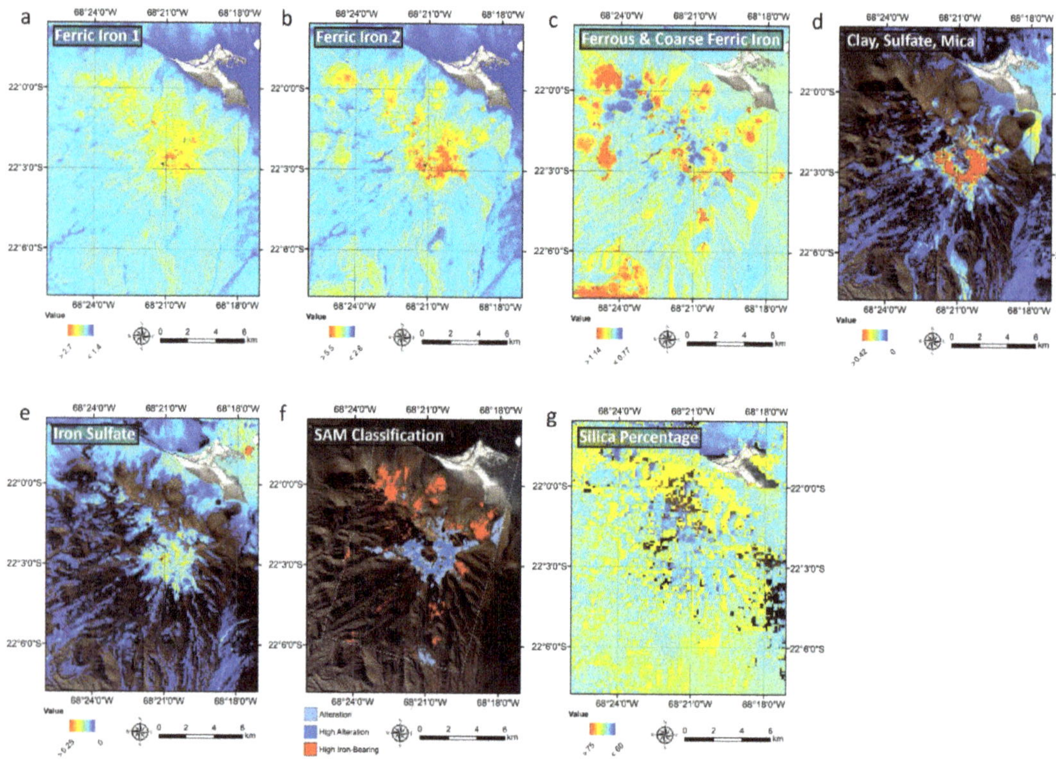

Figure 3. Maps of Cerro Carcanal and vicinity (21°57′36′′S, 68°25′48′′W to 22°7′48′′S, 68°17′24′′W). (**a**) Landsat 8 OLI ferric iron 1 "redness" band ratio index. (**b**) Landsat 8 OLI ferric iron 2 band ratio index. (**c**) Landsat 8 OLI ferrous/coarse-ferric iron band ratio index. (**d**) Landsat 8 OLI clay-sulfate-mica band ratio index. (**e**) Landsat 8 OLI iron-sulfate band ratio index. (**f**) EO-1 Hyperion SAM classification result. "Alteration" and "High Alteration" groups are based on montmorillonite and alunite/kaolinite minerals, respectively. "High-Iron-Bearing" group is based on high hematite spectral signature. (**g**) Estimated silica percentage from ASTER imagery.

3.1.3. Silica Percentage

Overall, the ASTER silica percentage map (Figures 2f and 3g) indicated that Carcanal volcano is enriched in silica compared to the volcanoes in its immediate surroundings, San Pedro (21°53′17′′S, 68°23′32′′W), San Pablo (21°52′41′′S, 68°21′49′′W), and Paniri, which suggests that Carcanal volcano experienced greater alteration (silicic and argillic) than its neighbors. This is consistent with the fact that Carcanal is older (Miocene) than others (Quaternary) [69] and thus has had more time for alteration to advance.

3.2. ALMA/Toconao Village Region

3.2.1. Alteration Minerals

The second region of interest lies in the area near Atacama Large Millimeter Array (ALMA) and Toconao village (Figure 2; 22°53′0′′S, 68°0′0′′W to 23°18′0′′S, 67°39′0′′W). Cerro Toco (22°57′36′′S, 67°46′12′′W) showed high Landsat clay-sulfate-mica and iron-sulfate mineral index values (Figure 5d,e). Hyperion analysis also indicated the presence of the argillic alteration minerals, montmorillonite, kaolinite, and alunite (Figure 5f).

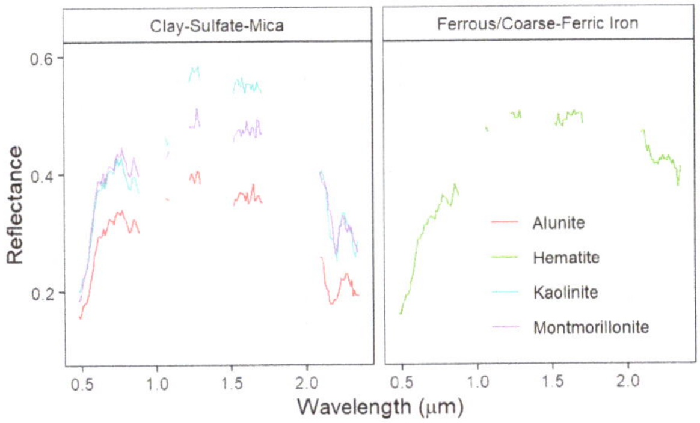

Figure 4. Hyperion spectra of alunite, kaolinite, montmorillonite, and hematite found at Cerro Carcanal and vicinity. The identification of minerals is the overall result of SAM and SFF methods in ENVI Spectral Analyst tool. The entire 0.48–2.5 μm wavelength range was used for hematite spectrum identification, and both 0.48–2.5 and 2.0–2.5 μm wavelength ranges were used for alunite/kaolinite/montmorillonite spectrum identification.

3.2.2. Iron-Bearing Minerals

This region was characterized by large extents of moderately-high-to-very-high iron-bearing minerals. The Landsat ferrous/coarse-ferric iron mineral index highlighted a few volcanic centers (23°5′24″S, 67°46′12″W; 23°10′12″S, 67°40′48″W; 23°14′24″S, 67°45′0″W), which were further found to possess hematite and hornblende with Hyperion hyperspectral analysis and classified as "High-Iron-Bearing." Similar to the iron-rich basaltic-andesite features at the Cerro Carcanal area, these iron-bearing features are likely oxidized basaltic-andesite volcanic deposits as well. The Landsat ferric iron 1 "redness" index (Figure 5a) was not high. The geological map (Figure A1b) indicates that these features are Pliocene volcanic centers with andesite, dacite, basalt, and basaltic-andesite composition, indicating younger age compared to the similar features at Cerro Carcanal.

The Hyperion hyperspectral analysis also identified pixels with hematite spectra but to a lower degree of agreement than the "High-Iron-Bearing" class, which were labeled as "Iron-Bearing" (Figure 5f). The Iron-Bearing pixels corresponded to moderately high Landsat ferrous/coarse-ferric iron index values (Figure 5c), and on the geological map, these regions were indicated as ignimbrites and pyroclastic flow deposits (Figure A1b). Furthermore, Landsat ferric iron index 1 and 2 also showed moderate-to-very-high values. Not all parts of the ignimbrites and pyroclastic flow deposits exhibited overall high iron-bearing mineral spectral signatures, however. The ignimbrites in the northern part of this location (23°0′36″S,″67°48′36″W) showed low values of Landsat ferric iron 1 and 2 index and very low values of ferrous/coarse-ferric iron index, and hematite (and other iron-bearing mineral) spectra were not identified with the Hyperion analysis.

3.2.3. Silica Percentage and Feldspar Minerals

The parts of ignimbrites and pyroclastic flow deposits that have high iron-bearing mineral spectral signatures were mostly more consistent with higher silica percentages than the parts with low iron-bearing mineral spectral signatures (Figure 5h). However, "High-Iron-Bearing" basaltic-andesite volcanic deposits exhibited low silica percentages (Figure 5h).

High values of Hyperion feldspar mineral index were found at Dacite Dome D of the Purico Complex (22°56′60″S,″67°43′12″W; Figure 5g [71,72]). For this dome, the ASTER silica percentage map indicated mostly average values (~62 wt.% SiO_2) but sometimes ranged up to nearly 70 wt.% SiO_2. Low feldspar index values were also found in the Purico ignimbrite located to the south of El Cerrillo dome.

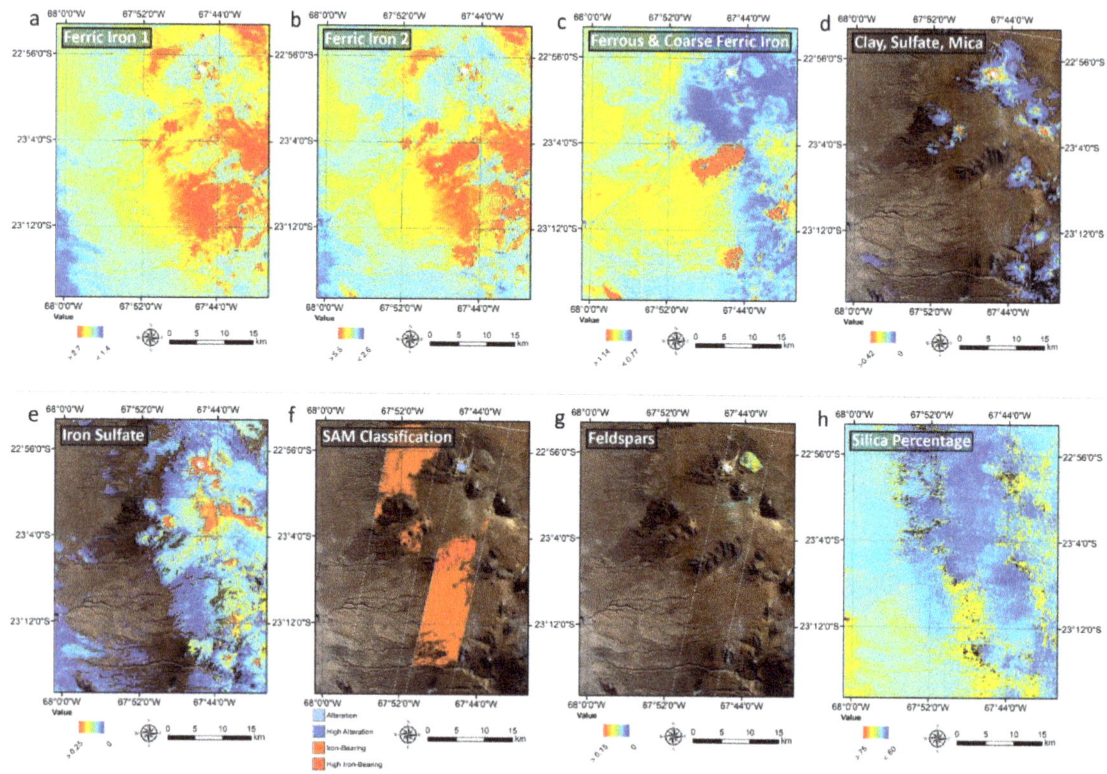

Figure 5. Maps of the region surrounding ALMA and Toconao Village (22°53′0″S, 28°0′0″W to 23°18′0″S, 67°39′0″W). (**a**) Landsat 8 OLI ferric iron 1 "redness" band ratio index. (**b**) Landsat 8 OLI ferric iron 2 band ratio index. (**c**) Landsat 8 OLI ferrous/coarse-ferric iron band ratio index. (**d**) Landsat 8 OLI clay-sulfate-mica band ratio index. (**e**) Landsat 8 OLI iron-sulfate band ratio index. (**f**) EO-1 Hyperion SAM classification result. "Alteration" and "High Alteration" groups are based on montmorillonite and alunite/kaolinite minerals, respectively. "Iron-Bearing" and "High-Iron-Bearing" groups are based on moderate hematite and high hematite/hornblende spectral signatures, respectively. (**g**) EO-1 Hyperion feldspar band depth index. (**h**) Estimated silica percentage from ASTER imagery.

3.3. *Lake Tuyajto/Salar de Talar Region*

3.3.1. Alteration Minerals

The third area of interest covers the vicinity of Lake Tuyajto and Salar de Talar (Figure 2; 23°48′36″S, 67°47′60″W to 24°13′12″S, 67°26′24″W). A similar assemblage of mineral groups to that seen at the previous two sites was found here. The volcanoes Incahuasi Sur (24°1′12″S, 67°31′12″W), Tuyajto (23°51′0″S, 67°35′60″W), and Cordón Purichare (24°8′60″S, 67°49′48″W) were highlighted by very high Landsat clay-sulfate-mica mineral index values (Figure 6d). They also showed high iron-sulfate mineral index values (Figure 6e), which is similar to the ALMA/Toconao village area but different from the Cerro Carcanal area. The flow deposits on the slopes and base of the volcanoes also showed high clay-sulfate-mica and iron-sulfate index values. The Hyperion analysis indicated that these areas contain montmorillonite, kaolinite, and alunite minerals. The flow materials of the Tuyajto volcano showed strong spectral signatures of alunite and kaolinite, suggesting advanced argillic alteration to greater extents (Figure 6d,f).

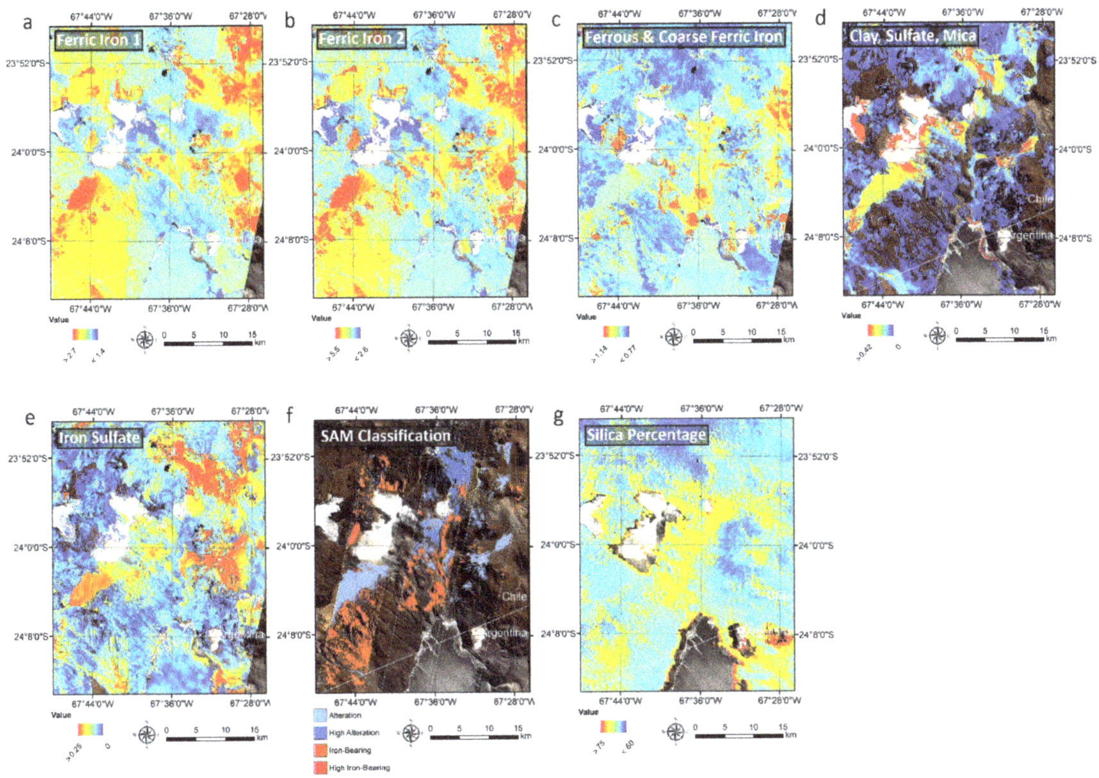

Figure 6. Maps of the region near Lake Tuyajto and Salar de Talar (67°47′60″S, 23°48′36″W to 67°26′24″S, 24°13′12″W). (**a**) Landsat 8 OLI ferric iron 1 "redness" band ratio index. (**b**) Landsat 8 OLI ferric iron 2 band ratio index. (**c**) Landsat 8 OLI ferrous/coarse-ferric iron band ratio index. (**d**) Landsat 8 OLI clay-sulfate-mica band ratio index. (**e**) Landsat 8 OLI iron-sulfate band ratio index. (**f**) EO-1 Hyperion SAM classification result. "Alteration" and "High Alteration" groups are based on montmorillonite and alunite/kaolinite minerals, respectively. "Iron-Bearing" and "High-Iron-Bearing" groups are based on moderate hematite and high hematite/hornblende spectral signatures, respectively. (**g**) Estimated silica percentage from ASTER imagery.

3.3.2. Iron-Bearing Minerals

Volcanic deposits with very high Landsat ferrous/coarse-ferric iron indices and strong Hyperion hematite spectral signatures interpreted as basaltic-andesite deposits at the previous two locations were also present ("High-Iron-Bearing" class, e.g., 23°58′48″S, 67°43′48″W). The Hyperion analysis also determined a class with a moderate hematite spectral signature (labeled "Iron-Bearing"). As observed in the ignimbrites and pyroclastic flow deposits of the ALMA/Toconao village region, the "Iron-Bearing" class showed moderate-to-very-high values of Landsat ferric iron 1 and 2 index values. On the geological map, "High-Iron-Bearing" deposits corresponded to ignimbrites, volcanic centers, and partially eroded volcanic centers Miocene to Pliocene in age (Figure A1b). The "Iron-Bearing" class corresponded largely to ignimbrites and volcanic centers but also to partially eroded volcanic centers and secondary sedimentary deposits (e.g., alluvial). As seen previously, the Landsat ferrous/coarse-ferric iron index did not highlight the "Iron-Bearing" mineral class, but it was highlighted in the ferric iron 1 and 2 indices (Figure 6a–c).

4. Discussion

4.1. Evaluation of Mapping Results

4.1.1. Iron-Bearing and Alteration Minerals

The distribution of iron-bearing and alteration mineral groups highlighted by the Landsat band ratio indices corresponded to the mineral maps made from the ALI-ASTER fused images by Hubbard and Crowley [27] at the area around Azufre Volcano (21°47′0″S, 68°13′0″W). The argillic alteration minerals represented by Landsat index 4 (clay-sulfate-mica) showed high abundance values at Azufre Volcano and the flow deposits surrounding the volcano (Figure 2d). These areas also showed high abundance values of index 5, iron sulfates. The ferric and ferrous iron mineral indices (index 1, 2, and 3; Figure 2a–c) also highlighted areas indicated as hematite in Hubbard and Crowley [27]. A spectrum of a sample of red scoria at El Rojo III (22°0′5″S, 68°18′46″W) showed consistency in general shape and ferrous/coarse-ferric iron index with the Landsat spectrum from a pixel containing these coordinates (Figure 7a), and Spectral Analyst confirmed the presence of hematite. The geochemical analysis at CRPG of El Rojo III confirmed nearly 10 wt.% Fe_2O_3, which was the highest iron oxide content among 155 APVC silicate rocks compiled from published data and CRPG measurements (Supplementary Material).

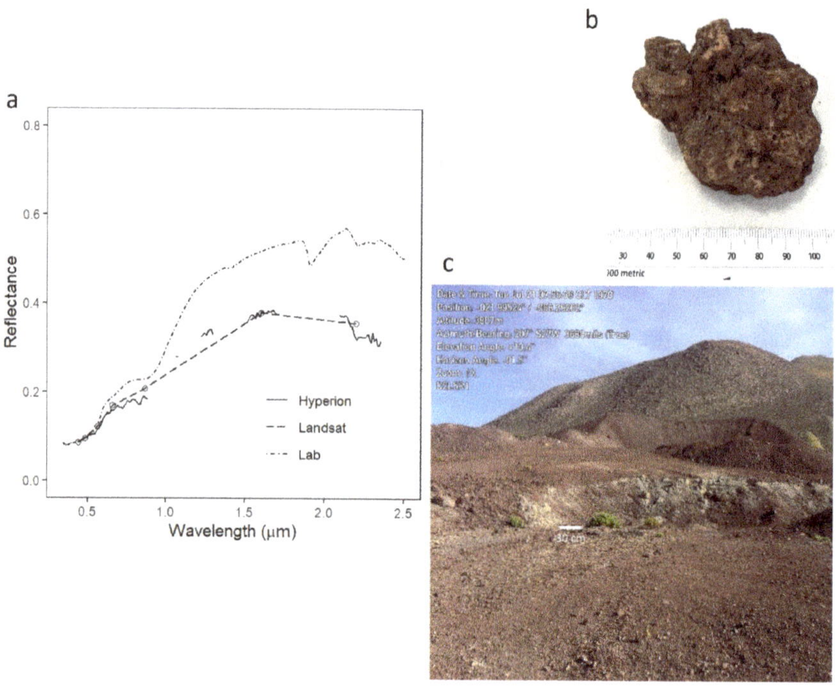

Figure 7. (**a**) Comparison of remote sensing spectra of El Rojo III location to a spectrum acquired in the laboratory of El Rojo III red scoria sample. (**b**) Image of the sample used for spectral measurement. (**c**) The view of El Rojo III scoria cone in the field (22°0′5″S, 68°18′46″W). Minor amounts of white minerals in (**c**) are alteration minerals, which contributed to the discrepancy of spectra in (**a**), in which the sample lacks this white mineral.

The Hyperion SAM classification map at Cerro Carcanal for iron-bearing and alteration minerals also indicated consistency with the results of Hubbard and Crowley [27], who identified hematite, aluminous clays, and alunite (mixed). Despite the low signal-to-noise ratio of Hyperion, the distinct absorption features of alunite, kaolinite, and montmorillonite in the 2.0–2.5 μm wavelength region were captured (Figure 4) and allowed further constraining of the mineral groups within the alteration mineral group, which was not possible with the Landsat clay-sulfate-mica index. Within the alteration mineral group, we distinguished between pixels dominated by montmorillonite and by alunite/kaolinite, indicating argillic alteration or advanced argillic alteration, respectively. The spatial correlation of the alteration minerals identified in the Hyperion analysis was in visually good agreement with the Landsat clay-sulfate-mica index (Figures 3d,f, 5d,f and 6d,f). Similarly, for iron-bearing minerals, the distinction between the strong and weak presence of iron was possible with Hyperion; pixels with a strong hematite spectral signature were classified as "High-Iron-Bearing" class (Figures 3f, 5f and 6f), and pixels with a weaker hematite spectral signature were classified as "Iron-Bearing" class (Figures 5f and 6f). Both of these iron-bearing mineral classes showed high spatial correlation with the Landsat ferrous/coarse-ferric iron index (Figures 3c,f, 5c,f and 6c,f).

4.1.2. Silica Percentage

Overall, the silica percentages derived from the ASTER GED highlighted roughly two main groups: secondary sedimentary deposits and primary volcanic deposits/centers high and low in silica, respectively. The absolute error in ASTER GED is reported as 1.24% [40,73], and using this, we estimated that uncertainties in the weight percentage of silica, calculated as in Section 2.3, were generally 0.5 wt.% SiO_2 or lower for over 90% of pixels in both categories (high and low silica content).

We gathered the published silica contents of samples of the APVC and also measured the silica content of the available APVC rock samples at the CRPG geochemistry laboratory for ground-truth comparison to the silica content derived from the ASTER GED (Appendix D and the Supplementary Material). In total, the locations of 251 samples corresponded to the ASTER silica content pixels, but four mafic samples with silica content of 50% SiO_2 or below were removed from further consideration (247 total). Assuming that the samples are representative of the surface seen from orbit, the ASTER GED silica content estimations were within a 6 wt.% SiO_2 difference for about two-thirds of the samples and within a 3 wt.% SiO_2 difference for 40% of the samples (Table 2). We found that there was a trend in the absolute differences as a function of silica content, according to which samples above approximately 60 wt.% SiO_2 had lower differences than those below 60 wt.% SiO_2 (Figure 8). Two groups, with high (group 1) and low (group 2) silica content, were distinguished using the k-means clustering method. Group 1 had a center point at 63 wt.% SiO_2 with a 2.5 wt.% SiO_2 difference and group 2 had a center point at 57 wt.% SiO_4 with an 8.1 wt.% SiO_2 difference (Figure 8). Approximately 90% of the measurements in group 1 and group 2 were below 5 and 12 wt.% SiO_2 difference, respectively. The silica content derived from ASTER GED was lower than that of the samples (i.e., underestimation) for 15% of the samples, and furthermore, these underestimations only occurred in group 1, which accounts for one-quarter of the group 1 samples.

Table 2. Summary of the differences in silica content derived from ASTER GED using Equation (3) and measured in geochemical laboratories. Bold numbers highlight the percentage of best estimations.

SiO_2 Abs. Diff.	0–3	3–6	6–9	9–12	+12
% of samples	**40**	**28**	18	10	3
no. samples at CRPG	5	3	2	1	2

Based on the ground-truth comparison, we believe that silica content estimation with ASTER GED is reasonable for deriving the silica content of APVC rocks within roughly 2–3 wt.% SiO_2 of the true value if they are above 60 wt.% SiO_2 (group 1). For a study dealing with rocks below 60 wt.% SiO_2 (group 2) where an average difference of 8.1 wt.% SiO_2 is not sufficient, we suggest a simple qualitative approach, e.g., high vs. low silica content comparison, which can still be valuable for remote locations.

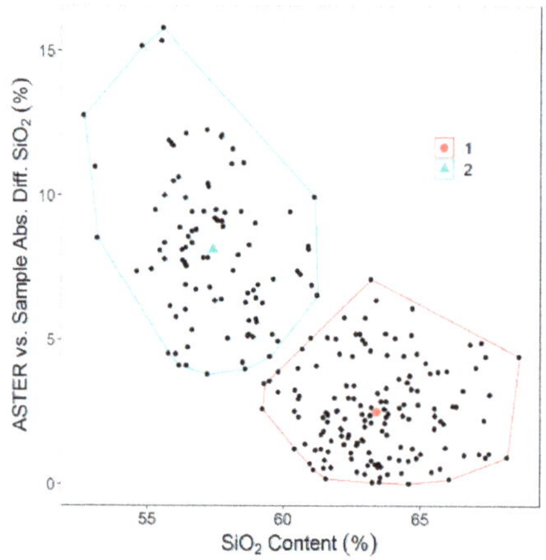

Figure 8. Absolute differences in silica content between ASTER and sample measurements as a function of silica content of the samples. Group 1 and group 2 from k-means clustering are indicated in red and blue boundaries, respectively. Center points of each group are shown in red circle and blue triangle for group 1 and group 2, respectively.

We suspected that the larger deviations of the ASTER-derived silica content from those of the samples could be attributed to Equation (3), where we assumed the exclusion of ultramafic rocks. Thus, we derived the silica content from the ASTER GED using the four-term Gaussian parameters of Hook et al. [45] that includes ultramafic rocks as

$$y = -10.548\ A_1 + 160.25 \tag{7}$$

where A_1 is from Equation (2). The result did not yield a significant trend in ASTER vs. sample differences as a function of silica content, and the overall agreement was lower than Equation (3), which excluded ultramafics (Table A3). For this reason, estimations based on Equation (3) are preferred. Although beyond the scope of this study, measuring the thermal emissivity spectra of the APVC samples and deriving the APVC's unique relationship between the wavelength of the minimum of the reststrahlen bands and the silica content of the rocks instead of using Equation (3) could lead to improvements. Additionally, the samples should ideally come from spatially representative locations rather than spatially biased locations (Figure A1a); however, the conditions of the APVC did not easily permit this (i.e., the purpose of using remote sensing techniques).

4.1.3. Feldspar Minerals

The feldspar spectra detected with Hyperion near the Chanka dome (21°45′0″S, 68°18′58″W) matched the spectra of a sample of feldspar-bearing rock collected at this location (Figure 9a). A visual inspection of the sample and optical microscope observations of a thin section indicated the presence of optically visible feldspar grains with sizes approximately 3 cm and smaller (Figure 9b–d). Additionally, the space between distinguishably large grains in the microscope image appeared to be composed of very small (roughly < 1 μm) feldspar and other mineral grains (Figure 9c,d). These observations give confidence to feldspar detections based on the 1.3 μm spectral feature.

For Dacite Dome D, which showed strong feldspar spectral signatures in the ALMA/ Toconao village region (22°56′0″S, 67°43′12″W; Figure 5g), the computed feldspar index values (>0.4) substantially exceeded the typical values for the Chanka dome (0.03). Samples of Dacite Dome D analyzed by Schmitt et al. [72] indicated that plagioclase feldspars accounted for 66–80% of the minerals (33–40% of the entire sample total when accounting for the presence of glass). According to Schmitt et al. [72], dacite pumice from the Purico ignimbrite surrounding Dacite Dome D closely resembles that of Dacite Dome D, with similar abundances of plagioclase feldspars, but curiously, we observed that only a small portion of the Purico ignimbrite to the South of El Cerrillo dome showed low feldspar index values.

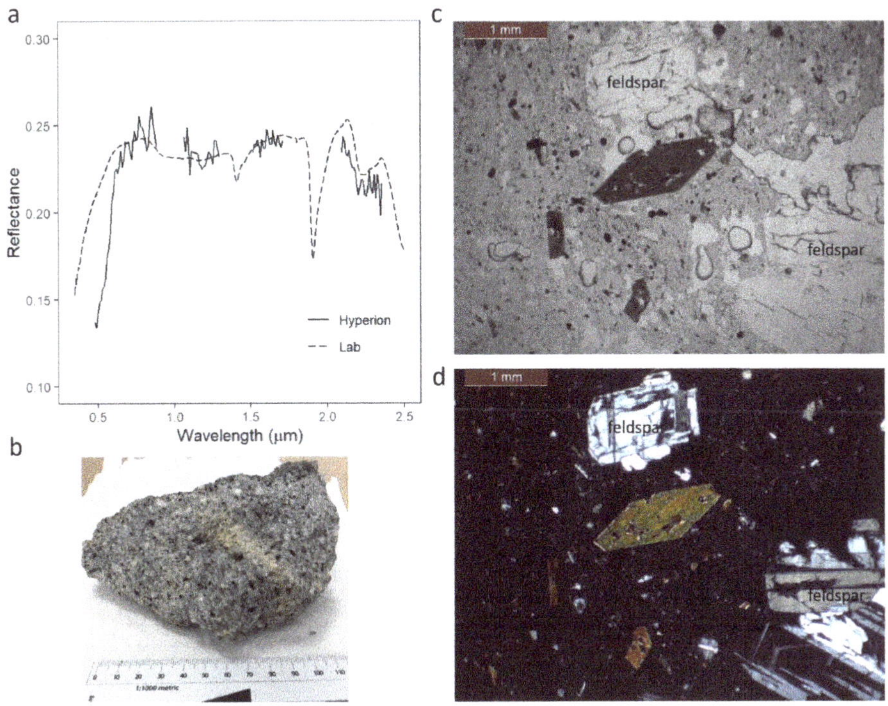

Figure 9. (**a**) Comparison of an EO-1 Hyperion spectrum to a spectrum acquired in the laboratory of feldspar-bearing rock from the Chanka dome (21°45′0′′S, 68°18′58′′W). (**b**) Image of the sample used for the measurement. White grains are feldspar. (**c**) Microscope image of a thin section of the sample. Feldspar grains are labeled. (**d**) Microscope image of a thin section of the sample in cross-polarized light. Twinning typical of plagioclase feldspars can be observed on feldspar grains.

4.2. Assessment of Different Datasets and Mapping Techniques

First, the Landsat 8 OLI DCS map, which is not quantitative but is computationally fast and suited for the initial evaluation of a large area, captured the diversity of the volcanic features, namely, the relative differences between the ignimbrites, low-silica volcanic centers, silicic domes, alteration areas, and sedimentary deposits (Figure A1c). Compositional variations were also observed within volcanic centers (e.g., Carcanal volcano versus the neighboring San Pedro and San Pablo volcanoes) or within ignimbrite bodies, which may appear in tones of yellow, orange, green, or purple in the Landsat DCS image (Figure A1c). Although the geological map did not distinguish them, possible differences may have arisen from the iron content, iron oxidation level, weathering, and compactness of the sediments.

After grasping the general pattern of the study site with the DCS technique, we applied semi-quantitative techniques to measure the presence of selected minerals. The band ratio indices for six mineral groups (Table 1) were computed with Landsat 8 OLI, which was fast and particularly effective at mapping a semi-quantitative distribution of key mineral groups, thereby capturing evidence of volcanic activities, such as iron-bearing and alteration minerals. A caveat of the band depth index approach, however, is that it targets a known mineral with a known absorption feature, which may in fact vary depending on its exact composition, structure, or environmental parameters. Therefore, the band depth index approach is not suitable for finding unexpected/unknown minerals. However, this limitation can be remedied to some extent by being aware of the variabilities shown with the DCS map and other sources of information, such as the geologic map. Another limitation is that the mineral groups were not precise enough, particularly for the clay-sulfate-mica group. The low spectral resolution of the Landsat 8 OLI does not allow the distinction among different minerals within each group. On the other hand, hyperspectral images are able to capture narrow spectral features and are better suited for more precise mineral characterizations; thus, we utilized EO-1 Hyperion hyperspectral imagery to obtain more definitive mineral characterizations. The spectral hourglass technique was used to derive mineral identification and distribution from the scene, and it distinguished alunite, kaolinite, and montmorillonite spectra within the Landsat clay-sulfate-mica group. This is a complementary approach to the spectral indices used with the Landsat dataset, and it is a means to achieve more precise mineral characterizations and cross-check the results from the Landsat analysis. In spite of that, precise distinctions were still not easy, since the minerals had overlapping or close spectral features, and spectral features were masked by other minerals. For this reason, we did not attempt to map alunite, kaolinite, and montmorillonite separately but rather distinguished only between alunite and a group of kaolinite and montmorillonite.

Band-depth indices can also be computed from Hyperion. This was necessary for the feldspar identification, where the diagnostic spectral feature is weak, and the spectral feature range is not adequately captured by Landsat 8 OLI (or other multispectral datasets). For weak spectral features such as those of feldspars, masking and inadequate absorption depth are especially problematic, even for hyperspectral datasets. We detected high feldspar index values from Dacite Dome D, but the Chanka dome had much lower index values and the felsic ignimbrites did not have significant values at all, even though they were expected to have feldspars. The spectral signature of feldspars is influenced by multiple parameters, including the mineral assemblage, the mineral's composition, its grain size, and the rock texture [64,66,67]. Spectral unmixing methods could resolve at least some of these problems.

Hyperion and other hyperspectral imaging techniques are usually spatially limited, only covering a fraction of Landsat 8 OLI imagery, with considerable spatial gaps among available images. For this reason, Hyperion is better suited to the study of small regions of interest already identified with Landsat or other datasets.

The ASTER GED in the TIR was highly complementary to the mineral analysis performed with the Landsat and Hyperion dataset in the VNIR-SWIR. Initially, the ASTER DCS map (Figure A1d) qualitatively revealed the relative distribution of high silica content (red), low silica content (blue), and non-silicates (green). Next, the quantitative silica percentages derived from the wavelengths of the minimum of reststrahlen bands followed the general spatial pattern seen in the ASTER DCS map and showed reasonable, albeit variable, errors when compared to the ground-truth samples. The mapping of the silica percentage of the silicate rocks at the APVC in the order of 10,000 s to 100,000 km^2 spatial extents is beneficial for volcanological studies of the APVC, and for our future work, it could define the lithological contexts of minerals and mineral groups identified with VNIR-SWIR spectra, especially feldspars.

4.3. Re-Emphasis of the APVC for Planetary Analog Use

Previously, water-related features, such as salars and hydrothermal fields, have been popular targets for planetary analog studies at the APVC. However, the remote sensing survey conducted in this study revealed other geological features, to large spatial extents, relevant for comparisons to those on other solar system bodies. Notably, the detection of the feldspar spectral signatures of non-anorthosite rocks with Hyperion could impact on the direction of research in Mars volcanism, in which recent works (e.g., Barthez et al. [31], Flahaut et al. [32]) have demonstrated alternative views to the hypothesis that feldspar spectral signature detections are likely anorthosites (which is the hypothesis of Carter and Poulet [62] and others). The detection of strong feldspar spectral signatures at Dacite Dome D in the ALMA/Toconao village region offers the potential for deeper insights into this problem, and it also is relevant for comparisons to silicic domes on the Moon (e.g., Flahaut et al. [21], Glotch et al. [74]). Questions remain over this dome as the ASTER silica percentage estimation yielded a range of 62–69 wt.% SiO_2 and the majority of nearby Purico ignimbrite did not show high feldspar spectral index values, despite it being mineralogically and chemically similar to Dacite Dome D. Often, the preferred approach for identifying feldspars or other silicate minerals is to detect the diagnostic reststrahlen bands in high-spectral-resolution TIR spectra (e.g., Lyon [42]); however, for both the Earth and Mars, orbital TIR hyperspectral imagery with sufficiently high spatial resolution (10′s of m/pixel) does not currently exist. The combination of feldspar detections with the 1.3 μm spectral feature in hyperspectral VNIR-SWIR imagery and the silica percentages derived from multispectral TIR imagery is currently viewed as a robust approach (e.g., Rogers and Nekvasil [64]). The outcome of the preferred interpretations of feldspar detections on Mars has a major impact on our understanding of Martian geological history; thus, follow-up investigations of the feldspar detections demonstrated in this work are crucial. We plan to send a geological field expedition to the Dacite Dome D area to gather further ground-truth data.

Ignimbrites in the ALMA/Toconao village region are another feature worthy of more detailed investigations. Parts of ignimbrites were highlighted by high silica percentages (~70%) according to ASTER, and the ignimbrite samples contained feldspar grains; however, although we did not find feldspar spectral signatures with Hyperion, we found moderate iron spectral signatures. In addition, the detection of high-iron-bearing mineral deposits at all three presented regions that were interpreted as basaltic-andesites is remarkable for research on mafic volcanism in the APVC [30,71,75].

Abundant alteration minerals were identified at the APVC, particularly in the Cerro Carcanal and Lake Tuyajto/Salar de Talar regions. Alteration analog studies have been important for understanding alterations on other planetary bodies (e.g., Mars; Carter et al. [76], Murchie et al. [77]), and they have also been conducted at other volcanic regions on Earth (e.g., Hawaii, Italy). The Landsat clay-sulfate-mica index and Hyperion alteration mineral maps contribute to this large-scale effort and will be useful for continued analog studies of alterations; moreover, they could play a major role in mineral resource assessments at the APVC, as argillic alterations are often associated with economically valuable minerals.

The methodology adopted in this work is commonly used in Earth and planetary remote sensing studies. We first used Landsat and ASTER, which are multispectral instruments and have large spatial coverage but low spectral resolution, to identify smaller regions of interest. Next, the minerals within those smaller regions were further analyzed with Hyperion hyperspectral imagery. Although it is not new, we emphasize that the dataset and mapping approach adopted in this work resemble those available for the Moon and Mars, and thus, our work is transferable to the analysis of the analogous terrains of these planetary bodies. The three satellite imagery datasets of the Earth, Landsat 8 OLI, Terra ASTER, and EO-1 Hyperion, are comparable to the Mars Odyssey (MO) Thermal Emission Imaging System (THEMIS) VNIR bands [78], the Mars Reconnaissance Orbiter Compact Reconnaissance Imaging Spectrometer for Mars (CRISM) [79], and MO THEMIS TIR bands [78]. For the Moon, they are comparable to Clementine multispectral imaging cameras (ultraviolet/visible camera, near-infrared camera) [80], the SELENE/Kaguya Multiband Imager [81], the Chandrayaan-1 Moon Mineralogy Mapper (M^3) [82], and the Lunar Reconnaissance Orbiter (LRO) Diviner [83]. Figure 1 shows the comparison of the instruments. Future orbiter missions, such as the Lunar Trailblazer [84], will carry similar but enhanced spectral instruments operating in both the VNIR-SWIR and TIR ranges.

Some limitations to this methodology, which is mainly based on the Earth's differences to other planetary bodies, must be understood. The atmosphere is a major factor in orbital remote sensing, and the Earth's atmosphere has strong absorptions arising from water, which is a dominant molecule on Earth, but not in other solar system bodies. This substantially impacts the usable wavelengths on Earth in both VNIR-SWIR and TIR and limits the constraining of some minerals, such as the clay-sulfate-mica group. Although Hyperion captured important spectral features in the 2.0–2.5 μm region, other major bands in the 1.0–2.0 μm would have helped with further constraining. On the other hand, the lack of atmosphere on some planetary bodies leads to spectral effects from space weathering in VNIR-SWIR [85] and thermal gradient in the TIR [86,87], which lead to difficulties in the interpretation of spectra. Additionally, the designs of existing instruments were intended for accomplishing the specific scientific objectives of a mission; thus, different instruments have different caveats, even if equivalent geological information can be retrieved. For instance, the LRO Diviner can retrieve silica content, but this is achieved by using the Christiansen feature of TIR spectra rather than reststrahlen bands, which was the case in this work [74,88,89].

5. Conclusions

The remote sensing data from various instruments (ASTER, Hyperion, Landsat) were used to map the mineral groups of mainly volcanic and secondary volcanic surfaces of the Altiplano-Puna Volcanic Complex, which are sound analogs of these types of surface on the Moon and Mars. Although mineral maps covering 10,000 s to 100,000 km^2 scales of the APVC are some of the fundamental resources used to conduct geological field work and planetary analog studies, they have not been created before. Using a combination of Earth observation instruments with comparable characteristics to lunar and Martian orbital instruments, we characterized, with different techniques, the alteration minerals, iron-bearing minerals, feldspar minerals, and silica percentages of bulk rocks. Three areas of particular interest for planetary analog studies were highlighted in this work, although other areas not mentioned in detail were also noteworthy for further investigations. The results of our maps were consistent with existing geological maps and a previous study performed locally at the Azufre volcano, and they also showed consistency, to varying degrees, with the geochemical analysis of samples compiled from past studies and measurements performed at CRPG. The benefits and limitations of each dataset and method were evaluated, and we observed that synthetically analyzing remote sensing datasets with different characteristics (e.g., wavelength range, spatial extent, spectral resolution) leads to more confident qualitative and quantitative interpretations of minerals and rocks over large areas.

One immediate impact of this work on the current state of planetary analog studies is the detection of feldspar spectral signatures on non-anorthositic rocks. Our feldspar detections at Chanka dome and Dacite Dome D influenced on-going efforts to refine the spectral detection of feldspar minerals on rocks other than anorthosites through laboratory measurements [31] and Mars CRISM hyperspectral imagery [32]. A laboratory analysis, by Barthez et al. [31], of entire feldspathic rocks of varied nature indicated that volcanic, plutonic, and metamorphic feldspathic rocks all show the spectral signature of plagioclase feldspars. Our observation of feldspar spectral signatures at Chanka dome (dacitic volcanic rock) and its sample analysis is consistent with their study. This supports the detection of the feldspar spectral signatures of non-anorthositic rocks on Mars by Flahaut et al. [32].

Supplementary Materials: The following supporting information can be downloaded at: https://www.mdpi.com/article/10.3390/rs14092081/s1.

Author Contributions: Conceptualization, J.F. and G.I.; methodology, J.F. and G.I.; software, V.P. and G.I.; validation, J.F., O.G.-M., B.G., M.B. and G.I.; formal analysis, J.F., O.G.-M., B.G., M.B. and G.I.; investigation, J.F., O.G.-M., B.G., M.B. and G.I.; resources, J.F., O.G.-M., B.G., M.B., V.P. and G.I.; data curation, J.F. and G.I.; writing—original draft preparation, J.F., O.G.-M., B.G., M.B. and G.I.; writing—review and editing, J.F., O.G.-M., B.G., M.B. and G.I.; visualization, J.F., M.B. and

G.I.; supervision, J.F.; project administration, J.F.; funding acquisition, J.F. All authors have read and agreed to the published version of the manuscript.

Funding: This research was funded by CNRS Momentum grant, the LUE future leader 'Viscous', and the ANR 'Mars-Spec' projects attributed to J. Flahaut. Financial support by FONDAP Folio 15090013 and 15200001 (Centro de Excelencia and Geotermia de los Andes) was attributed to B. Godoy.

Data Availability Statement: The remote sensing data listed in Table A1 are available on USGS EarthExplorer, https://earthexplorer.usgs.gov/, (accessed on 24 January 2022). The geochemistry analysis results are available in the Supplementary Material. The spectra of the samples are available on SSHADE, https://www.sshade.eu/db/mirabelle, (accessed on 24 January 2022).

Acknowledgments: The authors thank the NASA Hyperion team for acquiring dedicated targets prior to the instrument decommissioning. We are also thankful to Mélissa Martinot and Pascal Allemand for fruitful discussions. The authors are also grateful to the SARM Nancy, and in particular Delphine Yeghicheyan and Christophe Cloquet, for performing sample analyses. This is CRPG contribution N°2816.

Conflicts of Interest: The authors declare no conflict of interest.

Appendix A

Table A1. Summary of remote sensing-data used in this work.

Instrument	Data Type	ID [1]	Map
Landsat 8 OLI	Level 1 Radiance	LC08_L1TP_001075_20200727_20200807_01_T1 LC08_L1TP_001076_20200727_20200807_01_T1 LC08_L1TP_001077_20200727_20200807_01_T1 LC08_L1TP_233075_20200618_20200625_01_T1 LC08_L1TP_232076_20200729_20200807_01_T1 LC08_L1TP_233076_20200805_20200821_01_T1 LC08_L1TP_232077_20200729_20200807_01_T1 LC08_L1TP_233077_20200805_20200821_01_T1	Background imagery
	Level 2 Reflectance	LC08_L2SP_233075_20141125_20200910_02_T1 LC08_L2SP_233076_20141109_20200910_02_T1 LC08_L2SP_233077_20141109_20200910_02_T1	Iron-bearing minerals, alteration minerals
EO-1 Hyperion	Level 1 Radiance	EO1H2320772002361111PP EO1H2330752011245110KF EO1H2330762008075110P2 EO1H2330762012034110KF EO1H2330762016250110PF EO1H2330762016269110PF	Iron-bearing minerals, alteration minerals, feldspar minerals
Terra ASTER	GED AG 100	AG100.v003.-21.-068.0001.h5 AG100.v003.-21.-069.0001.h5 AG100.v003.-22.-068.0001.h5 AG100.v003.-22.-069.0001.h5 AG100.v003.-23.-068.0001.h5 AG100.v003.-23.-069.0001.h5 AG100.v003.-24.-068.0001.h5 AG100.v003.-24.-069.0001.h5	Silica percentage

[1] "Landsat Product Identifier", "Entity ID", and "Local Granule ID" for Landsat 8 OLI, EO-1 Hyperion, and Terra ASTER data, respectively.

Appendix B

Table A2. Wavelengths used for computing the feldspar band-depth index with EO-1 Hyperion. The last row shows the mean.

λ_s	λ_c		λ_l
0.783	1.003	1.225	1.548
0.793	1.013	1.235	1.558
0.803	1.023	1.245	1.568
0.813	1.033	1.255	1.578
0.824	1.044	1.266	1.588
0.844	1.064	1.286	1.609
0.854	1.074	1.296	1.619
0.864	1.084		1.629
0.875			1.639
0.884			1.649
			1.659
			1.669
			1.679
			1.689
			1.699
			1.709
0.834	**1.043**	**1.261**	**1.629**

Appendix C

Table A3. Summary of the differences in silica content derived from ASTER GED using Equation (7) and measured in geochemical laboratories.

SiO_2 **Abs. Diff.**	**0–5**	**5–10**	**10–15**	**15–20**	**+20**
% of samples	38	25	22	5	10
no. samples at CRPG	0	4	1	0	2

Appendix D

Rock samples from previous field campaigns were ground in an agate mortar. Major element analyses were performed by ICP-OES at the Service d'Analyse des Roches et des Minéraux (SARM) facility in Nancy, France (https://sarm.cnrs.fr/index.html, (accessed on 24 January 2022)). Data can be found in the Supplementary Material.

Appendix E

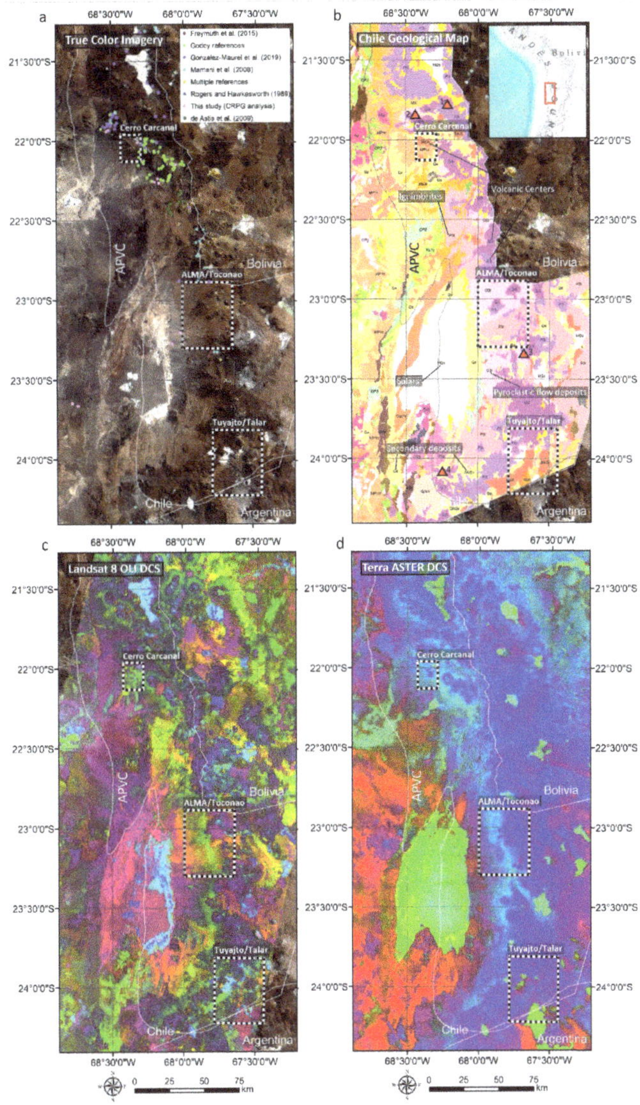

Figure A1. (**a**) Landsat 8 OLI radiance true color imagery of the APVC. Sampling locations used for comparison to ASTER silica content map are shown and grouped by source reference. The group "Godoy references" includes Godoy [90] and Godoy et al. [28,71,91]. The group "Multiple references" includes Francis et al. [92], Freymuth et al. [93], Mamani et al. [94], and Rogers and Hawkesworth [95]. (**b**) Geological map of Chile from Servicio Nacional de Geología y Minería [69]. Some geological units and features mentioned in the main text are indicated. Red triangles indicate volcanoes: 1 Azufre and Chanka, 2 San Pedro-San Pablo and La Poruña, 3 Lascar, 4 El Negrillar. Full descriptions of the geological units are extensive, and the reader is referred to [69]. (**c**) Landsat 8 OLI reflectance DCS using bands 7, 6, and 4 as red, green, and blue, respectively. (**d**) ASTER DCS using bands 14, 12, and 10 as red, green, and blue, respectively. On all four maps, three regions selected in this study are indicated by rectangles, and the approximate APVC boundary is shown.

References

1. Farley, K.A.; Williford, K.H.; Stack, K.M.; Bhartia, R.; Chen, A.; de la Torre, M.; Hand, K.; Goreva, Y.; Herd, C.D.; Hueso, R.; et al. Mars 2020 Mission Overview. *Space Sci. Rev.* **2020**, *216*, 142. [CrossRef]
2. Grasset, O.; Dougherty, M.K.; Coustenis, A.; Bunce, E.J.; Erd, C.; Titov, D.; Blanc, M.; Coates, A.; Drossart, P.; Fletcher, L.N.; et al. JUpiter ICy moons Explorer (JUICE): An ESA mission to orbit Ganymede characterise the Jupiter system. *Planet. Space Sci.* **2013**, *78*, 1–21. [CrossRef]
3. Howell, S.M.; Pappalardo, R.T. NASA's Europa Clipper—A mission to a potentially habitable ocean world. *Nat. Commun.* **2020**, *11*, 1311. [CrossRef] [PubMed]
4. International Space Exploration Coordination Group. The Global Exploration Roadmap. Washington, DC, USA. 2018. Available online: http://www.globalspaceexploration.org (accessed on 1 September 2021).
5. Carr, M.H. *The Surface of Mars*; Cambridge University Press: New York, NY, USA, 2006; pp. 43–76.
6. Chapman, M. *The Geology of Mars Evidence of Earth-Based Analogs*; Cambridge University Press: London, UK, 2009; p. 484.
7. Hörz, F.; Grieve, R.; Heiken, G.; Spudis, P.; Binder, A. Lunar Surface Processes. In *Lunar Sourcebook*; Heiken, G.H., Vaniman, D.T., French, B.M., Eds.; Cambridge University Press: New York, NY, USA, 1991; pp. 61–120.
8. Abercromby, A.F.J.; Gernhardt, M.L.; Jadwick, J. Evaluation of dual multi-mission space exploration vehicle operations during simulated planetary surface exploration. *Acta Astronaut.* **2013**, *90*, 203–214. [CrossRef]
9. Eppler, D.; Adams, B.; Archer, D.; Baiden, G.; Brown, A.; Carey, W.; Cohen, B.; Condit, C.; Evans, C.; Fortezzo, C.; et al. Desert Research and Technology Studies (DRATS) 2010 science operations: Operational approaches and lessons learned for managing science during human planetary surface missions. *Acta Astronaut.* **2013**, *90*, 224–241. [CrossRef]
10. Lim, D.S.S.; Abercromby, A.F.J.; Nawotniak, S.E.K.; Lees, D.S.; Miller, M.J. The BASALT research program: Designing and developing mission elements in support of human scientific exploration of Mars. *Astrobiology* **2019**, *19*, 245–261. [CrossRef]
11. Lim, D.S.S.; Heldmann, J.L. Exploring our Solar System through inter-disciplinary, analog research: An overview of the FINESSE, BASALT and SUBSEA programs. *Planet. Space Sci.* **2021**, *199*, 105187. [CrossRef]
12. Young, K.E.; Bleacher, J.E.; Rogers, A.D.; Schmitt, H.H.; McAdam, A.C.; Gary, W.B.; Whelley, P.L.; Scheidt, S.P.; Ito, G.; Knudson, C.A.; et al. The incorporation of field portable instrumentation into human planetary surface exploration. *Earth Space Sci.* **2018**, *5*, 697–720. [CrossRef]
13. Ruff, S.W.; Farmer, J.D. Silica deposits on Mars with features resembling hot spring biosignatures at El Tatio in Chile. *Nat. Commun.* **2016**, *7*, 13554. [CrossRef]
14. Flahaut, J.; Bishop, J.L.; Silvestro, S.; Tedesco, D.; Daniel, I.; Loizeau, D. The Italian Solfatara as an analog for Mars fumarolic alteration. *Am. Mineral.* **2019**, *104*, 1565–1577. [CrossRef]
15. Yant, M.; Young, K.E.; Rogers, A.D.; McAdam, A.C.; Bleacher, J.E.; Bishop, J.L.; Mertzman, S.A. Visible, near-infrared, and mid-infrared spectral characterization of Hawaiian fumarolic alteration near Kilauea's December 1974 flow: Implications for spectral discrimination of alteration environments on Mars. *Am. Mineral.* **2018**, *103*, 11–25.
16. Aerts, J.W.; van Spanning, R.J.M.; Flahaut, J.; Molenaar, D.; Bland, P.H.; Genge, M.J.; Ehrenfreund, P.; Martins, Z. Microbial communities in sediments from four mildly acidic ephemeral salt lakes in the Yilgarn Craton (Australia)—Terrestrial analogs to ancient Mars. *Front. Microbiol.* **2019**, *10*, 779. [CrossRef] [PubMed]
17. Aerts, J.W.; Riedo, A.; Melton, D.J.; Martini, S.; Flahaut, J.; Meierhenrich, U.J. Biosignature analysis of Mars soil analogs from the Atacama Desert: Challenges and implications for future missions to Mars. *Astrobiology* **2020**, *20*, 766–784. [CrossRef] [PubMed]
18. Dickson, J.L.; Head, J.W.; Levy, J.S.; Morgan, G.A.; Marchant, D.R. Gully formation in the McMurdo Dry Valleys, Antarctica: Multiple sources of water, temporal sequence and relative importance in gully erosion and deposition processes. In *Martian Gullies and Their Earth Analogues*; Conway, S.J., Carrivick, J.L., Carling, P.A., De Haas, T., Harrison, T.N., Eds.; The Geological Society of London: London, UK, 2018.
19. De Silva, S.L. Altiplano-Puna volcanic complex of the central Andes. *Geology* **1989**, *17*, 1102–1106. [CrossRef]
20. Flahaut, J.; Martinot, M.; Bishop, J.L.; Davies, G.R.; Potts, N. Remote sensing and in situ mineralogic survey of the Chilean salars: An analog to Mars evaporate deposits? *Icarus* **2017**, *282*, 152–173. [CrossRef]
21. Flahaut, J.; González-Maurel, O.; Godoy, B.; Martinot, M.; Guitreau, M. Rheology of the Andean domes as an analog for lunar silicic constructs. In Proceedings of the European Geoscience Union General Assembly, Online, 19–30 April 2021.
22. Hartley, A.J.; Chong, G. Late Pliocene age for the Atacama Desert: Implications for the desertification of western South America. *Geology* **2002**, *30*, 43–46. [CrossRef]
23. Placzek, C.; Quade, J.; Betancourt, J.L.; Patchett, P.J.; Rech, J.A.; Latorre, C.; Matmon, A.; Holmgren, C.; English, N.B. Climate in the dry central Andes over geologic, millennial, and interannual timescales. *Ann. Mo. Bot. Gard.* **2009**, *96*, 386–397. [CrossRef]
24. Wörner, G.; Schildgen, T.F.; Reich, M. The Central Andes: Elements of an extreme land. *Elements* **2018**, *14*, 225–230. [CrossRef]
25. González-Maurel, O.; Godoy, B.; Le Roux, P.; Rodríguez, I.; Marín, C.; Menzies, A.; Bertin, D.; Morata, D.; Vargas, M. Magmatic differentiation at La Poruña scoria cone, Central Andes, northern Chile: Evidence for assimilation during turbulent ascent processes, and genetic links with mafic eruptions at adjacent San Pedro volcano. *Lithos* **2019**, *338*, 28–140. [CrossRef]
26. De Silva, S.L.; Self, S.; Francis, P.W.; Drake, R.E.; Ramirez, C.R. Effusive silicic volcanism in the Central Andes: The Chao dacite and other young lavas of the Altiplano-Puna Volcanic Complex. *J. Geophys. Res.* **1994**, *99*, 17805–17825. [CrossRef]
27. Hubbard, B.E.; Crowley, J.K. Mineral mapping on the Chilean–Bolivian Altiplano using co-orbital ALI, ASTER and Hyperion imagery: Data dimensionality issues and solutions. *Remote Sens. Environ.* **2005**, *99*, 173–186. [CrossRef]

28. Godoy, B.; Lazcano, J.; Rodríguez, I.; Martínez, P.; Parada, M.A.; Le Roux, P.; Wilke, H.-G.; Polanco, E. Geological evolution of Paniri volcano, Central Andes, northern Chile. *J. S. Am. Earth Sci.* **2018**, *84*, 184–200. [CrossRef]
29. Godoy, B.; Taussi, M.; González-Maurel, O.; Renzulli, A.; Hernández-Prat, L.; Le Roux, P.; Morata, D.; Menzies, A. Linking the mafic volcanism with the magmatic stages during the last 1 Ma in the main volcanic arc of the Altiplano-Puna Volcanic Complex (Central Andes). *J. S. Am. Earth Sci.* **2019**, *95*, 102295. [CrossRef]
30. González-Maurel, O.; Le Roux, P.; Godoy, B.; Troll, V.R.; Deegan, F.; Menzies, A. The great escape: Petrogenesis of low-silica volcanism of Pliocene to Quaternary age associated with the Altiplano-Puna Volcanic Complex of northern Chile (21°10′–22°50′S). *Lithos* **2019**, *105162*, 346–347. [CrossRef]
31. Barthez, M.; Flahaut, J.; Ito, G.; Guitreau, M.; Pik, R. Near-infrared laboratory measurements of feldspathic rocks as a reference for hyperspectral Martian remote sensing data interpretation. In Proceedings of the Europlanet Science Congress, Online, 21 September–9 October 2020.
32. Flahaut, J.; Barthez, M.; Payet, V.; Fueten, F.; Guitreau, M.; Ito, G.; Allemand, P.; Quantin-Nataf, C. Identification and characterization of new feldspar-bearing rocks in the walls of Valles Marineris, Mars. In Proceedings of the Eleventh Moscow Solar System Symposium, Online, 5–9 October 2020.
33. Irons, J.R.; Dwyer, J.L.; Barsi, J.A. The next Landsat satellite: The Landsat Data Continuity Mission. *Remote Sens. Environ.* **2012**, *122*, 11–21. [CrossRef]
34. Hilsenrath, E.; Ward, A. Earth science mission operations, Part I: Flight operations—Orchestrating NASA's fleet of Earth observing satellites. *Earth Obs.* **2016**, *28*, 4–13.
35. Gillespie, A.R.; Kahle, A.B.; Walker, R.E. Color enhancement of highly correlated images. I. Decorrelation and HSI contrast stretches. *Remote Sens. Environ.* **1986**, *20*, 209–235. [CrossRef]
36. Rockwell, B.W.; Gnesda, W.R.; Hofstra, A.H. *Improved Automated Identification and Mapping of Iron Sulfate Minerals, Other Mineral Groups, and Vegetation Using Landsat 8 Operational Land Imager Data, San Juan Mountains, Colorado, and Four Corners Region*; U.S. Geological Survey Scientific Investigations Map 3466, 1 Sheet, Scale 1:325,000, 37-p. Pamphlet; US Geological Survey: Reston, VA, USA, 2021. [CrossRef]
37. Rockwell, B.W. *Automated Mapping of Mineral Groups and Green Vegetation from Landsat Thematic Mapper Imagery with an Example from the San Juan Mountains*; Colorado. U.S. Geological Survey Scientific Investigations Map 3252, 25-p. Pamphlet, 1 Map Sheet, Scale 1:325,000; US Geological Survey: Reston, VA, USA, 2012. [CrossRef]
38. Berk, A.; Conforti, P.; Kennett, R.; Perkins, T.; Hawes, F.; van den Bosch, J. MODTRAN6: A major upgrade of the MODTRAN radiative transfer code. In Proceedings of the SPIE 9088, Algorithms and Technologies for Multispectral, Hyperspectral, and Ultraspectral Imagery XX, 90880H, Baltimore, MA, USA, 13 June 2014.
39. Yamaguchi, Y.; Kahle, A.B.; Tsu, H.; Kawakami, T.; Pniel, M. Overview of Advanced Spaceborne Thermal Emission and Reflection Radiometer (ASTER). *IEEE Trans. Geosci. Remote Sens.* **1998**, *36*, 1062–1071. [CrossRef]
40. Hulley, G.C.; Hook, S.J.; Abbot, E.; Malakar, N.; Islam, T.; Abrams, M. The ASTER Global Emissivity Dataset (ASTER GED): Mapping Earth's emissivity at 100 meter spatial scale. *Geophys. Res. Lett.* **2015**, *42*, 7966–7976. [CrossRef]
41. Kruse, F.A.; Lefkoff, A.B. Knowledge-based geologic mapping with imaging spectrometers. *Remote Sens. Rev.* **1993**, *8*, 3–28. [CrossRef]
42. Lyon, R.J.P. *Evaluation of Infrared Spectrophotometry for Compositional Analysis of Lunar and Planetary Soils: Rough and Powdered Surfaces, Final Report, Part 2*; NASA contract NASr-49(04); Stanford Research Institute: Menlo Park, CA, USA, 1964.
43. Lyon, R.J.P. Analysis of rocks by spectral infrared emission (8–25 μm). *Econ. Geol.* **1965**, *60*, 715–736. [CrossRef]
44. Vincent, R.K.; Thomson, F. Spectral compositional imaging of silicate rocks. *J. Geophys. Res.* **1972**, *17*, 2465–2473. [CrossRef]
45. Hook, S.J.; Dmochowski, J.E.; Howard, K.A.; Rowan, L.C.; Karlstrom, K.E.; Stock, J.M. Mapping variations in weight percent silica measured from multispectral thermal infrared imagery—Examples from the Hiller Mountains, Nevada, USA and Tres Virgenes-La Reforma, Baja California Sur, Mexico. *Remote Sens. Environ.* **2005**, *95*, 273–289. [CrossRef]
46. Pan, C.; Rogers, A.D.; Michalski, J.R. Thermal and near-infrared analyses of central peaks of Martian impact craters: Evidence for a heterogeneous Martian crust. *J. Geophys. Res. Planets* **2015**, *120*, 662–688. [CrossRef]
47. Sabine, C.; Realmuto, V.J.; Taranik, J.V. Quantitative estimation of granitoid composition from thermal infrared multispectral scanner (TIMS) data, Desolation Wilderness, northern Sierra Nevada, California. *J. Geophys. Res.* **1994**, *99*, 4261–4271. [CrossRef]
48. Smith, M.R.; Bandfield, J.L.; Cloutis, E.A.; Rice, M.S. Hydrated silica on Mars: Combined analysis with near-infrared and thermal-infrared spectroscopy. *Icarus* **2013**, *223*, 633–648. [CrossRef]
49. Baldridge, A.M.; Hook, S.J.; Grove, C.I.; Rivera, G. The ASTER Spectral Library Version 2.0. *Remote Sens. Environ.* **2009**, *113*, 711–715. [CrossRef]
50. Meerdink, S.K.; Hook, S.J.; Roberts, D.A.; Abbott, E.A. The ECOSTRESS spectral library version 1.0. *Remote Sens. Environ.* **2019**, *230*, 111196. [CrossRef]
51. Payet, V.; Flahaut, J.; Ito, G.; Barthez, M.; Breton, S. Automated denoising for mineral identification on hyperspectral data. In Proceedings of the 11th Moscow International Symposium, Moscow, Russia, 5–9 October 2020. Available online: https://ms2020.cosmos.ru/docs/11M-S3-Abstract_book_full_version_04.pdf (accessed on 1 April 2021).
52. Middleton, E.M.; Ungar, S.G.; Mandl, D.J.; Ong, L.; Frye, S.W.; Campbell, P.E.; Landis, D.R.; Young, J.P.; Pollack, N.H. The Earth Observing One (EO-1) satellite mission: Over a decade in space. *IEEE J. Sel. Top. Appl. Earth Obs. Remote Sens.* **2013**, *6*, 243–256. [CrossRef]

53. Berk, A.; Bernstein, L.S.; Robertson, D.C. *MODTRAN®: A Moderate Resolution Model for LOWTRAN 7*; GL-TR-89-0122; Geophysics Laboratory: Hanscom Air Force Base, MA, USA, 1989.
54. Berk, A.; Bernstein, L.S.; Anderson, G.P.; Acharya, P.K.; Robertson, J.H.; Chetwynd, J.H., Jr.; Adler-Golden, S.M. MODTRAN® cloud and multiple scattering upgrades with application to AVIRIS. *Remote Sens. Environ.* **1998**, *65*, 367–375. [CrossRef]
55. Berk, A.; Acharya, P.K.; Bernstein, L.S.; Anderson, G.P.; Chetwynd, J.H., Jr.; Hoke, M.L. Reformulation of the MODTRAN® band model for finer spectral resolution. In Proceedings of the SPIE, Orlando, FL, USA, 27 March 2000; Volume 4049.
56. Morfitt, R.; Barsi, J.; Levy, R.; Markham, B.; Micijevic, E.; Ong, L.; Scaramuzza, P.; Vanderwerff, K. Landsat-8 Operational Land Imager (OLI) radiometric performance on-orbit. *Remote Sens.* **2015**, *7*, 2208–2237. [CrossRef]
57. Wolfe, J.D.; Black, S.R. Hyperspectral Analytics in ENVI Target Detection and Spectral Mapping Methods. 2018. Available online: http://www.spectroexpo.com/wp-content/uploads/2021/03/Hyperspectral_Whitepaper.pdf (accessed on 1 March 2021).
58. Green, A.A.; Berman, M.; Switzer, R.; Craig, M.D. A transformation for ordering multispectral data in terms of image quality with implications for noise removal. *IEEE Trans. Geosci. Remote Sens.* **1988**, *56*, 65–74. [CrossRef]
59. Boardman, J.W.; Kruse, F.A.; Green, R.O. Mapping target signatures via partial unmixing of AVIRIS data. In Summaries of the Fifth Annual JPL Airborne Geosciences Workshop. *Jet Propuls. Lab. Spec. Publ.* **1995**, *95*, 23–26.
60. Kokaly, R.F.; Clark, R.N.; Swayze, G.A.; Livo, K.E.; Hoefen, T.M.; Pearson, N.C.; Wise, R.A.; Benzel, W.M.; Lowers, H.A.; Driscoll, R.L.; et al. *USGS Spectral Library Version 7*; U.S. Geological Survey Data Series 1035; US Geological Survey: Reston, VA, USA, 2017; p. 61.
61. Kruse, F.A.; Lefkoff, A.B.; Boardman, J.W.; Heidebrecht, K.B.; Shapiro, A.T.; Barloon, P.J.; Goetz, A.F.H. The spectral image processing system (SIPS)—Interactive visualization and analysis of imaging spectrometer data. *Remote Sens. Environ.* **1993**, *44*, 145–163. [CrossRef]
62. Carter, J.; Poulet, F. Ancient plutonic processes on Mars inferred from the detection of possible anorthositic terrains. *Nat. Geosci.* **2013**, *6*, 1008–1012. [CrossRef]
63. Wray, J.J.; Hansen, S.T.; Dufek, J.; Swayze, G.A.; Murchie, S.L.; Seelos, F.P.; Skok, J.R.; Irwin, R.P., III; Ghiorso, M.S. Prolonged magmatic activity on Mars inferred from the detection of felsic rocks. *Nat. Geosci.* **2013**, *6*, 1013–1017. [CrossRef]
64. Rogers, A.D.; Nekvasil, H. Feldspathic rocks on Mars: Compositional constraints from infrared spectroscopy and possible formation mechanisms. *Geophys. Res. Lett.* **2015**, *42*, 2619–2626. [CrossRef]
65. Rogers, A.D.; Farrand, W.H. Spectral evidence for alkaline rocks and compositional diversity among feldspathic light-toned terrains on Mars. *Icarus* **2022**, *376*, 114883. [CrossRef]
66. Adams, J.B.; Goullaud, L.H. Plagioclase feldspars: Visible and near infrared diffuse reflectance as applied to remote sensing. In Proceedings of the 9th Lunar and Planetary Science Conference, 2901–2909, Houston, TX, USA, 13–17 March 1978.
67. Cheek, L.C.; Pieters, C.M. Reflectance spectroscopy of plagioclase-dominated mineral mixtures: Implications for characterizing lunar anorthosites remotely. *Am. Mineral.* **2014**, *99*, 1871–1892. [CrossRef]
68. Viviano-Beck, C.E.; Seelos, F.P.; Murchie, S.L.; Kahn, E.G.; Seelos, K.D.; Taylor, H.W.; Taylor, K.; Ehlmann, B.L.; Wiseman, S.M.; Mustard, J.F.; et al. Revised CRISM spectral parameters and summary products based on the currently detected mineral diversity on Mars. *J. Geophys. Res. Planets* **2014**, *119*, 1403–1431. [CrossRef]
69. Servicio Nacional de Geología y Minería. *Mapa Geológico de Chile: Versión Digital*; Publicación Geológica Digital, No. 7, Versión 1.0, 2004 [CD-ROM]; Servicio Nacional de Geología y Minería: Santiago, Chile, 2004.
70. Thompson, A.J.B.; Thompson, J.F.H. *Atlas of Alteration a Field and Petrographic Guide to Hydrothermal Alteration Minerals*; Geological Association of Canada: St. John's, NL, Canada, 1996.
71. Godoy, B.; Wörner, G.; Kojima, S.; Aguilera, F.; Simon, K.; Hartmann, G. Low-pressure evolution of arc magmas in thickened crust: The San Pedro–Linzor volcanic chain, Central Andes, Northern Chile. *J. South Am. Earth Sci.* **2014**, *52*, 24–42. [CrossRef]
72. Schmitt, A.K.; de Silva, S.L.; Trumbull, R.B.; Emmermann, R. Magma evolution in the Purico ignimbrite complex, northern Chile: Evidence for zoning of a dacitic magma by injection of rhyolitic melts following mafic recharge. *Contrib. Mineral. Petrol.* **2001**, *140*, 680–700. [CrossRef]
73. Hulley, G.C.; Hughes, C.G.; Hook, S.J. Quantifying uncertainties in land surface temperature and emissivity retrievals from ASTER and MODIS thermal infrared data. *J. Geophys. Res.* **2012**, *117*, D23113. [CrossRef]
74. Glotch, T.D.; Lucey, P.G.; Bandfield, J.L.; Greenhagen, B.T.; Thomas, I.R.; Elphic, R.C.; Bowles, N.; Wyatt, M.B.; Allen, C.C.; Donaldson Hanna, K.; et al. Highly silicic compositions on the Moon. *Science* **2010**, *329*, 1510–1513. [CrossRef]
75. Marín, C.; Rodríguez, I.; Godoy, B.; González-Maurel, O.; Le Roux, P.; Medina, E.; Bertín, D. Eruptive history of La Poruña scoria cone, Central Andes, Northern Chile. *Bull. Volcanol.* **2020**, *82*, 74. [CrossRef]
76. Carter, J.; Poulet, F.; Bibring, J.-P.; Mangold, N.; Murchie, S. Hydrous minerals on Mars as seen by the CRISM and OMEGA imaging spectrometers: Updated global view. *J. Geophys. Res. Planets* **2013**, *118*, 831–858. [CrossRef]
77. Murchie, S.; Mustard, J.F.; Ehlmann, B.L.; Milliken, R.E.; Bishop, J.L.; McKeown, N.K.; Noe Dobrea, E.Z.; Seelos, F.P.; Buczkowski, D.L.; Wiseman, S.M.; et al. A synthesis of Martian aqueous mineralogy after 1 Mars year of observations from the Mars Reconnaissance Orbiter. *J. Geophys. Res. Planets* **2009**, *114*, E00D06. [CrossRef]
78. Christensen, P.R.; Jakosky, B.M.; Kieffer, H.H.; Malin, M.C.; McSween, H.Y., Jr.; Nealson, K.; Mehall, G.L.; Silverman, S.H.; Ferry, S.; Caplinger, M.; et al. The Thermal Emission Imaging System THEMIS for the Mars 2001 Odyssey mission. *Space Sci. Rev.* **2004**, *110*, 85–130. [CrossRef]

79. Murchie, S.; Arvidson, R.; Bedini, P.; Beisser, K.; Bibring, J.-P.; Bishop, J.; Boldt, J.; Cavender, P.; Choo, T.; Clancy, R.T.; et al. Compact Reconnaissance Imaging Spectrometer for Mars (CRISM) on Mars Reconnaissance Orbiter (MRO). *J. Geophys. Res.* **2007**, *112*, E05S03. [CrossRef]
80. McEwen, A.S.; Robinson, M.S. Mapping of the Moon by Clementine. *Adv. Space Res.* **1997**, *19*, 1523–1533. [CrossRef]
81. Ohtake, M.; Haruyama, J.; Matsunaga, T.; Yokota, Y.; Morota, T.; Honda, C.; LISM Team. Performance and scientific objectives of the SELENE (KAGUYA) Multiband Imager. *Earth Planets Space* **2008**, *60*, 257–264. [CrossRef]
82. Green, R.O.; Pieters, C.; Mouroulis, P.; Eastwood, M.; Boardman, J.; Glavich, T.; Isaacson, P. The Moon Mineralogy Mapper (M3) imaging spectrometer for lunar science: Instrument description, calibration, on-orbit measurements, science data calibration and on-orbit validation. *J. Geophys. Res.* **2011**, *116*, E00G19. [CrossRef]
83. Paige, D.A.; Foote, M.C.; Greenhagen, B.T.; Schofield, J.T.; Calcutt, S.; Vasavada, A.R.; Preston, D.J.; Taylor, F.W.; Allen, C.C.; Snook, K.J.; et al. The Lunar Reconnaissance Orbiter Diviner Lunar Radiometer Experiment. *Space Sci. Rev.* **2010**, *150*, 125–160. [CrossRef]
84. Ehlmann, B.L.; Klima, R.I.; Bennett, C.L.; Blaney, D.; Bowles, N.; Calcutt, S.; Dickson, J.; Donaldson Hanna, K.; Edwards, C.S.; Green, R.; et al. Lunar Trailblazer: A pioneering smallsat for lunar water and lunar geology. In Proceedings of the 52nd Lunar and Planetary Science Conference, 2548, Virtual, 15–19 March 2021.
85. Pieters, C.M.; Noble, S.K. Space weathering on airless bodies. *J. Geophys. Res. Planets* **2016**, *121*, 1865–1884. [CrossRef]
86. Henderson, B.G.; Jakosky, B.M. Near-surface thermal gradients and their effects on mid-infrared emission spectra of planetary surfaces. *J. Geophys. Res.* **1994**, *99*, 19063–19073. [CrossRef]
87. Logan, L.M.; Hunt, G.R. Emission spectra of particulate silicates under simulated lunar conditions. *J. Geophys. Res.* **1970**, *75*, 6539–6548. [CrossRef]
88. Arnold, J.A.; Glotch, T.D.; Lucey, P.G.; Song, E.; Thomas, I.R.; Bowles, N.E.; Greenhagen, B.T. Constraints on olivine-rich rock types on the Moon as observed by Diviner and M3: Implications for the formation of the lunar crust. *J. Geophys. Res. Planets* **2016**, *121*, 1342–1361. [CrossRef]
89. Greenhagen, B.T.; Lucey, P.G.; Wyatt, M.B.; Glotch, T.D.; Allen, C.C.; Arnold, J.A.; Bandfield, J.L.; Bowles, N.E.; Donaldson Hanna, K.L.; Hayne, P.O.; et al. Global silicate mineralogy of the Moon from the Diviner Lunar Radiometer. *Science* **2010**, *329*, 1507–1509. [CrossRef]
90. Godoy, B. Evolución Petrológica de la Cadena Volcánica San Pedro-Linzor (21°30′S–22°10′S), Norte de Chile, y Su Relación con la Geodinámica Andina. Ph.D. Thesis, Universidad Católica del Norte, Antofagasta, Chile, 2014.
91. Godoy, B.; Wörner, G.; Le Roux, P.; de Silva, S. Sr- and Nd- isotope variations along the Pleistocene San Pedro—Linzor volcanic chain, N. Chile: Tracking the influence of the upper crustal Altiplano-Puna Magma Body. *J. Volcanol. Geotherm. Res.* **2017**, *341*, 172–186. [CrossRef]
92. Francis, P.W.; Roobol, M.J.; Walker, G.P.L.; Cobbold, P.R.; Coward, M. The San Pedro and San Pablo volcanoes of northern Chile and their hot avalanche deposits. *Geol. Rundsch.* **1974**, *63*, 357–388. [CrossRef]
93. Freymuth, H.; Brandmeier, M.; Wörner, G. The origin and crust/mantle mass balance of Central Andean ignimbrite magmatism constrained by oxygen and strontium isotopes and erupted volumes. *Contrib. Mineral. Petrol.* **2015**, *169*, 58. [CrossRef]
94. Mamani, M.; Tassara, A.; Wörner, G. Composition and structural control of crustal domains in the Central Andes. *Geochem. Geophys. Geosyst.* **2008**, *9*. [CrossRef]
95. Rogers, G.; Hawkesworth, C.J. A geochemical traverse across the North Chilean Andes: Evidence for crust generation from the mantle wedge. *Earth Planet. Sci. Lett.* **1989**, *91*, 271–285. [CrossRef]

remote sensing

MDPI

Article

Potential Applications of CE-2 Microwave Radiometer Data in Understanding Basaltic Volcanism in Heavily Ejecta-Contaminated Mare Frigoris

Jietao Lei [1,2], Zhiguo Meng [1,2], Yongzhi Wang [1,2,*], Shaopeng Huang [3], Jinsong Ping [4,5], Zhanchuan Cai [2] and Yuanzhi Zhang [4,5]

[1] College of Geoexploration Science and Technology, Jilin University, Changchun 130026, China; leijt19@mails.jlu.edu.cn (J.L.); mengzg@jlu.edu.cn (Z.M.)
[2] The State Key Laboratory of Lunar and Planetary Sciences, Macau University of Science and Technology, Macau 999078, China; zccai@must.edu.mo
[3] Institute of Deep Earth Science and Green Energy, Shenzhen University, Shenzhen 518060, China; shaopeng@szu.edu.cn
[4] Key Laboratory of Lunar and Deep Space Exploration, National Astronomical Observatory, Chinese Academy of Sciences, Beijing 100012, China; jsping@bao.ac.cn (J.P.); zhangyz@nao.cas.cn (Y.Z.)
[5] School of Space Science and Astronomy, University of Chinese Academy of Sciences, Beijing 100049, China
* Correspondence: wangyongzhi@jlu.edu.cn; Tel.: +86-431-8850-2362

Abstract: Mare Frigoris is the fifth largest and almost northernmost mare located on the near side of the Moon. Mare Frigoris has an elongated shape, with a length of approximately 1500 km and a width of approximately 200 km, which makes it susceptible to becoming contaminated by the impact ejecta from the nearby highlands. Comparatively speaking, microwave radiometer (MRM) data have good penetration capabilities. Therefore, the MRM data from Chang'e-2 satellite were employed to study the volumetric thermal emission features of basaltic deposits in Mare Frigoris. Combining the MRM data with the basaltic units with FeO and TiO_2 abundances identified using the small crater rim and ejecta probing (SCREP) methodology and with the gravity from Gravity Recovery and Interior Laboratory (GRAIL), the four potential conclusions that were obtained are as follows: (1) The MRM data are strongly related to the (FeO + TiO_2) abundance of pristine basalts and are less influenced by ejecta contamination; (2) in every quadrant of Mare Frigoris, the (FeO + TiO_2) abundance of the basalt decreases with an increase in age; (3) at least in Mare Frigoris, the main influencing factor regarding the brightness temperature remains the (FeO + TiO_2) abundance of surface deposits; (4) a warm microwave anomaly was revealed in the western-central and eastern-central areas of Mare Frigoris which has a strong relationship with the positive Bouguer gravity anomaly derived from GRAIL data in terms of spatial distribution. The results are significant in the context of improving our understanding the basaltic igneous rock and thermal evolution of the Moon using MRM data.

Keywords: Mare Frigoris; dielectric properties; basaltic units; Chang'e microwave radiometer; brightness temperature; Bouguer gravity

1. Introduction

Mare Frigoris is the fifth largest and northernmost lunar mare on the near side of the Moon. The mare extends from about 55°W to 45°E, with an average latitudinal extent of only 7.5° (Figure 1) [1–3]. In Mare Frigoris, there are light plain deposits, stress fractures, volcanic vents, rilles, pyroclastic deposits, cryptomare, and mare basalts, all of which have been identified mainly using optical remote sensing data [1–5]. With gravitational potential derived from the Gravity Recovery and Interior Laboratory (GRAIL) mission data, Andrews-Hanna et al. [6] suggested the existence of buried, ancient igneous intrusions in the mare. Thus, the study of Mare Frigoris can provide essential information on the basaltic volcanism and thermal evolution of the Moon.

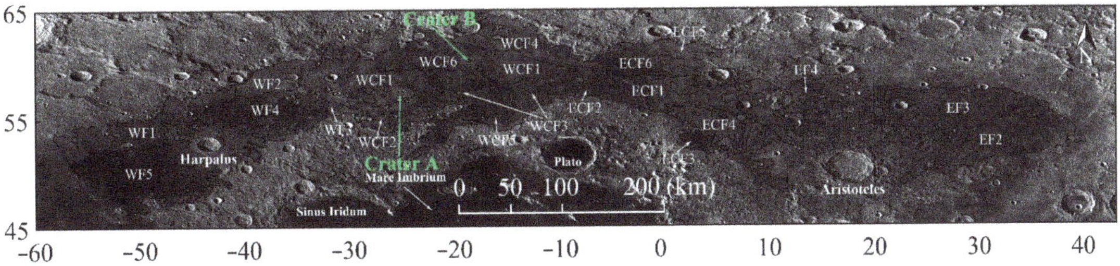

Figure 1. A wide-angle camera image of Mare Frigoris from the Lunar Reconnaissance Orbiter Camera (Equidistant Cylindrical projection), which was downloaded from https://wms.lroc.asu.edu/lroc/search (accessed on 20 June 2010). The black lines are the geologic boundaries and the abbreviations marked WCF5 in the west to EF2 in the east are the basaltic units of Mare Frigoris mapped by Kramer et al. [2]. The geologic boundaries were vectorized from the geologic map provided by Kramer et al. [2] and overlaid on this map and the following Figures 3–7, 9, 15, and 16. Crater A and Crater B are the two craters with brightness temperature anomalies, as discussed in Sections 3.2.2 and 4.2.1.

To accurately assess the basaltic volcanism of the Moon, it is necessary to characterize the pristine compositions of basaltic deposits [2]. Traditionally, basaltic units were mapped and their compositions estimated mainly by using ultraviolet, visible, and infrared data [1,7–11]. However, Mare Frigoris is narrow, and, therefore, ejecta from highland craters reach its central portions and obscure the basalt deposits more easily compared to large circular maria [1,2]. In addition, several large craters exist in Mare Frigoris, including the Harpalus crater (43.5°W, 52.7°N), which has a diameter of 40 km, and the Aristoteles crater (17.3°E, 50.2°N), which has a diameter of 88 km (Figure 1), the ejecta from which have heavily decreased the (FeO + TiO_2) abundance (FTA) of nearby mare deposits in the shallow layer [2]. This makes it difficult to infer the pristine compositions of the basaltic deposits of Mare Frigoris using optical data, as indicated by Figure 1.

To reduce the influence of the impact ejecta on the surface mafic contents of basaltic units, Kramer et al. [2,12,13] and Weider et al. [14] proposed the small crater rim and ejecta probing (SCREP) methodology, which collects composition information from small, immature impact craters that have penetrated the surface regolith. The SCREP methodology is considered the best approach to evaluate the pristine compositions of underlying basaltic units. Combined with the superposition of adjacent basaltic units and the FeO and TiO_2 abundances of the candidate 1533 SCREP craters, a new map of basaltic volcanism in Mare Frigoris was provided by Kramer et al. [2]. Using remote sensing data mainly from a Clementine Ultraviolet-Visible (UVVIS) camera and the Lunar Reconnaissance Orbiter (LRO) Wide Angle Camera (WAC), Kramer et al. [2] mapped 22 basaltic units, fewer than the 37 units previously mapped by Hiesinger et al. [10].

Generally, the penetration depth of optical remote sensing instruments is only several micrometers [15–17], which, combined with heavy surface contamination, has severely limited the application of optical data in the context of understanding basaltic volcanism in Mare Frigoris. Comparatively, the microwave radiometer (MRM) onboard the Chang'e (CE)-1 and -2 satellites, operated at microwave range and combined with the measured brightness temperature (TB), can reflect the dielectric properties of the lunar regolith at the corresponding penetration depths [15,17,18]. Moreover, MRM data have been proven to be highly correlated with the compositions of the regolith within the penetration depth [17,19]. At such depth, the deposits are less influenced by space weathering and less contaminated by the ejecta from nearby and distant impact events [20,21]. Thus, MRM data can provide a new potential choice to assess the basaltic units with heavy surface contamination in the maria on the Moon.

In this work, the MRM data from the CE-2 satellite were applied to assess the basaltic units in Mare Frigoris. Section 2 presents the process procedures for the CE-2 MRM data, the Clementine UVVIS data, and the GRAIL data. In Section 3, the microwave thermal emission features of the basaltic units are described, and the boundaries and ages of the basaltic units are assessed. Section 4 presents the potential application of the MRM data in understanding the basaltic volcanism and thermal evolution of Mare Frigoris. The conclusions are provided in Section 5.

2. Data Processing

The TB maps of Mare Frigoris were derived from the CE-2 MRM data. To test the TB anomaly and the mare's deep structure, the FeO, TiO_2, and Bouguer gravity anomaly maps were generated using the Clementine UVVIS data and GRAIL data, respectively.

2.1. CE-2 MRM Data Processing

From 15 October 2010 to 20 May 20 2011, the CE-2 MRM instrument measured the lunar surface for more than 5000 earth hours and operated at 3.0 GHz, 7.8 GHz, 19.35 GHz, and 37 GHz [22,23]. The integration time of the CE-2 MRM instrument is 200 ms, the observation angle is 0°, and the temperature sensitivity is approximately 0.5 K [22,24]. At an orbital altitude of approximately 100 km, the resolution of the MRM data is about 25 km at 3.0 GHz and 17.5 km at 7.8, 19.35, and 37 GHz. This study adopted the 2C level data after system calibration and geometric correction, obtained from the Lunar and Planetary Data Release System (https://moon.bao.ac.cn/ce5web/moonGisMap.search, accessed on 1 September 2019). For a description of the CE-2 MRM data, please see Zheng et al. [23] and Cai and Lan [25].

2.1.1. TB Maps Generation

The method of generating TB maps which was adopted in this study has been thoroughly described by Chan et al. [26], Cai and Lan [25], and Meng et al. [17,27]. According to a theoretical simulation using microwave radiative transfer equations, the TB of surface deposits is highly dependent on the surface temperature, which is dominated by the surface solar illumination and changes greatly with the lunation time [24,28]. Thus, the hour angle is introduced to describe the observed MRM points in terms of twenty-four lunation hours. However, for the special geometry of Mare Frigoris, it is difficult to generate TB maps with the sufficient MRM data points within one lunation hour. Thus, a new data selection method was proposed.

According to the geographical range of Mare Frigoris, 344,752 MRM data points was selected, and the local hours of every MRM data point were then calculated using the method provided by Chan et al. [26] and Zheng et al. [22,23]. Then, the histogram of the MRM data points in one lunation represented by the hour angles was counted (Figure 2a), which showed that the MRM data were concentrated around several time spans. After a comparison, we found that the data points from 12:26 to 13:08 and from 0:00 to 0:39 were enough to cover the whole of Mare Frigoris (Figure 2b,c). Therefore, the data points of the two time spans were extracted to generate TB maps at four channels using the linear interpolation method, and the spatial resolution was 0.25° × 0.25° (Figure 3, 37 GHz). The TB performances of the basaltic units at noon and night were represented by the observation from 12:26 to 13:08 and from 0:00 to 0:39, respectively. To clearly understand the thermophysical properties of basaltic deposits in Mare Frigoris, the boundaries of the basaltic units mapped by Kramer et al. [2] were vectorized and overlaid on the TB maps.

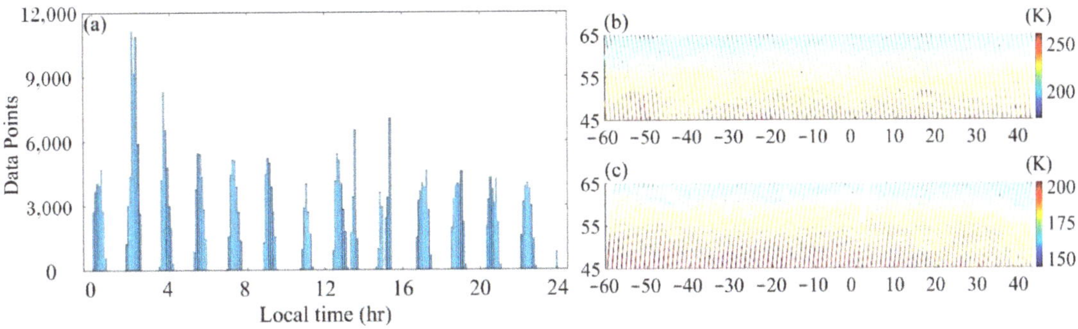

Figure 2. (**a**) The histogram of MRM data points in one lunation (24 h), (**b**) the scatter map of the MRM data points from 12:26 to 13:08, and (**c**) the scatter map of the MRM data points from 0:00 to 0:39 in Mare Frigoris (equidistant cylindrical projection).

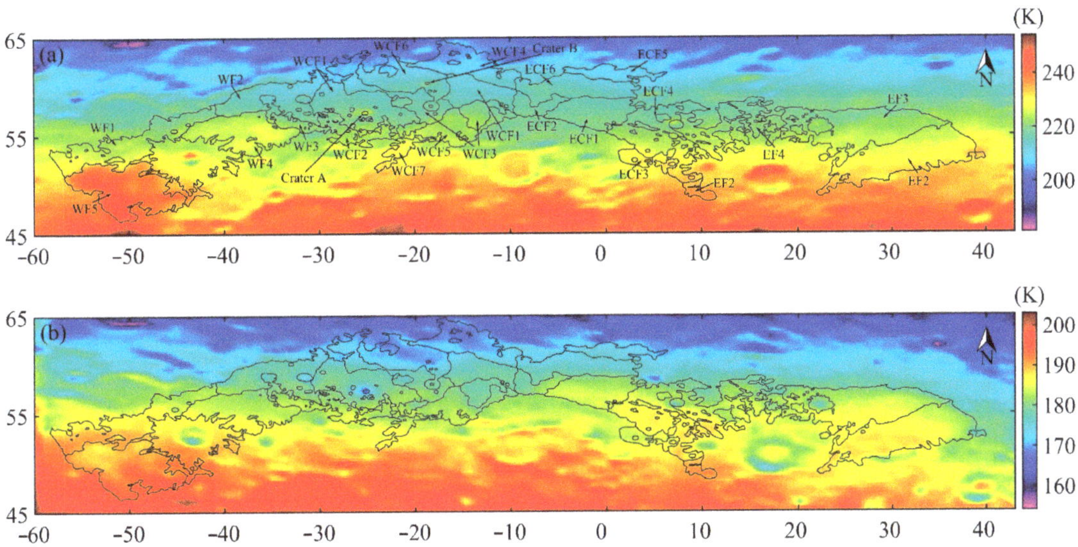

Figure 3. TB maps of Mare Frigoris at 37 GHz (equidistant cylindrical projection): (**a**) noon and (**b**) night. The black lines are the geologic boundaries of Mare Frigoris mapped by Kramer et al. [2].

The reliability of MRM data has been proven by Zheng et al. [22,23] and Cai and Lan [25], and our TB maps are identical to the previous results. Moreover, the TB maps in other hour spans were also generated and the TB performances were consistent in terms of the basaltic units, indicating the stability of the MRM data. What is more, Figure 3 shows a good correlation between the TB performances and the large craters, including Harpalus and Aristoteles. Particularly, the noon TB is enhanced in the southern slope of these craters, which are oriented towards solar illumination, and reduced in the northern slope. Furthermore, Figure 3 shows that the mare unit with high a FTA marked WF5 in the southwestern part of Mare Frigoris shows a high noon TB and low night TB, which is consistent with theoretical simulation result using microwave radiative transfer equations which show that the TB of deposits with a higher FTA will be higher at noon and lower at night when compared with deposits with a lower FTA [24,28]. Thus, the generated TB maps are rational.

2.1.2. nTB and TBD Maps Generation

Figure 3 shows an obvious N-S trend, demonstrating that the TB is higher in the south and lower in the north. Here, according to the latitude within and beyond Mare Frigoris, the change in TB is up to 40 K, which is much higher than the 12-K variation among the basaltic units within Mare Frigoris. This apparently decreases the dynamic range of the TB values in Mare Frigoris; thus, the following two steps were adopted to weaken the N-S trend of the TB and to enhance the TB performances of the mare units.

The first step was to exclude the nearby highlands by employing the traditional image cutting method according to the range of Mare Frigoris. Since the highest TB occurs in Sinus Iridum and the lowest TB occurs in the north highlands, such a process can, relatively speaking, enlarge the dynamic TB range in the cropped figures.

The second step was to operate normalized TB (nTB) mapping and TB difference (TBD) mapping, a method which was proposed by Meng et al. [17,27] to enhance the relationship between the TB and the surface materials.

To obtain the nTB maps, the standard TB values of every latitude were calculated according to the procedure provided in references [17,27]. The nTB maps at noon and night were generated by dividing the TB of every pixel by the standard TB of the corresponding latitude (Figures 4 and 5). The resulting nTB maps are considered to be sensitive to the compositions and thermal state of the surface deposits in the penetration depth [17,27].

The TBD is the difference between the interpolated noon and night TB of the same frequency before normalizing (Figure 6). The TBD maps are a good reflection of the compositions of the surface deposits in the penetration depth [17,27].

When compared with the TB maps, the microwave thermal emission features of surface deposits in Mare Frigoris are enhanced in Figures 4–6. The nTB and TBD maps agree well with the boundaries of the geologic units mapped by Kramer et al., especially in the units marked WF4, WF5, WCF6, and WCF7 [2]. In other words, the nTB and TBD maps can highlight mare deposits with different FTAs, which can be used to re-evaluate surface deposits in Mare Frigoris.

2.2. FeO and TiO_2 Abundance (FTA)

FTA is the decisive factor for the dielectric constant, which is the main influencing factor of the TB emitted from the lunar regolith [24,28,29]. In the microwave domain, the FTA value is always thought of as one of the important factors that can influence the loss tangent of the regolith [30].

Currently, FeO and TiO_2 abundances are also considered to be the most important index in the identification of basaltic units in different eras [1,2,10,31,32] with severe surface contamination, including Mare Frigoris. Thus, the FeO and TiO_2 abundances (Figure 7) were retrieved from the Clementine UVVIS data using the methods developed by Lucey et al. [33]. The Clementine UVVIS data can be downloaded from USGS Astrogeology Science Center (https://astrogeology.usgs.gov/search/map/Moon/Clementine/UVVIS/Lunar_Clementine_UVVIS_WarpMosaic_5Bands_200m, accessed on 9 February 2022). Figure 7 shows that the FeO and TiO_2 abundances are highest in the western region, moderate in the middle region, and lowest in the eastern region. These maps are used as an important reference to evaluate the TB performances of basaltic units.

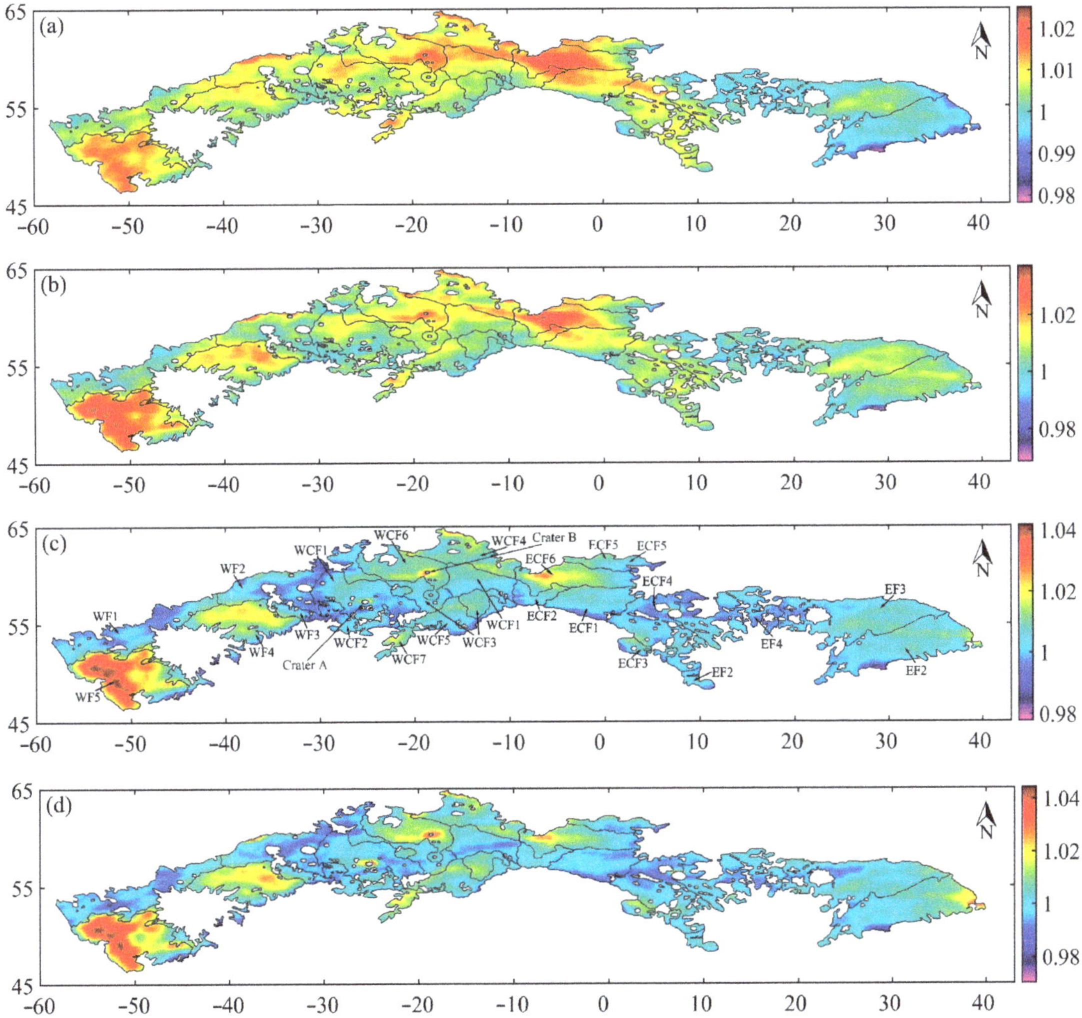

Figure 4. Normalized TB maps of Mare Frigoris at noon: (**a**) 3.0 GHz, (**b**) 7.8 GHz, (**c**) 19.35 GHz, and (**d**) 37 GHz (equidistant cylindrical projection). The black lines are the geologic boundaries of Mare Frigoris mapped by Kramer et al. [2]. Crater A and Crater B are two craters with TB anomalies, as discussed in Sections 3.2.2 and 4.2.1.

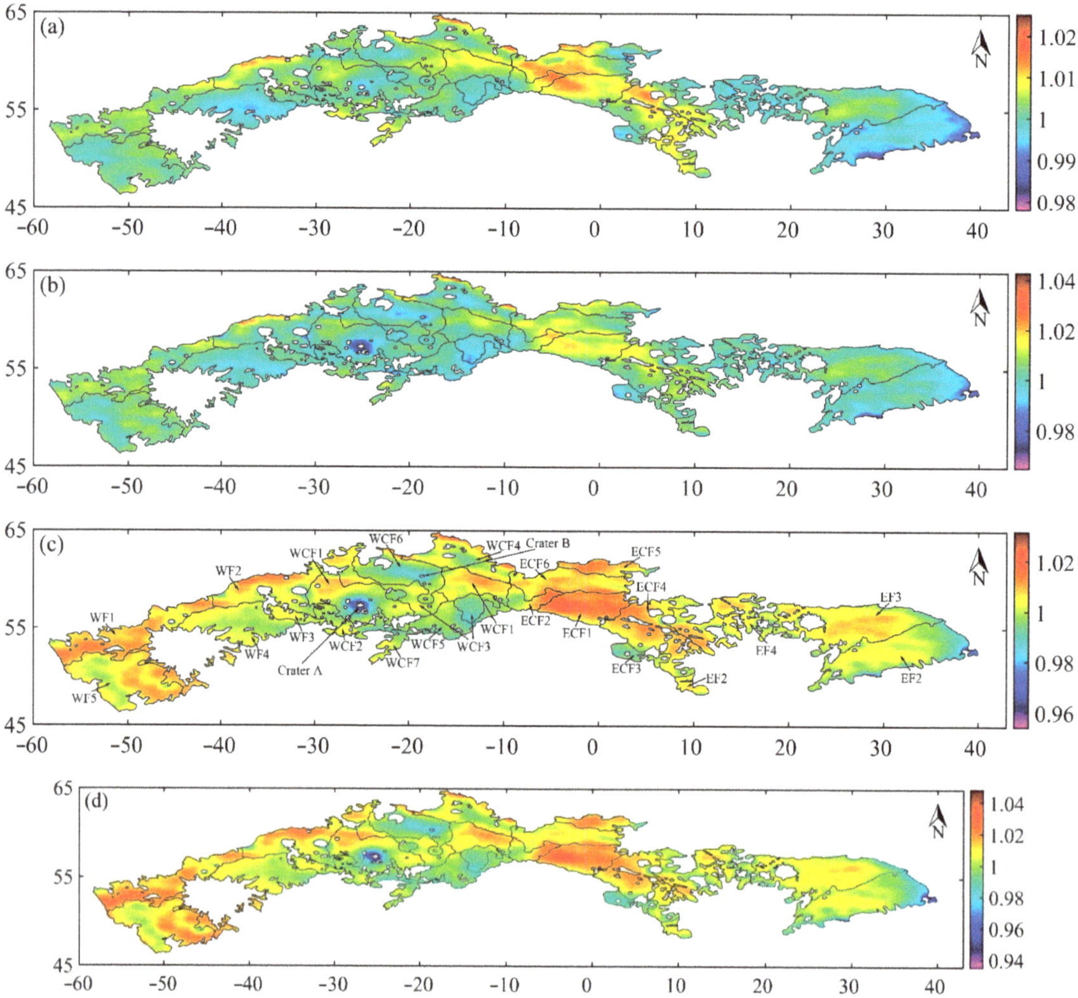

Figure 5. Normalized TB maps of Mare Frigoris at night: (**a**) 3.0 GHz, (**b**) 7.8 GHz, (**c**) 19.35 GHz, and (**d**) 37 GHz (equidistant cylindrical projection). The black lines are the geologic boundaries of Mare Frigoris mapped by Kramer et al. [2]. Crater A and Crater B are two craters with TB anomalies, as discussed in Sections 3.2.2 and 4.2.1.

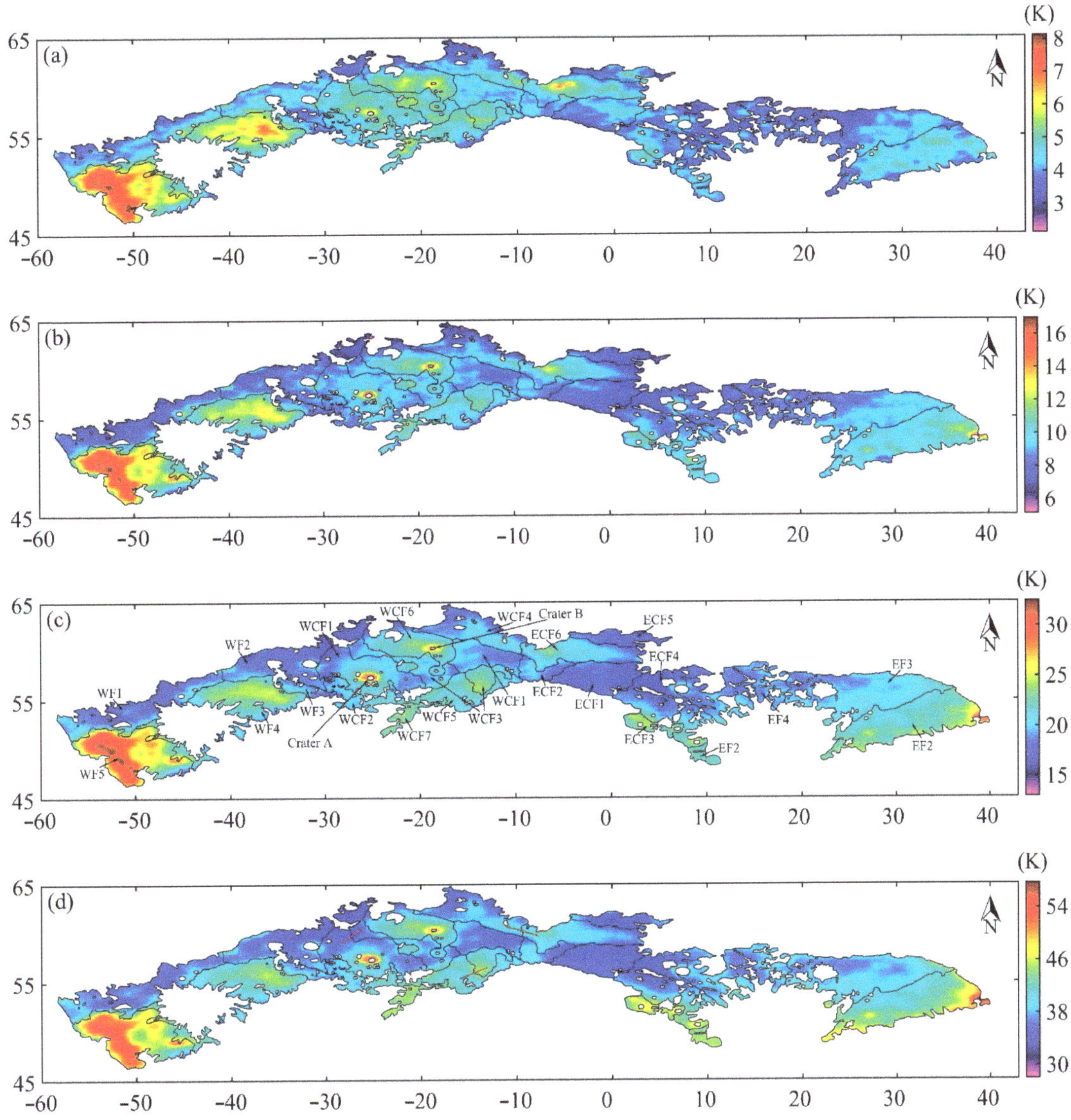

Figure 6. TB difference maps of Mare Frigoris: (**a**) 3.0 GHz, (**b**) 7.8 GHz, (**c**) 19.35 GHz, and (**d**) 37 GHz (equidistant cylindrical projection). The black lines are the geologic boundaries of Mare Frigoris mapped by Kramer et al. [2]. The orange lines are the profiles from the western WCF1 unit to the western WCF3 unit, from the WCF4 unit to the ECF6 unit, and from the WCF5 unit to the eastern the WCF3 unit. Crater A and Crater B are two craters with TB anomalies, as discussed in Sections 3.2.2 and 4.2.1.

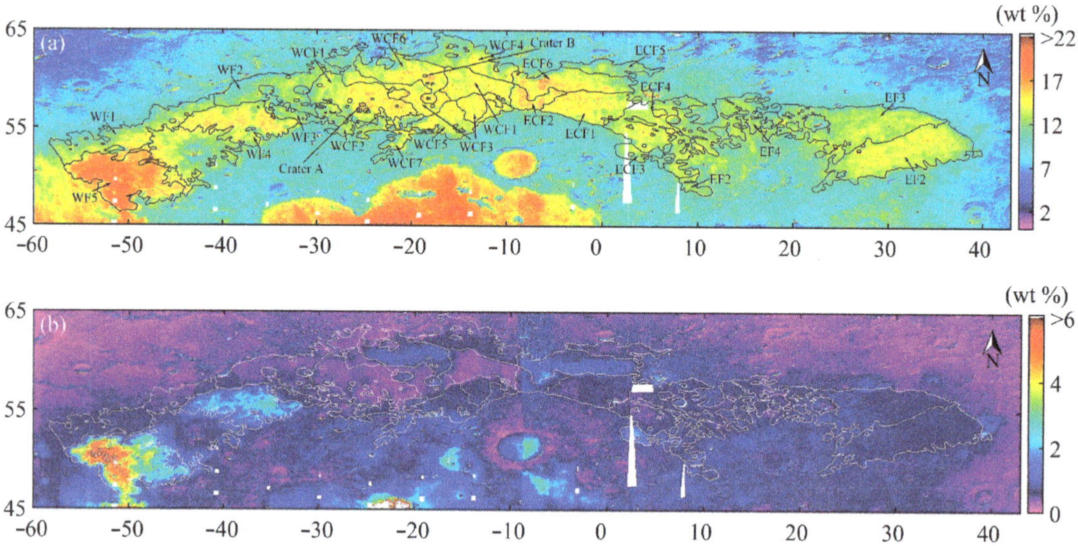

Figure 7. (**a**) FeO and (**b**) TiO_2 abundance maps of Mare Frigoris (equidistant cylindrical projection). The (**a**) black and (**b**) white lines are the geologic boundaries of Mare Frigoris mapped by Kramer et al. [2].

The TB of the Moon and its characteristics have been thoroughly theoretically investigated by a number of authors [22–25]. After evaluating the TB performances on a global scale, Meng et al. [34] proposed that the ($FeO + TiO_2$) abundance (FTA) is the controlling factor with regard to surface deposits on a local scale. As shown in Figure 8, the simulated TB using the radiative transfer model shows a clear change with respect to the FTA of the regolith. Compared to the deposits with lower FTAs, the TB is higher during the daytime (represented by surface temperatures of 390 K and 300 K) and lower at night (represented by surface temperatures 200 K and 100 K) in the deposits with higher FTAs. Correspondingly, the TBD will be much higher in FTA-rich deposits. This relationship between the TB (nTB) and TBD and the FTA of surface deposits is helpful in our understanding of the findings regarding the basaltic units in the following sections.

2.3. Bouguer Gravity Anomaly

The TB can reflect the dielectric properties of the regolith within the penetration depth and is sensitive to the substrate temperature [17,27]. Meanwhile, there is no air on the Moon, and the TB results likely represent thermal behaviors in the shallow lunar crust. Thus, the GRAIL data were introduced to unravel the potential relationship between the nTB results and the Bouguer gravity anomaly. The Bouguer gravity anomaly map was filtered from degree 42 to 600 and derived from GRAIL data (Figure 9), which can be downloaded from https://pds-geosciences.wustl.edu/grail/grail-l-lgrs-5-rdr-v1/grail_1001/shadr/ (accessed on 27 April 2016).

In Figure 9, there exists a considerable linear gravity anomaly centered on the ECF6 unit and several relatively low gravity anomalies to the north of the WCF1 unit, the junction of the WCF4 and WCF6 units, the junction of the eastern WCF3 and WCF5 units, the EF3 unit, and the EF4 unit. The map was used as a useful reference to understand the TB anomalies mentioned in the following sections.

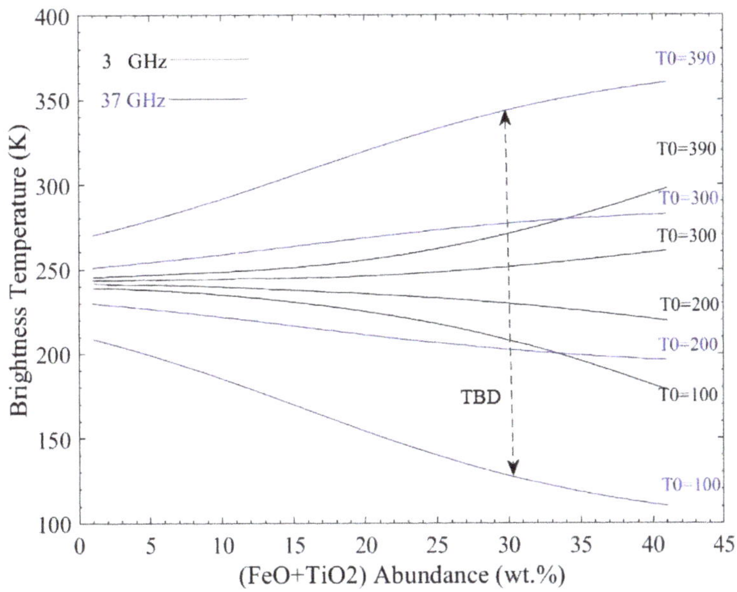

Figure 8. Change in simulated TB with (FeO + TiO_2) abundance and surface temperature [34,35]. T0 is the supposed surface temperature (unit: K), which is given as 390 K and 300 K (typical daytime temperatures) and 200 K and 100 K (typical nighttime temperatures).

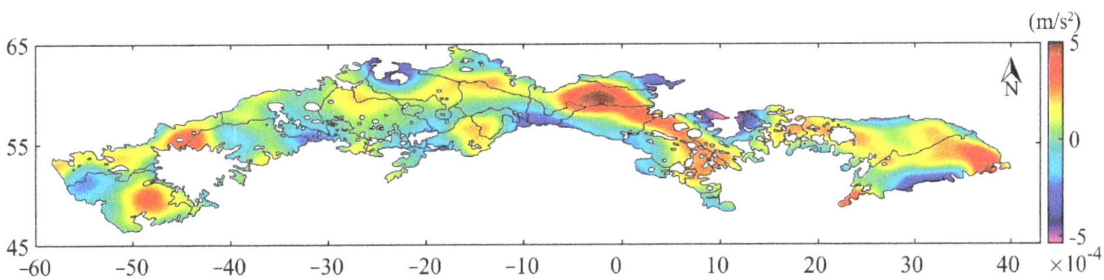

Figure 9. Bouguer gravity anomaly of Mare Frigoris (equidistant cylindrical projection). The black lines are the basaltic unit boundaries of Mare Frigoris mapped by Kramer et al. [2].

3. Results

SCREP-based results can reduce the influence of surface contamination on the pristine compositions of underlying materials. Thus, the basaltic units mapped by Kramer et al. [2] were used to evaluate the nTB and TBD results. Also, the basaltic units mapped by Hiesinger et al. [10] were used for a comparison between the nTB and TBD results, and the estimated ages were used to evaluate the ages of the basaltic units mapped by Kramer et al. [2].

3.1. Assessing the Basaltic Results

Regarding the weathering effects of the contamination of impact ejecta from other locations and space on the surface deposits, Kramer et al. [2] depended on the SCREP craters to probe the material beneath the surface regolith to determine the pristine compositions of the basaltic units.

Comparatively, a CE-2 microwave signal can penetrate surface deposits up to 10 to 20 times the wavelength of the used microwave [17,36]. Also, TBD maps have been proven to be sensitive to the FTA of deposits within the penetration depth of the microwave that is used [17,27]. Thus, the MRM data provided a new way to evaluate the pristine compositions of the basaltic units.

Figure 6 demonstrates how, in the study of the basaltic units on the lunar surface, the overlay of the geologic boundaries mapped by Kramer et al. [2] on TBD maps highlights the MRM data. Kramer et al. [2] divided Mare Frigoris into four quadrants comprising Western Frigoris (WF), West-central Frigoris (WCF), East-central Frigoris (ECF), and Eastern Frigoris (EF), which include 21 basaltic units in addition to the EF1 unit in Lacus Mortis, pyroclastic deposits, and cryptomare that are not mentioned by Hiesinger et al. [10]. Here, in the WF4, WF5, WCF4, WCF6, ECF3, and ECF5 units, the TBD results correlate with the basaltic unit mapped by Kramer et al. [2] (Figure 6). This correlation indicates that the TBD is highly related to the SCREP-based (FeO + TiO_2) abundance of pristine basalts, which is not as influenced by the ejecta contamination as the UVVIS-based FeO and TiO_2 abundances, as shown in Figure 7.

Moreover, the correspondence between the TBD results and the basaltic units is good in the WF, WCF, and ECF quadrants, while the correspondence is only poor in the EF quadrant (Figure 6). Two large craters exist in Mare Frigoris, the Harpalus crater in the WF quadrant, with a diameter of 40 km, and the Aristoteles crater in the EF quadrant, with a diameter of 88 km (Figure 1). Figure 6 shows that the influence of Aristoteles crater on the compositions of pristine basaltic units are complex. In the EF quadrant, the TBD in the southern area is higher than that in the northern area, which deserves to be further studied.

For comparative purposes, the basaltic units identified by Hiesinger et al. [10] were also vectorized and overlaid on the 37-GHz TBD map (Figure 10). Only four (the F1, F25, F27, and F31 units) of the 37 units (units F1–F37) show a good agreement with the TBD results. This suggests that the weathering effects of the contamination of the impact ejecta from other locations and space on the compositions of surfaces are fairly strong in Mare Frigoris.

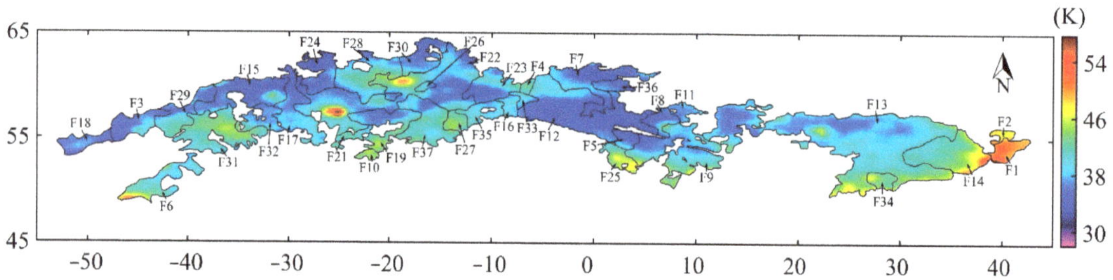

Figure 10. TB difference map of Mare Frigoris at 37 GHz (equidistant cylindrical projection). The black lines are the geologic boundaries determined by Hiesinger et al. [10].

3.2. *nTB and TBD Performances of Basaltic units*

According to the results in Mare Moscoviense and Mare Australe, Meng et al. [27,28] proposed that TBD values are sensitive to the compositions of the volumetric regolith, which are used in the following analysis. The nTB maps were adopted as references to obtain the dielectric properties of the regolith within the penetration depth of the used microwave.

3.2.1. Western Frigoris

Five basaltic units were mapped by Kramer et al. [2] in the WF quadrant, and these represent the total range in compositions and ages of the basaltic units in Mare Frigoris. Kramer et al. [2] estimated the pristine compositions of the five units, as shown in Table 1.

Table 1. The SCREP-based FeO, TiO_2, and FTA values mapped by Kramer et al. [2] and the four-channel TBDs of the WF1, WF2, WF3, WF4, and WF5 units.

Units	TBD (K)				FeO (wt %)	TiO_2 (wt %)	FTA (wt %)	Age (Ga)
	3.0 GHz	7.8 GHz	19.35 GHz	37 GHz				
WF1	4.0	7.9	17.8	35.8	13–14	1	14–15	3.72
WF2	4.3	8.5	18.4	36.0	12	<1	12–13	3.54
WF3	4.2	8.8	19.5	38.0	12	1	13	3.39
WF4	6.6	12.8	24.5	44.5	16	3–4	19–20	<3.39
WF5	7.3	15.5	30.9	53.2	18	6–7	24–25	<WF4

In the WF quadrant, the TBD is highest in the WF5 unit and second highest in the WF4 unit. Correspondingly, the nTB of the WF5 unit is highest at noon and almost lowest at night, followed by the nTB of the WF4 unit. Here, two issues should be pointed out. The first issue (named Issue 1) concerns the fact that the TBD and nTB results agree with the theoretical simulation finding that the nTB will be higher at noon but lower at night in deposits with higher FTAs compared to those with lower FTAs [24,28]. Correspondingly, the TBD will be considerably high in FTA-rich deposits. The second issue (named Issue 2) is that coincidences occur between the TBD and nTB results and the boundaries outlined by Kramer et al. [2] in the WF4 and WF5 units. Again, these verify that the MRM data are of potential significance in the identification of basaltic units. The two issues are significant in the context of understanding the relationship between TB results and basaltic units in the following analysis. The WF3 unit indicates a relatively low nTB at noon, a high nTB at night, and a low TBD compared with the WF4 unit. The boundary between the WF2 and WF3 units is clear at 19.35- and 37-GHz according to the nTB and TBD maps.

The WF1 and WF2 units are located in the northern section of the WF quadrant, and their TBDs are apparently lower than those of the other units in the WF quadrant. Here, we noticed an abnormal phenomenon: Table 1 shows that the TBD of the WF1 unit is 0.3 K, 0.6 K, 0.6 K, and 0.2 K lower than that of the WF2 unit at 3.0 GHz, at 3.0 GHz, 7.8 GHz, 19.35 GHz, and 37 GHz, respectively, but the FTA of the WF1 unit is 2–3 wt % higher than that of the WF2 unit. According to Issue 1 and the theoretical simulation [24,28], the FTA of the WF1 unit should be lower than that of the WF2 unit, which is opposite of the SCREP-based FTA result.

When referring to the ages estimated by Hiesinger et al. [10], the WF1 unit corresponds to the F3 unit (3.72 Ga) and the F18 unit (3.53 Ga), the WF2 unit mainly includes the F15 unit (3.54 Ga), and the WF3 unit corresponds to the middle of the F32 unit (3.39/4.00 Ga). Thus, combined with the relative ages provided by Kramer et al. [2], the WF1 unit is probably 3.72 Ga in age, older than the WF2 unit, at 3.54 Ga, and the WF3 unit should be younger still, at 3.39 Ga. Considering the superposition relationship of the stratum [2], the WF4 unit should be younger than 3.39 Ga, and the WF5 unit is the youngest in the WF quadrant.

3.2.2. West-Central Frigoris

The WCF1 unit consists of two parts: the western WCF1 and the eastern WCF1. The WCF1 unit is the only "very low-Ti" (VLT) basalt in the quadrant. Table 2 shows that the TBD of the western WCF1 is approximately 0.5 K higher than that of the eastern WCF1 at 3.0 GHz, except for the region surrounding Crater A (La Condamine S crater, 25.22°W, 57.34°N). In the other three channels, the TBD of the eastern WCF1 is consistent with that of the western WCF1, indicating the similarity between the two parts.

Table 2. The SCREP-based FeO, TiO_2, and FTA values mapped by Kramer et al. [2] and the four-channel TBDs of the WCF1, WCF2, WCF3, WCF4, WCF5, WCF6, and WCF7 units.

Units	TBD (K)				FeO (wt %)	TiO_2 (wt %)	FTA (wt %)	Age (Ga)
	3.0 GHz	7.8 GHz	19.35 GHz	37 GHz				
WCF1	4.5	8.8	19.0	35.9	13–14	<1	13–15	3.53
WCF1 (W)	4.7	9.1	19.1	36.3	13–14	<1	13–15	3.53
WCF1 (E)	4.2	8.3	18.8	35.4	13–14	<1	13–15	3.53
WCF2	4.5	9.0	20.5	40.0	12–13	1	13–14	3.49
WCF3	4.9	9.6	20.6	38.4	13–14	1	14–15	3.46
WCF3 (W)	4.9	9.1	19.5	36.4	13–14	1	14–15	3.46
WCF3 (E)	4.9	10.4	22.7	42.2	13–14	1	14–15	3.46
WCF4	4.8	9.6	20.6	39.0	13	1	14	3.43
WCF5	4.8	9.9	21.7	41.3	14	1	15	3.11
WCF6	5.1	10.4	22.3	42.7	14–15	2	16–17	<3.11
WCF7	4.9	10.3	23.0	44.6	15	2–3	17–18	<WCF6

Concerning the WCF2 unit, Table 2 shows that the TBD of the WCF2 unit is 0 K at 3.0 GHz and 0.2 K higher at 7.8 GHz than that of the WCF1 unit, while it is 1.5 K at 19.35 GHz and 4.1 K higher at 37 GHz than that of WCF1 unit, which indicates the complexity of the deposits in the vertical direction. Meanwhile, the same problem pointed out in the WF1 and WF2 units also occurs here: the SCREP-based FTA of the WCF2 unit is significantly lower than that of the WCF1 unit; however, the TBD of the WCF2 unit is higher than that of the WCF1 unit. Thus, theoretically, the FTA of the WCF2 unit should be higher than that of the WCF1 unit.

The WCF3 unit includes two parts: the western WCF3 and the eastern WCF3. The TBD maps show that the boundary between the western WCF3 and the nearby WCF6 unit in the north and that between the eastern WCF3 and its nearby WCF1 unit in the north are well mapped by the TBD results. However, the TBD of the eastern WCF3 is clearly higher than that of the western WCF3, and the differences are approximately 0.1 K at 3.0 GHz, 1.3 K at 7.8 GHz, 3.2 K at 19.35 GHz, and up to 5.8 K at 37 GHz (Table 2). Compared with the TBD results of the other units, the differences are considerably large, and, thus, the category of the eastern and western WCF3 units is problematic.

Moreover, the TBD maps show that the difference between the western WCF3 unit and the nearby WCF1 unit is not clear, where the differences are only 0.2 K at 3.0 GHz, 0 K at 7.8 GHz, 0.4 K at 19.35 GHz, and 0.1 K at 37 GHz. Furthermore, according to the TBD profiles (Figure 11), compared with the clear difference between the WCF5 unit and the eastern WCF3 unit, there is no obvious TBD boundary between the western WCF1 unit and the western WCF3. Therefore, the western WCF3 should be includes as part of the WCF1 unit, and they are both clearly different to the eastern WCF3 (Figure 12).

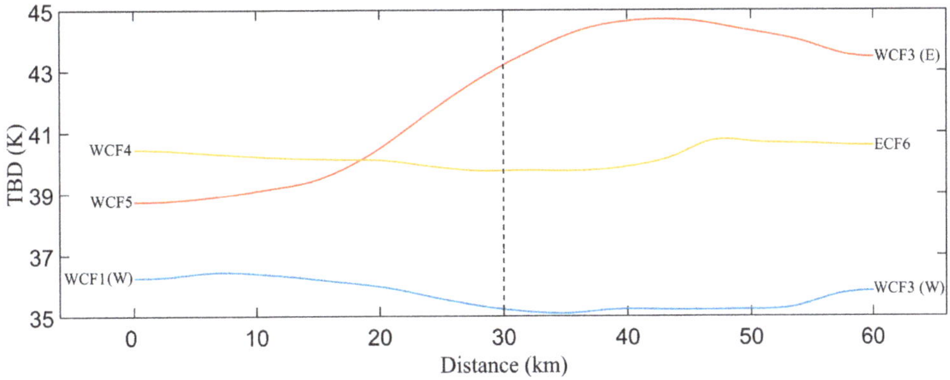

Figure 11. The profiles of the 37-GHz TBD from the western WCF1 (WCF1 (W)) to the western WCF3 (WCF3 (W)), from the WCF4 unit to the ECF6 unit, and from the WCF5 unit to the eastern WCF3 (WCF3 (E)). The black dashed line represents the boundary between the basaltic units mapped by Kramer et al. [2]. The locations of the profiles are shown in Figure 6d (orange lines).

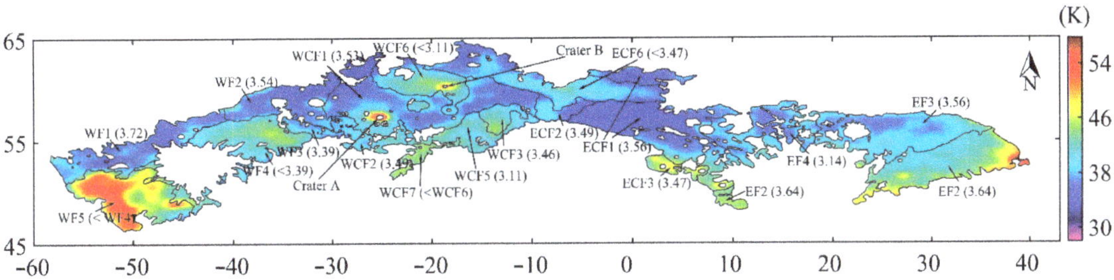

Figure 12. New geologic sketch of basaltic units in Mare Frigoris (equidistant cylindrical projection). The numbers after the unit names represent the ages in Ga. The base map is the TBD at 37 GHz.

In the case of the WCF4 unit, the four-channel TBDs are significantly lower than those of the WCF6 unit, but higher than those of the eastern WCF1. Compared with the WCF1 unit, the WCF4 unit exhibits a relatively high nTB at noon and a relatively low nTB at night. According to Issue 1 and the theoretical simulation [24,28], such nTB and TBD results indicate that the FTA of the WCF4 unit should be higher than that of the WCF1 unit, which would be inconsistent with the SCREP-based FTA provided by Kramer et al. [2].

One issue that should be noted is the fact that the TBD in the WCF4 unit increases gradually from the northwest to the east, and the TBD in the eastern area is close to that in the west of the ECF6 unit (Figure 6). Here, we ascribed the WCF4 unit to the ECF6 unit, which is discussed further in the following section.

The WCF5 unit is located in the southeast of the WCF quadrant, and the FeO and TiO_2 abundances of the pristine basalt were estimated to be 14 wt % and 1 wt % [2]. The WCF5 unit includes the F35 (3.11 Ga) and F37 (2.61 Ga) units mapped by Hiesinger et al. [10]. Here, the TBD maps show that the difference between the F35 and F37 units is almost zero in each channel. It should be noted that the TBD of the WCF5 unit is slightly lower than that of the eastern WCF3 (Figure 11). Meanwhile, the noon nTB is lower and the night nTB is slightly higher in the WCF5 unit compared to the eastern WCF3. Therefore, the FTA of the WCF5 unit should be lower than that of the eastern WCF3.

In WCF6 unit, the TBD is significantly higher than that of the surrounding WCF1, WCF3, and WCF4 units. Although the TBD clearly changes from the north to the southeast of this unit, no clear boundary exists. Meanwhile, the unit has the second highest noon nTB and the second lowest night nTB. These TBD and nTB results are consistent with the SCREP-based FTA of the WCF6 and WCF7 units.

The WCF7 unit has the highest TBD results at 19.35 and 37 GHz, corresponding to the highest FeO and TiO_2 abundances here.

Compared with the ages estimated by Hiesinger et al. [10], the ages of the basaltic units mapped by Kramer et al. [2] are complicated. The WCF1 unit is mainly located in the F17 (3.53 Ga), F19 (3.51 Ga), F22 (3.48 Ga), and F23 (3.48 Ga) units. Since the WCF1 unit is the oldest in the WCF quadrant, we set the age of the unit as 3.53 Ga. The WCF2 unit corresponds to the F17 (3.53 Ga) and F21 (3.49) units; therefore, it would be rational to set the WCF2 unit as 3.49 Ga. Most of the western WCF3 unit lies within the F19 (3.51/3.79 Ga), F22 (3.48/3.63 Ga), F24 (3.47/3.70 Ga), F28 (3.45 Ga), and F37 (2.61 Ga) units, while the eastern WCF3 unit is close to the F27 unit (3.46 Ga) and its age was set as 3.46 Ga. The WCF4 unit is located in the F22 (3.48/3.63 Ga), F23 (3.48 Ga), and F30 (3.43 Ga) units, so 3.43 Ga (F30) was set as the age for this unit. The WCF5 unit includes the F35 (3.11 Ga) and F37 (2.61 Ga) units. According to the relative ages provided by Kramer et al. [2], the WCF6 unit is much younger than the WCF5 unit, at 3.48 Ga (F22) and 3.45 Ga (F28). The WCF7 unit should be younger than the WCF6 unit, which is also younger than 3.6 Ga (F10) and 3.51 Ga/3.79 Ga (F19).

3.2.3. East-Central Frigoris

Figures 4–6 show that the nTB and TBD performances of the ECF quadrant are complex, or at least they are in the ECF1, ECF2, ECF5, and ECF6 units, all of which show relatively high 3.0-GHz nTBs both at noon and at night. According to previous studies [28], the warm microwave anomaly caused by the higher substrate temperature is likely to exist in the ECF quadrant, which will be discussed further in the next section.

The ECF1 unit has the lowest TBD in the ECF quadrant. The nTB maps show that the ECF1 unit has the lowest noon nTB and the highest night nTB in the ECF quadrant at 37 GHz (representing the mare deposits in the shallow layer). These nTB and TBD results are highly correlated with the lowest FeO and TiO_2 abundances in the ECF quadrant.

The TBD of the ECF2 unit is notably higher than the ECF1 unit and lower than the ECF6 unit. The noon nTB of the ECF2 unit is higher than that of the ECF1 unit but lower than that of the ECF6 unit at 37 GHz, and the opposite is true at night.

Compared with the ECF1 and ECF2 units, at 37 GHz, the TBD and the noon nTB of the ECF3 unit are higher but the night nTB is lower.

These nTB and TBD results in the ECF1, ECF2, and ECF3 units agree with their comparative SCREP-based FTA according to the theoretical simulation [24,28].

Table 3 shows that the TBD of the ECF4 unit differs very little from the ECF1 unit: only 0.1 K at 3.0 GHz, 0.4 K at 7.8 GHz, 0.5 K at 19.35 GHz, and 0.4 K at 37 GHz, indicating the homogeneity of mare deposits in the two units. Meanwhile, the ECF4 unit has a relatively low nTB at noon and a relatively high nTB at night at 37 GHz, which is also consistent with the ECF1 unit. Therefore, we combined the ECF4 unit with the ECF1 unit based on the nTB and TBD results (Figure 12). Hiesinger et al. [10] classified the west of the ECF4 unit as the ECF1 unit, which also confirmed the reliability of the nTB and TBD results.

The nTB of the ECF5 unit is very close to that of the ECF1 unit, and the TBD discrepancies between the two units are 0.4 K at 7.8 GHz and 0 K in the other three channels (Table 3). According to the strong relationship between the TBD and the compositions of the mare deposits, the FTA of the ECF5 unit should be consistent with the ECF1 unit. Therefore, considering that the FTA is an important basis for Kramer et al. [2] in the mapping of different units, the ECF1 and ECF5 units should be considered the same unit (Figure 12). The F7 unit mapped by Hiesinger et al. [10] comprises the ECF1 and ECF5 units, validating that the combination is reliable.

Table 3. The SCREP-based FeO, TiO_2, and FTA values mapped by Kramer et al. [2] and the four-channel TBDs of the ECF1, ECF2, ECF3, ECF4, ECF5, and ECF6 units.

Units	TBD (K)				FeO (wt %)	TiO_2 (wt %)	FTA (wt %)	Age (Ga)
	3.0 GHz	7.8 GHz	19.35 GHz	37 GHz				
ECF1	4.1	8.2	17.6	34	14	1	15	3.56
ECF2	4.1	9	19.3	38.2	15	1.5	16.5	3.54
ECF3	4.5	10.2	23.5	46.1	14	3	17	3.47
ECF4	4	7.8	17.1	34.4	13	1	14	3.56
ECF5	4.1	7.8	17.6	34	11–12	1	12–13	3.56
ECF6	5	10.2	21.2	39.6	16	2	18	<3.47

The ECF6 unit has the highest TBD, a relatively high nTB at noon, and a low nTB at night in the ECF quadrant, which is consistent with the highest FeO and TiO_2 abundances in the unit.

According to the nTB and TBD results, the ECF quadrant contains the new ECF1 unit (the combination of the ECF1, ECF4, and ECF5 units), the ECF2 unit, the ECF3 unit, and the ECF6 unit (Figure 12). The new ECF1 is mainly located in the F5 (3.71 Ga), F7 (3.63 Ga), F8 (3.62/3.76 Ga), F9 (3.62 Ga), F11 (3.58 Ga), and F12 (3.56/3.71 Ga) units. The ECF2 unit includes the southwestern area of the F12 (3.56/3.71 Ga) unit and the F16 (3.54 Ga) unit. The ECF3 unit corresponds to the F25 (3.47 Ga) unit, and the ECF6 unit is located in the F7 (3.63 Ga) and F12 (3.56/3.71 Ga) units. Combined with the relative ages provided by Kramer et al. [2], the ECF1 unit was set as 3.56 Ga, the ECF2 unit as 3.54 Ga, the ECF3 unit as 3.47 Ga, and the ECF6 unit is less than 3.47 Ga in age.

3.2.4. Eastern Frigoris

Kramer et al. [2] mapped the East Frigoris into four basaltic units, but the EF1 unit in Lacus Mortis was not discussed in the study. Compared with the WF quadrant, in which there is an excellent correspondence between WCF and ECF quadrants, with a good agreement between the TBD results and the basaltic units, this relationship is poor in the EF quadrant, and there are no obvious nTB and TBD boundaries between the three units.

Compared with the other basaltic units in the quadrant, the TBD of the EF2 unit is significantly higher and the nTB of the unit is higher at noon but lower at night (Table 4). The nTB of the EF3 unit is similar to that of the EF2 unit at noon, but it is significantly higher than that of the EF2 unit at night (Table 4). Accordingly, the TBD of the EF3 unit is significantly lower than that of the EF2 unit. Figure 4 and Table 4 show that the EF4 unit has the lowest TBD and noon nTB in the EF quadrant.

Table 4. The SCREP-based FeO, TiO_2, and FTA values mapped by Kramer et al. [2] and the four-channel TBDs of the EF2, EF3, and EF4 units.

Units	TBD (K)				FeO (wt %)	TiO_2 (wt %)	FTA (wt %)	Age (Ga)
	3.0 GHz	7.8 GHz	19.35 GHz	37 GHz				
EF2	4.5	10	21.7	41.3	14	1.5–2	15.5–16	3.64
EF3	4	9.1	19.9	37.8	13	1–1.5	14–14.5	3.56
EF4	3.8	8.1	18.5	36.2	13	1	14	3.14

Generally speaking, the nTB and TBD performances of the EF2, FE3, and EF4 units are consistent with the SCREP-based FTA results according to the theoretical simulation [24,28]. Therefore, the composition features obtained by Kramer et al. [2] in the EF quadrant can be confirmed by the nTB and TBD results, but the boundaries between these units are not clear.

The relative ages provided by Kramer et al. [2] are different from the aging results obtained by Hiesinger et al. [10] and Wilhelms [8]. The EF2 unit is located in the F13 (3.56 Ga), F14 (3.56 Ga), and F34 (3.14 Ga) units, the EF3 unit includes the northeast of the F13 unit (3.56 Ga), and the EF4 unit corresponds to the F9 (3.62 Ga) and F13 (3.56 Ga) units. Boyce [7] and Wilhelms [8] estimated the age of the ECF4 unit to be 3.4 Ga and late Imbrium. Therefore, based on the nTB and TBD results and the age estimated by Hiesinger et al. [10], Boyce [7], and Wilhelms [8], the EF4 unit is older (approximately 3.64 Ga), the EF3 unit is about 3.56 Ga, and the EF2 unit is about 3.14 Ga.

3.3. Redrawing Geologic Sketch of Basaltic Units and Remaining Problems

It is difficult to renew fine geologic maps when only using TB maps because the pixel size of MRM data is 25 km at 3.0 GHz and 17.5 km at other channels. In general, the pristine compositions exposed by SCREP-based craters, particularly FeO and TiO_2 abundances, were the most important references for Kramer et al. [2] in their re-evaluation of the basaltic units of Mare Frigoris, and the FeO and TiO_2 abundances were also the key regolith parameters of the MRM data when the influence of the surface temperature was weakened in this study. Thus, based on the analysis described above, the basaltic units in Mare Frigoris mapped by Kramer et al. [2] were rearranged to obtain a new geologic sketch (Figure 12). Compared to the geologic map provided by Kramer et al. [2], differences only occur in the following basaltic units: the western WCF3 unit was included in the WCF1 unit, the ECF4 and ECF5 units were combined with the ECF1 unit, and the WCF4 unit was included in the ECF6 unit.

According to the nTB and TBD results for Mare Frigoris, we can find the following three problems:

(1) The spatial distributions of the nTB and TBD values are consistent with the boundaries of the basaltic units in the WF, WCF, and ECF quadrants mapped by Kramer et al. [2], as shown in Figures 4–6. This comparison hints at the fact that the MRM data have the potential capability to probe the pristine basalt under ejecta, such as that reflected by the SCREP craters.
(2) In most of the basaltic units mapped by Kramer et al. [2], the TBD results have a strong relationship with the FTA. Whether this relationship is valid or not should be verified for the whole of Mare Frigoris.
(3) In most quadrants, the TBD results have a good agreement with the basaltic units. However, there were abnormal 3.0-GHz nTB results for the ECF1, ECF2, ECF5, and ECF6 units, which are relatively high both at noon and at night. The penetration depth of a 3.0-GHz microwave is about 1 m to 2 m [15], and the substrate temperature at this depth is slightly influenced by the surface temperature [30]. The causes of the TB anomaly, which will be discussed further in Section 4.2.

4. Discussion

Air and water are absent on the Moon, and because of this, microwaves can penetrate the regolith up to 10 to 20 times their wavelength [15,16], that is, about 1–2 m at 3.0 GHz, 38.5 cm to 75 cm at 7.8 GHz, 15.5 cm to 31 cm at 19.35 GHz, and 8.1 cm to 16.2 cm at 37 GHz. Therefore, the MRM data provide a fresh view of the surface deposits in Mare Frigoris when compared to the optical observations.

4.1. New Views of Basaltic Volcanism in Mare Frigoris

Although Mare Frigoris has suffered from severe contamination due to impact crater ejecta, the TB results of the basaltic units provide an opportunity to study the lunar basaltic volcanism of the maria.

4.1.1. Representation of TB to Basaltic Units

Mare Frigoris is one of the northmost maria on the lunar nearside, and its special elongated shape means that it has suffered more from severe contamination due to impact crater ejecta than other maria (Figure 1). Boyce et al. [7] and Hiesinger et al. [10] used Lunar Orbiter IV photographs and Clementine UVVIS data, respectively, to divide the basaltic units of Mare Frigoris, and both used the crater size-frequency distribution (CSFD) to determine the ages of the divided basaltic units. Kramer et al. [2] re-evaluated the basaltic units of Mare Frigoris by using the pristine compositions exposed by SCREP craters and the superposition relationship of the stratum. Generally, the results of these studies differ greatly from each other.

Based on the nTB and TBD maps, this study evaluated the basaltic unit result mapped by Kramer et al. [2] with reference to Hiesinger et al. [10]. The nTB and TBD results are consistent with the basaltic units mapped by Kramer et al. [2] (Figures 4–6). Among all 21 units (excluding the EF1 unit in Lacus Mortis), most unit boundaries (in the case of 17 units) agree with the TBD results, but the boundaries of the WCF1, WCF3, ECF1, and ECF4 units do not (Figure 6). In particular, the TBD results for the WF4, WF5, WCF4, WCF6, ECF3, and ECF5 units strongly agree with the basaltic units obtained by Kramer et al. [2] (Figure 6). These indicate that the SCREP methodology and MRM data are consistent when it comes to characterizing basaltic units contaminated by impact crater ejecta.

Figure 13 shows the SCREP-based FeO abundances, TiO_2 abundances, and FTA estimated by Kramer et al. [2] and the four-channel TBDs of all units. Here, the TBD is strongly correlated with the SCREP-based FTA of the basaltic units, that is, a higher TBD is directly related to a higher FTA.

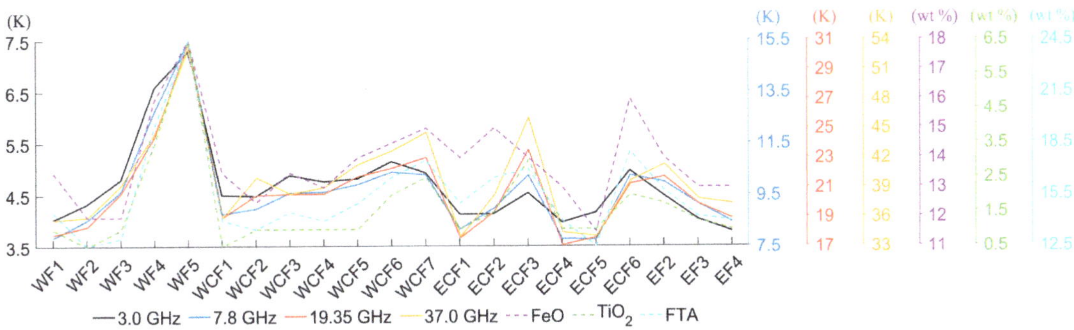

Figure 13. The SCREP-based FeO, TiO_2, and FTA values mapped by Kramer et al. [2] and the four-channel TBDs of all units.

The Pearson correlation coefficient (PCC) is one of the most widely used measures of relationships, and it can measure the linear correlation of two random variables [37,38]. To further evaluate the consistency of the pristine basalt compositions reflected by the MRM data and SCREP craters, we calculated the PCC between the TBDs and the SCREP-based FeO abundances, the TiO_2 abundances, and the FTAs estimated by Kramer et al. [2] (Figure 14). The PCC of two sets of data, X ($X_1, X_2, X_3, \ldots, X_n$) and Y ($Y_1, Y_2, Y_3, \ldots, Y_n$), is defined as the ratio between the covariance of the two sets of data and the product of their standard deviations, which is expressed as follows:

$$r_{xy} = \frac{\sum(x_i - \overline{x})\sum(y_i - \overline{y})}{\sqrt{\sum(x_i - \overline{x})^2}\sqrt{\sum(y_i - \overline{y})^2}} \tag{1}$$

where $\overline{x} = \frac{1}{n}\sum_{i=1}^{n} x_i$ is the mean value of data X and $\overline{y} = \frac{1}{n}\sum_{i=1}^{n} y_i$ is the mean value of data Y. The coefficient r_{xy} ranges from −1 to 1. If the two sets of data are directly related, r_{xy}

is positive; otherwise, it is negative. The closer the absolute value of r_{xy} is to 1, the stronger the correlation between the two sets of data, and the two sets of data do not correlate when $r_{xy} = 0$ [37,38].

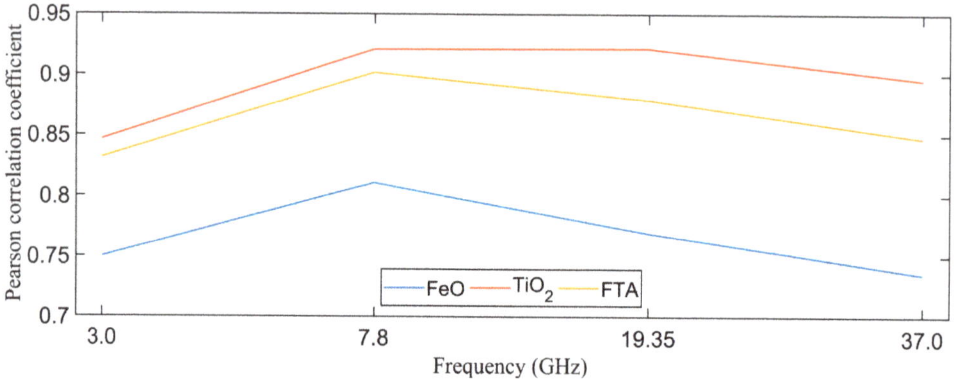

Figure 14. Pearson correlation coefficients between the TBD and the SCREP-based FeO, TiO_2, and FTA values of the basaltic units mapped by Kramer et al. [2].

In Figure 14, the average PCC is 0.77 between the four-channel TBDs and the FeO abundances of all units, 0.90 between the four-channel TBDs and the TiO_2 abundances of all units, and 0.86 between the four-channel TBDs and the FTAs of all units. Generally, the PCC results indicate that there is a very strong correlation between the TBD and the pristine basalt compositions reflected by SCREP craters.

Therefore, compared with the optical detection results, the TB, especially the TBD, is highly related to the FTAs of the surface deposits and can represent the composition and distribution features of pristine basaltic units contaminated by impact crater ejecta.

4.1.2. New Views of Basaltic Volcanism

By combining the TB results, the SCREP-based FTA, and the superposition relationship of the stratum, two interesting findings related to the basaltic volcanism in Mare Frigoris were provided.

(1) In every quadrant, the younger basaltic units have the larger FTAs.

In the WF quadrant, according to the above discussions, the FTA of the WF1 unit should be lower than that of the WF2 unit. Thus, one conclusion can be obtained: the FTA gradually increases from the oldest WF1 unit to the youngest WF5 unit.

Meanwhile, the same conclusion was not obtained in whole WCF quadrant. As mentioned in Section 3.2.2, the FTA of the WCF1 unit was overestimated and should be smaller than that of the WCF2 unit. Thus, from the younger WCF1 unit to the older WCF3 unit, the FTA gradually increased, and from the younger WCF5 unit to older WCF7 unit, this trend was also detectable. The issue exists from the WCF3 unit to the WCF5 unit, where the FTA shows a trend of initially decreasing and subsequent increasing. Here, the relationship between the compositions and age of the WCF4 unit is problematic in the WCF quadrant. Figures 11 and 12 show that there is no obvious TBD boundary between the WCF4 unit and the ECF6 unit. Therefore, if the WCF4 unit was removed from the WCF quadrant and included in the ECF6 unit, the remaining units in the WCF quadrant would satisfy observation regarding the basaltic volcanism of maria that the younger the age of the unit, the larger the FTA.

In the ECF quadrant, the new ECF1, ECF2, ECF3, and ECF6 units support the same conclusion, that is, the FTAs of older units are apparently lower than those of younger units. In the EF quadrant, if the ages estimated by Hiesinger et al. [10] are adopted, the ages of the three units still satisfy the observation that the younger the age of the basaltic unit, the larger the FTA. Therefore, in every quadrant, the FTA (($FeO + TiO_2$) abundances) of the basaltic units increases the younger they are.

(2) SCREP-based FTA results cannot completely avoid the influence of ejecta contamination. As mentioned in Section 3.2, combined with the TB results and the ages obtained by Hiesinger et al. [10] and Kramer et al. [2], we simply counted the age of each unit as shown in Tables 1–4. However, when combining the age and SCREP-based FTA results of all quadrants together, it was difficult to obtain a similar basaltic volcanism in every quadrant. In this regard, the accuracy of the estimated SCREP-based FTA was still debatable, and this can be explained thanks to the following three aspects.

(i) Tables 1 and 3 show that the difference in terms of the SCREP-based FTA between the WF3 unit and the ECF6 unit is 5 wt %, while the difference in terms of the 37-GHz TBDs is only 0.4 K. Moreover, the difference in terms of the SCREP-based FTA between the EF1 unit and the ECF1 unit is only 1.5 wt %, but the difference in terms of the 37-GHz TBDs is as much as 4.9 K (Tables 3 and 4). The same problem also exists in the case of the ECF6 and WF4 units, the EF1 and ECF1 units, and the WCF1 and WCF2 units.

(ii) In the above discussions, the TBD of the WF1 unit is lower than that of the WF2 unit, while Kramer et al. [2] showed that the FTA of the WF1 is higher than that of the WF2 unit. This circumstance also occurs in cases of the WCF1, WCF5, ECF4, and ECF5 units. Based on the theoretical simulation [24,28] and the aforementioned PCC between the TBD and the FTA, a possible reason for this issue is that Kramer et al. [2] overestimated the FTAs of the WF1, WCF1, and WCF5 units and underestimated the FTAs of the ECF4 and ECF5 units.

(iii) It has been noted that there are two large craters in Mare Frigoris: the Harpalus and Aristoteles craters. Apparently, the Harpalus crater has a great impact on the surrounding WF4 and WF5 units, which is reflected in the gradually decreasing TBD of the WF5 unit from west to east. The complex relationship between the TBDs, SCREP-based FTAs, and ages of the EF units is likely brought about by the ejecta from the Aristoteles craters.

Therefore, the TBD results suggest that the SCREP-based FTA results cannot completely avoid the influence of ejecta contamination. The combination of TBD and SCREP methodology likely provides a new way to accurately track the FTAs of basaltic deposits in the context of studying lunar basaltic volcanism in the future.

4.2. TB Anomalies in Mare Frigoris

There are two main types of TB anomalies in Mare Frigoris: the cold microwave anomaly and the warm microwave anomaly.

4.2.1. Cold Microwave Anomaly

Figures 4 and 5 show two obvious nTB anomalies in the WCF1 and WCF6 units. In the WCF1 unit, Crater A indicates low nTB values, 3.0 and 7.8 GHz at noon and at the four channels at night, while Crater B in the WCF6 unit shows low nTB values of 7.8, 19.35, and 37 GHz at night and high nTB values at the four channels at noon. They both present high TBD values. Such TB behaviors are also found in the Necho, King, Giordano Bruno, and Vavilov craters [39], and we have called this the cold microwave anomaly, which is brought about by the existence of surface rocks.

Thus, this brought about a new problem in this study: what was the influence of surface rocks on our understanding of the relationship between the nTB and TBD results and the basaltic units?

This problem was also pointed out by Bugiolacchi et al. [40], who observed that the TBD features of interest are linked to impact craters and their ejecta in the western lunar farside, resulting in a strong relationship between the size of the anomaly and the target's physical properties. Thus, a comparison between the TB behaviors in Crater A (the La Condamine S crater, 25.22°W, 57.34°N) and Crater B (an unnamed crater, 18.65°W, 60.33°N) provided a new way to evaluate the problem.

To better understand the relationship between the cold microwave anomaly and surface rocks, a rock abundance map was generated with the Diviner Radiometer data onboard the Lunar Reconnaissance Orbiter (LRO) satellite [41]. Figure 15 shows the rock abundance map of Mare Frigoris, which indicates abundant rocks appearing around Craters A and B. When combined with the findings in the Hertzsprung basin with the MRM data and the numerical simulation based on the microwave radiative transfer equations [39], the nTB and TBD results for Crater A are related to the existence of abundant rocks. Therefore, the cold microwave anomaly, which is related to the rocks in lunar highlands, is also relevant to the maria and should be taken into consideration in the study of mare deposits using MRM data.

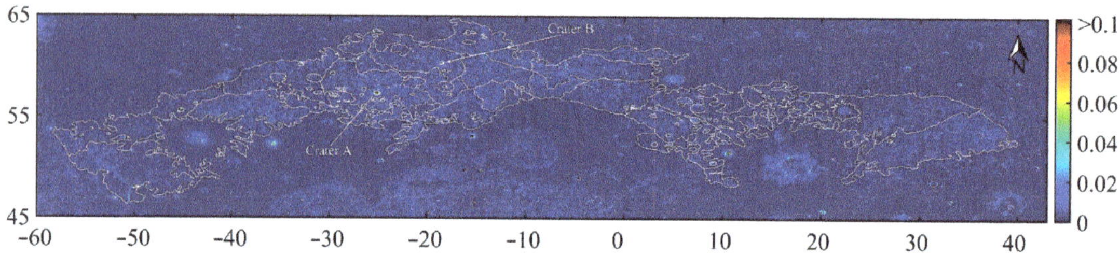

Figure 15. Rock abundance map of Mare Frigoris (equidistant cylindrical projection). The white lines are the geologic boundaries of Mare Frigoris mapped by Kramer et al. [2]. The rock abundance data were downloaded from https://ode.rsl.wustl.edu/moon/index.aspx (accessed on 21 February 2013).

However, the nTB and TBD results for Crater B are rather complicated. Here, Figure 15 shows that the rock abundance in Crater B is high, while Figures 4 and 5 show an abnormally high 3.0-GHz nTB at night. In addition to this channel, others point towards a high nTB at noon and low nTB at night, which are consistent with the high FTA of the surface deposits in the region [24,28].

Moreover, in addition to the two craters, many craters occur with high rock abundances, but few of them can be identified by the nTB and TBD maps. Therefore, at least in Mare Frigoris, the main influencing factor behind the TB remains the FTAs of surface deposits. This finding also confirms the reliability of the new results regarding the basaltic volcanism of Mare Frigoris based on nTB and TBD performances.

Even so, the influence of rocks on the TB behaviors of surface deposits, as explained by Bugiolacchi et al. [40], should be considered for younger craters and will be further studied on a global scale in future.

4.2.2. Warm Microwave Anomaly

In the above discussion, we mentioned an abnormally high 3.0-GHz nTB in Crater B at night. At 3.0 GHz, high nTB anomalies exist also in the case of the ECF1, ECF6, WCF4, and WCF6 units, which was outlined in Figure 16a. Meanwhile, the 3.0-GHz nTB is still high at noon for these units, which is defined the warm microwave anomaly in this study. In the above section, we provided the abnormal 3.0-GHz nTB results for Crater B. Figure 15 shows that rocks are abundant in Crater B, and, theoretically, the 3.0-GHz nTB should be cold both at noon and at night [39]. However, the 3.0-GHz nTB is actually high both at noon and at night in the observation maps (Figures 4 and 5), validating the rationality of the warm microwave anomaly in the case of Crater B and Mare Frigoris.

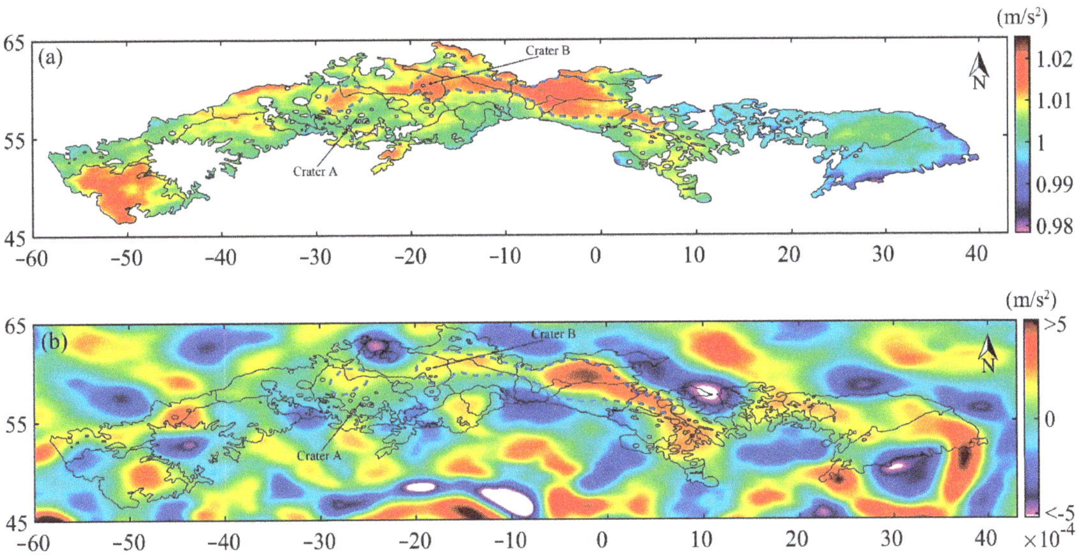

Figure 16. (**a**) 3.0-GHz normalized TB map at noon and (**b**) Bouguer gravity anomaly map of Mare Frigoris (equidistant cylindrical projection). The blue dash lines indicate the range of the warm microwave anomaly in Mare Frigoris. The black lines are the geologic boundaries of Mare Frigoris mapped by Kramer et al. [2].

The warm microwave anomaly also occurs in Mare Nubium and Mare Moscoviense [28]. Through the use of numerical simulations, Meng et al. [28] eliminated the influences of the FTA of the regolith, surface topography, and surface rocks on the warm microwave anomaly. They hypothesized that the warm microwave anomaly is probably caused by the high substrate temperature, which was always set as a constant in previous studies [18,28]. To determine the cause of the warm microwave anomaly is priority when using passive microwave data [28,42]. The distribution of the warm microwave anomaly in Mare Frigoris provides some useful information related to this work.

Generally, in Mare Frigoris, the high nTB anomaly is mainly distributed in the ECF6 unit and extends from the boundary between the ECF1 and ECF4 units, through the ECF6 and WCF4 units, to the WCF6 unit, with a linear pattern in the NWW direction. When studying the gravity map of Mare Frigoris (Figure 16b), the positive Bouguer gravity anomaly has a clear relationship with the warm microwave anomaly in terms of spatial distribution. Figure 16b shows the Bouguer gravity anomaly derived from the Gravity Recovery and Interior Laboratory (GRAIL) mission, where a positive gravity anomaly extends from the boundary between the ECF1 and ECF4 units, through the ECF6 and WCF4 units, to the WCF6 unit in a linear pattern, with this agreeing with the distribution of the warm microwave anomaly. In particular, the positive Bouguer gravity anomaly occurs at the junction of eastern WCF1 unit and WCF4 unit with a nearly east-west distribution, coinciding with the warm microwave anomaly here. In addition, a positive Bouguer gravity anomaly also occurs in the north of WCF1 unit, which also indicates the high 3.0-GHz nTB anomaly.

Interestingly, Andrews-Hanna et al. [6] indicated the existence of the linear positive gravity anomaly in Mare Frigoris and interpreted it as a part of the huge magma plumbing system located in the Procellarum KREEP Terrane (Figure 16b). Moreover, the TB can reflect the dielectric properties of the regolith within the penetration depth and it is sensitive to the substrate temperature at 3.0 GHz [17,27]. Meanwhile, because there is no air on the

Moon, abnormal TB performances are directly related to the thermal state of the shallow lunar crust and are significant in the study of the thermal evolution of the Moon.

Though the relationship between the positive Bouguer gravity anomaly, the magma pluming system, and the warm microwave anomaly is not clear, this possible relationship provides a new, noteworthy avenue of study to pursue in the context of determining the cause of the warm microwave anomaly and its potential geologic significance using the MRM data.

5. Conclusions

In this paper, the microwave thermal emission features of the basaltic units in Mare Frigoris, in addition to their (FeO + TiO_2) abundances (FTAs) and ages, were studied using normalized brightness temperature (nTB) and brightness temperature difference (TBD) maps generated from Chang'e-2 microwave radiometer (MRM) data and a Bouguer gravity anomaly map derived from GRAIL data. The main conclusions are as follows:

(1) The nTB and TBD results agree with the basaltic units based on the small crater rim and ejecta probing (SCREP) methodology. This indicates that MRM data are highly related to the (FeO + TiO_2) abundance of pristine basalts and are less influenced by ejecta contamination when compared to SCREP-based FeO and TiO_2 abundances.

(2) Based on the nTB and TBD results, the SCREP-based (FeO + TiO_2) abundance, and the superposition relationship of the stratum, it was found that, in most quadrants of Mare Frigoris, the (FeO + TiO_2) abundances of basaltic units were higher at younger ages.

(3) A comparison of the rock abundance map and the nTB and TBD results for Craters A and B indicated that, at least in Mare Frigoris, the (FeO + TiO_2) abundances of surface deposits remains the main influencing factor behind the TB.

(4) The warm microwave anomaly was revealed in the case of the ECF1, ECF6, WCF4, and WCF6 units, and showed a similar distribution to the Bouguer gravity anomaly derived from GRAIL data. This provides useful information in the context of determining the cause of the warm microwave anomaly.

Generally, in Mare Frigoris, in the context of heavy contamination due to impact crater ejecta, the MRM data verified the discoveries of Kramer et al. [2] concerning pristine basaltic deposits using SCREP craters, which are difficult to locate directly using optical data. Moreover, the MRM data provided a new description of the (FeO + TiO_2) abundances of basaltic units and indicated their possible relationship with the deep information revealed by the Bouguer gravity anomaly. More work should be undertaken to further understand the relationship between brightness temperature performances and surface deposits on the Moon.

Author Contributions: Conceptualization, Z.M. and J.L.; methodology, Z.M. and Y.W.; software, J.L.; validation, Y.W., S.H. and J.P.; formal analysis, Z.M. and Z.C.; data curation, Z.M. and Y.W.; writing—original draft preparation, J.L. and Z.M.; writing—review and editing, J.L., Z.M., Y.W., S.H., J.P., Z.C. and Y.Z.; visualization, Z.M. and Y.Z.; supervision, Y.W. and S.H.; project administration, Z.M.; funding acquisition, Z.M. and Y.W. All authors have read and agreed to the published version of the manuscript.

Funding: This research was funded by the National Key R & D Program of China (No. 2021YFA0715104), the National Natural Science Foundation of China (No. 42071309), the Strategic Priority Research Program of the Chinese Academy of Sciences (No. XDB 41000000), the Opening Fund of the State Key Laboratory of Lunar and Planetary Sciences (Macau University of Science and Technology), Macau FDCT grant number 119/2017/A3, and the Science and Technology Development Fund (FDCT) of Macau (No. 0059/2020/A2).

Data Availability Statement: In this study, the MRM data were downloaded from https://moon.bao.ac.cn/ce5web/moonGisMap.search (accessed on 1 September 2019), the Clementine UVVIS data were downloaded from https://astrogeology.usgs.gov/search/map/Moon/Clementine/UVVIS/Lunar_Clementine_UVVIS_WarpMosaic_5Bands_200m (accessed on 9 February 2022), the WAC images were downloaded from https://wms.lroc.asu.edu/lroc/search (accessed on 20 June 2010), the GRAIL data were downloaded from https://pds-geosciences.wustl.edu/grail/grail-l-lgrs-5-rdr-v1/grail_1001/shadr/ (accessed on 27 April 2016), and the Rock abundance data were downloaded from https://ode.rsl.wustl.edu/moon/index.aspx (accessed on 21 February 2013).

Acknowledgments: The authors express their thanks to the people who aided this work, the providers of the data used in this article, and acknowledge the valuable suggestions provided by the peer reviewers.

Conflicts of Interest: The authors declare no conflict of interest regarding the publication of this paper.

References

1. Whitford-Stark, J.L. The volcanotectonic evolution of Mare Frigoris. In Proceedings of the Lunar and Planetary Science Conference, Houston, TX, USA, 12–16 March 1990; pp. 175–185.
2. Kramer, G.; Jaiswal, B.; Hawke, B.; Öhman, T.; Giguere, T.; Johnson, K. The basalts of mare Frigoris. *J. Geophys. Res. Planets* **2015**, *120*, 1646–1670. [CrossRef]
3. Williams, N.R.; Bell, J.F., III; Watters, T.R.; Banks, M.E.; Daud, K.; French, R.A. Evidence for recent and ancient faulting at Mare Frigoris and implications for lunar tectonic evolution. *Icarus* **2019**, *326*, 151–161. [CrossRef]
4. Giguere, T.; Hawke, B.R.; Gaddis, L.; Gustafson, J.; Lawrence, S.; Stopar, J.; Mattson, S.; Robinson, M. Geologic studies of volcanic constructs in eastern Mare Frigoris. In Proceedings of the Lunar and Planetary Science Conference, The Woodlands, TX, USA, 17–21 March 2014; p. 1318.
5. Albin, E.; Harris, R. Volcanism Within Mare Frigoris: Further Evidence for Recent Eruptions on the Moon. In Proceedings of the Lunar and Planetary Science Conference, The Woodlands, TX, USA, 16–20 March 2015; p. 2633.
6. Andrews-Hanna, J.C.; Besserer, J.; Head, J.W., III; Howett, C.J.; Kiefer, W.S.; Lucey, P.J.; McGovern, P.J.; Melosh, H.J.; Neumann, G.A.; Phillips, R.J. Structure and evolution of the lunar Procellarum region as revealed by GRAIL gravity data. *Nature* **2014**, *514*, 68–71. [CrossRef]
7. Boyce, J.M. Ages of flow units in the lunar nearside maria based on Lunar Orbiter IV photographs. In Proceedings of the Lunar and Planetary Science Conference, Houston, TX, USA, 15–19 March 1976; pp. 2717–2728.
8. Wilhelms, D.E.; John, F.; Trask, N.J. *The Geologic History of the Moon*; USGS: Reston, VA, USA, 1987; pp. 2330–7102.
9. Staid, M.I.; Pieters, C.M. Integrated spectral analysis of mare soils and craters: Applications to eastern nearside basalts. *Icarus* **2000**, *145*, 122–139. [CrossRef]
10. Hiesinger, H.; Head, J., III; Wolf, U.; Jaumann, R.; Neukum, G. Ages and stratigraphy of lunar mare basalts in Mare Frigoris and other nearside maria based on crater size-frequency distribution measurements. *J. Geophys. Res. Planets* **2010**, *115*, E03003. [CrossRef]
11. Morota, T.; Haruyama, J.; Ohtake, M.; Matsunaga, T.; Honda, C.; Yokota, Y.; Kimura, J.; Ogawa, Y.; Hirata, N.; Demura, H. Timing and characteristics of the latest mare eruption on the Moon. *Earth Planet. Sci. Lett.* **2011**, *302*, 255–266. [CrossRef]
12. Kramer, G.Y. Characterizing bedrock lithologies using small crater rims and ejecta probing (SCREP). *Adv. Space Res.* **2010**, *45*, 1257–1267. [CrossRef]
13. Kramer, G.Y.; Combe, J.P.; Harnett, E.M.; Hawke, B.R.; Noble, S.K.; Blewett, D.T.; McCord, T.B.; Giguere, T.A. Characterization of lunar swirls at Mare Ingenii: A model for space weathering at magnetic anomalies. *J. Geophys. Res. Planets* **2011**, *116*, E04008. [CrossRef]
14. Weider, S.Z.; Crawford, I.A.; Joy, K.H. Individual lava flow thicknesses in Oceanus Procellarum and Mare Serenitatis determined from Clementine multispectral data. *Icarus* **2010**, *209*, 323–336. [CrossRef]
15. Campbell, B.A.; Carter, L.M.; Hawke, B.R.; Campbell, D.B.; Ghent, R.R. Volcanic and impact deposits of the Moon's Aristarchus Plateau: A new view from Earth-based radar images. *Geology* **2008**, *36*, 135–138. [CrossRef]
16. Fang, T.; Fa, W. High frequency thermal emission from the lunar surface and near surface temperature of the Moon from Chang'E-2 microwave radiometer. *Icarus* **2014**, *232*, 34–53. [CrossRef]
17. Meng, Z.; Hu, S.; Wang, T.; Li, C.; Cai, Z.; Ping, J. Passive microwave probing mare basalts in mare imbrium using CE-2 CELMS data. *IEEE J. Sel. Top. Appl. Earth Obs. Remote Sens.* **2018**, *11*, 3097–3104. [CrossRef]
18. Jin, Y.; Yan, F.; Liang, Z. Simulation of brightness temperature from the lunar surface using multi-channels microwave radiometers. *Chin. J. Radio Sci.* **2003**, *18*, 477–486.
19. Hu, G.; Chan, K.L.; Zheng, Y.; Tsang, K.T.; Xu, A. Comparison and evaluation of the Chang'E microwave radiometer data based on theoretical computation of brightness temperatures at the Apollo 15 and 17 sites. *Icarus* **2017**, *294*, 72–80. [CrossRef]
20. Mustard, J.F.; Head, J.W. Buried stratigraphic relationships along the southwestern shores of Oceanus Procellarum: Implications for early lunar volcanism. *J. Geophys. Res. Planets* **1996**, *101*, 18913–18925. [CrossRef]

21. Xie, M.; Xiao, Z.; Zhang, X.; Xu, A. The provenance of regolith at the Chang'e-5 candidate landing region. *J. Geophys. Res. Planets* **2020**, *125*, e2019JE006112. [CrossRef]
22. Zheng, Y.; Tsang, K.; Chan, K.; Zou, Y.; Zhang, F.; Ouyang, Z. First microwave map of the Moon with Chang'E-1 data: The role of local time in global imaging. *Icarus* **2012**, *219*, 194–210. [CrossRef]
23. Zheng, Y.; Chan, K.L.; Tsang, K.T.; Zhu, Y.; Hu, G.P.; Blewett, D.T.; Neish, C. Analysis of Chang'E-2 brightness temperature data and production of high spatial resolution microwave maps of the Moon. *Icarus* **2019**, *319*, 627–644. [CrossRef]
24. Fa, W.; Jin, Y. A primary analysis of microwave brightness temperature of lunar surface from Chang-E 1 multi-channel radiometer observation and inversion of regolith layer thickness. *Icarus* **2010**, *207*, 605–615. [CrossRef]
25. Cai, Z.; Lan, T. Lunar brightness temperature model based on the microwave radiometer data of Chang'E-2. *IEEE Trans. Geosci. Remote Sens.* **2017**, *55*, 5944–5955. [CrossRef]
26. Chan, K.L.; Tsang, K.T.; Kong, B.; Zheng, Y. Lunar regolith thermal behavior revealed by Chang'E-1 microwave brightness temperature data. *Earth Planet. Sci. Lett.* **2010**, *295*, 287–291. [CrossRef]
27. Meng, Z.; Chen, S.; Zheng, Y.; Cheng, W.; Zhu, Y.; Cai, Z.; Zhang, Y.; Cao, W.; Hou, L. Mare Deposits Identification and Feature Analysis in Mare Australe Based on CE-2 CELMS Data. *J. Geophys. Res. Planets* **2020**, *125*, e2019JE006330. [CrossRef]
28. Meng, Z.; Chen, S.; Wang, Y.; Wang, T.; Cai, Z.; Zhang, Y.; Zheng, Y.; Hu, S. Reevaluating Mare Moscoviense and its vicinity using Chang'E-2 microwave sounder data. *Remote Sens.* **2020**, *12*, 535. [CrossRef]
29. Hu, G.; Chan, K.L.; Zheng, Y.; Xu, A. A rock model for the cold and hot spots in the Chang'E microwave brightness temperature map. *IEEE Trans. Geosci. Remote Sens.* **2018**, *56*, 5471–5480. [CrossRef]
30. Heiken, G.H.; Vaniman, D.T.; French, B.M. *Lunar Sourcebook, a User's Guide to the Moon*; Cambridge University Press: New York, NY, USA, 1991.
31. Whitford-Stark, J.L.; Fryer, R. Origin of Mare Frigoris. *Icarus* **1975**, *26*, 231–242. [CrossRef]
32. Hiesinger, H.; Head, J., III; Wolf, U.; Jaumann, R.; Neukum, G. Ages and stratigraphy of mare basalts in oceanus procellarum, mare nubium, mare cognitum, and mare insularum. *J. Geophys. Res. Planets* **2003**, *108*, 5065. [CrossRef]
33. Lucey, P.G.; Blewett, D.T.; Jolliff, B.L. Lunar iron and titanium abundance algorithms based on final processing of Clementine ultraviolet-visible images. *J. Geophys. Res. Planets* **2000**, *105*, 20297–20305. [CrossRef]
34. Meng, Z.; Yang, G.; Ping, J.; Cai, Z.; Gusev, A.; Osei, E.M. Influence of (FeO+ TiO_2) abundance on the microwave thermal emissions of lunar regolith. *Sci. China Earth Sci.* **2016**, *59*, 1498–1507. [CrossRef]
35. Meng, Z.; Lei, J.; Qian, Y.; Xiao, L.; Head, J.W.; Chen, S.; Cheng, W.; Shi, J.; Ping, J.; Kang, Z. Thermophysical Features of the Rümker Region in Northern Oceanus Procellarum: Insights from CE-2 CELMS Data. *Remote Sens.* **2020**, *12*, 3272. [CrossRef]
36. Campbell, B.A.; Hawke, B.; Campbell, D.B. Surface morphology of domes in the Marius Hills and Mons Rümker regions of the Moon from Earth-based radar data. *J. Geophys. Res. Planets* **2009**, *114*, E01001. [CrossRef]
37. Lee Rodgers, J.; Nicewander, W.A. Thirteen ways to look at the correlation coefficient. *Am. Stat.* **1988**, *42*, 59–66. [CrossRef]
38. Zhou, H.; Deng, Z.; Xia, Y.; Fu, M. A new sampling method in particle filter based on Pearson correlation coefficient. *Neurocomputing* **2016**, *216*, 208–215. [CrossRef]
39. Meng, Z.; Lei, J.; Xiao, Z.; Cao, W.; Cai, Z.; Cheng, W.; Feng, X.; Ping, J. Re-Evaluating Influence of Rocks on Microwave Thermal Emission of Lunar Regolith Using CE-2 MRM Data. *IEEE Trans. Geosci. Remote Sens.* **2022**, *60*, 1–12. [CrossRef]
40. Bugiolacchi, R.; Hu, G.P. The Moon in the Microwave: Shedding New Light on the Lunar Farside. *J. Geophys. Res. Planets* **2022**, *127*, e2021JE007039. [CrossRef]
41. Bandfield, J.L.; Ghent, R.R.; Vasavada, A.R.; Paige, D.A.; Lawrence, S.J.; Robinson, M.S. Lunar surface rock abundance and regolith fines temperatures derived from LRO Diviner Radiometer data. *J. Geophys. Res. Planets* **2011**, *116*, E00H02. [CrossRef]
42. Keihm, S.J. Interpretation of the lunar microwave brightness temperature spectrum: Feasibility of orbital heat flow mapping. *Icarus* **1984**, *60*, 568–589. [CrossRef]

remote sensing

MDPI

Article

Spectroscopic and Petrographic Investigations of Lunar Mg-Suite Meteorite Northwest Africa 8687

Lang Qin [1,2], Xing Wu [1,*], Liying Huang [1,2], Yang Liu [1,3] and Yongliao Zou [1]

1 State Key Laboratory of Space Weather, National Space Science Center, Chinese Academy of Sciences, Beijing 100190, China; qinlang@nssc.ac.cn (L.Q.); huangliying21@mails.ucas.ac.cn (L.H.); yangliu@nssc.ac.cn (Y.L.); zouyongliao@nssc.ac.cn (Y.Z.)
2 University of Chinese Academy of Sciences, Beijing 100049, China
3 Center for Excellence in Comparative Planetology, Chinese Academy of Sciences, Hefei 230026, China
* Correspondence: wuxing@nssc.ac.cn

Abstract: Magnesian suite (Mg-suite) rocks represent plutonic materials from the lunar crust, and their global distribution can provide critical information for the early magmatic differentiation and crustal asymmetries of the Moon. Visible and near-infrared (VNIR) spectrometers mounted on orbiters and rovers have been proven to be powerful approaches for planetary mineral mapping, which are instrumental in diagnosing Mg-suite rocks. However, due to the scarcity and diversity of Mg-suite samples, laboratory measurements with variable proportions of minerals are imperative for spectral characterization. In this study, spectroscopic investigation and petrographic study were conducted on lunar Mg-suite meteorite Northwest Africa 8687. We classify the sample as a pink spinel-bearing anorthositic norite through spectral and petrographic characteristics. The ground-truth information of the Mg-suite rock is provided for future exploration. Meanwhile, the results imply that the VNIR technique has the potential to identify highland rock types by mineral modal abundance and could further be applied in extraterrestrial samples for primary examination due to its advantage of being fast and non-destructive.

Keywords: meteorite; Moon; spectroscopy; Mg-suite; anorthositic norite

1. Introduction

Lunar meteorites and samples returned by the Apollo and Luna missions and the recent Chang'E-5 program provide significant insights into the formation and evolution of the Moon (e.g., [1–3]). Beyond that, large-scale remote-sensing observations have greatly revolutionized our understanding of the unsampled place (e.g., [4]). Early evolutionary models based on sample studies predicted the existence of a globally distributed layer enriched in incompatible elements (e.g., potassium (K), rare earth elements (REE), and phosphorus (P); referred to as KREEP) produced by the fractional crystallization of the lunar magma ocean (LMO) [5,6]. However, global remotely sensed data reveal that only a restrictive area on the lunar nearside shows KREEPy enrichment. This distinct province is the so-called Procellarum KREEP Terrane (PKT), since the region is coincidentally located in the Procellarum and Imbrium impact basin [4,7,8]. The lateral extent and distribution of KREEP are controversial, and the crustal asymmetry would be essential for understanding the planetary differentiation at large scales.

Ancient crustal samples bear critical information on lunar magmatic differentiation [3]. These nonmare igneous rocks, which record the early history of the Moon, mainly include the ferroan anorthosites (FANs) and magnesian suite (Mg-suite) rocks [9–11]. The FANs formed the primary crust by accumulating from the global magma ocean and are the main components (probably 80%) of lunar highlands [12]. Mg-suite rocks, also known as Mg-rich plutonic rocks, likely represent materials from the lower lunar crust [3]. Due to their deep crustal origin, few outcrops are present on the lunar surface, and the Mg-rich rocks account for merely a tiny proportion of lunar samples (around 0.7% of 381.7 kg samples returned by the Apollo programs [10]). Despite this, the global distribution of Mg-suite rocks is crucial for constraining its petrogenesis and the last evolutionary stage proposed by the LMO hypothesis [13]. Early research on the Apollo returned sample suggested that Mg-suite magmatism was strongly controlled by KREEPy magma [3,13,14]. Namely, the distribution should be closely associated with the geochemical anomalous PKT [4]. Nevertheless, the trace elemental gap between lunar meteorites and the Apollo samples [15–17], along with new laboratory models [18], indicate that high magnesian rocks can be crystallized from KREEP-free melts; that is, the distribution of Mg-suite rocks could be Moon-wide. Whether Mg-suite rocks are widespread and their petrogenetic link to the last dregs of magma ocean holds the keys to unlocking the mysteries of their origin and asymmetries on the planetary body, while it remains an unsolved enigma in lunar science.

The possible Mg-suite materials on the lunar surface can be identified by discerning varied pyroxene- and olivine-bearing assemblages and Mg-spinel [13,19–21]. Visible and near-infrared (VNIR) spectroscopy has the potential to differentiate various minerals [13,22]; hence, it has attracted increasing attention over the past decades. Several spectral surveys identified Mg-suite clusters in the eastern South Pole–Aitken (SPA) and the northern rim of the Imbrium Basin [13]. Cahill et al. [23] detected Mg-suite exposures in central peaks of craters. However, the identical spectra could lead to different interpretations. For instance, the mineral compositions of lunar rocks in the SPA decoded from the VNIR spectra of the Yutu-2 rover yielded significant differences [24,25]. It is challenging to validate the mapping results. The ground-truth information is essential in calibrating and interpreting spectral features, while extensive laboratory measurements of variable proportions of minerals are lacking, due to the scarcity and diversity of Mg-suite rocks. Furthermore, previous laboratory spectral studies have mainly focused on bulk samples in powder form, which is difficult to interpret the spectral characteristics of individual clasts precisely [26].

In contrast to returned samples from limited areas, lunar meteorites represent random samplings of the lunar surface [1]. This allows meteoritic samples to contribute to a more diverse lithological dataset. Moreover, due to the rarity of extraterrestrial samples, non-destructive methods need to be explored for future analysis. To achieve these aims, we followed the efforts of [26–28], and conducted a spectral survey on lunar meteorite Northwest Africa (NWA) 8687. To well-characterize the sample, a careful petrographic study was performed. Some of the highlights of this study are as follows:

- A new compilation of mineral chemistry of NWA 8687 is reported. Based on detailed analyses of mineral composition and spectral investigation, the sample NWA 8687 is confirmed as a lunar Mg-suite anorthositic norite.
- The coordinated reflectance spectra of NWA 8687 are acquired. Unlike the VNIR spectra of bulk samples collected in powder form, the measurement of NWA 8687 was performed on a chip, which facilitates spectral analysis of individual rocks and clasts.
- The VNIR spectroscopy is helpful for both the in situ characterization of lunar rocks and non-destructive laboratory characterization of returned samples and lunar meteorites.

2. Materials and Methods

2.1. Lunar Meteorite Northwest Africa 8687

Northwest Africa (NWA) 8687 was found in a hot desert within Morocco in April 2014, and it was broken into 5 pieces, weighing a total of 563 g [29]. Based on textural and chemical similarities, NWA 8687 is likely to be paired to the lunar granulitic anorthositic troctolite breccia NWA 5744 clan (including paired stones NWA 5744, NWA 8599, NWA 10140, NWA 10178, NWA 10318, NWA 10401, and NWA 11252) [30–33].

The saw-cut surface of the sample used for point spectroscopy analysis in this study is greenish-white, with large mineral grains and dark impact melt veins visible (Figure 1). After obtaining the spectral data, a 1-inch polished thin section made with the same chip was prepared for petrological analysis, and the area for research is around 16 × 14 mm.

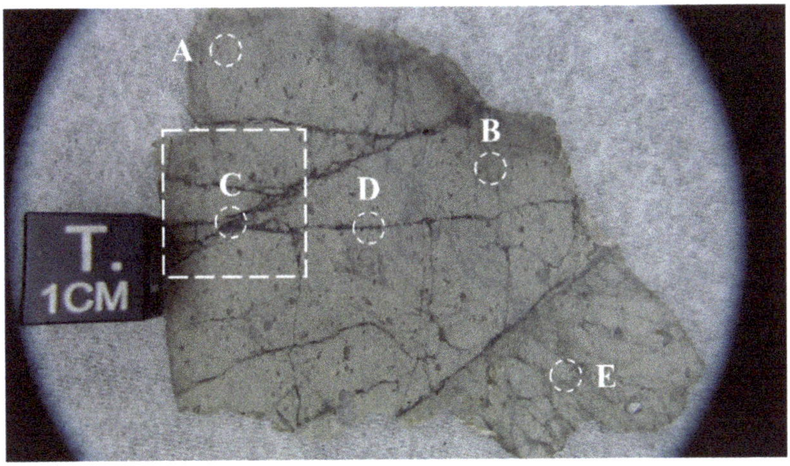

Figure 1. Photograph of lunar meteorite Northwest Africa 8687 and the location of point spectral measurements (**A**–**E**) in this study. The frame shows the rough location of the polished thin section for the petrographic study.

2.2. Visible and Near-Infrared Reflectance Spectroscopy

The VNIR spectra were acquired by the PSR-3500 portable spectrometer (Spectral Evolution Inc., Lawrence, MA, USA), which covers a spectral range of 350–2500 nm, and offers a spectral resolution of 3.5 nm at 700 nm, 10 nm at 1500 nm, and 7 nm at 2100 nm [34]. The number of spectral bands is 1024. The spectralon produced by Labsphere was used to provide the absolute reflectance factor for spectral measurements. A 50 W halogen lamp mounted 60 cm above the samples served as the light source in the darkroom. To facilitate comparison with other lunar samples' spectra, i.e., spectra from the Reflectance Experiment Laboratory (RELAB) database, we set the incidence and emergence angles as 30° and 0°, respectively. A total of 10 scans were averaged for each spectrum in order to improve the signal-to-noise ratio. The fields of view of five spectral measurements are shown in Figure 1.

2.3. Petrography and Mineralogical Analysis

2.3.1. Scanning Electron Microscope

The backscattered electron (BSE) image of the thin section of NWA 8687 was captured by the Thermofisher Apreo field emission scanning electron microscope (FE-SEM) at the Institute of Geology and Geophysics, Chinese Academy of Sciences (IGGCAS). The instrument is equipped with a QUANTAX energy dispersive spectrometer (EDS) for obtaining Kα X-ray element maps under 15 kV and 13 nA operating conditions. The X-ray acquisition time is 8 ms per pixel, and the spacing is 10 μm.

Mineral modal abundance is achieved by automated mineral identification based on X-ray maps via the software NanoMin. To compare the result, we used the software ImageJ to count the gray values of the BSE image, based on the principle that the values are related to chemical compositions, and can be used as proxies for minerals.

2.3.2. Electron Probe Micro Analyzer

The quantitative analysis of main mineral phases was carried out by electron probe microanalysis (EPMA) using the JEOL JXA-8100 at the Institute of Geology, Chinese Academy of Geological Sciences (CAGS). The acceleration voltage of the EPMA is 15 kV, and the beam current is 20 nA. The focused beams measured 1 to 5 μm, adjusted by the size of the mineral grains. Natural minerals of known compositions were used as standards for the corresponding minerals, and the standard used for impact melt veins was consistent with plagioclase. All data were corrected using the ZAF method.

The bulk composition was calculated by the weight percentage of mineral phases and their average point analysis, using the equation:

$$Bulk = \frac{\sum \chi \rho c}{\sum \chi \rho} \quad (1)$$

where χ = mineral phase abundance; ρ = mineral phase density; c = average element concentration.

2.4. Spectral Analysis

2.4.1. Spectral Parameters

The acquired spectrum was first smoothed using the Savitzky–Golay filter with a window size of 11 bands and a polynomial order of 2 [35]. The continuum removal was conducted to highlight absorption features while removing the baseline [36]. The band center and area are widely used to differentiate the mafic minerals [37,38]. To estimate the exact band center and area, we employed a sixth-order polynomial fitting to the absorption regions around ~1 μm and ~2 μm of the continuum-removed spectra [39]. The band center is defined as the wavelength with the minimum value according to the polynomial function. The band area is defined as the area covered by the absorption band of a given absorption feature. The band area ratio (BAR) is the ratio of the Band II (2 μm) area to the Band I (1 μm) area.

2.4.2. Spectral Similarity Measurement

We collected all available RELAB reflectance spectra of lunar samples and searched for the best matches between NWA 8687 spectra and the library spectra. The spectral angle mapper (SAM) is a tool that permits the rapid mapping of spectral similarity. The SAM algorithm (Equation (2)) determines the similarity of two spectra by calculating the angle parameter θ (in degree) between them [40]. Smaller spectral angles indicate a higher similarity between the NWA 8687 spectra and the laboratory spectra. The reason for choosing the SAM technique is that it concentrates on the spectral shape and is insensitive to absolute reflectance associated with illumination or albedo effects.

$$\theta = \cos^{-1}\left(\frac{\sum_{i=1}^{L} x_i y_i}{\sqrt{\sum_{i=1}^{L} x_i^2}\sqrt{\sum_{i=1}^{L} y_i^2}}\right) \quad (2)$$

where x and y are spectra from the RELAB database and NWA 8687, respectively.

As the spectra of NWA 8687 and the RELAB database were measured using different spectrometers, we performed the spectral resampling first to ensure the success of the comparison using SAM. A total of 884 RELAB spectra of returned samples and lunar meteorites have been investigated. We ranked the matching results in ascending order of the spectral angle. The spectrum with the smallest spectral angle was selected as the best match.

2.4.3. Spectral Unmixing

The signal collected by a spectrometer is often a mixture of light scattered by substances located in the field of view [41]. To separate the observed spectrum into a collection of pure endmember spectra and their respective abundances, both the linear and nonlinear mixing models were proposed. The reflectance spectra of meteorite chip surfaces can be regarded as linear mixtures, since minerals are separated from one another on a much larger scale than the average grain size and wavelength, so that reflected light from different minerals do not interact significantly [27]. The observed spectrum $\boldsymbol{y}$ can be represented as follows:

$$\boldsymbol{y} = \boldsymbol{M\alpha} + \boldsymbol{n} \tag{3}$$

where $\boldsymbol{y} \in R^{L\times 1}$ is the spectral vector with L bands, $\boldsymbol{M} \in R^{L\times q}$ is the endmember matrix containing q pure endmembers, $\boldsymbol{\alpha} \in R^{q\times 1}$ is the abundance vector, and $\boldsymbol{n} \in R^{L\times 1}$ denotes the additive perturbation due to spectral noise and modeling error. The abundance vector should follow two physical constraints, namely the abundance nonnegativity constraint (ANC) and the abundance sum-to-one constraint (ASC):

$$\boldsymbol{\alpha} \geq 0 \tag{4}$$

$$\mathbf{1}^T \boldsymbol{\alpha} = 1 \tag{5}$$

where $\mathbf{1}^T$ is a line vector of 1's compatible with $\boldsymbol{\alpha}$.

In this work, we used the linear unmixing model with two physical constraints [42] to derive mineral abundances. Plagioclase, pyroxene, olivine, and ilmenite were used as endmembers, for the reason that they are the four major minerals of the Moon [43]. The pyroxene contains clinopyroxene and orthopyroxene, and the plagioclase includes plagioclase glass (maskelynite). The endmembers were manually selected based on the petrographic results and are exhibited in Figure 2. The spectra of endmembers were collected in powder form, while that of NWA 8687 were in chip form. For meteorite chips, the reflectance of the left shoulder of 1 μm absorption is greater than the right shoulder of 2 μm absorption. Thus, measuring the samples as chips rather than particulates led to a pronounced negative continuum slope [26]. The continuum removal was performed on both the endmember and NWA 8687 spectra first, and then the linear unmixing was conducted on the continuum removed spectra.

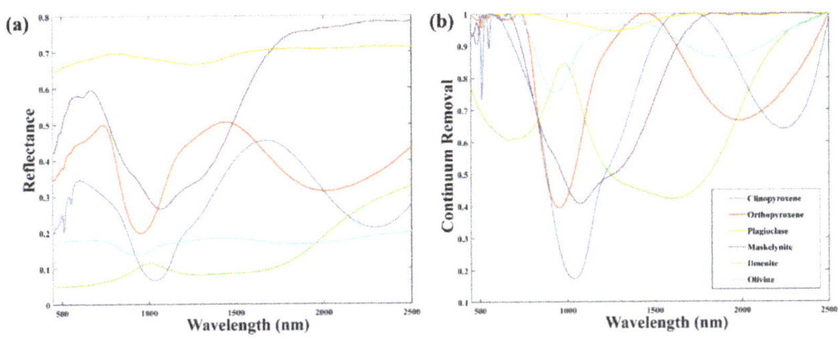

Figure 2. (**a**) Visible and near-infrared (VNIR) reflectance spectra and (**b**) spectra after continuum removal of endmembers used for spectral unmixing. Spectral ID: clinopyroxene (C1DL84A), orthopyroxene (C1LR209), plagioclase (C1LR223), maskelynite (C1LR222), ilmenite (C1LR222), and olivine (C2PO81).

3. Results

3.1. Petrography and Mineral Modal Abundance

The meteorite NWA 8687 shows a porphyritic texture, with large phenocrysts embedded in the fine-grained matrix and impact melt veins crosscutting the samples (Figure 3). Few clasts are present in the sample, indicating that NWA 8687 is a monomict breccia.

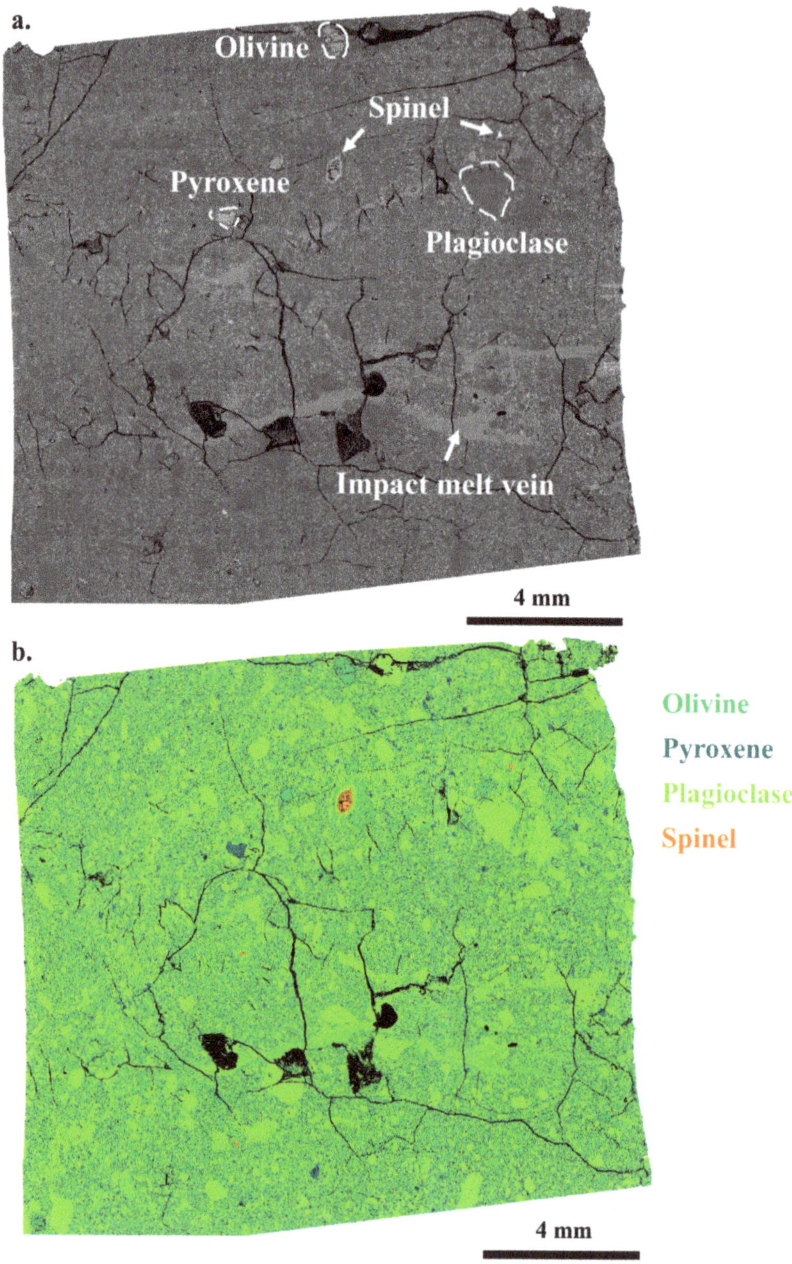

Figure 3. (**a**) backscattered electron (BSE) image of a polished section of NWA 8687. (**b**) Distribution of main minerals in NWA 8687 based on X-ray element maps.

The sample in this study mainly consists of plagioclase (or maskelynite; 67.1–71.8 vol.%), pyroxene (16.5–20.8 vol.%), and olivine (11.6–11.8 vol.%), along with minor spinel (0.1–0.2 vol.%) (Table 1). Accessory minerals include Fe–Ni metals (0.07 vol.%), troilites

(0.04 vol.%), ilmenites (0.04 vol.%), and baddeleyites (0.0002 vol.%). Compared with pairs NWA 5744 and NWA 10401 [15,32], NWA 8687 has more pyroxene and less olivine.

Table 1. Mineral modal abundance in the lunar meteorites NWA 8687 compared to pairs NWA 5744 and NWA 10401 (in vol. %).

Method [3]	NWA 8687		NWA 5744 [1]	NWA 10401 [2]	
	NanoMin	BSE	BSE	XMapTool	IDRISI Selva
Plagioclase *	67.10	71.8	67	59	65
Olivine	11.80	11.6	26.3	26	23
Total pyroxene *	20.77	16.5	6.6	15	12
High-Ca pyroxene	1.03	-	-	-	-
Low-Ca pyroxene	19.74	-	-	-	-
Spinel/chromite	0.18	0.1	0.1	0.001	0.001
Metal	0.07	-	-	-	-
Troilite	0.04	-	-	-	-
Ilmenite	0.04	-	-	-	-
Baddeleyite	0.0002	-	-	-	-
Total	100	100	100	100	100

* Plagioclase includes plagioclase glass (maskelynite). Total pyroxene is the sum of high-Ca pyroxene and low-Ca pyroxene. [1] Data from Ref. [32]. [2] Data from Ref. [15]. [3] Abbreviations represent modal abundance calculated by NanoMin, XMapTool, IDRISI software, or by backscattered electron (BSE) image grayscale statistics.

3.2. Mineral Chemistry

Major elements of the four major mineral phases in NWA 8687 were analyzed in this study. The representative analyses of individual minerals and estimated bulk composition are given in Table 2 and Supplementary Tables S1–S5.

Table 2. Representative EPMA analysis of olivine, pyroxene, plagioclase (or maskelynite), spinel, and impact melt vein, and estimated bulk composition of lunar meteorite NWA 8687 (in wt.%).

	Olivine			Pyroxene		Plagioclase		Spinel		Vein	Bulk [1]
n	Grain	Matrix	Vein	Low-Ca	High-Ca	Grain	Matrix	Mg-Al	Cr-Fe		
	17	9	5	14	12	8	6	5	4	6	
SiO_2	38.73	38.69	38.65	53.96	51.31	45.03	44.78	0.46	0.45	46.03	46.16
TiO_2	0.04	0.08	0.06	0.69	1.30	0.02	0.03	0.46	5.41	0.29	0.20
Al_2O_3	0.08	0.12	0.02	1.34	2.09	34.84	35.04	57.36	19.53	23.41	22.36
Cr_2O_3	0.04	0.04	0.04	0.40	0.67	0.01	0.04	8.93	37.78	0.15	0.17
V_2O_3	-	-	-	-	-	-	-	0.23	0.51	-	0.001
FeO	20.98	21.15	20.60	12.52	7.77	0.13	0.18	15.59	26.85	5.40	5.82
MnO	0.23	0.23	0.23	0.23	0.18	0.01	0.01	0.13	0.30	0.08	0.09
MgO	39.26	38.94	39.83	27.47	16.31	0.12	0.08	16.42	8.44	10.28	11.66
CaO	0.12	0.17	0.08	2.64	19.81	19.17	19.03	0.07	0.12	13.65	12.85
NiO	0.01	0.01	0.01	-	-	-	-	-	-	-	0.001
CoO	0.04	0.04	0.04	-	-	-	-	-	-	-	0.01
Na_2O	0.01	0.02	0.02	0.02	0.08	0.36	0.37	-	-	0.27	0.24
K_2O	0.01	0.01	0.01	0.01	0.01	0.01	0.02	-	-	0.01	0.01
P_2O_5	0.01	0.02	0.01	0.02	0.02	0.01	0.02	-	-	0.03	0.02
Total	99.56	99.53	99.62	99.31	99.55	99.72	99.61	99.66	99.39	99.59	99.57

[1] Bulk composition is estimated by Equation (1).

3.2.1. Pyroxene

Pyroxene is the most abundant mafic mineral in NWA 8687. It appears throughout the sample as orthopyroxene (OPX; 19.74 vol.%) and a small amount of clinopyroxene (CPX; 1.03 vol.%). These two phases belong to low-Ca pyroxenes and high-Ca pyroxene, respectively. The CPX presented as phenocrysts in the slab, along with OPX, whereas no exsolution texture was seen. In the matrix, low-Ca pyroxene and olivine usually grew together, surrounded by anhedral plagioclase glass (maskelynite) (Figure 4c).

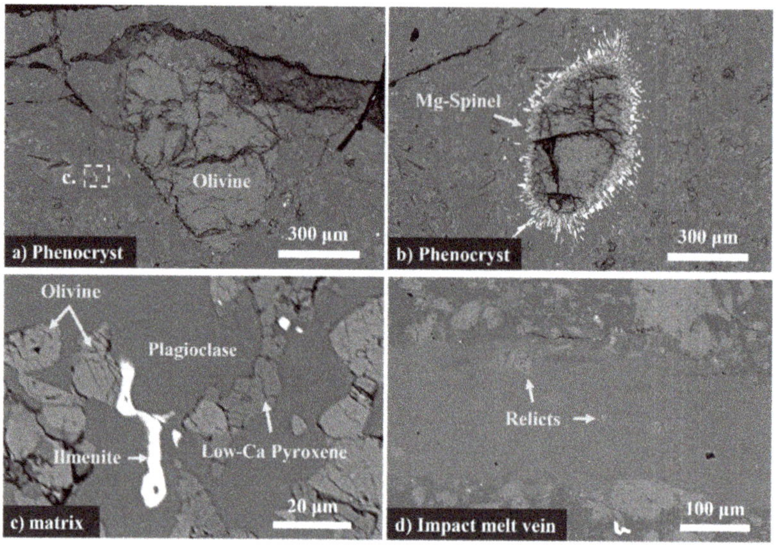

Figure 4. Backscattered electron (BSE) images of (**a**) an olivine phenocryst. The frame shows the location of figure (**c**). (**b**) A Mg-spinel in the matrix. (**c**) Fine-grained matrix, and (**d**) some mineral relicts are visible in the impact melt vein.

The composition of OPX in NWA 8687 is Mg-rich (En_{65-78} Fs_{17-20} Wo_{3-20}; $n = 14$), while the CPX has a compositional range of En_{43-50} Fs_{10-16} Wo_{38-43} ($n = 12$) (Figure 5a). The Mg number (Mg#; molar Mg/(Mg + Fe) × 100) in pyroxene ranges from 72.8 to 83.0, with an average of 79.3. The pyroxene composition of NWA 8687 is consistent with that of NWA 5744 and NWA 10401 [15,32], apart from a slightly lower calcium content.

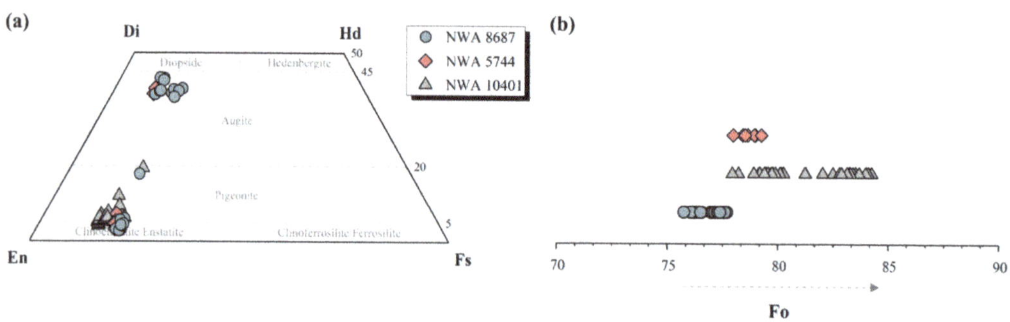

Figure 5. Pyroxene (**a**) and olivine (**b**) composition for NWA 8687, compared with a counterpart from pairs NWA 5744 and NWA 10401 [15,32]. Di—diopside, Hd—hedenbergite, En—enstatite, Fs—ferrosilite, and Fo—forsterite.

3.2.2. Olivine

Phenocryst olivine was distributed occasionally in the matrix, with grain sizes up to 700 μm (Figure 4a). Fine-grained olivine was also ubiquitous in the matrix, with varying sizes of several to tens of micrometers (Figure 4c). Cracks were on the surface of each olivine grain.

The composition of olivine in NWA 8687 is identical and forsteritic, with a high Mg# of 76.9 on average. The forsterite (Fo) contents of the 31 selected olivine grains varied between 75.8 to 77.7 mol%, with an average of 76.9 ± 0.6 mole%. However, compared to the NWA 5744 clan, olivine in NWA 8687 is relatively depleted in Mg (Figure 5b).

3.2.3. Plagioclase and Maskelynite

The feldspar in NWA 8687 is predominantly plagioclase (Ca-rich feldspar), with a narrow range of compositions ($An_{95·6–97·5}$ $Ab_{2·4–4·2}$ $Or_{0–0·2}$) (n = 14). Some of them exist in the sample as phenocryst (plagioclase) (Figure 3), and some are devitrified to glass (maskelynite). This article will not distinguish these two phases particularly.

3.2.4. Spinel

Spinel is a general term for a series of oxide minerals with diverse compositions, which can be represented by the spinel multicomponent prism [44]. It is common in mare basalts, and most lunar spinel is iron-rich, with components falling between chromite and ilmenite [43]. Remote-sensing observations have shown that Mg-rich spinel (also known as pink spinel) are ubiquitous on the lunar surface [20,21], while spinel of this composition is rare in lunar samples, most of which are highland rocks (e.g., pink spinel troctolite) [43].

A magnesian spinel surrounded by chromites and ilmenites in NWA 8687 is shown in Figure 4b. The Cr# (molar Cr/(Cr +Al) × 100) and Mg# (molar Mg/(Mg + Fe) × 100) of the Mg-spinel are in the range of pink spinel (Figure 6), although it still does not reach the component observed by the M^3 (Mg# > 90, Cr# < 5) [21].

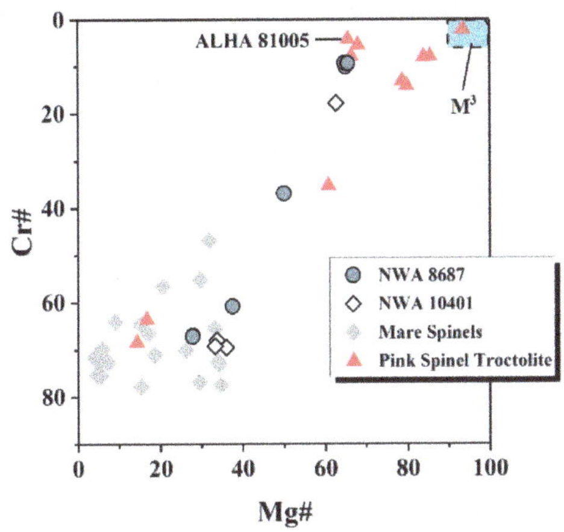

Figure 6. Spinel composition in NWA 8687 relative to compositional range from M^3 (after Ref. [21]); also compared with spinel in pair stone NWA 10401 [15], as well as in mare basalts [45] and pink spinel troctolites [46–54].

3.2.5. Impact Melt Veins

Several impact melt veins presented as crosscutting the NWA 8687, and there were mineral relicts in veins that were not completely melted (Figure 4d). A total of six spots were analyzed in the study, and the result was shown to be chemically homogenous. Notably, the average composition of these impact melt veins is almost consistent with that of the entire rock (Table 2), with an Mg number of 78.

3.3. Minerlogy Inferred from VNIR Spectroscopy

3.3.1. Qualitative Analysis

As demonstrated in Figure 7, the VNIR spectra of NWA 8687 are characterized by two broad and strong absorptions at ~900 nm and ~1900 nm, demonstrating the presence of low-Ca pyroxene. The weak absorption at ~1250 nm indicates the occurrence of abundant plagioclase in NWA 8687. This feature can be easily muted by mafic minerals. For example, the spectral mixture model shows that only 2 vol.% pyroxene could mask the ~1250 nm absorption in plagioclase [55]. The weak absorption at ~1300 is indicative of olivine.

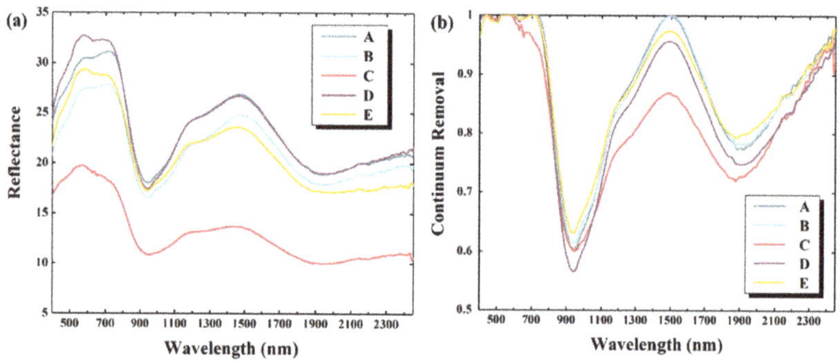

Figure 7. (**a**) Visible and near-infrared (VNIR) reflectance spectra of NWA 8687 chip surface spots marked in Figure 1. (**b**) VNIR spectra after continuum removal.

The best-matched spectra are S3LS07 and CFLM34 (spectrum ID), with the spectral angle of 1.91° and 3.18°, taken separately from the returned sample 60019.214, and lunar meteorite Allan Hills A81005 (ALHA 81005) (Figure 8a). Sample 60019 was returned by the Apollo 16 mission, and it has been classified as an ancient regolith breccia [56]. According to the geochemistry data documented by RELAB, the Mg number of sample 60019.214 is 69.2, together with an Al_2O_3 content of 26.3%, indicating that the sample has the characteristics of FANs instead of Mg-suite rocks. The ALHA 81005 is a regolith breccia with various clasts, such as basalts, FANs, and Mg-suite rocks [57], whereas the mineral modal abundance and composition of the specified sample used for collected spectra are not available.

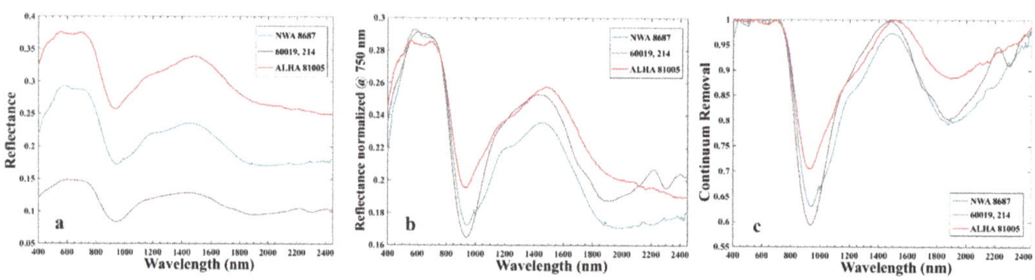

Figure 8. Comparisons of NWA 8687 and the best-matched RELAB reflectance spectra of returned lunar sample 60019, 214, and lunar meteorite ALHA 81005 using SAM algorithm. (**a**) Reflectance. (**b**) Spectra normalized at 750 nm, and (**c**) spectra after continuum removal. Spectrum ID for sample 60019.214 and ALHA 81005 is S3LS07 and CFLM34, respectively.

For visual comparison, the NWA 8687 spectrum and the RELAB spectra were normalized at 750 nm (Figure 8b). The continuum removal was performed (Figure 8c) to compare the band centers and depths. Since the reflectance of the left shoulder of 1 μm absorption is greater than the right shoulder at 2 μm absorption, the continuum shows a negative slope. Similar absorption features near ~900 nm and ~1900 nm were demonstrated in three reflectance spectra, while NWA 8687 has deeper absorptions. Additionally, ALHA 81005 and NWA 8687 show absorption at ~1250 nm, indicating that they are rich in plagioclase.

Figure 9a illustrates the Band I and Band II centers of NWA 8687 spectra in the background of those produced from laboratory synthetic pyroxene [58]. Generally, adding

Ca into pyroxene crystal makes Band I and Band II centers shift to longer wavelengths [38]. The OPX cluster is at the lower left corner, while the CPX is mainly distributed in the upper right corner. Moreover, the high Mg (Mg# > 70) pyroxene is annotated in the shaded region, based on laboratory results. Band centers of NWA 8687 are in the range of OPX, and are located in the high Mg region, indicating that the composition of pyroxene is low in calcium and high in magnesium, which is in agreement with the petrography.

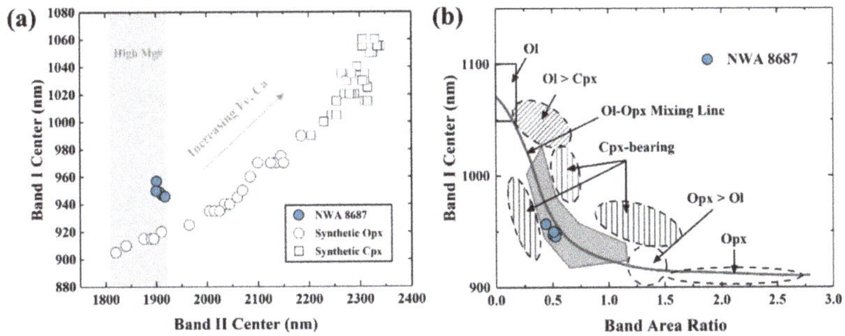

Figure 9. (**a**) Plot of the Band I center versus the Band II center. The laboratory pyroxene spectra are from Ref. [58]. (**b**) Plot of the Band Area Ratio versus the Band I center. The olivine–orthopyroxene mixing line is represented by the solid line. Different spectral regions are defined by Refs. [37,59].

The band area ratio is inversely proportional to olivine abundance [59], and its relationship to Band I center is useful for estimating the relative abundance of olivine and OPX [37,59]. The NWA 8687 spectra allocate on the Ol–OPX mixing line (Figure 9b), suggesting that NWA 8687 contains both OPX and olivine, while the proportions of the two-phase are not well distinguished.

3.3.2. Quantitative Analysis

From the results of spectral unmixing (Table 3), the NWA 8687 is dominated by OPX and plagioclase (including meskelynite), with an average abundance of 54.02% and 34.05%, respectively. Olivine follows as the third most abundant constituent (7.08% on average). There are tiny amounts of CPX and ilmenite in the meteorite, consistent with the petrography (Table 1). Since OPX easily masks the spectrum of plagioclase, the derived OPX abundance is higher than that of plagioclase.

Table 3. Mineral abundances are estimated from NWA 8687 spectra (in %).

	A	B	C	D	E
Clinopyroxene	0	1.51	0	0	1.71
Orthopyroxene	58.20	50.33	55.30	63.55	42.73
Plagioclase	35.24	28.49	20.83	24.82	24.53
Olivine	6.08	6.12	9.58	7.55	6.08
Ilmenite	0.49	0.50	14.28	4.07	1.69
Maskelynite	0	13.05	0	0	23.27
Total	100	100	100	100	100

Once the abundance vector is derived, the modeled spectrum can be obtained through the endmember matrix multiple by the abundance vector. The measured and modeled spectra generally fit well, with the exception of the shift around 1900 nm (Figure 10). The shift could be the difference in chemical compositions (mainly in Ca, Fe, and Mg) between the pyroxene in NWA 8687 and the pyroxene endmembers used in the unmixing process.

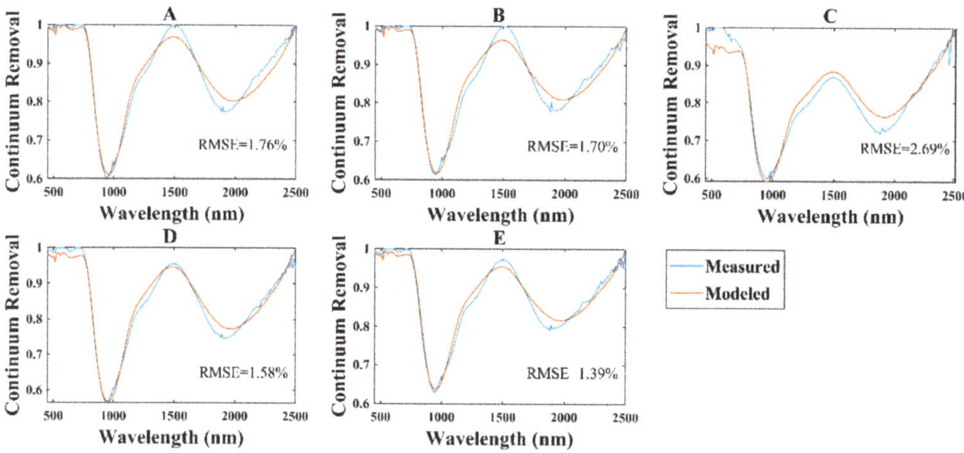

Figure 10. Comparisons between the measured point spectra (**A–E**) (in blue) and their modeled spectra (in red) of the NWA 8687 chip.

4. Discussion

4.1. Lunar Origin

Basalt magmatism is a basic geological process of the Moon, Earth, and other terrestrial planets. As probes to planetary interiors, basaltic silicate minerals record distinct igneous histories and signatures of their planetary bodies [60,61]. Figure 11 illustrates that each planet has a unique Mn/Fe^{2+} ratio in pyroxene and olivine, and the ratio increases with the heliocentric distance, probably due to heating events in the early solar system [62,63]. The practically constant values are widely used as a "fingerprint" of meteorite origins (e.g., [15,64]). NWA 8687 has an average Mn/Fe^{2+} ratio of 0.021 in pyroxene and 0.011 in olivine; both fall into the range of lunar samples (Figure 11). Therefore, we confirm that the sample in this study has a lunar origin.

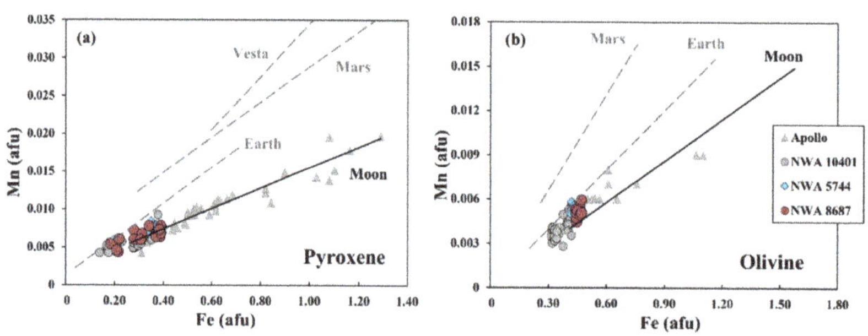

Figure 11. Mn versus Fe^{2+} atoms per formula unit (afu) in pyroxene (**a**) and olivine (**b**) of NWA 8687, compared with pairs NWA 5744 [32] and NWA 10401 [15]. Apollo data and best-fit trend lines are from Ref. [60] (pyroxene) and Ref. [61] (olivine).

4.2. Classification of NWA 8687: A Mg-Suite Anorthositic Norite?

For mineral compositions, NWA 8687 is dominated by calcic plagioclase, with a high content of anorthite ($An_{95.6–97.5}$) coexisting with Mg-rich mafic silicates (Mg# between 75.8 and 77.7) (Figure 12a). This mineral assemblage is consistent with the Mg-suite rocks (e.g., [65,66]). For bulk chemistry, Mg number versus Al_2O_3 content of NWA 8687 and other lunar samples are plotted in Figure 12b (after Refs. [15,16,32]). The figure clearly shows that Mg-rich rocks are distinct from other lunar samples due to higher Mg# and lower Al_2O_3 (e.g., [67]), and NWA 8687 is located ideally in the Mg-suite region. Thus, NWA 8687 can be considered to have the characteristics of magnesian rocks.

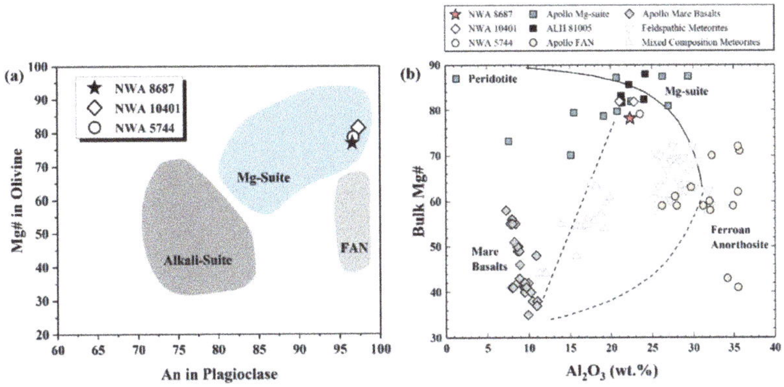

Figure 12. (**a**) Mg# in olivine versus An in plagioclase for NWA 8687 and pairs NWA 5744 and NWA 10401 (after Ref. [13]). (**b**) Bulk Mg# versus Al_2O_3 content for NWA 8687 and other lunar samples (after Refs. [15,16,32] and references therein). Lines show mixtures among four endmembers: a hypothetical magnesian peridotite, Mg-suite rocks, ferroan anorthosite, and mare basalts. Data source: NWA 5744 [32], NWA 10401 [15], ALH 81005 [16], and other lunar meteorites [33]; Apollo Mg-suite [68], FAN, and mare basalts [9].

For the convenience of academic communication, Stöffler et al. [69] developed a classification scheme to allocate lunar highland rocks, based on their modal compositions (Figure 13). The Mg-suite rock comprises norites, troctolites, ultramafic rocks, and closely related lithologies (e.g., anorthositic norite) in this scheme [15]. According to mineral modal abundances (Table 1; here, maskelynite is considered part of plagioclase), this particular sample, NWA 8687, is better described as an anorthositic norite instead of a troctolite, reported by previous studies [29,30,70]. Similarly, NWA 5744, examined by Ref. [31], also has noritic lithology differing from other studies. Although the meteorite and pairs NWA 5744 group have slight differences in classification, they are broadly the same in texture (Figures 3 and 4) and mineral chemistry (Figures 5, 6, and 12a), as well as whole-rock composition (Figure 12b); thus, NWA 8687 is still intimately connected with the NWA 5744 clan.

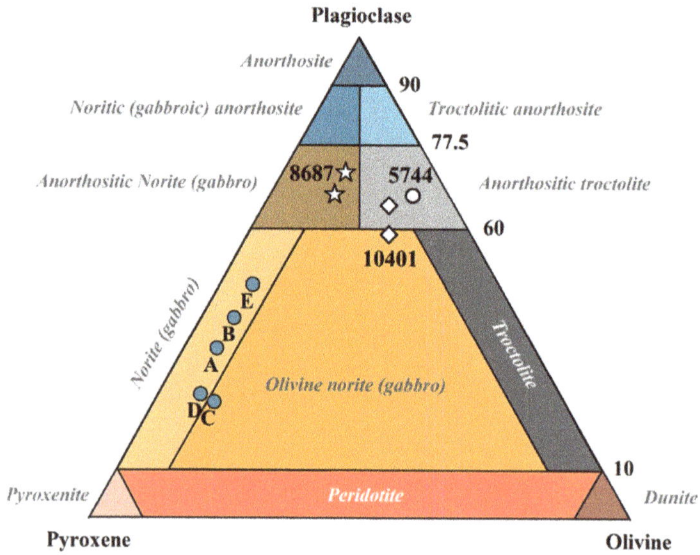

Figure 13. Classification of lunar meteorite NWA 8687 and comparison with pairs NWA 5744 and NWA 10401. Modal abundance (in vol.%) obtained by petrography (star) and by spectroscopy (blue circle) is expressed in the plot. This figure is modified from Figure 2 in Ref. [69].

4.3. Connections between Lab Results and Remote-Sensing Interpretations

The mineral abundances derived from spectral unmixing (Table 3) were compared with the petrographic ground-truth (Table 1). The average differences between the derived abundances and the ground-truth are −0.004 (CPX), 0.34 (OPX), −0.33 (plagioclase), and −0.05 (olivine), respectively. The result shows that the VNIR spectra can accurately determine the abundances of CPX and Olivine. As the OPX overwhelms the plagioclase spectral features, the meteorite spectral features are dominated by the OPX. Specifically, the VNIR spectra overestimate the OPX abundance (more than 2.5 times the actual fraction of OPX) and underestimate the plagioclase abundance (about half the actual fraction of plagioclase). This phenomenon could also happen during lunar orbital observations.

Nonetheless, as shown in the pyroxene–olivine–plagioclase ternary diagram (Figure 13), both the petrographic and spectral results are located at the left edge of the triangle, indicating that the NWA 8687 is noritic instead of troctolitic. In lunar geology, norite and almost all of the anorthositic norite represent material from the lower lunar crust [71]. Thus, the deviation in classification based on petrography and spectroscopy does not lead to considerable differences in regional geologic interpretations.

Our work shows that the VNIR spectroscopy has the potential to crudely classify lunar rock types without sample preparation. The orbital hyperspectral image (HSI) can be used to search for Mg-Suite exposures across the Moon. However, the spectral interpretation of the lunar surface would be complicated, due to both the intrinsic (e.g., grain size, crystalline) and external (e.g., space weathering, imaging conditions, and surface relief) factors. For example, the pervasive space weathering effects on an airless body, such as the Moon, dampens the absorption features. The instrumental noise also impairs the spectral features. These factors affect the determination of band parameters and mineral abundances. Therefore, the spectral data covering less mature surfaces (e.g., fresh craters) are preferred for extracting the mineralogical information. For spectral unmixing, a priori knowledge of endmembers is sometimes challenging to obtain, or unavailable in practice. The unmixing performance would be impaired when using imprecise endmembers. Extracting in-scene endmembers will alleviate this issue to some extent. Using an imaging spectrometer to acquire meteorite HSI and test the endmember extraction algorithms will be the focus in our feature research.

5. Conclusions

We have investigated the lunar meteorite Northwest Africa 8687 by integrated petrographic and spectral approaches. Our conclusions are the following:

1. NWA 8687 is an anorthositic norite mainly composed of plagioclase (67 vol.%) and mafic silicates (21 vol.% for pyroxene and 12 vol.% for olivine), with minor spinel and accessory minerals, such as Fe–Ni metals, troilites, ilmenites, and baddeleyites.
2. The mineral composition is relatively uniform in NWA 8687, with plagioclase dominated by plagioclase (Ca-rich feldspar; $An_{95.6-97.5}$), and olivine dominated by forsterite (Mg-rich olivine; $Fo_{75.8-77.7}$). The main component of pyroxene is enstatite (low-Ca pyroxene; En_{65-78} Fs_{17-20} Wo_{3-20}), followed by augite (high-Ca pyroxene; En_{43-50} Fs_{10-16} Wo_{38-43}). A Mg-spinel (pink spinel) occurs in the sample, with Mg# > 65 and Cr# < 10.
3. Mineral compositions and bulk chemistry imply that NWA 8687 has characteristics of lunar Mg-suite rock, and is likely paired with the NWA 5744 clan, although their petrological classification is slightly different.
4. The mafic mineral compositions inferred from VNIR reflectance are consistent with the petrological result. However, it is challenging to determine plagioclase, since mafic minerals easily mask its spectral features.
5. The mineral modal abundance of the NWA 8687 chip derived from linear unmixing confirms the noritic characteristics of the sample. It shows that spectral investigation has the potential to classify lunar highland rocks, although extended works of diverse lithologies are warranted.

Supplementary Materials: The following supporting information can be downloaded at: https://www.mdpi.com/article/10.3390/rs14122952/s1, Tables S1–S5: Major element abundances of minerals in NWA 8687.

Author Contributions: Conceptualization, L.Q. and X.W.; methodology, L.Q., X.W. and L.H.; software, L.Q., X.W. and L.H.; validation, L.Q. and X.W.; formal analysis, L.Q. and X.W.; investigation, L.Q. and X.W.; resources, Y.Z.; data curation, L.Q., X.W. and L.H.; writing—original draft preparation, L.Q. and X.W.; writing—review and editing, L.Q., X.W. and Y.L.; visualization, L.Q., X.W. and L.H.; supervision, X.W., Y.L. and Y.Z.; project administration, Y.L. and Y.Z.; funding acquisition, Y.L. and Y.Z. All authors have read and agreed to the published version of the manuscript.

Funding: This research was funded by National Key R&D Program of China (No. 2020YFE0202100) and the pre-research project on Civil Aerospace Technologies funded by China National Space Administration (CNSA) (Grants No. D020201 and D020203). Xing Wu also acknowledges the support from the China Postdoctoral Science Foundation (No. 2021M700149).

Data Availability Statement: The RELAB spectral database is available on PDS Geosciences Node Spectral Library (https://pds-speclib.rsl.wustl.edu/ (accessed on 1 May 2022)).

Acknowledgments: We acknowledge Yuchen Xu for providing the sample, and we are grateful to Yazhou Yang for his valuable discussions. Many thanks to Xiaohong Mao (Chinese Academy of Geological Sciences) for her kind assistance with the EPMA.

Conflicts of Interest: The authors declare no conflict of interest.

References

1. Korotev, R.L. Lunar geochemistry as told by lunar meteorites. *Geochemistry* **2005**, 65, 297–346. [CrossRef]
2. Mitchell, R.N. Chang'E-5 reveals the Moon's secrets to a longer life. *Innovation* **2021**, 2, 100177. [CrossRef] [PubMed]
3. Shearer, C.K.; Hess, P.C.; Wieczorek, M.A.; Pritchard, M.E.; Parmentier, E.M.; Borg, L.E.; Longhi, J.; Elkins-Tanton, L.T.; Neal, C.R.; Antonenko, I. Thermal and magmatic evolution of the Moon. *Rev. Mineral. Geochem.* **2006**, 60, 365–518. [CrossRef]

4. Jolliff, B.L.; Gillis, J.J.; Haskin, L.A.; Korotev, R.L.; Wieczorek, M.A. Major lunar crustal terranes: Surface expressions and crust-mantle origins. *J. Geophys. Res. Planets* **2000**, *105*, 4197–4216. [CrossRef]
5. Elkins-Tanton, L.T. Magma oceans in the inner solar system. *Annu. Rev. Earth Planet. Sci.* **2012**, *40*, 113–139. [CrossRef]
6. Warren, P.H.; Wasson, J.T. The origin of KREEP. *Rev. Geophys.* **1979**, *17*, 73–88. [CrossRef]
7. Gillis, J.J.; Jolliff, B.L.; Korotev, R.L. Lunar surface geochemistry: Global concentrations of Th, K, and FeO as derived from lunar prospector and Clementine data. *Geochim. Cosmochim. Acta* **2004**, *68*, 3791–3805. [CrossRef]
8. Lawrence, D.; Elphic, R.; Feldman, W.; Prettyman, T.; Gasnault, O.; Maurice, S. Small-area thorium features on the lunar surface. *J. Geophys. Res. Planets* **2003**, *108*, 5102. [CrossRef]
9. Papike, J.J.; Ryder, G.; Shearer, C.K. Lunar samples. In *Planetary Materials*; De Gruyter: Berlin, Germany, 1998; pp. 719–952.
10. Wieczorek, M.A.; Jolliff, B.L.; Khan, A.; Pritchard, M.E.; Weiss, B.P.; Williams, J.G.; Hood, L.L.; Righter, K.; Neal, C.R.; Shearer, C.K.; et al. The Constitution and Structure of the Lunar Interior. *Rev. Mineral. Geochem.* **2006**, *60*, 221–364. [CrossRef]
11. Taylor, S.R.; Norman, M.D.; Esat, T.M. The Mg-suite and the highland crust: An unsolved enigma. In Proceedings of the Lunar and Planetary Science Conference, Houston, TX, USA, 15–19 March 1993.
12. Dowty, E.; Prinz, M.; Keil, K. Ferroan anorthosite: A widespread and distinctive lunar rock type. *Earth Planet. Sci. Lett.* **1974**, *24*, 15–25. [CrossRef]
13. Shearer, C.K.; Elardo, S.M.; Petro, N.E.; Borg, L.E.; McCubbin, F.M. Origin of the lunar highlands Mg-suite: An integrated petrology, geochemistry, chronology, and remote sensing perspective. *Am. Mineral.* **2015**, *100*, 294–325. [CrossRef]
14. Shearer, C.K.; Papike, J. Early crustal building processes on the moon: Models for the petrogenesis of the magnesian suite. *Geochim. Cosmochim. Acta* **2005**, *69*, 3445–3461. [CrossRef]
15. Gross, J.; Hilton, A.; Prissel, T.C.; Setera, J.B.; Korotev, R.L.; Calzada-Diaz, A. Geochemistry and Petrogenesis of Northwest Africa 10401: A New Type of the Mg-Suite Rocks. *J. Geophys. Res. Planets* **2020**, *125*, e2019JE006225. [CrossRef]
16. Treiman, A.H.; Maloy, A.K.; Shearer, C.K., Jr.; Gross, J. Magnesian anorthositic granulites in lunar meteorites Allan Hills A81005 and Dhofar 309: Geochemistry and global significance. *Meteorit. Planet. Sci.* **2010**, *45*, 163–180. [CrossRef]
17. Takeda, H.; Yamaguchi, A.; Bogard, D.; Karouji, Y.; Ebihara, M.; Ohtake, M.; Saiki, K.; Arai, T. Magnesian anorthosites and a deep crustal rock from the farside crust of the moon. *Earth Planet. Sci. Lett.* **2006**, *247*, 171–184. [CrossRef]
18. Prissel, T.C.; Gross, J. On the petrogenesis of lunar troctolites: New insights into cumulate mantle overturn & mantle exposures in impact basins. *Earth Planet. Sci. Lett.* **2020**, *551*, 116531.
19. Klima, R.L.; Pieters, C.M.; Boardman, J.W.; Green, R.O.; Head, J.W., III; Isaacson, P.J.; Mustard, J.F.; Nettles, J.W.; Petro, N.E.; Staid, M.I.; et al. New insights into lunar petrology: Distribution and composition of prominent low-Ca pyroxene exposures as observed by the Moon Mineralogy Mapper (M^3). *J. Geophys. Res. Planets* **2011**, *116*, E00G06. [CrossRef]
20. Pieters, C.M.; Besse, S.; Boardman, J.; Buratti, B.; Cheek, L.; Clark, R.; Combe, J.; Dhingra, D.; Goswami, J.; Green, R. Mg-spinel lithology: A new rock type on the lunar farside. *J. Geophys. Res. Planets* **2011**, *116*, E00G08. [CrossRef]
21. Prissel, T.C.; Parman, S.W.; Jackson, C.R.M.; Rutherford, M.J.; Hess, P.C.; Head, J.W.; Cheek, L.; Dhingra, D.; Pieters, C.M. Pink Moon: The petrogenesis of pink spinel anorthosites and implications concerning Mg-suite magmatism. *Earth Planet. Sci. Lett.* **2014**, *403*, 144–156. [CrossRef]
22. Pieters, C. Compositional diversity and stratigraphy of the lunar crust derived from reflectance spectroscopy. In *Remote Geochemical Analysis Elemental and Mineralogical Composition*; Cambridge University Press: Cambridge, UK, 1993; pp. 309–339.
23. Cahill, J.; Lucey, P.; Wieczorek, M. Compositional variations of the lunar crust: Results from radiative transfer modeling of central peak spectra. *J. Geophys. Res. Planets* **2009**, *114*, E09001. [CrossRef]
24. Li, C.; Liu, D.; Liu, B.; Ren, X.; Liu, J.; He, Z.; Zuo, W.; Zeng, X.; Xu, R.; Tan, X. Chang'E-4 initial spectroscopic identification of lunar far-side mantle-derived materials. *Nature* **2019**, *569*, 378–382. [CrossRef] [PubMed]
25. Huang, J.; Xiao, Z.; Xiao, L.; Horgan, B.; Hu, X.; Lucey, P.; Xiao, X.; Zhao, S.; Qian, Y.; Zhang, H. Diverse rock types detected in the lunar South Pole–Aitken Basin by the Chang'E-4 lunar mission. *Geology* **2020**, *48*, 723–727. [CrossRef]
26. Isaacson, P.; Liu, Y.; Patchen, A.; Pieters, C.; Taylor, L. Spectroscopy of lunar meteorites as constraints for ground truth: Expanded sample collection diversity. In Proceedings of the 41st Annual Lunar and Planetary Science Conference, The Woodlands, TX, USA, 1–5 March 2010; p. 1927.
27. Hiroi, T.; Kaiden, H.; Yamaguchi, A.; Kojima, H.; Uemoto, K.; Ohtake, M.; Arai, T.; Sasaki, S. Visible and near-infrared spectral survey of lunar meteorites recovered by the National Institute of Polar Research. *Polar Sci.* **2016**, *10*, 476–496. [CrossRef]
28. Isaacson, P.; Liu, Y.; Patchen, A.; Pieters, C.; Taylor, L. Integrated analyses of lunar meteorites: Expanded data for lunar ground truth. In Proceedings of the 40th Annual Lunar and Planetary Science Conference, The Woodlands, TX, USA, 23–27 March 2009; p. 2119.
29. Ruzicka, A.; Grossman, J.; Bouvier, A.; Agee, C.B. The meteoritical bulletin, No. 103. *Meteorit. Planet. Sci.* **2017**, *52*, 1014. [CrossRef]
30. Zhou, Q.; Zhang, G.; Zhu, X.; Zhang, X.; Xiong, Y.; Li, C. Petrography and Mineralogy of Northwest Africa 8687. In Proceedings of the Lunar and Planetary Science Conference, The Woodlands, TX, USA, 18–22 March 2019; p. 2372.
31. Robinson, K.; Kring, D. The Northwest Africa 5744 Group: A glimpse into Schrödinger-like lithologies? In Proceedings of the Lunar and Planetary Science Conference, The Woodlands, TX, USA, 19–23 March 2018; p. 1635.
32. Kent, J.J.; Brandon, A.D.; Joy, K.H.; Peslier, A.H.; Lapen, T.J.; Irving, A.J.; Coleff, D.M. Mineralogy and petrogenesis of lunar magnesian granulitic meteorite Northwest Africa 5744. *Meteorit. Planet. Sci.* **2017**, *52*, 1916–1940. [CrossRef]

33. Korotev, R.L.; Irving, A.J. Lunar meteorites from northern Africa. *Meteorit. Planet. Sci.* **2021**, *56*, 206–240. [CrossRef]
34. Sun, W.; Zhang, X.; Sun, X.; Sun, Y.; Cen, Y. Predicting nickel concentration in soil using reflectance spectroscopy associated with organic matter and clay minerals. *Geoderma* **2018**, *327*, 25–35. [CrossRef]
35. Savitzky, A.; Golay, M.J. Smoothing and differentiation of data by simplified least squares procedures. *Anal. Chem.* **1964**, *36*, 1627–1639. [CrossRef]
36. Clark, R.N.; Roush, T.L. Reflectance spectroscopy: Quantitative analysis techniques for remote sensing applications. *J. Geophys. Res. Solid Earth* **1984**, *89*, 6329–6340. [CrossRef]
37. Gaffey, M.J.; Bell, J.F.; Brown, R.H.; Burbine, T.H.; Piatek, J.L.; Reed, K.L.; Chaky, D.A. Mineralogical variations within the S-type asteroid class. *Icarus* **1993**, *106*, 573–602. [CrossRef]
38. Cloutis, E.A.; Gaffey, M.J. Pyroxene spectroscopy revisited: Spectral-compositional correlations and relationship to geothermometry. *J. Geophys. Res. Planets* **1991**, *96*, 22809–22826. [CrossRef]
39. Yang, Y.; Lin, H.; Liu, Y.; Lin, Y.; Wei, Y.; Hu, S.; Yang, W.; Xu, R.; He, Z.; Zou, Y. The effects of viewing geometry on the spectral analysis of lunar regolith as inferred by in situ spectrophotometric measurements of Chang'E-4. *Geophys. Res. Lett.* **2020**, *47*, e2020GL087080. [CrossRef]
40. Kruse, F.A.; Lefkoff, A.; Boardman, J.; Heidebrecht, K.; Shapiro, A.; Barloon, P.; Goetz, A. The spectral image processing system (SIPS)—interactive visualization and analysis of imaging spectrometer data. *Remote Sens. Environ.* **1993**, *44*, 145–163. [CrossRef]
41. Bioucas-Dias, J.M.; Plaza, A.; Dobigeon, N.; Parente, M.; Du, Q.; Gader, P.; Chanussot, J. Hyperspectral unmixing overview: Geometrical, statistical, and sparse regression-based approaches. *IEEE J. Sel. Top. Appl. Earth Obs. Remote Sens.* **2012**, *5*, 354–379. [CrossRef]
42. Heinz, D.; Chang, C.-I.; Althouse, M.L. Fully constrained least-squares based linear unmixing [hyperspectral image classification]. In Proceedings of the IEEE 1999 International Geoscience and Remote Sensing Symposium. IGARSS'99 (Cat. No. 99CH36293), Hamburg, Germany, 28 June–2 July 1999; pp. 1401–1403.
43. Lucey, P.; Korotev, R.L.; Gillis, J.J.; Taylor, L.A.; Lawrence, D.; Campbell, B.A.; Elphic, R.; Feldman, B.; Hood, L.L.; Hunten, D. Understanding the lunar surface and space-Moon interactions. *Rev. Mineral. Geochem.* **2006**, *60*, 83–219. [CrossRef]
44. Haggerty, S. Luna 24-Systematics in spinel mineral chemistry in the context of an intrusive petrogenetic grid. In Proceedings of the Mare Crisium: The View from Luna 24, Houston, TX, USA, 1–3 December 1977; pp. 523–535.
45. Papike, J.; Hodges, F.; Bence, A.; Cameron, M.; Rhodes, J. Mare basalts: Crystal chemistry, mineralogy, and petrology. *Rev. Geophys.* **1976**, *14*, 475–540. [CrossRef]
46. Keil, K.; Prinz, M.; Bunch, T. Mineral chemistry of lunar samples. *Science* **1970**, *167*, 597–599. [CrossRef]
47. Anderson, A. The texture and mineralogy of lunar peridotite, 15445, 10. *J. Geol.* **1973**, *81*, 219–226. [CrossRef]
48. Ridley, W.; Hubbard, N.; Rhodes, J.; Weismann, H.; Bansal, B. The petrology of lunar breccia 15445 and petrogenetic implications. *J. Geol.* **1973**, *81*, 621–631. [CrossRef]
49. Prinz, M.; Dowty, E.; Keil, K.; Bunch, T. Spinel troctolite and anorthosite in Apollo 16 samples. *Science* **1973**, *179*, 74–76. [CrossRef]
50. Baker, M.B.; Herzberg, C.T. *Spinel Cataclasites in 15445 and 72435: Petrology and Criteria for Equilibrium*; Pergamon Press: Oxford, UK, 1980.
51. Marvin, U.B.; Carey, J.W.; Lindstrom, M.M. Cordierite-spinel troctolite, a new magnesium-rich lithology from the lunar highlands. *Science* **1989**, *243*, 925–928. [CrossRef] [PubMed]
52. Snyder, G.; Ruzicka, A.; Taylor, L.; Patchen, A. Journey to the center of the regolith: A spinel troctolite and other clasts from drive tube 68001. In Proceedings of the Lunar and Planetary Science Conference, Houston, TX, USA, 16–20 March 1998; p. 1144.
53. Daubar, I.J.; Kring, D.A.; Swindle, T.D.; Jull, A.T. Northwest Africa 482: A crystalline impact-melt breccia from the lunar highlands. *Meteorit. Planet. Sci.* **2002**, *37*, 1797–1813. [CrossRef]
54. Gross, J.; Treiman, A.H. Unique spinel-rich lithology in lunar meteorite ALHA 81005: Origin and possible connection to M3 observations of the farside highlands. *J. Geophys. Res. Planets* **2011**, *116*, E10009. [CrossRef]
55. Cheek, L.C.; Pieters, C.M. Reflectance spectroscopy of plagioclase-dominated mineral mixtures: Implications for characterizing lunar anorthosites remotely. *Am. Mineral.* **2014**, *99*, 1871–1892. [CrossRef]
56. McKay, D.; Bogard, D.; Morris, R.; Korotev, R.; Johnson, P.; Wentworth, S. Apollo 16 regolith breccias: Characterization and evidence for early formation in the mega-regolith. *J. Geophys. Res. Solid Earth* **1986**, *91*, 277–303. [CrossRef]
57. Treiman, A.H.; Drake, M.J. Origin of lunar meteorite ALHA 81005: Clues from the presence of Terrae clasts and a very low-titanium mare basalt clast. *Geophys. Res. Lett.* **1983**, *10*, 783–786. [CrossRef]
58. Klima, R.L.; Dyar, M.D.; Pieters, C.M. Near-infrared spectra of clinopyroxenes: Effects of calcium content and crystal structure. *Meteorit. Planet. Sci.* **2011**, *46*, 379–395. [CrossRef]
59. Cloutis, E.A.; Gaffey, M.J.; Jackowski, T.L.; Reed, K.L. Calibrations of phase abundance, composition, and particle size distribution for olivine-orthopyroxene mixtures from reflectance spectra. *J. Geophys. Res. Solid Earth* **1986**, *91*, 11641–11653. [CrossRef]
60. Karner, J.; Papike, J.; Shearer, C. Comparative planetary mineralogy: Pyroxene major-and minor-element chemistry and partitioning of vanadium between pyroxene and melt in planetary basalts. *Am. Mineral.* **2006**, *91*, 1574–1582. [CrossRef]
61. Karner, J.; Papike, J.; Shearer, C. Olivine from planetary basalts: Chemical signatures that indicate planetary parentage and those that record igneous setting and process. *Am. Mineral.* **2003**, *88*, 806–816. [CrossRef]
62. Taylor, S.R. *Solar System Evolution: A New Perspective*; Cambridge University Press: Cambridge, UK, 2001.

63. Papike, J.J. Comparative planetary mineralogy; chemistry of melt-derived pyroxene, feldspar, and olivine. *Rev. Mineral. Geochem.* **1998**, *36*, 701–712.
64. Papike, J.; Karner, J.; Shearer, C. Determination of planetary basalt parentage: A simple technique using the electron microprobe. *Am. Mineral.* **2003**, *88*, 469–472. [CrossRef]
65. Warner, J.; Simonds, C.; Phinney, W. Genetic distinction between anorthosites and Mg-rich plutonic rocks: New data from 76255. In Proceedings of the Lunar and Planetary Science Conference, Houston, TX, USA, 15–19 March 1976.
66. Haskin, L.; Shih, C.-Y.; Bansal, B.; Rhodes, J.; Wiesmann, H.; Nyquist, L. Chemical evidence for the origin of 76535 as a cumulate. In Proceedings of the Lunar and Planetary Science Conference Proceedings, Houston, TX, USA, 18–22 March 1974; pp. 1213–1225.
67. Lindstrom, M.M.; Lindstrom, D.J. Lunar granulites and their precursor anorthositic norites of the early lunar crust. *J. Geophys. Res. Solid Earth* **1986**, *91*, 263–276. [CrossRef]
68. Warren, P.H. Lunar anorthosites and the magma-ocean plagioclase-flotation hypothesis; importance of FeO enrichment in the parent magma. *Am. Mineral.* **1990**, *75*, 46–58.
69. Stöffler, D.; Knöll, H.-D.; Marvin, U.; Simonds, C.; Warren, P. Recommended classification and nomenclature of lunar highland rocks-a committee report. In Proceedings of the Lunar Highlands Crust, Houston, TX, USA, 1 October 1980; pp. 51–70.
70. Korotev, R.; Irving, A. Not quite keeping up with the lunar meteorites—2016. In Proceedings of the Lunar and Planetary Science Conference, The Woodlands, TX, USA, 21–25 March 2016; p. 1358.
71. Wieczorek, M.A.; Zuber, M.T. The composition and origin of the lunar crust: Constraints from central peaks and crustal thickness modeling. *Geophys. Res. Lett.* **2001**, *28*, 4023–4026. [CrossRef]

remote sensing

MDPI

Article

High-Resolution Regional Digital Elevation Models and Derived Products from MESSENGER MDIS Images

Madeleine R. Manheim [1,*], Megan R. Henriksen [1], Mark S. Robinson [1], Hannah R. Kerner [2], Bradley A. Karas [1], Kris J. Becker [3,4], Matthew Chojnacki [5], Sarah S. Sutton [4] and David T. Blewett [6]

1 School of Earth and Space Exploration, Arizona State University, 1100 South Cady Mall, Tempe, AZ 85287, USA; mhenriksen@ser.asu.edu (M.R.H.); robinson@ser.asu.edu (M.S.R.); bradley.karas@asu.edu (B.A.K.)
2 School of Computing and Augmented Intelligence, Arizona State University, 1100 South Cady Mall, Tempe, AZ 85287, USA; hkerner@umd.edu
3 Astrogeology Science Center, United States Geological Survey, 2255 North Gemini Dr., Flagstaff, AZ 86001, USA; kbecker@orex.lpl.arizona.edu
4 Lunar and Planetary Laboratory, University of Arizona, Tucson, AZ 85721, USA; ssutton@lpl.arizona.edu
5 Planetary Science Institute, Lakewood, CO 80401, USA; mchojnacki@psi.edu
6 Planetary Exploration Group, Johns Hopkins Applied Physics Laboratory, Laurel, MD 20723, USA; david.blewett@jhuapl.edu
* Correspondence: mmanheim@ser.asu.edu

Abstract: The Mercury Dual Imaging System (MDIS) on the Mercury Surface, Space ENvironment, GEochemistry, and Ranging (MESSENGER) spacecraft has provided global images of Mercury's surface. A subset of off-nadir observations acquired at different times resulted in near-global stereo coverage and enabled the creation of local area digital elevation models (DEMs). We derived fifty-seven DEMs covering nine sites of scientific interest and tied each to a geodetic reference derived from Mercury Laser Altimeter (MLA) profiles. DEMs created as part of this study have pixel scales ranging from 78 m/px to 500 m/px, and have vertical precisions less than the DEM pixel scale. These DEMs allow detailed characterizations of key Mercurian features. We present a preliminary examination of small features called "hollows" in three DEM sites. Depth measurements from the new DEMs are consistent with previous shadow and stereo measurements.

Keywords: Mercury; digital elevation models; MESSENGER; MDIS; stereophotogrammetry; hollows; high resolution

1. Introduction

The Mercury Dual Imaging System (MDIS) onboard the Mercury Surface, Space ENvironment, GEochemistry, and Ranging (MESSENGER) spacecraft has provided a variety of observations to investigate the origin and evolution of the innermost planet. One of the primary objectives of the MDIS experiment was the acquisition of high-resolution images of significant landforms [1]. The MDIS experiment consisted of a Narrow-Angle Camera (NAC) and a Wide-Angle Camera (WAC) [1,2]. Although MDIS is not a stereo camera, off-nadir observations acquired at different times resulted in near-global stereo coverage [1,3]. In addition, specific high-priority targets were imaged in stereo at a higher resolution than the global campaign, enabling the creation of local area Digital Elevation Models (DEMs).

The orbit of the spacecraft was close to polar and highly eccentric, with a periapsis that varied from 200 to 506 km and an apoapsis of 10,000 to 15,000 km; the periapsis occurred within 30° to 6° from the north pole [4–6]. Its orbital period was 12 h from entry into orbit around Mercury on 8 March 2011 until March 2012, when the orbit was lowered, resulting in an 8 h period. Then, beginning in mid-June 2014, a series of maneuvers resulted in periapsis varying from 15 to 200 km until MESSENGER impacted Mercury on 30 April

2015 [4,5]. These later mission periods enabled the acquisition of higher-resolution NAC observations (~1 m to ~300 m pixel scales) of regions of interest [2,6,7].

This paper describes the production of high-resolution regional MDIS NAC DEMs with an uncertainty analysis. It also demonstrates how these products can be helpful for scientific investigation with a topographic analysis of hollows, an enigmatic landform found only on Mercury [8–10].

2. Data Sources

The DEMs were computed from MDIS NAC images, with elevation profiles from the Mercury Laser Altimeter (MLA) [11] serving as a geodetic reference frame. When no MLA data directly covered the NAC DEMs, MDIS WAC images controlled to MLA profiles bridged the gaps in coverage.

2.1. Mercury Dual Imaging System (MDIS)

The NAC was an off-axis reflector with a 1.5° field of view (FOV), providing images with a pixel scale of 5 m at an altitude of 200 km. The WAC was a four-element refractor with a 10.5° FOV and 12-position filter wheel, providing images with a pixel scale of 36 m at an altitude of 200 km. Each camera had an identical 1024 × 1024 charge-coupled device (CCD) detector. The two cameras were mounted and co-aligned on a pivoting platform with a 90° range of motion about the +z axis (the boresight of most of the MESSENGER instruments). This alignment allowed off-nadir image acquisition without affecting the pointing of other instruments or misalignment of the sunshade, enabling the acquisition of stereo observations [1,12].

2.2. Mercury Laser Altimeter (MLA)

We used altimetry obtained from the MLA to increase the absolute accuracy of NAC DEMs. MLA was a time-of-flight altimeter that measured the shape of Mercury by using pulse detection and pulse edge timing to precisely determine the range from the spacecraft to the surface [11]. The MLA pulse width was ~6 ns (FWHM), taken at a rate of 8 Hz with a 20 mJ laser pulse energy at a wavelength of 1064.3 nm [5]. MLA ranging observations were only available for latitudes between 90°N and 18°S due to MESSENGER's highly elliptical orbit and periapsis at high northern latitudes, with measurements becoming sparser with decreasing latitude. At the time of this work, the best available MLA profiles were the version archived in volume MESSMLA_2001; we downloaded these from the PDS and filtered them to use only channel 0 [13]. These version 1 tracks have not been crossover-corrected. While these MLA measurements' horizontal accuracies are not well constrained, they had a radial precision of <1 m and a radial accuracy of <20 m with respect to Mercury's center of mass [14]. Additionally, the spacecraft position was known to within a few tens of meters [15].

3. Methodology

High-resolution regional WAC and local-scale NAC DEM processing were completed using a combination of Integrated Software for Imagers and Spectrometers (ISIS) [16–18] and SOCET SET from BAE Systems [19,20]. Although there are important differences (e.g., framing vs. pushbroom camera systems), the DEM processing for MDIS DEMs is similar to the process used for the Lunar Reconnaissance Orbiter Camera (LROC) NAC DEM production described by Henriksen et al. [21] and for the High-Resolution Imaging Science Experiment (HiRISE) DEM production of Mars reported by Kirk et al. [22] and Sutton et al. [23].

Other regional MDIS DEMs at comparable resolutions have been created using various methods. Fassett [24] generated 96 DEMs at 45–245 m/pixel using the Ames Stereo Pipeline. Ostrach and Dundas [25] created 11 DEMs using SOCET SET (made by USGS Astrogeology). Tenthoff et al. [26] adapted a shape-from-shading method for use on MDIS images to extract eight DEMs with pixel scales as low as 3.3 m; in addition, Weirich et al. [27] used stereophotoclinometry to create four DEMs at 20–120 m/px. At the km scale, Preusker et al. [28]

divided the Mercurian surface into 15 quadrangles and is producing 222 m/pixel DEMs by photogrammetrically processing thousands of images. To date, 4 quadrangles have been published [29]. We compare our DEMs to the available co-located products in Section 4.3.

DEM production is described in the following sections and summarized in Figure 1. After determining appropriate stereo pairs based on acquisition conditions and image overlap, images were calibrated and formatted for ingestion into SOCET SET using ISIS. Once imported into SOCET SET, each image was corrected for relative orientation to all other images in the stereo model and then for absolute orientation using MLA profiles. When the stereo model bundle-adjustment achieved a root mean square error (RMSE) < 0.5 pixels, we extracted a final DEM for each stereo pair. DEMs were manually edited, as needed, to minimize artifacts. Geometrically correct orthorectified maps, or orthophotos, were then generated from the edited DEM. Final orthophotos and DEMs were exported from SOCET SET and post-processed using ISIS for release into the Planetary Data System. Uncertainty analysis was performed, and additional products were created and released.

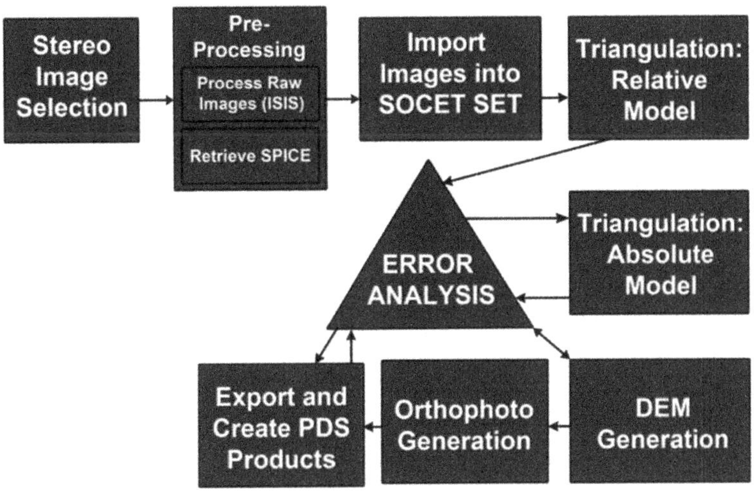

Figure 1. Workflow diagram for processing MESSENGER high-resolution stereo pairs to DEMs (adapted from Henriksen et al. [21]).

3.1. Stereo Image Selection

We selected stereo images via a two-step process. First, we filtered all the images over the desired target area for favorable acquisition conditions and then identified specific stereo pairs. As proposed by Becker et al. [30], we adapted eight criteria, covering illumination conditions, resolution, parallax angle, and image overlap. The search was performed using the USGS *isisminer* application [16,17]. We initially limited the searches to NAC images; if the site had sufficiently complete NAC stereo coverage, we checked for MLA profile overlap. If NAC-MLA overlap did not exist, we then searched for WAC stereo images that crossed MLA profiles. Due to MESSENGER's highly elliptical orbit, NAC images used for DEM production could range in pixel scale from less than 5 m to over 150 m, and image illumination conditions and spacecraft geometries varied greatly. The amount of overlap and the actual footprint of the DEMs were affected by the topography and acquisition parameters.

We first sorted the images covering the target region to ensure adequate lighting, viewing conditions, and spatial resolution. The criteria used in this sorting process are summarized in Figure 2a,b. The incidence angle is between the Sun and the normal surface vector (ideally between 40° and 65°). The emission angle is between the normal surface vector and the camera boresight intercept point; we selected images with emission angles between 0° and 40° to include images acquired during the so-called nadir (emission = 0°) mosaic imaging campaign [30]. Additionally, the native pixel scale of the images was taken into consideration as any final DEMs would have a ground sampling distance of at least three times the coarsest pixel scale of the stereo pair. For the DEMs we created, images had pixel scales of ~150 m or better.

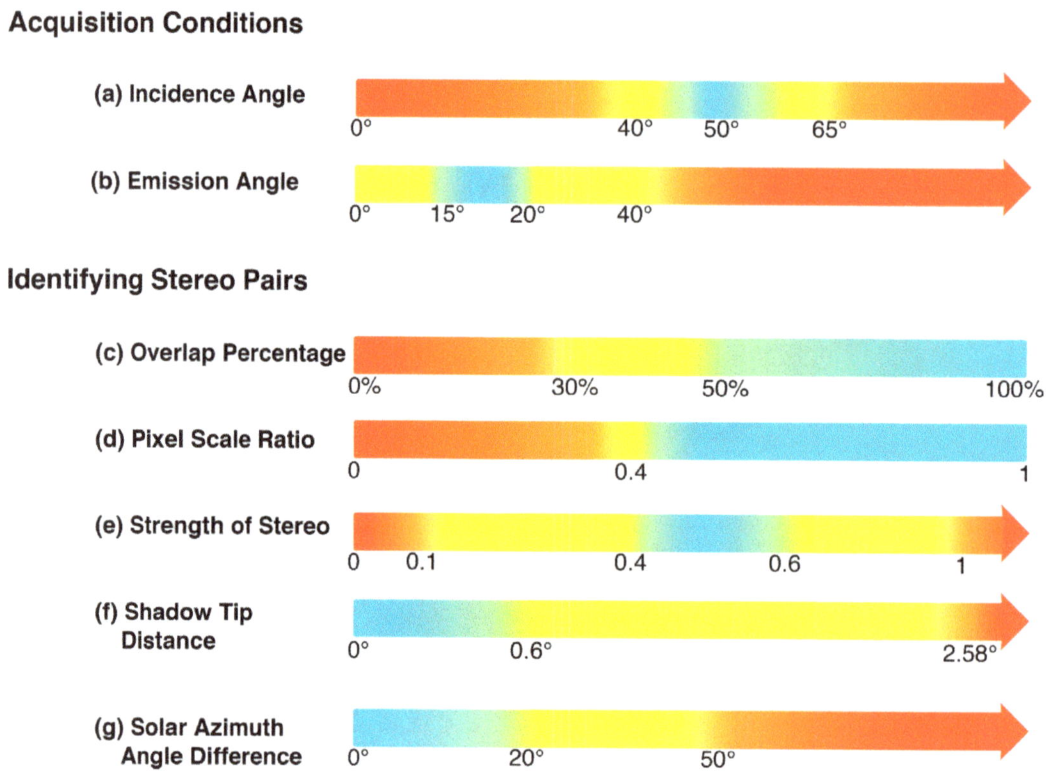

Figure 2. Criteria for choosing NAC and WAC images to produce stereo pairs (blue indicates optimal conditions, yellow is acceptable, and red is not serviceable). The values shown here are adapted from Becker et al. [30].

After vetting the images on acquisition parameters, images were evaluated for suitability as part of a stereo pair (Figure 2c–g). Image pairs were first filtered to those that shared at least a 30% overlap of common surface coverage. These overlapping regions were then assessed based on pixel scale ratio, stereo strength, and illumination compatibility. Strength of stereo is calculated as a ratio of the parallax between the two images over unit height and illumination compatibility is obtained by measuring the shadow tip distance, or the distance between the tips of the shadows in two images for a hypothetical post of unit height [30]. Additionally, the absolute difference in the solar azimuth angle was required to be less than 50° (optimally $\leq$ 20°) [30].

Ultimately, we created DEMs covering nine different sites (see Appendices A and B). Among the stereo pairs we used to create these DEMs, images had pixel scales ranging from 19 m–150 m. Incidence angles ranged from 30°–64° (all Catullus crater, Cunningham crater, and Kuiper crater images had incidence angles between 30°–40°; images used for Kertész crater, Paramour Rupes, and Sander crater DEMs all had incidence angles > 41°). Emission angles ranged from 1°–50°. Except for those used in the Sander DEM, all pairs had stereo strength values between 0.37 and 0.65 (Table 1).

Table 1. Image selection criteria statistics for the stereo images used in this study.

	Acquisition Conditions		Identifying Stereo Pairs				
	Incidence Angle (°)	Emission Angle (°)	Overlap Percentage	Pixel Scale Ratio	Strength of Stereo	Shadow Tip Distance (°)	Solar Az. Angle Diff. (°)
Minimum	30	1	0.45	0.40	0.37	0.00	0.00
Maximum	64	50	100	0.99	1.39	0.02	32.65
Average	45	26	56.86	0.66	0.80	0.15	8.99
Median	44	32	51.81	0.68	0.72	0.11	6.60

After all the acceptable stereo pairs in a region were identified, we then selected pairs that would give us the desired coverage for the site (Figure 3). These pairs were controlled together and ultimately mosaicked in ISIS to create the final DEM. Because some of the pairs were acquired serendipitously rather than systematically, often from substantially different orbits, the coverage in regions of interest could vary significantly in resolution and overlap between pairs [31]. We selected stereo pairs to provide optimal coverage of the region; however, because we sampled our DEMs at 3× the pixel scale of the lowest resolution image, and because controlling images with drastically different pixel scales is more difficult, we selected stereo pairs with similar pixel scales wherever possible. Individual images were frequently used in more than one stereo pair to maximize coverage while minimizing production time.

3.2. Image Pre-Processing

The selected stereo pairs were ingested into ISIS, radiometrically calibrated, and initialized with position and orientation parameters stored in SPICE kernels using the planetary kernel pck00010_msgr_v21.tpc with a reference radius of 2439.4 km. This kernel was used by all the products archived in the PDS volume MESSDEM1001 [32]. The kernel parameters were then converted to a format compatible with SOCET SET 5.6 and imported into SOCET SET. Finally, the images were converted to a 16-bit Tagged Interchange File Format (TIFF) and imported into SOCET SET using an off-axis framing camera sensor model developed by the USGS Astrogeology Science Center to accommodate the MDIS NAC off-axis refractor [16,21].

3.3. Relative Orientation

An element of uncertainty is always present in spacecraft positioning and pointing; therefore, it is necessary to correct for relative positional and pointing offsets, or the orientation of one image compared to that of the other images being used [33]. All overlapping images were linked together with at least nine equally distributed "tie" points, manually selected by matching pixels between stereo images. We added additional points for large areas or those with a complex topography. For each stereo pair, we started by putting in nine points, and then added more until we had sufficient coverage to extract a precise model that accurately followed the topography. For mosaics, where multiple stereo pairs were used to build up coverage, we also added points to tie together each area of stereo pair overlap. The images were then bundle-adjusted using a multi-sensor triangulation (MST) algorithm to align the images in the Mercury body-fixed system [33,34].

A priori spacecraft and camera parameters were used as inputs into the MST algorithm (Table 2) [21]. These parameters are loose constraints to help optimize the bundle-adjustment solution. They were empirically selected based on estimates of the spacecraft uncertainties and refined by iteratively adjusting values and examining the detailed bundle adjustment output. For MDIS images, we used a set of three parameters for camera position and three for camera pointing. These parameters were applied to every image except the nadir-most image, which was temporarily held constant to prevent an overly large (km-scale) shift from occurring due to the bundle adjustment. Following bundle-adjustment, we examined individual tie point residuals and adjusted the points until an adequate overall RMSE was reached (ideally less than 0.2 pixels, although under 0.5 pixels was acceptable) [21].

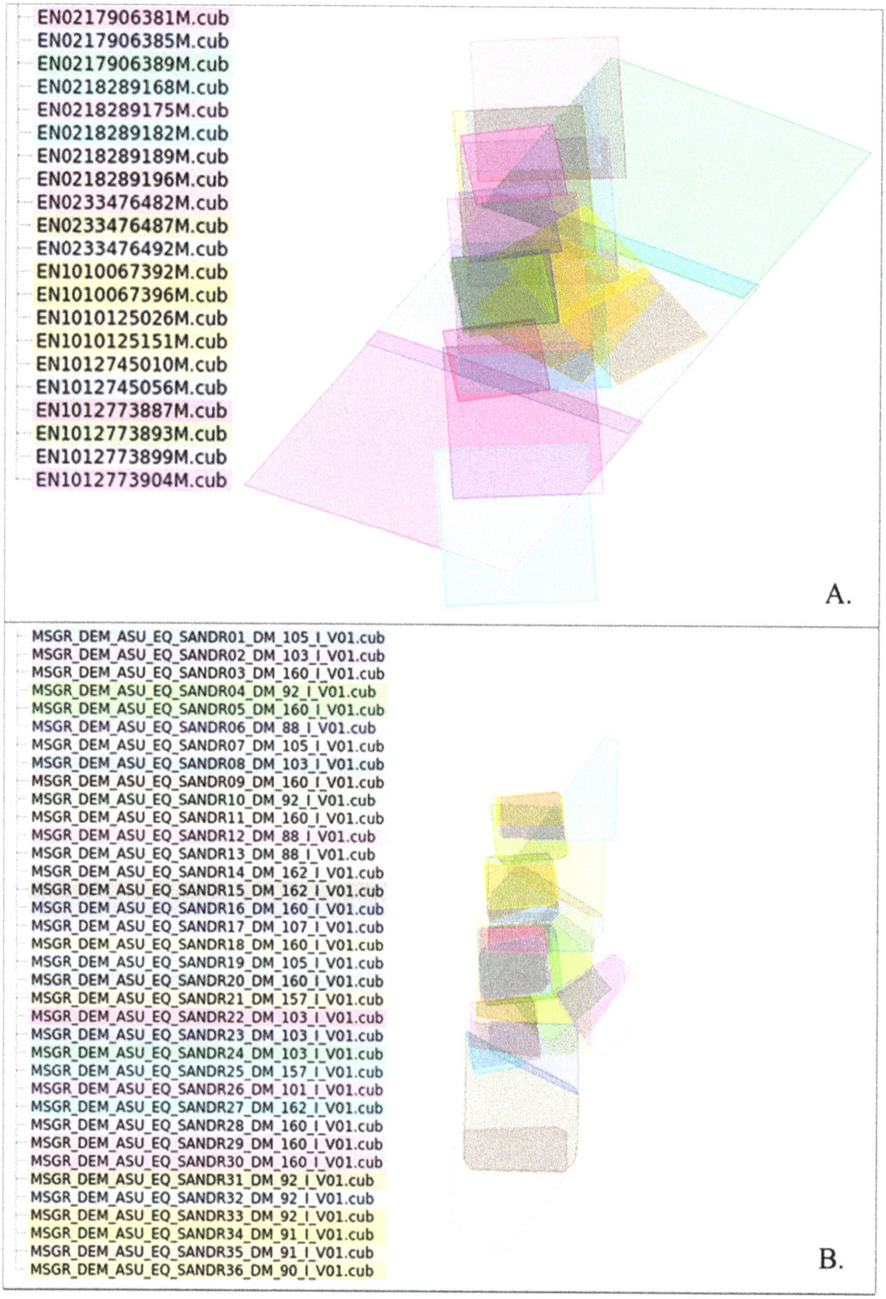

Figure 3. Stereo pairs used to create the Sander crater DEM mosaic: (**A**) shows the footprints of the 21 images chosen, while (**B**) shows the footprints of the 36 individual DEMs made from those images.

Table 2. A priori bundle adjustment parameters.

MST Parameters	Values
Camera ΔX	100 m
Camera ΔY	100 m
Camera ΔZ	10 m
Attitude ΔOmega *	0.1°
Attitude Δphi *	0.1°
Attitude Δkappa *	0.1°

* These three attitude parameters are functionally similar to roll, pitch, and yaw, although not exactly the same. Roll, pitch, and yaw describe the rotation of a body while phi, omega, and kappa describe the rotation of an image relative to a projected coordinate system.

3.4. Absolute Orientation

Absolute orientation describes the relationship between the features in the image and their corresponding locations on the surface of Mercury. Here, absolute control was achieved by adding ground points specifying exact ground (x, y, z) coordinates for given pixel locations in the image. For MDIS DEMs, we extracted these ground coordinates from MLA profiles covering the same areas.

Once an acceptable bundle adjustment solution for relative orientation was found, the NAC images could be controlled directly to the MLA profiles. Ground control points were created by identifying one or more points along individual profiles (which can be overlaid over the images in SOCET SET by importing them as shapefiles) and visually identifying their corresponding locations in the images. Using a stereo monitor to view the images and profiles in 3D, users can select these corresponding locations by identifying distinct topographic features that appear in both datasets. Ground points were added or adjusted iteratively until we reached a state where we could visually observe that the MLA profiles aligned well with the DEM topography. Therefore, while relatively few ground control points were added (3 to 10), the rest of the overlying profiles acted as visual check points.

We then ran a new bundle adjustment that included both the ground control and tie points to triangulate absolute orientation. We used the same parameters as the initial bundle adjustment but with the nadir image no longer held constant. Because the bundle adjustment now also included ground control points, we now had RMSE values for latitude, longitude, and elevation in addition to the overall RMSE. These latitude, longitude, and elevation RMSEs were considered acceptable when they fell within the known accuracies of the MLA profiles.

Ideally, the NAC images would be co-registered directly to overlapping MLA profiles. However, MLA profiles are relatively sparse, especially at southern latitudes. Occasionally, a NAC image had insufficient or no MLA coverage (a minimum of three control points are necessary to triangulate ground). For our Catullus crater and Paramour Rupes DEM mosaics, where the sparsity of MLA points prevented direct control, a WAC stereo pair was chosen that overlapped with the NAC stereo pair and MLA profiles. The WAC images had resolutions ranging from 235 to 475 m/px, and were used to create DEMs of 1337 m/px and 1589 m/px. Following the method described in Section 3.3, the two WAC images were tied to each other and to the NAC images, and a bundle adjustment solution was performed. This solution was evaluated by the same standards described in Section 3.3, with an additional check to ensure the vertical offset between the WAC and NAC stereo models was less than the expected precision for a DEM created from the WAC images. Then, ground control points were added to control the WAC images to MLA profiles. Finally, by bundle-adjusting the WACs and NACs together with these ground control points, the NAC DEM could be indirectly controlled to the MLA profiles (Figure 4).

3.5. DEM Extraction

The sensor model contained a bug that resulted in errors when calculating extents for image overlap. Therefore, it could not be used to directly extract a DEM. As a work-around, we generated a cubic rational polynomial (CRP) approximation of the sensor model for each controlled image; these approximations have residual errors of less than 0.01 pixels. We used the Next Generation Automatic Terrain Extraction (NGATE) program in SOCET SET to extract elevations pixel by pixel from the CRP images; the program then downsampled the results to create a DEM at a pixel scale that is a factor of three greater than the largest pixel scale of an image in a stereo pair [35,36]. NGATE can be customized to a certain degree using a strategy file that defines several correlation parameters, window size parameters, and ways to filter the correlation results. We used strategy files customized for the MDIS images. See [21] for more details about NGATE.

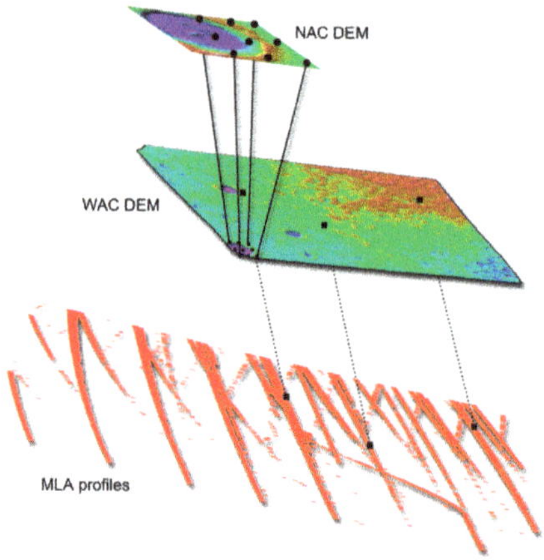

Figure 4. Tie points connecting Kertész NAC DEM (**top**; displayed as a color-shaded relief map) to an overlapping WAC DEM (**middle**; made from image pair EW0218288557G, EW0233348623G), which is controlled to MLA profiles (**bottom**) projected to Mercury's surface. Black circles represent tie points between the NAC and WAC images, while black squares represent control points in the WAC DEM from MLA points.

Despite the robust image matching techniques used by NGATE, the automatically extracted DEM can contain some artifacts, or blunders, especially near edges, areas of steep topography, or particularly homogenous areas (shadows can also cause DEM artifacts; however, our images were selected to avoid shadows). Artifacts were identified visually and from the confidence map generated by SOCET SET for the DEM and were removed using SOCET SET interactive tools that allow the user to cut out sections of the DEM. Then, using NGATE, new DEM "patches" were generated to cover these missing areas and were merged back into the original DEM. These small DEM patches were regenerated using boundary coordinates in stereo that finely delineated the overlap between the images and accounted for local changes in topography. Constraining the stereo algorithm to a small portion of each image as defined by these boundaries forced the algorithm to compare identical locations and yielded good correlation results. SOCET SET also offers interactive tools that allow a user to manually adjust the heights of individual points, but we did not use these tools for these DEMs [21]. The final confidence map product maps the per-pixel results of the NGATE correlations to a small number of integer values (1–15) to enable user evaluation of the DEM. A look-up table is provided in the Readme for each DEM.

The final DEM underwent a last comparison to MLA to ensure that there were no remaining systematic offsets. If any were detected, we then went back to the bundle adjustment step to adjust the ground control and tie points.

3.6. Orthophoto Generation

Once a final DEM was corrected and checked for errors, it was used to orthorectify the 16-bit NAC images used in its creation. This process, performed within SOCET SET, uses the elevations given by the DEM to correct image distortion by calculating and correcting for shifts in pixel location due to camera obliquity and terrain relief [34]. The resulting orthophotos are free from distortion (within the limits of the DEM accuracy) caused by topography and camera viewing angle, allowing direct and accurate distance measurements between pixels [37]. Orthophotos were generated at both the DEM pixel scale and the native image pixel scale.

3.7. Planetary Data System (PDS) Products and Derived Products

In addition to the DEM, which was converted to IMG format using ISIS, several derived products were created and made available through the PDS (Figure 5). A confidence map and orthophotos of each image in the stereo pair are provided at both the pixel scale of the DEM and the largest native pixel scale from the stereo pair. A terrain-shaded relief map, a color-shaded relief map, a slope map, and corresponding legends are also provided at the pixel scale of the DEM in the EXTRAS directory of the PDS in GeoTIFF format, as well as a 32-bit GeoTIFF of the DEM. GeoTIFF products were created using the Geospatial Data Abstraction Library (GDAL) [38]. For sites where multiple stereo pairs were used to build up coverage, the overlapping products were mosaicked using the ISIS program *automos* and released in addition to the individual products.

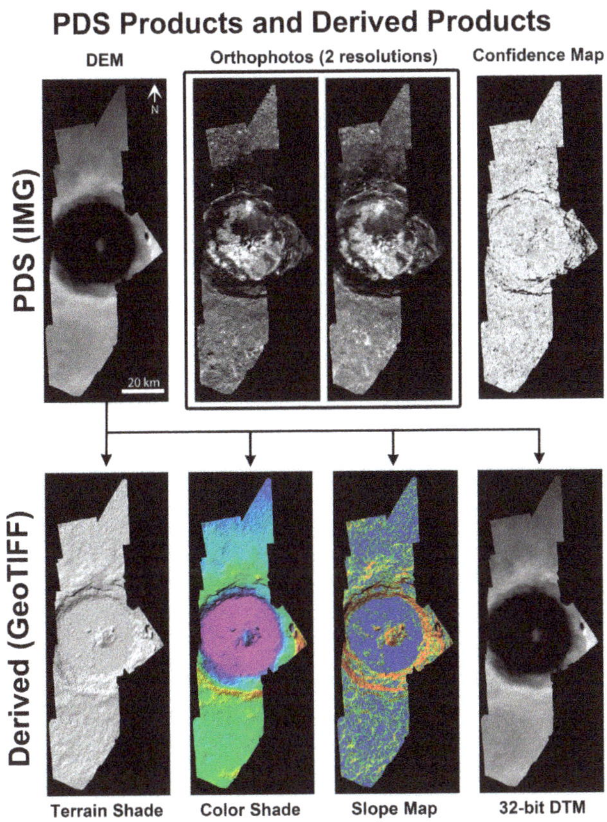

Figure 5. PDS and derived products from the Sander crater DEM mosaic (42.42°N, 154.64°E), created from 21 images forming 36 stereo pairs. The DEM mosaic, orthophoto mosaics at two resolutions, and a confidence map are available in PDS format, as well as individual DEMs, four orthophotos, and confidence maps for each stereo pair. In addition, terrain-shaded relief, color-shaded relief, and slope maps are derived from each DEM and are released as 8-bit GeoTIFFs along with a 32-bit GeoTIFF of the DEM. See Appendix A for stereo pair details.

4. Results: Uncertainty Analysis

The DEMs were evaluated based on qualitative and quantitative error analyses. Contour intervals created from the DEMs were compared to the images using a stereo viewer to confirm a close match with the terrain. The overlapping stereo pairs were also compared to the available MLA elevation points to ensure that there was no tilt present in the DEM and that the profiles closely aligned with the images in stereo. Here, we report quantitative metrics for precision and accuracy for each site (Table 3), and compare them to other regional DEMs.

Table 3. Summary data and error analysis for completed regions.

Site Name	# of Stereo Pairs	Center Latitude, Longitude (°N, °E)	DEM Mosaic Pixel Scale (m)	Relative Linear (Vertical) Error (m)	# of MLA Points for Comparison	MLA Mean (Vertical) Offset (m)	MLA Standard Deviation (m)
Catullus Crater *	1	21.88°, 292.5°	84	84	777	−255	188
Cunningham Crater	9	30.40°, 157.0°	105	85	127	2	58
Degas Crater	1	36.86°, 232.6°	97	70	125	−1	109
Scarp in NW Rembrandt Basin	1	−31.16°, 82.5°	500	383	-	-	-
Kertész Crater	2	31.45°, 146.3°	120	98	27	−9	51
Kuiper Crater	1	−11.3°, 329.1°	270	161	-	-	-
Paramour Rupes *	3	−5.07°, 145.1°	450	261	138	−5	185
Raditladi Hollows	3	15.20°, 120.2°	180	137	154	−44	52
Sander Crater	36	42.50°, 154.7°	162	70	756	−44	65

* MLA offsets are obtained from WAC DEMs tied to the NAC DEMs.

4.1. Relative Linear Error

The SOCET SET software calculates precision as a relative error at a 90% confidence level, meaning that 90% of the errors in elevation measurements will be equal to or less than the reported value [34,37]. The vertical precision for each DEM is reported as relative linear error (Table 3) and is expected to be less than the pixel scale of the DEM [21,39]. The horizontal precision of the DEM is reported to be equal to the pixel scale of the DEM, as the pixel scale is consistently greater than the relative circular (or horizontal) error at 90% confidence, as reported by SOCET SET [34,37].

4.2. Offsets from MLA

Positional accuracy was evaluated by comparing DEM elevations with MLA elevations. Wherever MLA profiles directly crossed the DEM, the mean, median, and standard deviation of the offsets were evaluated (Figure 6). However, if, due to the highly elliptical orbit of MESSENGER and the sparse MLA coverage, no co-located profiles were available, then offsets were reported from the broader area WAC DEMs (controlled to MLA profiles). Some DEMs in the southern hemisphere had no MLA coverage; these products were not absolutely controlled, and for these we report only the relative precision.

With the range accuracy of MLA better than 20 m, ideally the measured differences between the DEMs and MLA profiles would also be <20 m. However, despite ongoing refinement of both our error analysis and DEM processing methods, consistently obtaining this level of accuracy remained challenging; manual alignment in stereo was made more difficult by varying image resolutions within stereo pairs. The seven DEMs we controlled to MLA had a median mean offset of −9 m. These offsets are also affected by horizontal errors due to uncertainty in position and orientation for MLA profiles [28]. Nevertheless, in all but one case, we were able to achieve mean offsets less than the pixel scale of the DEM (Table 3).

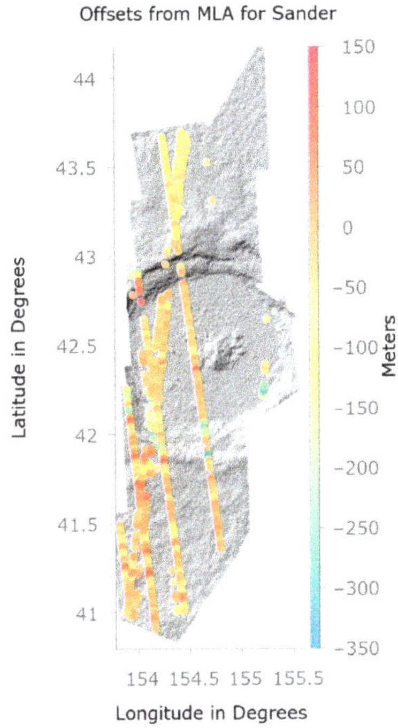

Figure 6. Plot showing the difference between MLA profiles and the Sander crater regional DEM. The DEM mosaic consists of 36 stereo pairs with a pixel scale of 162 m.

4.3. Comparison to Other Regional Topographic Products

The sparse MLA data does not allow us to fully evaluate the level of detail captured in the DEMs; therefore, we compared our products to overlapping DEMs from other publicly available regional datasets with overlapping DEMs and similar pixel scales (Table 4; Fassett [24]; Ostrach and Dundas [25]; and Preusker et al. [28]). Of the other high-resolution MDIS DEM datasets, we shared no overlapping DEMs with Tenthoff et al. [26], and Weirich et al.'s [27] DEMs are not yet publicly available.

Table 4. Regional MDIS DEM datasets used in this comparison.

Dataset	Method	Resolution Range (m/px)	Number of DEMs	ASU Site (# DEMs from Dataset That Overlap)
Manheim et al. (this study) ("ASU")	SOCET SET	78–500	9 mosaics; 57 individual DEMs	-
Fassett [24] ("FAS")	Ames Stereo Pipeline	45–245	96	Catullus (2), Degas (1), Raditladi (3), Sander (1)
Ostrach and Dundas [25] ("OST")	SOCET SET	25–120	11	Degas (1), Sander (1), Kertész (2)
Preusker et al. [28] ("PRU")	Custom stereophotogrammetry	222	4 quadrangles	Catullus (1), Degas (1), Kuiper (1)

Because the Ostrach and Dundas [25] ("OST") and Fassett [24] ("FAS") DEMs used reference radii of 2440.0 km, we added 600 m to these DEMs to adjust them to the 2439.4 km radii used by us ("ASU") and by Preusker et al. [28] ("PRU"). The FAS and OST DEMs were also produced using version 2 (crossover-corrected) of the MLA profiles, whereas only version 1 of the MLA profiles was available when our DEMs were created. In the regions we examined, there was a km-scale shift between the two versions of the MLA. To account for any remaining offsets, we applied translations produced by running the pc_align tool from the Ames Stereo Pipeline [40].

To collect profiles for comparison, all the DEMs were loaded into QGIS [41]. The QGIS Profile Tool plugin was used to simultaneously collect profiles across the overlapping DEMs [42]. These profiles were then exported for analysis.

The translations from pc_align were predominantly horizontal (see Figure 7b,c). The horizontal offsets originate at least partially from differences between the two versions of MLA data used; other minor factors are the sparseness of the MLA profiles and the fact that each product was controlled to different profiles and numbers of MLA points.

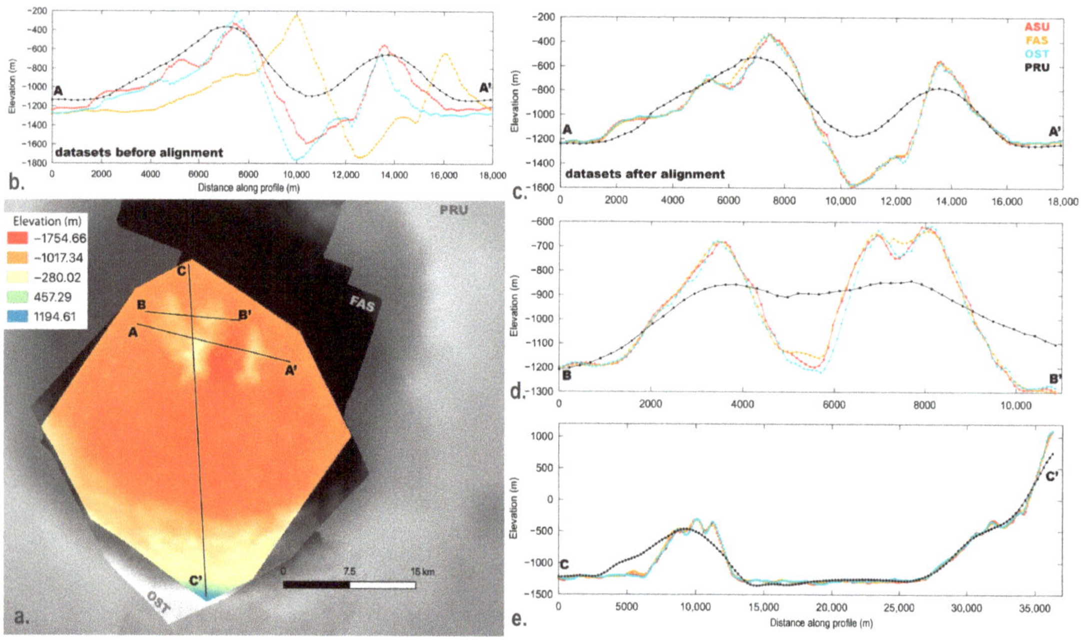

Figure 7. Degas crater profiles. (**a**) Colorized elevation map of ASU (97 m/px) shown over OST DEM (90 m/px), FAS DEM (95 m/px), and PRU H03 quadrangle (222 m/px). (**b**) Elevations for profile A for each DEM before alignment. (**c**) Elevations for profile A after alignment. (**d**,**e**) Post-alignment elevations for profiles B and C.

The remaining differences between the DEMs are largely due to the differing pixel scales, as expected. The 222 m/px PRU DEMs are significantly smoother and shallower than the ASU (90–97 m) Degas DEMs when observing small features (e.g., the central peaks; Figure 7c,d), but elevation ranges agree well over longer baselines (e.g., crater diameter; Figure 7e). The detail captured by our Degas DEM most closely matches the detail from OST, which was produced using the same method. Minor differences between ASU and FAS likely originated from the different extraction algorithms, as well as some remaining misalignment.

Profiles over Sander crater's central peak show that ASU, FAS, and OST DEMs capture similar features, but that the ASU mosaic, with approximately twice the pixel scale of FAS and OST, has more outliers and captures less fine detail (Figure 8). For the Sander site, we also viewed the DEMs as contour lines over an ASU DEM orthophoto (Figure 8b). For this small area, features in FAS are somewhat more smoothed out than in OST and ASU, while the ASU, as expected, contains more noise (see Figure 8c).

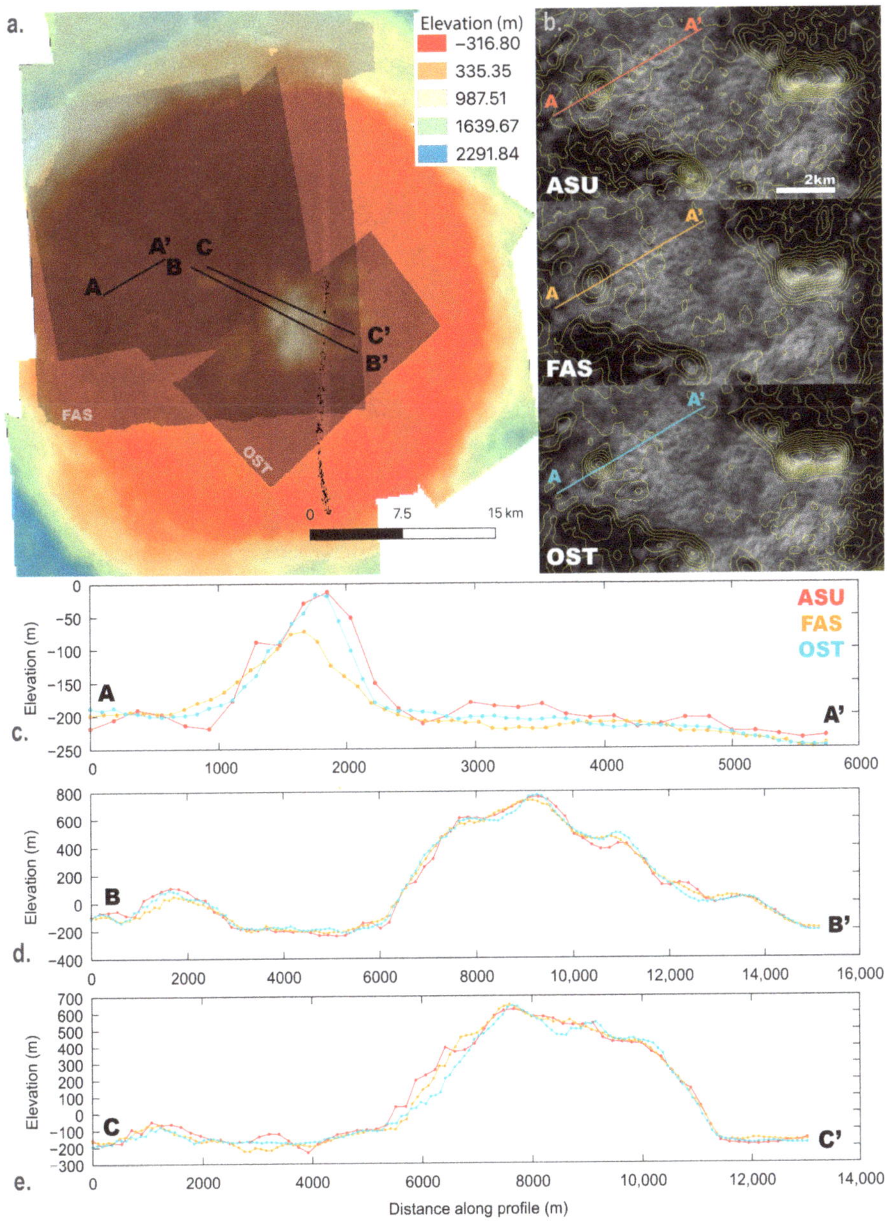

Figure 8. Sander crater topographic profiles. (**a**) Colorized elevation map of ASU (SANDRMS; 162 m/px). Gray polygons show the extents of FAS (90 m/px) and OST (80 m/px); both have been aligned to ASU. (**b**) ASU, FAS, and OST displayed as 30 m contours overlaid on an ASU orthophoto. (**c**–**e**) DEM elevations for profiles A, B, and C.

Our DEM of Kuiper crater was not controlled to MLA at all, as MLA becomes extremely sparse south of the equator. However, it does overlap with the PRU H06 quadrangle, which extends from 22.5°S to 22.5°N and therefore has MLA coverage over its northern half; its RMS vertical offset from MLA is 88 m [28]. We used pc_align to co-register the ASU Kuiper DEM to H06 and found a 934.4 m vertical offset, but a relatively good horizontal alignment (the optimal horizontal translation was <1 px for the ASU DEM) (Figure 9). At 270 m/px, the ASU DEM is a lower resolution than the 222 m/px quadrangle; however, more fine detail is captured in the ASU profile, likely due to differences in methodology and product coverage.

Figure 9. Kuiper crater profiles. Here, ASU DEM has been vertically translated by subtracting 934.4 m; no other alignment was performed. (**a**) ASU 90 m contour map overlaid on ASU orthophoto, shown over PRU DEM. (**b**) Elevations from ASU (270 m/px) and PRU (222 m/px) for profile A.

To more comprehensively evaluate the residual offsets between the DEMs, we created difference maps by subtracting the pc_aligned DEMs produced by FAS, OST, and PRU from the overlapping ASU DEMs. Table 5 shows the statistical summaries of these maps. The mean differences are within 4 m with the exception of three outliers; all are within 20 m. The means of the absolute differences are all less than 65 m. Larger mean absolute differences are associated with larger differences in pixel scales between the two DEMs. The standard deviations indicate fairly tight distributions; however, the minimums and maximums can be quite large, with total ranges between 285 m and 1226 m.

Following Bland et al., 2021 [43], we also created histograms of the difference maps to examine the distribution of offsets between the DEMs. They were approximately normally distributed, but with abnormally large tails. As a generally representative example, we show difference maps and histograms of Sander crater in Figure 10. Figure 10a,c show that the largest residual offsets are concentrated at the areas of most extreme topography: along the crater walls and at the steepest points of the central peak. Both FAS and OST here have an approximately 2× finer pixel scale than ASU, which may account for some of the residual offsets.

Table 5. Summary of differences between ASU DEMs and overlying DEMs.

ASU DEM	Subtracted DEM	Standard Deviation	Mean Diff. (m)	Std Dev. of Abs. Diff.	Mean Abs. Diff.	Minimum (m)	Maximum (m)
Catullus	FAS [a]	39.97	2.31	29.51	27.06	−417.34	284.83
Catullus	FAS [b]	62.27	−0.06	50.61	36.29	−873.00	352.96
Catullus	PRU	56.83	−6.71	41.25	39.66	−404.20	263.67
Degas	FAS	17.78	−0.18	13.39	11.69	−161.41	123.54
Degas	OST	19.77	0.81	14.51	13.46	−207.45	185.70
Degas	PRU	84.62	−10.45	66.54	53.31	−571.25	430.98
Kertész	OST [c]	20.23	0.22	14.67	13.93	−193.86	201.48
Kertész	OST [d]	37.57	−2.30	28.82	24.20	−256.25	315.43
Kuiper	DLR	55.77	0.93	37.21	41.55	−552.13	620.03
Raditladi	FAS [e]	85.43	19.87	59.39	64.55	−397.21	559.83
Raditladi	FAS [f]	31.30	2.46	22.49	21.90	−248.57	246.98
Raditladi	FAS [g]	31.79	2.97	23.24	21.89	−256.89	594.35
Sander	FAS	38.54	−0.60	28.07	26.42	−367.55	366.59
Sander	OST	35.70	−3.60	25.75	24.99	−237.77	236.10

Note: ASU DEM mosaics were used where available instead of individual pairs. Where ASU DEMs are compared to multiple DEMs from one other dataset, we disambiguate by providing the DEM names: [a] 293E22N_0, [b] 293E22N_1, [c] DEM_100m_MSGR_Kertesz_crater_2_isis3, [d] DEM_70m_MSGR_Kertesz_crater_M2015, [e] 118E26N_0, [f] 121E27N_0, [g] 121E27N_1.

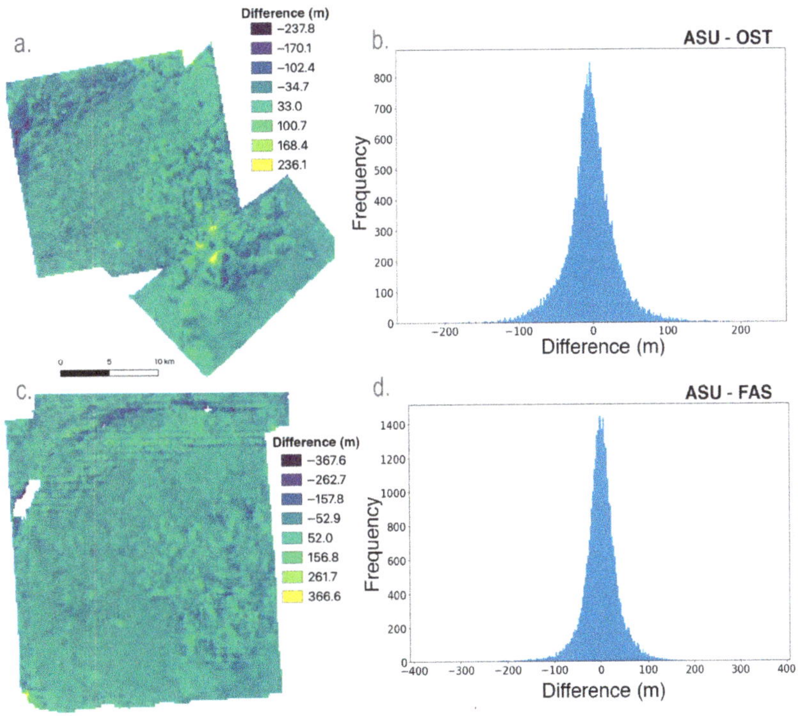

Figure 10. (**a**,**c**) Difference maps of Sander crater: OST and FAS DEMs subtracted from ASU DEM. (**b**,**d**) Histograms of the difference values.

Our comparisons with other regional DEMs show that once large offsets are removed, products have comparable elevation ranges. Most discrepancies in the small-scale features captured are attributable to differences in DEM pixel scale, although minor horizontal misalignment and noise are also factors. The ASU, FAS, and OST DEMs, which have similar areas of coverage and numbers of input images, capture similar amounts of detail when they have roughly the same pixel scales.

5. Discussion: Scientific Application

5.1. Anomalous High-Reflectance Regions in Kertész and Sander Craters

Blewett et al. [8,10,44,45] characterized distinctive features called "hollows" composed of relatively blue (433 nm to 559 nm ratio) and high-reflectance material. Hollows are shallow, rimless, irregularly shaped depressions, are distributed globally, and appear to form preferentially on crater floors, central peaks, and peak rings. The new DEMs of Sander crater, Kertész crater, and Raditladi basin presented here, all of which host hollows, provide the means to investigate formation theories proposed by previous studies.

5.2. Formation Hypotheses

Current hypotheses proposed to explain hollow formation agree that a volatile component is being lost, resulting in a collapse of the weakened matrix [10]. Several possibilities for the loss mechanism and the source of volatiles have been proposed. Vaughan et al. [46] suggested that sulfides or chlorides in differentiated impact melt could sublimate at surface temperatures. Blewett et al. [8,44] proposed that space weathering (sputtering by solar-wind ions, bombardment by micrometeoroids, exposure to solar heating, and ultraviolet fluxes) could cause the destruction and loss of volatile-bearing phases, such as sulfides. Alternately, Blewett et al. [8,44] noted that sequestered magmatic volatiles or fumarolic minerals from volcanic activity could later be exposed by impacts and lost by solar heating. Thomas et al. [9] suggested that endogenic heating of volatiles in upper crustal rocks could lead to sublimation. Phillips et al. [47] proposed a model in which thermal decay of sulfide minerals resulted in fumarolic systems that altered the near-surface and drove hollow formation. Measurements of hollow depths in high-resolution images have shown that hollows consistently have depths of tens of meters [8,9,24,45,46]. The mechanisms proposed generally posit that hollows grow in depth until either the material is exhausted or a lag deposit reaches sufficient thickness to halt the volatile loss [10].

For Kertész and Sander craters, we investigate three distinct floor units that are not observed in Raditladi: smooth sheet, pitted sheet, and bright knob/deep floor materials (Figures 11–13). Vaughan et al. [46] defined these units and proposed that they represent the progression of hollow formation from smooth sheet to pitted sheet to the final stage, where the lowest elevation of the deep floor represents the maximum depth of the hollow-forming layer. Because the Kertész hollows are larger and more sharply defined than those in Sander crater, and the Kertész DEM, at 120 m/px, has a finer pixel scale than Sander (162 m/px) or Raditladi (180 m/px), we focused our investigation primarily on Kertész crater.

5.3. Findings: Kertész Crater

Kertész is a 33 km diameter crater with a depth of 2.2 km. It has two central peaks, with heights of 550 m and 600 m from the surrounding crater floor. A slump is present on the south wall, and abutting the crater wall on the southwest side of the floor just west of the slump is a pit ~360 m deeper than the surrounding crater floor. Hollows are distributed over the entire crater floor, extending in some areas to slumped units in the crater walls and onto lower regions of the central peaks (Figure 11).

In a sample of Kertész's most well-defined hollows (six on the north/northwest side of the crater, four on the east/southeast sides; both regions defined by Vaughan et al. [46] as bright knob/deep floor), we calculated depth by subtracting minimum elevation on the hollow floor from the average hollow rim elevation. Measurements were taken in SOCET SET from the stereo pair released as KERTS01 using a stereo viewer (see Appendix C for red–blue anaglyphs of Kertész, Sander, and Raditladi stereo pairs). The images had pixel scales of 33 m and 40 m/px. The southeast hollows ranged from 20–62 m deep, while the significantly larger pits on the northwest side ranged from 35–101 m deep; the overall average depth was 51 m or a median depth of 48 m. Measurements taken from the DEM were somewhat shallower, averaging 30 m deep on the southeast side and 48 m deep on the northwest side. The stereo measurements, while involving a more ambiguous amount of uncertainty due to the user intervention required, are able to take advantage of the full resolution of the images; therefore, a stereo viewer may be required to extract the most data out of the stereo model. The depths found in stereo somewhat exceed the ~30 m maximum depth found by Vaughan et al. [46], possibly because that study used higher-resolution images and therefore was able to measure smaller hollows. The less well-defined hollows ("pitted sheet") were too small to measure in our DEM, although both profiles and measurements taken in stereo in SOCET SET suggest depths of <10 m. Our measurements do fall within the general ranges of typical hollow depths found by previous studies (tens of meters).

We also noted several large contiguous regions where the Kertész smooth sheet was not pitted (or very minimally pitted), but was concave to at least the depth of the surrounding deep floor units; first, in a 4 km by 1 km sheet to the southwest, bordering the deep pit mentioned earlier; and secondly, in a smaller (1 × 1 km) region adjacent to the taller of the central peaks. At their highest points, these two smooth sheets are raised above the surrounding deep floor units to approximately the same height as the tops of the bright knobs, but they sink at their centers to depths of 27 m for the southwest sheet and 72 m for the central sheet. For the southwest sheet, the deepest point is approximately the same elevation as the bottom of the surrounding pits. For the central sheet, the lowest point of the concavity is at least 40 m lower than the average floor of the surrounding bright knob/deep floor unit. We discuss these smooth sheet regions in our analysis in Section 5.6.

5.4. Findings: Sander Crater

Sander crater is larger than Kertész, with a diameter of 52 km and a depth of 2.5 km. The floor of Sander crater is dominated by the unit designated as pitted sheet; these hollows are generally too small to be reliably measured at the DEM pixel scale (Figure 12). However, five distinct bright knob/deep floor hollows were identified and measured from the stereo pair used to create SANDR18. These hollows had an average depth of 31 m when measured in stereo (range of 9–65 m).

5.5. Findings: Raditladi Basin

Hollows in Raditladi Basin surround the peak ring and are found on peak mountains (Figure 13). Unlike the hollows in Sander and Kertész, only one type of occurrence is found; the hollows are clearly defined and most closely resemble the bright knob/deep floor unit. However, the hollows are small enough (<1 km in diameter) that, in most cases, the DEM pixel scale (180 m/px) was not fine enough for precise measurements of individual hollows. Furthermore, the Raditladi DEM for the stereo pair measured (RADIT02) has a relative linear error of 137 m, significantly worse than either Kertész or Sander. We measured five of the largest hollows in stereo to find an average depth of 28 m (range of 24–32 m). This value differs from the average shadow-length depth for this area (44 m) reported by Blewett et al. [8]; however, our calculated depth likely represents a very conservative estimate due to the DEM's coarser resolution.

5.6. Analysis

Our results for the depth of the pitted regions, knobby regions, and individual hollows in Sander and Kertész are consistent with the depth of hollows found by other analyses. Blewett et al. [45] used shadow-length measurements on 565 high-resolution images to determine the depths of hollows and found an average depth of 24 m. Other shadow-length depth estimates of hollows [8,9,45,46] generally agree with the findings described here.

Fassett [24] used MDIS DEMs produced with the Ames Stereo Pipeline to examine hollows and found an average depth of 32 m.

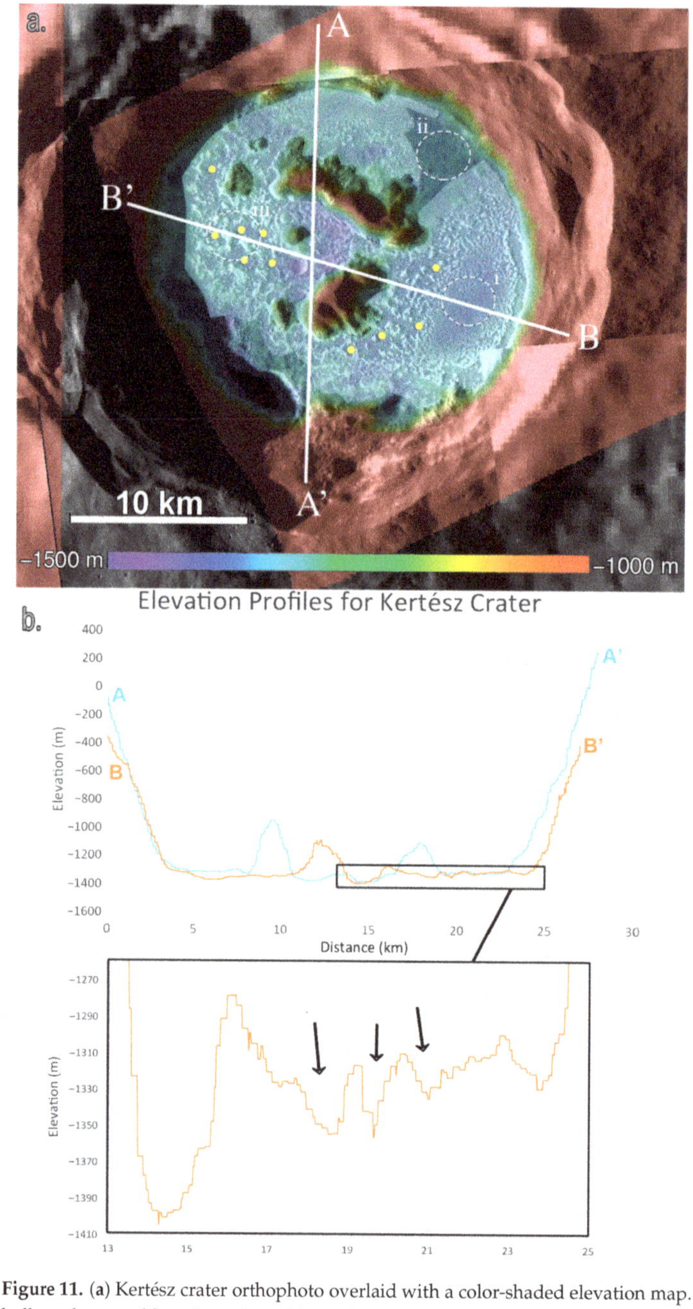

Figure 11. (**a**) Kertész crater orthophoto overlaid with a color-shaded elevation map. Polygons show hollowed areas of floor investigated here; dotted circles mark characteristic units of pitted terrain (i), smooth floor (ii), and bright knob/deep floor (iii). Yellow dots indicate ten hollows measured for this study. See Appendix A for stereo pair details. (**b**) Elevation profiles. Inset profile shows a detail of bright knob/deep floor area; arrows indicate distinct hollows.

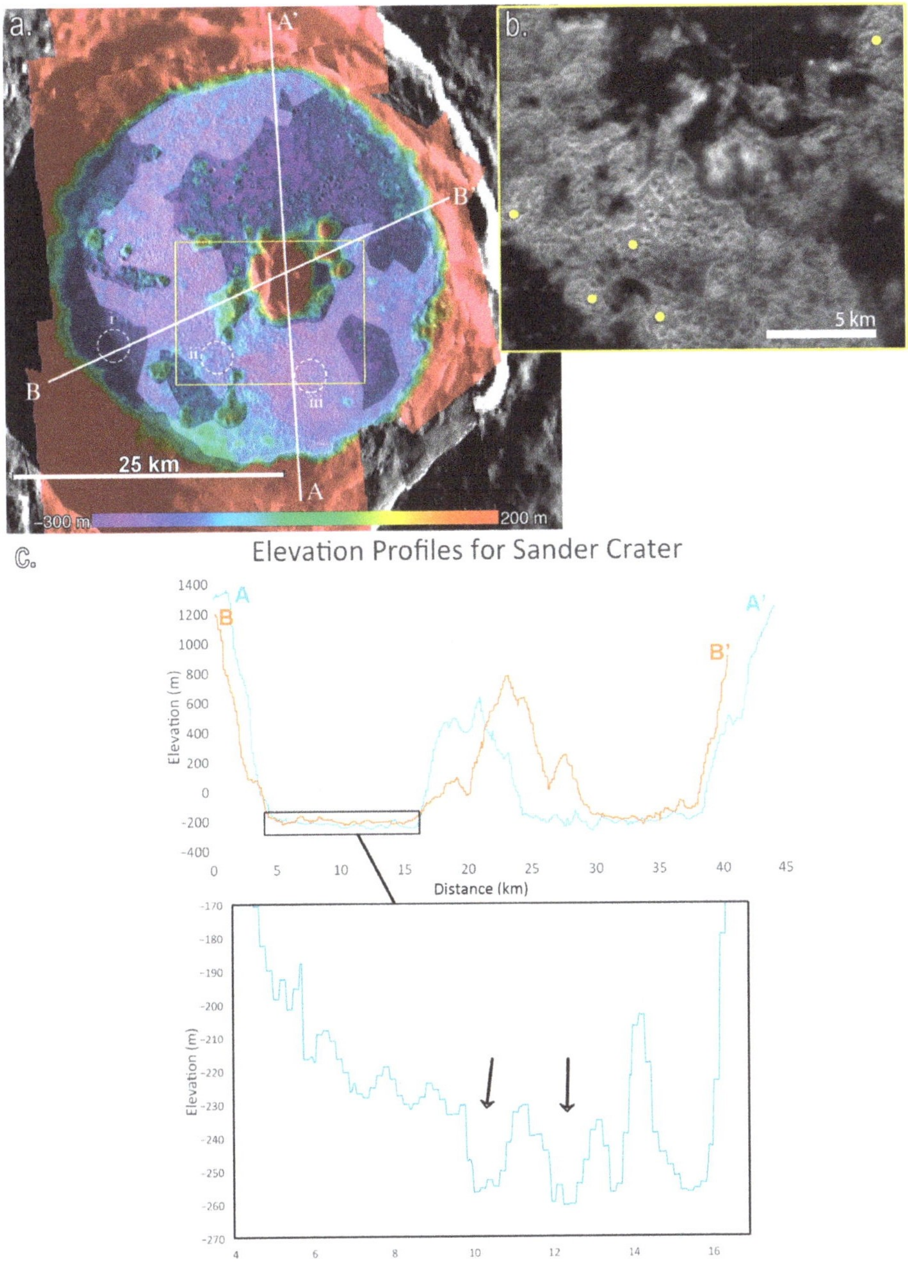

Figure 12. (**a**) Sander crater overlaid with a color-shaded elevation map. Polygons show hollowed areas of floor investigated here; dotted circles mark characteristic units of pitted terrain (i), smooth floor (ii), and bright knob/deep floor (iii). (**b**) Detail of SANDR18 orthophoto showing a portion of the crater floor and central peak. Yellow dots indicate the hollows measured for this study. See Appendix A for stereo pair details. (**c**) Elevation profiles. Inset profile shows a detail of bright knob/deep floor area; arrows indicate distinct hollows.

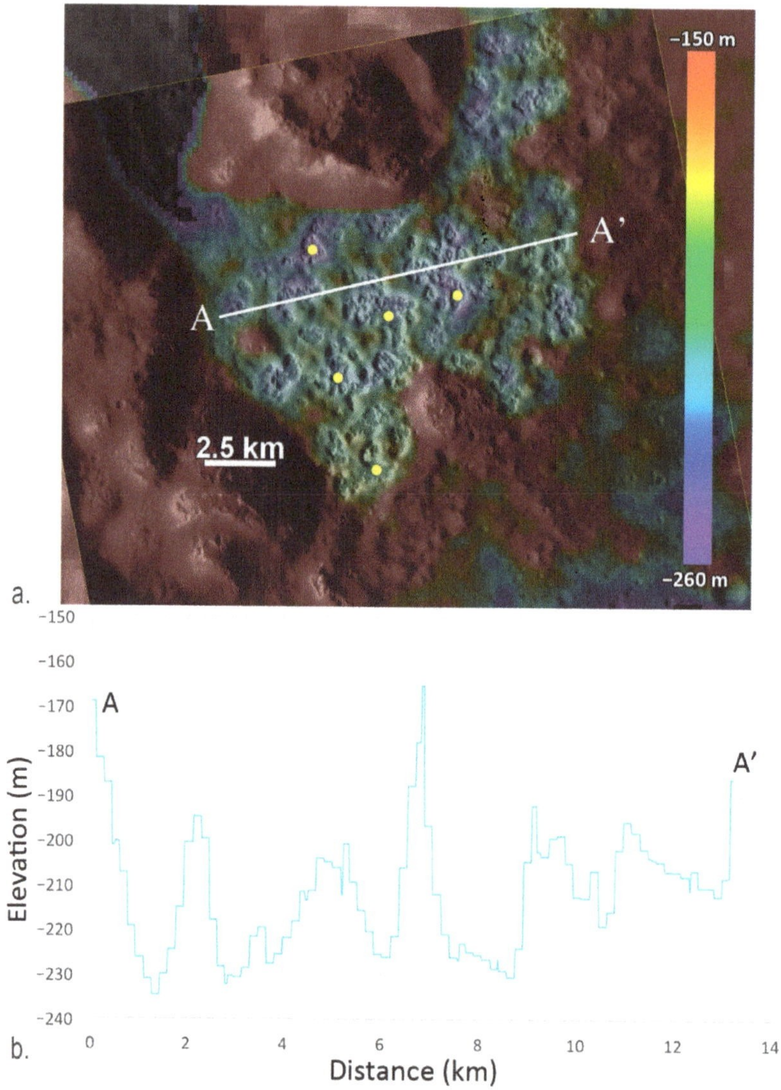

Figure 13. (**a**) Hollows along the peak ring of Raditladi Basin in color-shaded elevation map overlaid on associated orthophoto mosaic. The white line (A–A') indicates the extent of the profile taken across the hollowed region. Yellow dots indicate five individual hollows measured for this study. See Appendix A for stereo details. (**b**) Elevation profile.

Blewett et al. [45] interpreted the typical depth of several tens of meters to indicate that hollows increase in depth until a sufficient lag deposit of devolatilized material develops and protects the underlying material from further loss of volatiles. Our observations of Kertész, Sander, and Raditladi are consistent with this hypothesis. The presence of a deep, smooth, concave sheet in Kertész crater is also consistent with Vaughan et al. [46]. They proposed that the hollows in Kertész formed in a massive sulfide or chloride deposit that resulted from differentiation and immiscibility in the impact melt that pooled in the crater floor. The concavities revealed by the topography may be related to volatile loss from such a thick, relatively pure volatile-bearing layer. Hollows found elsewhere commonly occur in places with steep topographic slopes (e.g., crater central peaks and walls) where pooling and differentiation of impact melt seems unlikely [10]. Thus, formation of hollows within impact melt pools (as at Kertész) may differ from that in material containing a lower abundance of a volatile-bearing phase.

6. Conclusions

Nine science sites consisting of 57 stereo pairs were completed (Appendix A; Appendix B) and archived to the NASA PDS in December 2016 as part of the MESSENGER DEM PDS Archive [29] (https://pdsimage2.wr.usgs.gov/archive/mess-h-mdis-5-dem-elevation-v1.0/MESSDEM_1001/DEM/REGIONAL/IMG/, accessed on 20 July 2022). Created from 49 NAC images, these DEMs are some of the highest-resolution, most precise, and most accurate topographic products of Mercury currently available. However, these represent only a few of the thousands of stereo pairs acquired with pixel scales of 13–150 m. As well as producing more DEMs, future work could include using global MESSENGER DEM products as additional sources of absolute control to improve local DEM accuracy and resolve issues with sparse MLA data at lower latitudes. We could also further automate several of the steps in our methodology (adding tie points, registering to MLA). That said, one of the benefits of our method over other, faster, more automated methods is that we use manual intervention and stereo viewing to visually inspect and validate the entire model [21,43]. More precise, accurate DEMs will provide necessary data for geologic studies of unique features on Mercury's surface and complement observations by the SIMBIO-SYS instrument on the BepiColumbo spacecraft scheduled to enter orbit in 2025 [48].

Author Contributions: Conceptualization, M.S.R., M.R.H. and M.R.M.; methodology, M.R.H., M.R.M., S.S.S., M.C., K.J.B. and B.A.K.; software, K.J.B.; validation, M.R.H., M.R.M., S.S.S. and M.C.; formal analysis, M.R.M., M.R.H., H.R.K. and D.T.B.; investigation, H.R.K., B.A.K., M.R.M., M.R.H. and D.T.B.; resources, M.S.R. and K.J.B.; data curation, M.R.H. and M.R.M.; writing—original draft preparation, M.R.M., M.R.H. and H.R.K.; writing—review and editing, M.R.M., M.R.H., M.S.R., M.C., D.T.B., S.S.S. and K.J.B.; visualization, M.R.M., M.R.H. and H.R.K.; supervision, M.S.R.; project administration, M.S.R.; funding acquisition, M.S.R. All authors have read and agreed to the published version of the manuscript.

Funding: The majority of the work presented here was funded by NASA MESSENGER Project grant DTM3250022. The OSIRIS-REx Mission provided support for M. Chojnacki and S. Sutton. The contribution of D. Blewett was supported by NASA MESSENGER Participating Scientist grant NNX08AN29G.

Data Availability Statement: The MDIS NAC and WAC images, MLA data, and SPICE kernels used to produce the DEMs in this work are freely available through the MESSENGER PDS Archive at: http://pds-geosciences.wustl.edu/missions/messenger/index.htm (accessed on 20 July 2022). The DEMs themselves are archived in the PDS at https://doi.org/10.17189/1520282 (accessed on 20 July 2022).

Acknowledgments: We are deeply indebted to Annie Howington-Kraus, who shared with us the preliminary methodology, documentation, and pipelines for this work. We are grateful to Nick Aldana, Sarah Deitrick, Chris Hansen, Kerry Martin, and Corrine Rojas who worked to produce the products described here, and to Emerson Speyerer for his encouragement and thoughtful review.

Conflicts of Interest: The authors declare no conflict of interest. The funders had no role in the design of the study; in the collection, analyses, or interpretation of data; in the writing of the manuscript, or in the decision to publish the results.

Appendix A

List of DEMs created for each site and the images providing stereo for each DEM, DEM pixel scales, and relative linear errors. A DEM mosaic (indicated by a name ending in "MS") was created for each site with more than one associated DEM.

Site	DEM Name	Image 1	Image 2	Pixel Scale (m)	Relative Linear Error (m)
Vent in Catullus Crater	CATLS01	EN1020925599M	EN1020925715M	84	83.56
Cunningham Crater	CNGHM01	EN0215633679M	EN1012716022M	78	48.97
"	CNGHM02	EN0215633679M	EN1012716026M	78	48.00
"	CNGHM03	EN0215633673M	EN1025248438M	105	53.67
"	CNGHM04	EN0215633679M	EN1025248438M	105	53.28
"	CNGHM05	EN0215633673M	EN1012716026M	78	48.51
"	CNGHM06	EN0215633679M	EN1025248434M	102	52.89
"	CNGHM07	EN0215633685M	EN1025248434M	102	52.64
"	CNGHM08	EN1010038716M	EN1012716018M	78	82.29
"	CNGHM09	EN1010038716M	EN1012716022M	78	84.69
"	CNGHMMS	-	-	105	84.69
Degas Crater	DEGAS01	EN1008944217M	EN1024240075M	97	69.26
Scarp in NW Rembrandt Basin	HYNEK01	EN1015859904M	EN1015861167M	500	383.01
Kertész Crater	KERTS01	EN1010096377M	EN1025335005M	120	71.46
"	KERTS02	EN1010096377M	EN1010096528M	120	97.34
"	KERTSMS	-	-	120	97.34
Kuiper Crater	KUIPR01	EN0223659984M	EN1002982634M	270	160.37
Paramour Rupes	PARMR01	EN1045386684M	EN1045387902M	450	256.82
"	PARMR02	EN1045386684M	EN1045387906M	432	256.94
"	PARMR03	EN1045386688M	EN1045387906M	432	260.93
"	PARMRMS	-	-	450	260.93
Raditladi Basin	RADIT01	EN1015425730M	EN1015426008M	180	134.99
"	RADIT02	EN1015425734M	EN1015426008M	180	136.62
"	RADIT03	EN1015425734M	EN1015426012M	180	134.74
"	RADITMS	-	-	180	136.62
Sander Crater	SANDR01	EN1012745010M	EN0218289182M	105	31.95
"	SANDR02	EN1012745010M	EN0218289189M	103	31.11
"	SANDR03	EN1012745010M	EN0217906385M	160	56.26
"	SANDR04	EN1012745010M	EN0233476482M	92	21.43
"	SANDR05	EN1010125026M	EN0217906385M	160	45.96
"	SANDR06	EN1012745056M	EN1010125151M	88	66.78
"	SANDR07	EN1012745056M	EN0218289182M	105	37.46
"	SANDR08	EN1012745056M	EN0218289189M	103	36.23
"	SANDR09	EN1012745056M	EN0217906385M	160	56.36
"	SANDR10	EN1012745056M	EN0233476482M	92	23.73
"	SANDR11	EN1010125151M	EN0217906385M	160	50.19
"	SANDR12	EN1010125151M	EN1010067392M	88	67.22
"	SANDR13	EN1010125151M	EN1010067396M	88	69.59
"	SANDR14	EN0218289168M	EN0217906381M	162	40.46
"	SANDR15	EN0218289175M	EN0217906381M	162	39.96
"	SANDR16	EN0218289175M	EN0217906385M	160	40.05
"	SANDR17	EN0218289175M	EN1012773887M	107	44.74
"	SANDR18	EN0218289182M	EN0217906385M	160	39.62
"	SANDR19	EN0218289182M	EN1012773893M	105	44.19

Site	DEM Name	Image 1	Image 2	Pixel Scale (m)	Relative Linear Error (m)
"	SANDR20	EN0218289189M	EN0217906385M	160	39.11
"	SANDR21	EN0218289189M	EN0217906389M	157	39.09
"	SANDR22	EN0218289189M	EN1012773893M	103	42.36
"	SANDR23	EN0218289189M	EN1012773899M	103	43.55
"	SANDR24	EN0218289189M	EN1012773904M	103	44.59
"	SANDR25	EN0218289196M	EN0217906389M	157	38.6
"	SANDR26	EN0218289196M	EN1012773904M	101	42.77
"	SANDR27	EN0217906381M	EN1012773887M	162	48.77
"	SANDR28	EN0217906385M	EN1010067392M	160	54.88
"	SANDR29	EN0217906385M	EN1012773887M	160	48.64
"	SANDR30	EN0217906385M	EN1012773893M	160	48.35
"	SANDR31	EN0233476482M	EN1012773887M	92	25.61
"	SANDR32	EN0233476482M	EN1012773893M	92	26.36
"	SANDR33	EN0233476482M	EN1012773899M	92	27.14
"	SANDR34	EN0233476487M	EN1012773899M	91	26.22
"	SANDR35	EN0233476487M	EN1012773904M	91	26.99
"	SANDR36	EN0233476492M	EN1012773904M	90	26.11
"	SANDRMS	-	-	162	69.59

Appendix B

Color-shaded relief maps of all completed regional MESSENGER DEMs (nine sites).

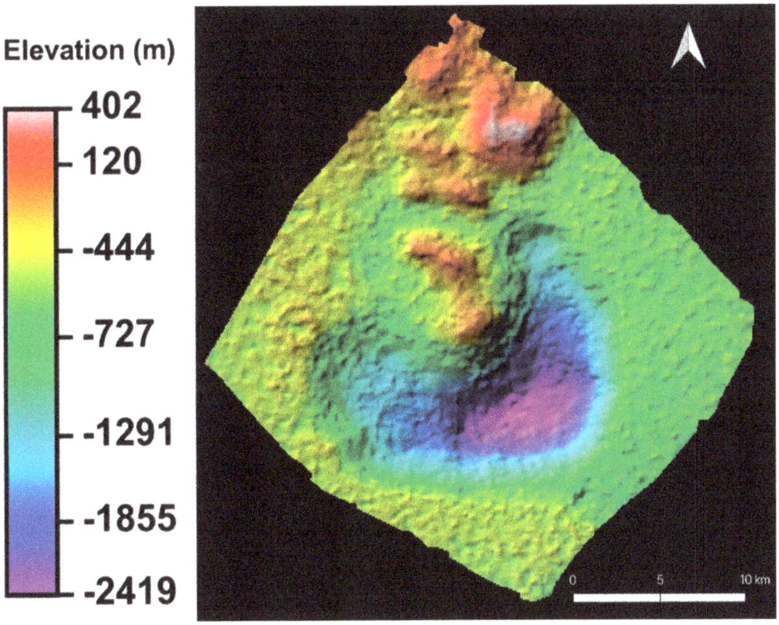

Figure A1. Vent in Catullus Crater.

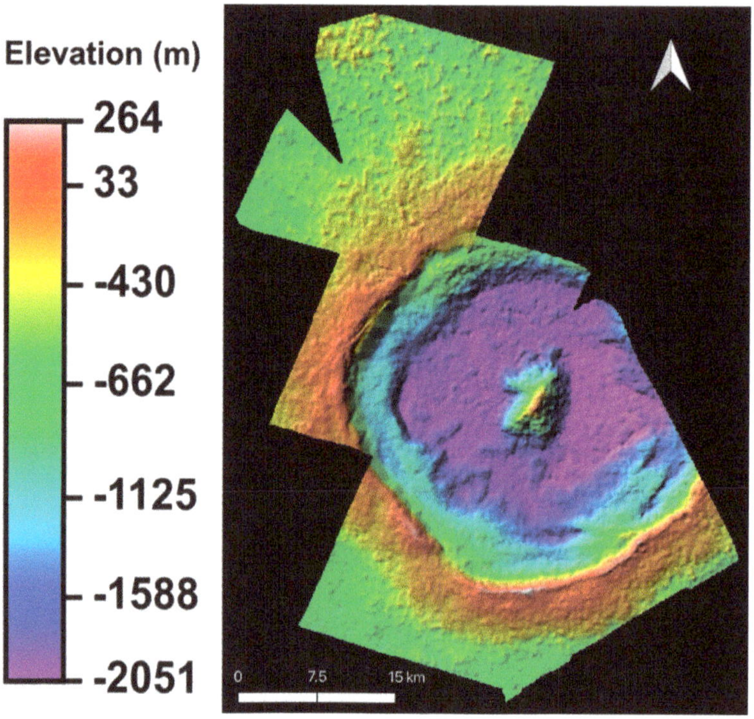

Figure A2. Cunningham Crater.

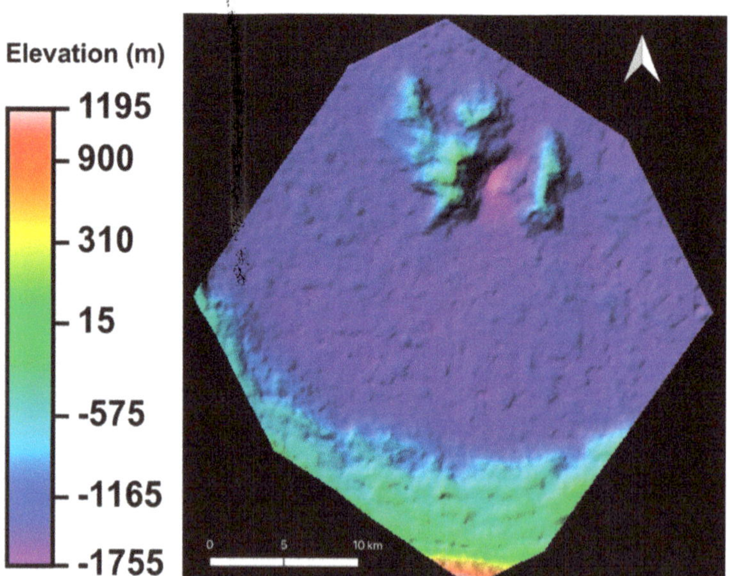

Figure A3. Degas Crater.

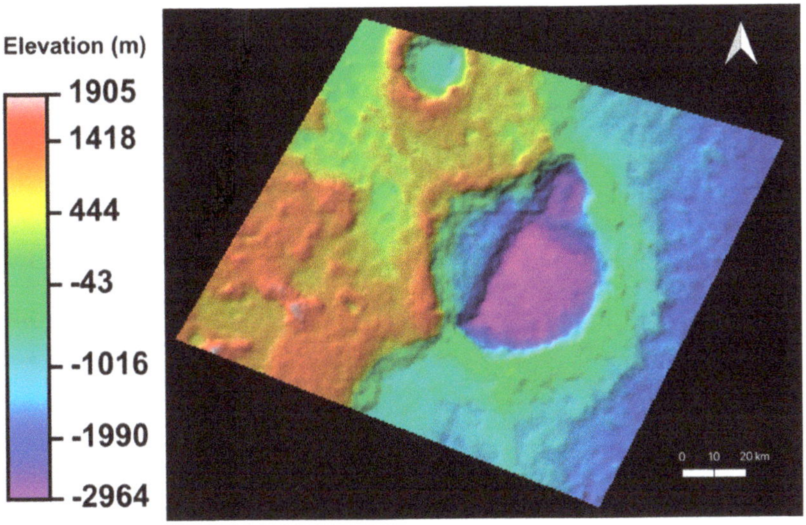

Figure A4. Scarp in northwest Rembrandt Basin.

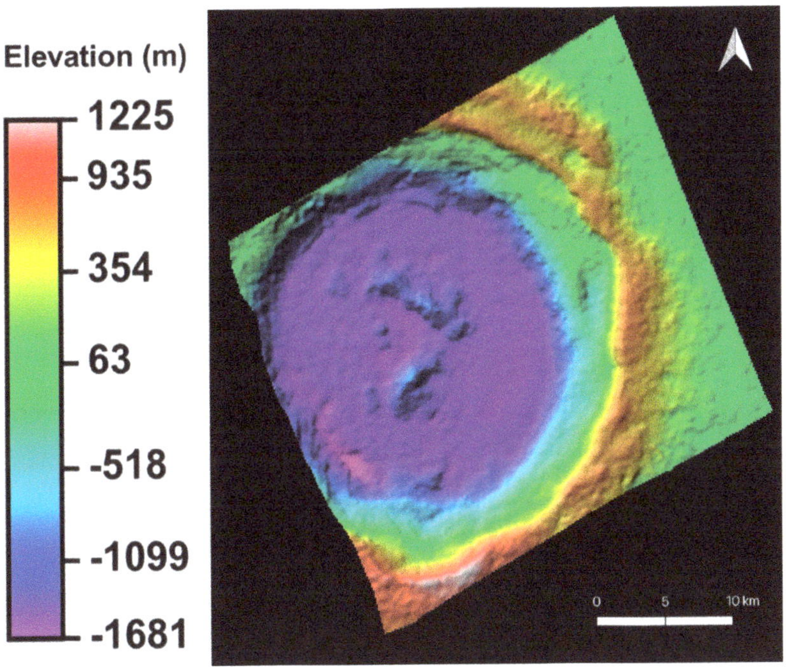

Figure A5. Kertész Crater.

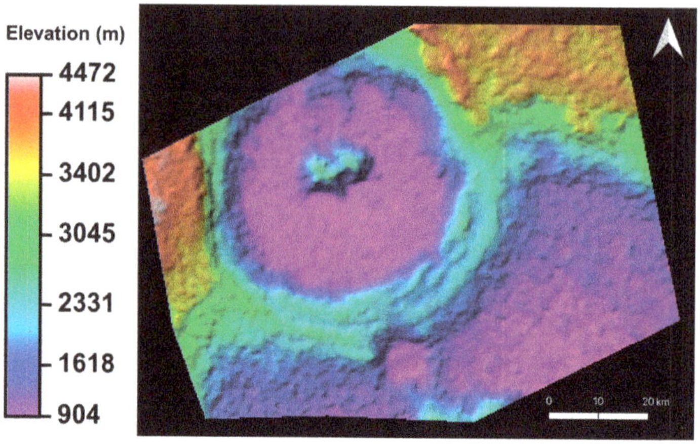

Figure A6. Kuiper Crater.

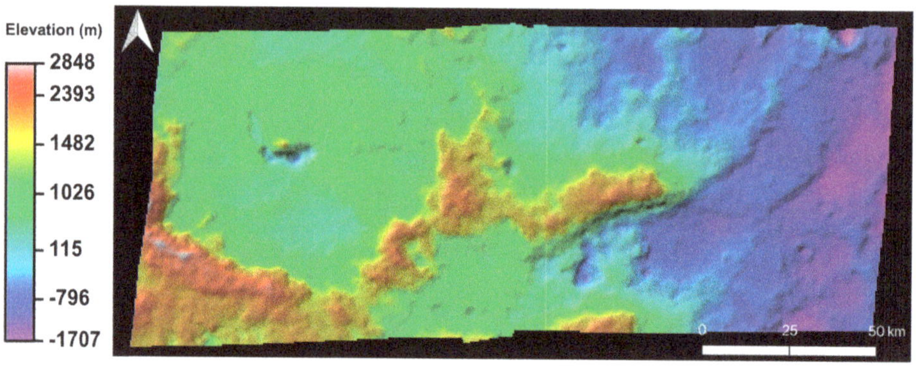

Figure A7. Paramour Rupes.

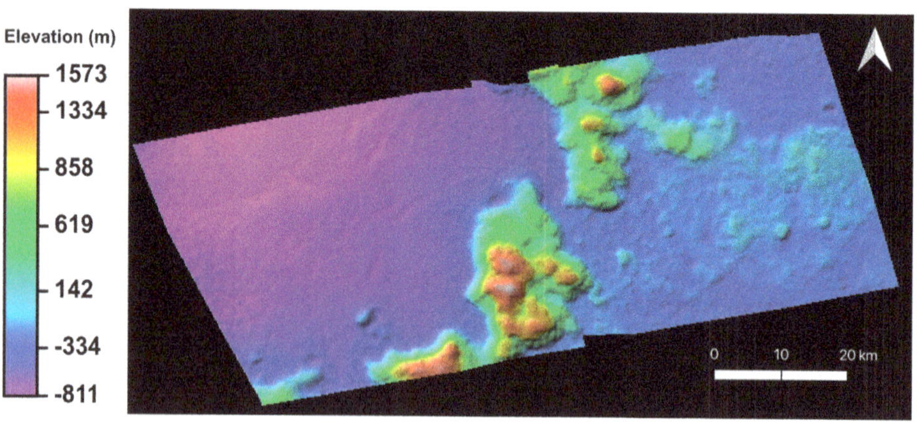

Figure A8. Hollows in Raditladi Basin.

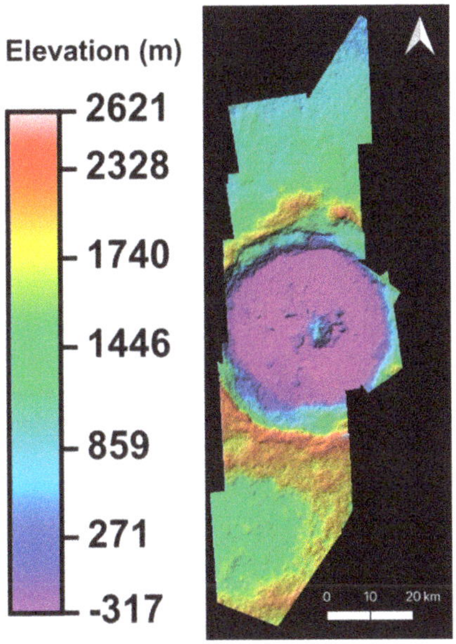

Figure A9. Sander Crater.

Appendix C

Anaglyphs of DEMs containing hollowed regions. Note that anaglyphs are vertically exaggerated by approximately 10× the true vertical distance.

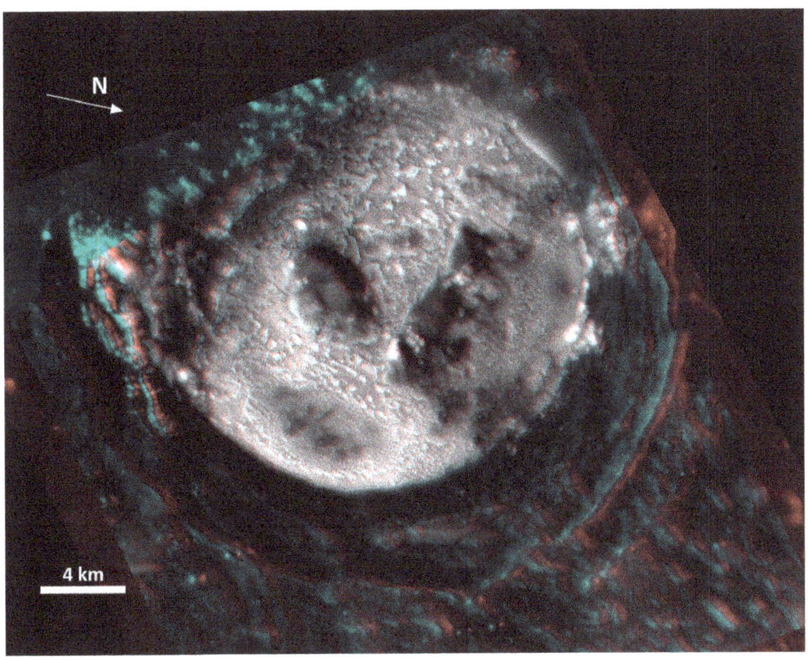

Figure A10. Kertész Crater. Images: EN1010096377M, EN1010096528M.

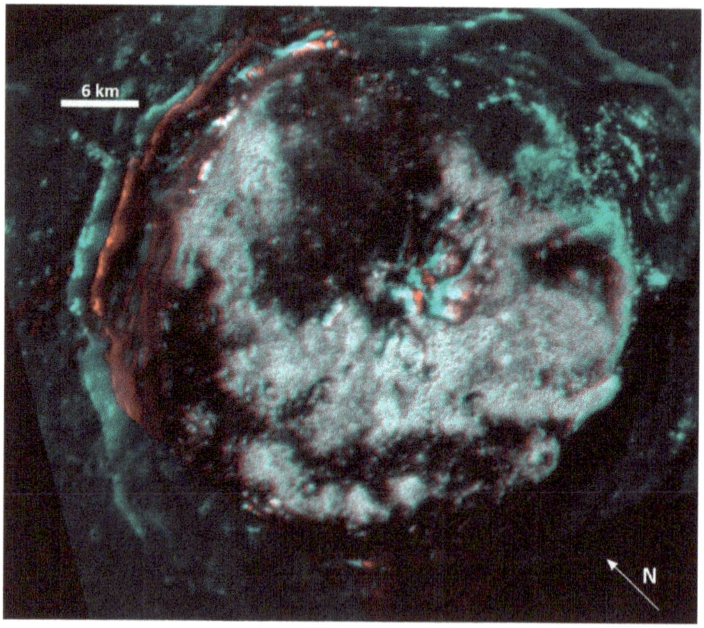

Figure A11. Sander Crater. Images: EN0218289182M, EN0217906385M.

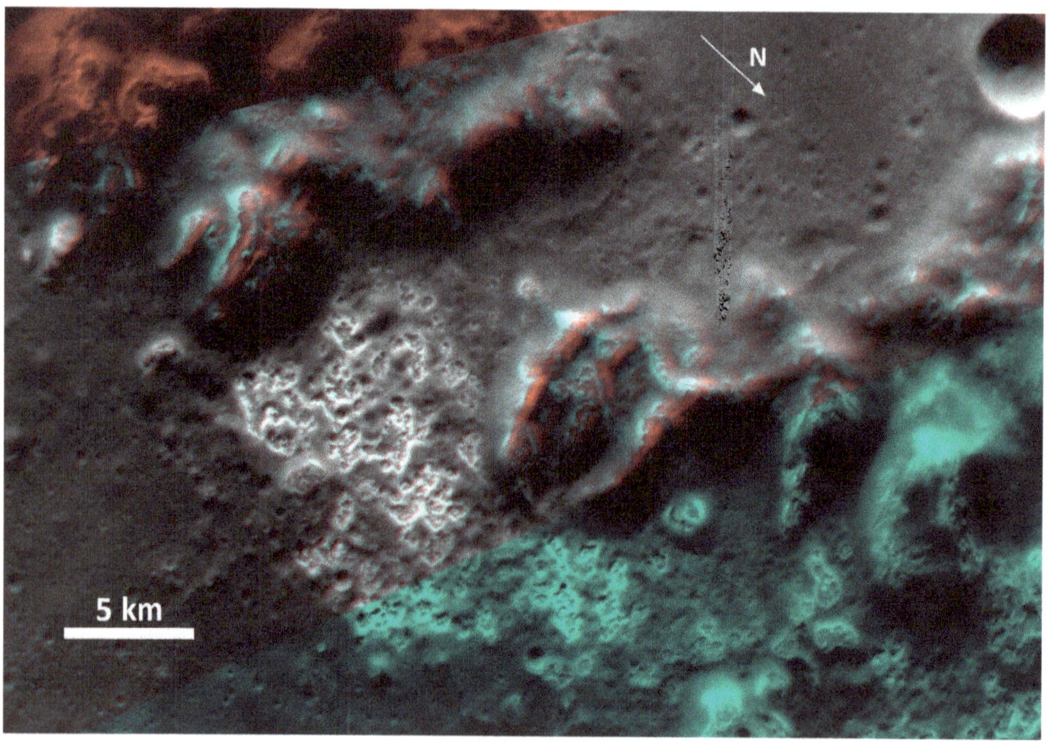

Figure A12. Raditladi Basin. Images: EN1015426008M, EN1015425734M.

References

1. Hawkins, S.E., III; Boldt, J.D.; Darlington, E.H.; Espiritu, R.; Gold, R.E.; Gotwols, B.; Grey, M.P.; Hash, C.D.; Hayes, J.R.; Jaskulek, S.E.; et al. The Mercury Dual Imaging System on the MESSENGER Spacecraft. *Space Sci. Rev.* **2007**, *131*, 247–338. [CrossRef]
2. Hawkins, S.E., III; Murchie, S.L.; Becker, K.J.; Selby, C.M.; Turner, F.S.; Noble, M.W.; Chabot, N.L.; Choo, T.H.; Darlington, E.H.; Denevi, B.W.; et al. In-flight performance of MESSENGER's Mercury Dual Imaging System. *SPIE Opt. Eng. Appl.* **2009**, *7441*, 74410Z. [CrossRef]
3. Chabot, N.L.; Denevi, B.W.; Murchie, S.L.; Hash, C.D.; Ernst, C.M.; Blewett, D.T.; Nair, H.; Laslo, N.R.; Solomon, S.C. Mapping Mercury: Global imaging strategy and products from the MESSENGER mission. In Proceedings of the 47th Annual Lunar and Planetary Science Conference, The Woodlands, TX, USA, 21–25 May 2016. Abstract #1256.
4. Moessner, D.P.; McAdams, J.V. Design, implementation, and outcome of MESSENGER's trajectory from launch to impact. In Proceedings of the Astrodynamics Specialist Conference, American Astronautical Society (AAS 15-608), Vail, CO, USA, 9–13 August 2015.
5. Sun, X.; Neumann, G.A. Calibration of the Mercury Laser Altimeter on the MESSENGER Spacecraft. *IEEE Trans. Geosci. Remote Sens.* **2015**, *53*, 2860–2874. [CrossRef]
6. Solomon, S.C.; Anderson, B.J. The MESSENGER Mission: Science and Implementation Overview. In *Mercury: The View after MESSENGER*; Cambridge University Press: Cambridge, UK, 2018; pp. 1–29. [CrossRef]
7. Bedini, P.D.; Solomon, S.C.; Finnegan, E.J.; Calloway, A.B.; Ensor, S.L.; McNutt, R.L.; Anderson, B.J.; Prockter, L.M. MESSENGER at Mercury: A mid-term report. *Acta Astronaut.* **2012**, *81*, 369–379. [CrossRef]
8. Blewett, D.T.; Chabot, N.L.; Denevi, B.W.; Ernst, C.M.; Head, J.W.; Izenberg, N.R.; Murchie, S.L.; Solomon, S.C.; Nittler, L.R.; McCoy, T.J.; et al. Hollows on Mercury: MESSENGER Evidence for Geologically Recent Volatile-Related Activity. *Science* **2011**, *333*, 1856–1859. [CrossRef]
9. Thomas, R.J.; Rothery, D.A.; Conway, S.J.; Anand, M. Hollows on Mercury: Materials and mechanisms involved in their formation. *Icarus* **2014**, *229*, 221–235. [CrossRef]
10. Blewett, D.T.; Ernst, C.M.; Murchie, S.L.; Vilas, F. Mercury's Hollows. In *Mercury: The View after MESSENGER*; Cambridge University Press: Cambridge, UK, 2018; pp. 324–345. [CrossRef]
11. Cavanaugh, J.F.; Smith, J.C.; Sun, X.; Bartels, A.E.; Ramos-Izquierdo, L.; Krebs, D.J.; McGarry, J.F.; Trunzo, R.; Novo-Gradac, A.M.; Britt, J.L.; et al. The Mercury Laser Altimeter Instrument for the MESSENGER Mission. *Space Sci. Rev.* **2007**, *131*, 451–479. [CrossRef]
12. Solomon, S.C.; McNutt, R.L.; Gold, R.E.; Domingue, D.L. MESSENGER Mission Overview. *Space Sci. Rev.* **2007**, *131*, 3–39. [CrossRef]
13. MESSENGER MLA Team. MESSENGER MLA Calibrated (CDR/RDR) Data Bundle, Version 1. 2018. Available online: https://doi.org/10.17189/1518580 (accessed on 20 July 2022).
14. Zuber, M.T.; Smith, D.E.; Phillips, R.J.; Solomon, S.C.; Neumann, G.A.; Hauck, S.A., II; Peale, S.J.; Barnouin, O.S.; Head, J.W.; Johnson, C.L.; et al. Topography of the Northern Hemisphere of Mercury from MESSENGER Laser Altimetry. *Science* **2012**, *336*, 217–220. [CrossRef]
15. MESSENGER MLA Team. MESSENGER MLA CDR/RDR Data Set Description, Version 1. Lidvid urn:nasa:pds:mess_mla_calibrated:document:mess_mla_cdr_rdr_ds::1.0. 2018. Available online: https://pds-geosciences.wustl.edu/messenger/mess-e_v_h-mla-3_4-cdr_rdr-data-v1/messmla_2001/document/mess_mla_cdr_rdr_ds.txt (accessed on 14 July 2022).
16. Anderson, J.A.; Sides, S.C.; Soltesz, D.L.; Sucharski, T.L.; Becker, K.J. Modernization of the Integrated Software for Imagers and Spectrometers. In Proceedings of the 35th Lunar and Planetary Science Conference, Houston, TX, USA, 15 March 2004. Abstract #2039.
17. Keszthelyi, L.; Becker, T.; Sides, S.; Barrett, J.; Cook, D.; Lambright, S.; Lee, E.; Milazzo, M.; Oyama, K.; Richie, J.; et al. Support and future vision for the Integrated Software for Imagers and Spectrometers. In Proceedings of the 44th Lunar and Planetary Science Conference, The Woodlands, TX, USA, 6–8 March 2013. Abstract #2546.
18. Adoram-Kershner, L.; Berry, K.; Lee, K.; Laura, J.; Mapel, J.; Paquette, A.; Rodriguez, K.; Sanders, A.; Sides, S.; Weller, L.; et al. USGS Astrogeology/ISIS3: ISIS3.10.1 Public Release. 2020. Available online: https://zenodo.org/record/3697255 (accessed on 20 July 2022).
19. Miller, S.B.; Walker, A.S. Further developments of Leica digital photogrammetric systems by Helava. *ACSM/ASPRS Annu. Conv. Expo.* **1993**, *3*, 256–263.
20. Miller, S.B.; Walker, A.S. Die Entwicklung der digitalen photogrammetrischen Systeme von Leica und Helava. *Z. Photogramm. Fernerkund.* **1995**, *63*, 4–16.
21. Henriksen, M.R.; Manheim, M.R.; Burns, K.N.; Seymour, P.; Speyerer, E.J.; Deran, A.; Boyd, A.K.; Howington-Kraus, E.; Rosiek, M.R.; Archinal, B.A.; et al. Extracting accurate and precise topography from LROC Narrow Angle Camera stereo observations. *Icarus* **2017**, *283*, 122–137. [CrossRef]
22. Kirk, R.L.; Howington-Kraus, E.; Rosiek, M.R.; Anderson, J.A.; Archinal, B.A.; Becker, K.J.; Cook, D.A.; Galuszka, D.M.; Geissler, P.E.; Hare, T.; et al. Ultrahigh resolution topographic mapping of Mars with MRO HiRISE stereo images: Meter-scale slopes of candidate Phoenix landing sites. *J. Geophys. Res. Planets* **2008**, *113*. [CrossRef]

23. Sutton, S.S.; Chojnacki, M.; McEwen, A.S.; Kirk, R.L.; Dundas, C.M.; Schaefer, E.I.; Conway, S.J.; Diniega, S.; Portyankina, G.; Landis, M.E.; et al. Revealing Active Mars with HiRISE Digital Terrain Models. *Remote Sens.* **2022**, *14*, 2403. [CrossRef]
24. Fassett, C.I. Ames stereo pipeline-derived digital terrain models of Mercury from MESSENGER stereo imaging. *Planet. Space Sci.* **2016**, *134*, 19–28. [CrossRef]
25. Ostrach, L.R.; Dundas, C.M. Topographic assessment of hollows on Mercury: Distinguishing among formation hypotheses. In Proceedings of the Lunar and Planetary Science Conference XLVIII, The Woodlands, TX, USA, 20–24 March 2017. Abstract #1656.
26. Tenthoff, M.; Wohlfarth, K.; Wöhler, C. High Resolution Digital Terrain Models of Mercury. *Remote Sens.* **2020**, *12*, 3989. [CrossRef]
27. Weirich, J.R.; Domingue, D.L.; Palmer, E.E.; Rodriguez, A. High-Resolution Topography and Photometric Data Cubes for Mercury and the Moon. In Proceedings of the Lunar and Planetary Science Conference LIII, The Woodlands, TX, USA, 7–11 March 2022. Abstract #1115.
28. Preusker, F.; Stark, A.; Oberst, J.; Matz, K.-D.; Gwinner, K.; Roatsch, T.; Watters, T.R. Toward high-resolution global topography of Mercury from MESSENGER orbital stereo imaging: A prototype model for the H6 (Kuiper) quadrangle. *Planet. Space Sci.* **2017**, *142*, 26–37. [CrossRef]
29. Becker, K.; Robinson, M.; Pruesker, F. MESSENGER MDIS DEM V1.0, NASA Planetary Data System. 2015. Available online: https://doi.org/10.17189/1520282 (accessed on 20 July 2022).
30. Becker, K.J.; Archinal, B.A.; Hare, T.H.; Kirk, R.L.; Howington-Kraus, E.; Robinson, M.S.; Rosiek, M.R. Criteria for automated identification of stereo image pairs. In Proceedings of the Lunar and Planetary Science Conference XLVI, The Woodlands, TX, USA, 16–20 March 2015. Abstract #2703.
31. McAdams, J.V.; Bryan, C.G.; Bushman, S.S.; Calloway, A.B.; Carranza, E.; Flanigan, S.H.; Kirk, M.N.; Korth, H.; Moessner, D.P.; O'Shaughnessy, D.J.; et al. Engineering MESSENGER's grand finale at Mercury—The low-altitude hover campaign. In Proceedings of the Astrodynamics Specialist Conference, American Astronautical Society (AAS 15-634), Vail, CO, USA, 9–11 August 2015.
32. Becker, K.J.; Edmundson, K.; Weller, L.; Isbell, C.; Bowman-Cisneros, E.; Speyerer, E.; Henriksen, M.; Preusker, F.; Ensor, S.; Reid, M.; et al. MESSENGER Digital Elevation Model Software Interface Specification, Version 1.6. 2017. Available online: https://pdsimage2.wr.usgs.gov/archive/mess-h-mdis-5-dem-elevation-v1.0/MESSDEM_1001/DOCUMENT/MSGR_DEM_SIS.PDF (accessed on 20 July 2022).
33. Forstner, W.; Wrobel, B.; Paderes, F.; Fraser, C.S.; Dolloff, J.; Mikhail, E.M.; Rujikietgumjorn, W. Analytical photogrammetric operations. In *Manual of Photogrammetry*, 6th ed.; McGlone, J.C., Ed.; American Society for Photogrammetry and Remote Sensing: Bethesda, MD, USA, 2013; pp. 785–955.
34. BAE Systems. SOCET SET User's Manual; Version 5.6; BAE Systems. 2011. Available online: https://www.geospatialexploitationproducts.com/content/ (accessed on 20 July 2022).
35. BAE Systems. Next-Generation Automatic Terrain Extraction (NGATE). White Paper. 2007. Available online: https://www.geospatialexploitationproducts.com/wp-content/uploads/2016/01/wp_ngate.pdf (accessed on 20 July 2022).
36. Zhang, B. Towards a higher level automation in softcopy photogrammetry: NGATE and LIDAR processing in SOCET SET. In Proceedings of the GeoCue Corporation 2nd Annual Technical Exchange Conference, Nashville, TN, USA, 26–27 September 2006.
37. Miller, S. Photogrammetric products. In *Manual of Photogrammetry*, 6th ed.; McGlone, J.C., Ed.; American Society for Photogrammetry and Remote Sensing: Bethesda, MD, USA, 2013; pp. 1009–1043.
38. Warmerdam, F. The Geospatial Data Abstraction Library. In *Open Source Approaches in Spatial Data Handling*; Hall, G.B., Leahy, M.G., Eds.; Springer: Berlin/Heidelberg, Germany, 2008; pp. 87–104. [CrossRef]
39. Burns, K.N.; Speyerer, E.J.; Robinson, M.S.; Tran, T.; Rosiek, M.R.; Archinal, B.A.; Howington-Kraus, E.; the LROC Science Team. Digital Elevation Models and Derived Products from LROC NAC stereo observations. *ISPRS Int. Arch. Photogramm. Remote Sens. Spat. Inf. Sci.* **2012**, *39*, 483–488. [CrossRef]
40. Beyer, R.A.; Alexandrov, O.; McMichael, S. The Ames Stereo Pipeline: NASA's Open Source Software for Deriving and Processing Terrain Data. *Earth Space Sci.* **2018**, *5*, 537–548. [CrossRef]
41. QGIS.org. QGIS Geographic Information System. QGIS Association. 2022. Available online: http://www.qgis.org (accessed on 6 June 2022).
42. Jurgiel, B.; Verchere, P.; Tourigny, E.; Becerra, J. Profile Tool, v4.1.7. PANOimagen S. L. Available online: https://github.com/PANOimagen/profiletool (accessed on 20 July 2022).
43. Bland, M.T.; Kirk, R.L.; Galuszka, D.M.; Mayer, D.P.; Beyer, R.A.; Fergason, R.L. How Well Do We Know Europa's Topography? An Evaluation of the Variability in Digital Terrain Models of Europa. *Remote Sens.* **2021**, *13*, 5097. [CrossRef]
44. Blewett, D.T.; Vaughan, W.M.; Xiao, Z.; Chabot, N.L.; Denevi, B.W.; Ernst, C.M.; Helbert, J.; D'Amore, M.; Maturilli, A.; Head, J.W.; et al. Mercury's hollows: Constraints on formation and composition from analysis of geological setting and spectral reflectance. *J. Geophys. Res. Planets* **2013**, *118*, 1013–1032. [CrossRef]
45. Blewett, D.T.; Stadermann, A.C.; Susorney, H.C.; Ernst, C.M.; Xiao, Z.; Chabot, N.L.; Denevi, B.W.; Murchie, S.L.; McCubbin, F.M.; Kinczyk, M.J.; et al. Analysis of MESSENGER high-resolution images of Mercury's hollows and implications for hollow formation. *J. Geophys. Res. Planets* **2016**, *121*, 1798–1813. [CrossRef]
46. Vaughan, W.M.; Helbert, J.; Blewett, D.T.; Head, J.W.; Murchie, S.L.; Gwinner, K.; McCoy, T.J.; Solomon, S.C. Hollow-forming layers in impact craters on Mercury: Massive sulfide or chloride deposits formed by impact melt differentiation? In Proceedings of the Lunar and Planetary Science Conference XLIII, The Woodlands, TX, USA, 7–11 March 2022. Abstract #1187.

47. Phillips, M.S.; Moersch, J.E.; Viviano, C.E.; Emery, J.P. The lifecycle of hollows on Mercury: An evaluation of candidate volatile phases and a novel model of formation. *Icarus* **2021**, *359*, 114306. [CrossRef]
48. Benkhoff, J.; Murakami, G.; Baumjohann, W.; Besse, S.; Bunce, E.; Casale, M.; Cremosese, G.; Glassmeier, K.-H.; Hayakawa, H.; Heyner, D.; et al. BepiColombo—Mission Overview and Science Goals. *Space Sci. Rev.* **2021**, *217*, 90. [CrossRef]

remote sensing

MDPI

Article

Population of Degrading Small Impact Craters in the Chang'E-4 Landing Area Using Descent and Ground Images

Teng Hu [1,2], Ze Yang [1,2], Zhizhong Kang [1,2,*], Hongyu Lin [3], Jie Zhong [4], Dongya Zhang [1,2], Yameng Cao [1,2] and Haomin Geng [1,2]

1 Department of Remote Sensing and Geo-Information Engineering, School of Land Science and Technology, China University of Geosciences, Xueyuan Road, Beijing 100083, China; huteng@cugb.edu.cn (T.H.); yangze1204@cugb.edu.cn (Z.Y.); 2012170008@email.cugb.edu.cn (D.Z.); 2012170007@email.cugb.edu.cn (Y.C.); 2012170009@email.cugb.edu.cn (H.G.)

2 Subcenter of International Cooperation and Research on Lunar and Planetary Exploration, Center of Space Exploration, Ministry of Education of The People's Republic of China, No. 29 Xueyuan Road, Haidian District, Beijing 100083, China

3 Beijing Institute of Space Mechanics and Electricity, 104 Youyi Road, Beijing 100094, China; linhy_bisme@spacechina.com

4 Institute of Optics and Electronics, Chinese Academy of Sciences, Chengdu 610209, China; zhongjie@ioe.ac.cn

* Correspondence: zzkang@cugb.edu.cn

Abstract: The landing camera (LCAM) of Chang'e-4 lander provides a series of low (46 cm/pixel) to high (2.3 cm/pixel) resolution images, which are suitable for centimeter-scale craters. In this paper, we analyze the degradation of those small-sized craters to provide detailed information on the local geological evolution of the lunar surface. From the mosaicked descent image, 6316 craters were extracted and classified into four degradation levels based on their morphology on the image: fresh, slightly degraded, moderately degraded, and severely degraded. The ground terrain camera (TCAM) image and the DEM of the Yutu-2 panoramic camera (PCAM) validate the crater degradation levels from a qualitative and quantitative perspective, respectively. The results show that the smaller the size of the craters, the more easily they are degraded. The crater populations in equilibrium in the four study areas indicate that the cumulative size–frequency distribution (SFD) slope is different from previous research results, and the smaller the craters, the more difficult to reach an equilibrium state (for craters smaller than a given size, the production rate is exactly balanced by the removal rate), which may be due to secondary cratering and surface resurfacing caused by the burial of ejecta from neighboring craters.

Keywords: meter to centimeter size impact craters; Yutu-2 panoramic camera; landing camera; terrain camera; crater populations in equilibrium; surface resurfacing

1. Introduction

The Chang'e-4 mission was the world's first soft landing on the far side of the Moon and is also China's second lunar surface soft landing mission [1–3]. The lander is equipped with a variety of cameras, capturing in situ field images at different resolutions and positions and providing imagery of impact craters down to centimeter scale. These high-resolution data from the Chang'e-4 landing area provide the possibilities for small-scale (especially centimeter-scale) geomorphological features, such as impact craters and boulders [4,5], enabling a more detailed investigation of the morphological features associated with them in the landing area [5,6]. Therefore, the analysis of the centimeter-scale geomorphological features of the Chang'e-4 landing area on the lunar far side, especially the impact crater population, is valuable for obtaining statistical characteristics of the centimeter-scale impact crater distribution, and detailed information on their geomorphology.

Craters are among the most common features on the lunar surface and reveal significant information primarily regarding relative and absolute surface chronologies and erosion processes [7–9]. Craters after the formation of impacts are affected by the micrometeorite bombardment, emplacement of ejecta from distal larger craters, and stresses associated with geological deformation [10–13]. Consequently, the spatial morphology changes and gradually degenerates [14]. The first lunar image captured by Ranger 7 confirmed the degradation of the lunar surface. Among the factors influencing the degradation of impact craters, direct micrometeorite bombardment and ejecta from other impact craters result in incomplete craters, disappearance of the crater rims, and the infilling of the crater cavities. Additionally, internal geological processes such as geomorphological processes (avalanches, talus slides, landslides, and creep) influence the morphology of impact craters, especially small craters [10,13].

From formation to obliteration, impact craters experience different stages of degradation with various morphological characteristics. The degradation of a population of craters reflects the age of the local geological units, providing insights into the processes of lunar geological evolution. The relationship between the degradation of impact crater populations and the age of geological units has been the focus of study. Craddock and Howard [11] studied the crater degradation process and described the relative age of geological units with the concept of degradation parameters; the degradation pattern of impact craters and the impact history of the region have also been studied in much detail in recent years [15–17]. Meanwhile, the evaluation of the impact degradation level is also of interest in the studies. Basilevsky [10], Wood et al. [18], and Ivanov and Basilevsky [19] distinguished the degradation level according to crater morphology; Li [20] used craters of typical lunar maria acquired from Chang'e-2 digital orthophoto map (DOM) and digital elevation model (DEM) to study crater shape and degradation, and crater depth was used as the critical parameter to evaluate crater degradation. Wood, Head, and Cintala [19] Ivanov and Basilevsky [20] Li [16–18,21,22]. The image-based ray brightness, rim, and terrace sharpness, texture, and digital elevation model (DEM) data-based impact crater depth, profile, and slope are the key elements of impact crater degradation analysis in existing studies. However, considering the high resolution of the Landing Camera (LCAM, cm-level resolution at the center of the landing area), in the absence of a high-resolution DEM, we analyze the degradation of small impact craters mainly using the features obtained from the LCAM images [13,15–17,21–23].

Some craters, which we call fresh impact craters, have sharp edges and bright ejecta in optical images, while in the case of other craters, the edges are softened or incomplete, and their walls are even collapsed and discontinuous. The bottoms of these craters, which we call degraded impact craters, are filled, and their reflectance is consistent with the surrounding background [11,13,24]. To better understand crater degradation, an image-based statistical method based on the degradation classification of impact craters is developed. Chapman [25] defined four levels of lunar impact crater degradation. Based on this work, we classify the degradation of the crater population into four classes based on the features shown in the images, namely (1) the presence of ray patterns around the craters, (2) the clarity of the crater rims, and (3) the contrast between the bright and shadowed regions of the crater interiors. Impact craters that are clear and complete, with sharp contrasting bright and shadowed interiors, are considered fresh impact craters. The impact craters are classified into the following types according to the degree of degradation: fresh craters (C1), slightly degraded craters (C2), moderately degraded craters (C3), and severely degraded craters (C4). Overall, empirical measures of crater degradation levels based on orbiter image data are highly dependent on the appearance of impact craters, and studies of small, degraded craters are subject to the image resolution of the orbital data.

Existing impact crater degradation studies have generally concluded that the smaller the crater is, the more likely it is to be filled or destroyed by subsequent impacts, while large craters are relatively stable and less prone to degradation [11,13,26–28]. However, few studies have focused on the population degradation of small craters, especially those smaller than one meter. Therefore, in this study, we focus on the analysis of population degradation statistics for impact craters with sizes of centimeters to one meter, which has not been addressed in previous studies. In addition, we do not analyze the effect of secondary impact craters on the degradation of impact craters because the sizes of the craters involved in this study are small and the distribution is dense; thus, it is currently not possible to distinguish primary and secondary craters by their spatial distribution patterns. The main contributions include the following:

(1) Impact craters of C1, C2, C3, and C4 degradation levels at the centimeter to meter scale in the Chang'e-4 landing area were acquired based on LCAM mosaic imagery.
(2) The impact crater degradation level classification was qualitatively verified from a side-view perspective using TCAM images, and the crater depth-to-diameter ratio calculated by DEM using PCAM of Yutu-2 was used to quantitatively verify the impact crater degradation level.
(3) The equilibrium state of the centimeter to meter impact crater population in the Chang'e-4 landing area was evaluated. Moreover, the local surface resurfacing was analyzed.

Degradation analysis of lunar impact craters extends and deepens our understanding of crater morphology, crater identification, lunar cratering history, and lunar chronology [12]. Degradation analysis of small-sized impact crater populations with sizes on the meter to centimeter scale can provide detailed information on the local geological evolution of the lunar surface relative to larger-sized impact craters (hundreds of meters), providing the possibility of obtaining localized surface resurfacing of the lunar surface.

2. Methodology

In this paper, high-resolution descent images are used to explore the distribution and degradation of small impact crater populations in the landing area. The descent images, however, have no geolocation information. Thus, we needed to mosaic and register them based on existing georeferenced data. All the craters in the mosaic images were manually identified based on the crater textures, as the accuracy and completeness of the automatic identification techniques are limited. Craters were manually classified to different degradation levels to permit a quantitative analysis of degradation in the crater population, a multi-vote process to compensate for human subjective factors. The impact crater degradation level classification was verified qualitatively and quantitatively using the side-view morphology of the impact crater from the images of the ground terrain camera (TCAM) and the depth-to-diameter ratio of the DEM acquired by the Yutu-2 panoramic camera (PCAM). The specific flow is shown in Figure 1. A detailed description of the method is described below.

2.1. The Mosaic and Registration of the Images from the LCAM

Typically, we used the structure from motion (SfM) algorithm [29] to estimate the photo position and attitude to obtain the normalized image; however, the final stage of the Chang'e-4 lander was the vertical landing, and the images were acquired along the main optical axis direction, leading to large errors in the position and attitude calculated by the SfM algorithm [30,31]. Therefore, this paper adopted the mosaic method.

Since the descent images lacked geo-referencing, we registered the descent images before mosaicking, making them have the same coordinate reference as Chang'e-2 DOM. However, the resolution of Chang'e-2 DOM is 7 m/pixel, which is quite different from the first descent image (46 cm/pixel). Therefore, we considered using the lunar reconnaissance orbiter camera (LROC) Narrow-Angle Camera (NAC) image to register with the Chang'e-2 DOM image first and then registered the descent images based on the georeferenced LROC NAC image.

We used ArcGIS for the image registration and applied the spline transformation model to transfer the images. For the registration of the LROC NAC image and the Chang'e-2 DOM, we selected 300 pairs of corresponding points and ensured that they were evenly distributed. For the registration of the descent images and the LROC NAC image, we needed to register 25 images. For each image pair, 100 pairs of corresponding points were selected. Experimentally, 100 pairs of random points were selected to test the registration error and the results showed that the error was less than 2 pixels. Furthermore, the experiments selected larger impact craters (with diameters larger than 50 pixels) and compared their diameters to the LROC NAC image to determine the difference in crater diameter. The maximum diameter difference was no more than 2 pixels. This finding suggested that image distortion due to image registration had little effect on subsequent crater size acquisition. The resolution of the georeferenced LCAM mosaic image is 0.023 m/pixel (center) to 0.46 m/pixel (edge).

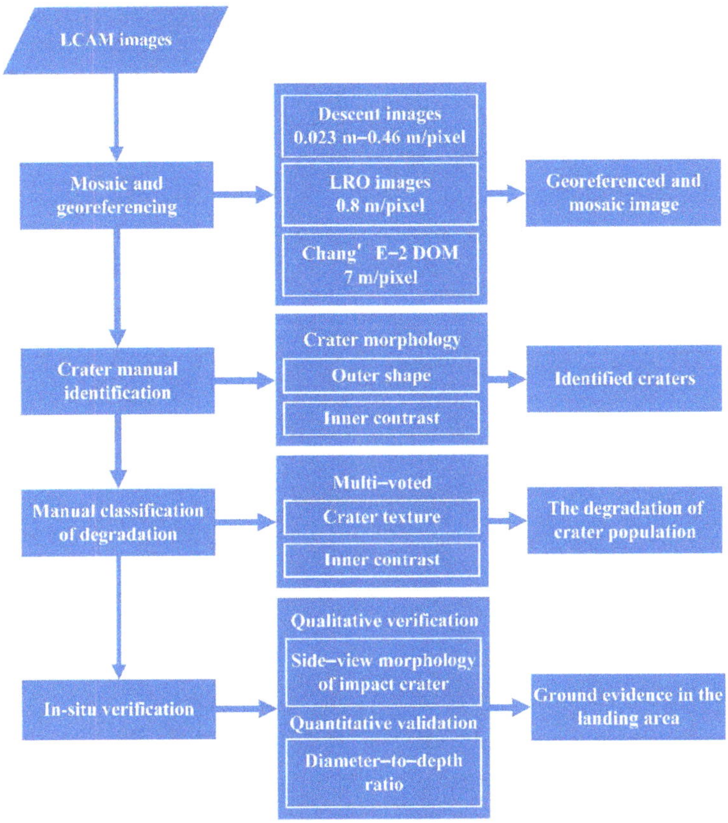

Figure 1. Flow chart of the analysis of impact crater population degradation.

2.2. Identification of Impact Craters

As impact craters are the most typical geomorphic structure on the lunar surface, they provide important information on past and present geological processes and have also been applied to age dating of geologic surfaces [32,33]. In this paper, impact craters were extracted around the landing site of Chang'e-4 using manual visual identification, as shown in Figure 2a,b. Figure 2a illustrates the contrast between the shadowed areas (dark areas, blue outline) and the illuminated areas (light areas, red outline) presented internally along the illumination direction of the impact crater in the image. Figure 2b shows the circular shape of craters (yellow arrows mark the impact crater boundaries); thus, they can be distinguished from the background.

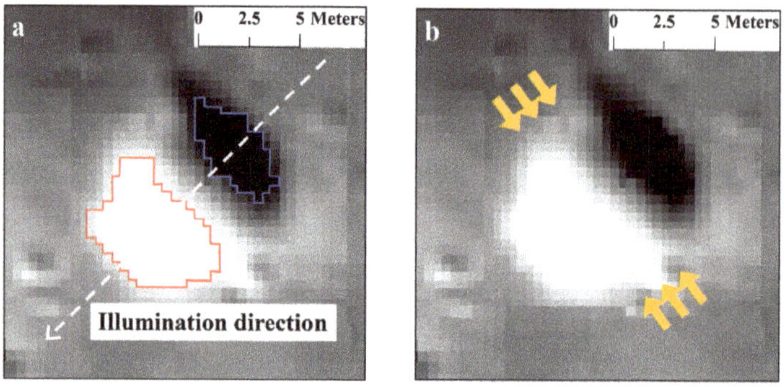

Figure 2. Visual identification criteria of impact crater features. (**a**) Crescent-like bright and dark regions. (**b**) Circular shape, yellow arrows mark the impact crater boundaries.

Therefore, craters can be identified through the combination of contrast and shape attributes. Identifying craters manually means craters can be detected, marked, and verified through human eyes, decreasing the likelihood of incomplete and mistaken identification evident in currently deployed automatic methods.

For the identification of impact craters, a minimum crater diameter of 5 pixels (e.g., Figure 3a,b) or more is needed, but for the degradation analysis of impact craters, a minimum crater diameter of 10 pixels (e.g., Figure 3c,d) is required due to the detailed information on the shape of the crater, the contrast between bright and dark and the rays around the crater.

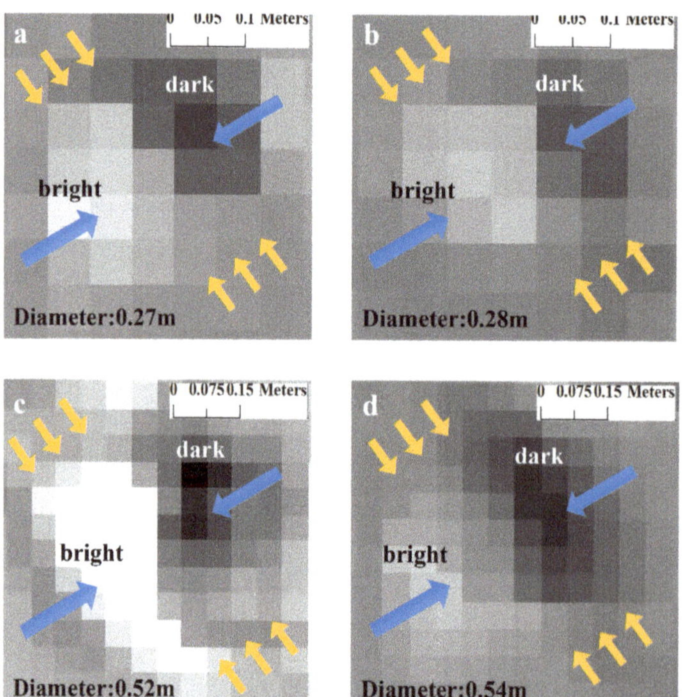

Figure 3. Diagram of impact craters with diameters of 5 pixels (**a**,**b**) and 10 pixels (**c**,**d**), with (**c**) a C2 type impact crater and (**d**) a C3 type impact crater. Yellow arrows mark the impact crater boundaries.

2.3. Classification of Degradation Levels of Impact Craters

Once formed, impact craters begin to degrade gradually. As the degradation of the impact crater becomes more severe, the distal ejecta rays (if any) around the fresh impact crater gradually disappear, and the contrast between the bright and dark areas of the impact crater is no longer sharp, while the boundary gradually becomes indistinguishable and nearly blends in the surrounding lunar surface [2,13,28,34,35]. Statistical studies of morphological and other related characteristics of their degradation level are important to understand the correlation between the evolution of crater morphology and the age of the corresponding geological units. Unlike existing studies that have targeted larger impact craters (meters and even hundreds of meters), most craters in our study are smaller than 2 m in diameter. Unfortunately, there is no DEM with a resolution higher than 0.5 m/pixel to calculate the exact value of the quantitative classification indicators for craters. Thus, a qualitative classification was applied in this paper based on the traditional method [25]. As shown in Figure 4, we used a qualitative identification method to define and determine the degradation levels of impact craters. The latter were divided into four types (C1–C4):

(1) C1 type: fresh craters, with obvious rays or clear edges, and a sharp contrast between bright and dark.
(2) C2 type: slightly degraded craters, with clear edges, and a relatively clear contrast between bright and dark.
(3) C3 type: moderately degraded craters, which show clear signs of degradation with small differences between bright and dark, raised crater floors due to filling, and fuzzy circular outlines.
(4) C4 type: severely degraded craters with unclear outlines, shallow crater floors, and a circular outline that can be distinguished when the image is stretched.

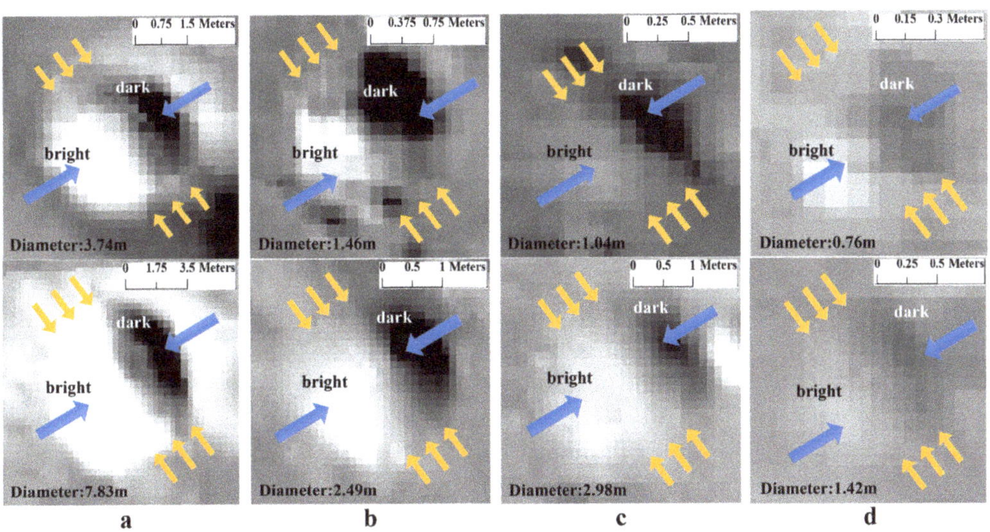

Figure 4. Examples of craters at four degradation levels, where (**a**–**d**) correspond to impact craters of types C1, C2, C3, and C4, respectively. Yellow arrows mark the impact crater boundaries. The diameters of the impact craters are written in the figures.

The manual qualitative classification of craters usually introduces human subjective factors, so judgment errors are inevitable. For example, the C3-type craters in Figure 4 are like C4-type craters and may be misclassified; thus, a multi-voting process for all craters was adopted. In the voting process, first, all impact craters were individually classified by three researchers into degradation levels. Then, we summarize the obtained impact crater degradation levels. For a given impact crater, the degradation level of the crater is confirmed only when the number of votes is greater than or equal to 2. Otherwise, the crater is considered to have an uncertain degradation level and needs to be reclassified. Finally, the impact crater degradation level will not be determined until all impact crater degradation levels have obtained 2 or more equal votes. In this process, the number of impact craters with uncertain degradation levels was less than 15% when researchers first independently classified the crater degradation levels, and after two rounds of reclassifying craters with uncertain degradation levels, the degradation levels of all impact craters were determined. The specific crater degradation levels classification results can be downloaded from the Science Data Bank of Beijing, China.

Nevertheless, we are concerned that the very small number of C3 and C4 impact craters may still be misclassified. Therefore, in the subsequent analysis, we also combined the statistical analysis of C3 and C4 impact craters. In addition, the qualitative and quantitative methods for classifying impact crater degradation levels are described in the next section.

2.4. Qualitative and Quantification Analysis of the Degradation Level of Impact Craters Using Ground TCAM Images and DEMs Acquired by Yutu-2 PCAM

We compared craters in images from both the TCAM and LCAM. We manually identified the craters in both types of images and analyzed the degradation of the crater population in the landing area. Then, we verified our results by comparing the identified craters on the LCAM mosaic images and the TCAM images. Figure 5 shows two impact craters with different degradation levels in the LCAM mosaic image and TCAM image (red circles are C3 type impact craters and blue circles are C4 type impact craters).

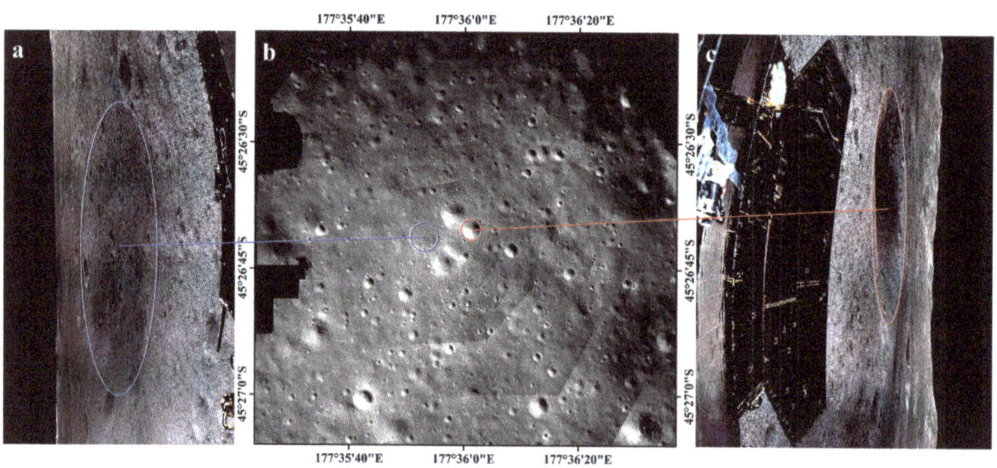

Figure 5. Comparison of impact craters on the LCAM mosaic image (**b**) and the TCAM images (**a**,**c**: cylindrical projection). Circled in red and blue are two impact craters with different degradation levels (red circles are C3 type impact craters and blue circles are C4 type impact craters) shown in LCAM mosaic image and in TCAM images, respectively.

As shown in Figure 5, we show two pairs of corresponding impact craters on TCAM images and LCAM mosaic images, where the blue circles are C4 level impact craters, and the red circles are C3 level impact craters. The TCAM images are obtained from a side view, for the C3 impact crater compared to the C4 impact crater, part of the crater rim can be identified, while the concave interior of the C3 impact crater is also more obvious. We carefully compared the LCAM mosaic image and the corresponding impact crater on the TCAM images from different angles for close-up verification. In the central part of the landing area, the degradation levels and actual spatial forms of craters were checked one by one at close range. This comparison helped us to understand the actual morphology of craters in the descent image.

For the PCAM image of Yutu-2, we used the Context software program to obtain the DOM and DEM of the region. After that, the LCAM mosaic image was used as the reference image for the registration of the DOM and DEM by using ArcGIS.

To quantitatively verify the degradation level of impact craters acquired from LCAM images, a crater diameter-to-depth ratio was used. Considering the variation in the depth-to-diameter ratio of different types and sizes of impact craters, we selected simple impact craters of the similar size (diameter difference less than 1 m) for the calculation of the depth-to-diameter ratio. For the calculation of the impact crater depth, we used the average of the elevations of the eight directions at the edge of the crater to represent the topographic elevation and obtained the crater depth by subtracting this value from the depth at the center of the DEM impact crater.

3. Results and Analysis

The difference in the resolution of different regions of the mosaic image makes the minimum diameter of the impact crater that can be extracted different for each part. Therefore, we divided the study area into four study areas based on the minimum diameter of the impact crater that can be extracted from each area. In addition, the degradation statistics of crater populations in each study area were analyzed. Finally, TCAM and LCAM images were used to analyze the impact crater degradation classification qualitatively and quantitatively.

3.1. Study Area and Data

Chang'e-4 landed on the floor of the Von Kármán crater (186 km in diameter), located in the central part of the northwestern South Polar Aitken (SPA) basin. Recent geological mapping surveys show that this crater is approximately 3.6–3.7 Gyr old (pre-Nectarian) [36]. The landing site (45.457°S, 177.588°E) [37] is approximately 350 km northwest of the SPA ellipse center (53°S, 169°W) [38]. During the landing process, the images captured by the LCAM of Chang'e-4 provided an unprecedented opportunity to study the degradation of the population of small impact craters with diameters of 10 m or less in the area.

We selected 25 images captured at different landing altitudes (600 to 30 m). The ground resolution of the captured images increased gradually as the camera descended, and impact craters of sub-meter size can gradually be identified. Because of their high spatial resolution, descent images are ideal for interpreting the small geological structures in landing areas and analyzing the geological evolution of planetary surfaces. An accurate geographic reference is the foundation for mapping and interpretation of planetary geology using descent images. The images from the LCAM, however, do not have geographic coordinates. Therefore, we must introduce images with geographic references to register descent images.

The Chang'e-2 DOM (7 m/pixel) is the lunar global DOM with the highest ground resolution and is thus used as the reference image when registering the descent images. The great difference in spatial resolution between the Chang'e-2 DOM and the descent images, however, presented a serious challenge when selecting corresponding pixels. Thus, we used the LROC NAC image (M1298916428LR) to bridge the gap in resolutions [39].

We used full-color mosaic images, captured by the lander, and published by China's Lunar and Deep Space Exploration (CLEP), containing a high-resolution TCAM image of the landing zone displayed by azimuth and cylindrical projections, for the qualitative comparison of craters found in LCAM and TCAM images. To quantitatively compare the results of the impact crater degradation classification, a DEM was acquired from the Yutu-2 rover's PCAM image to calculate the diameter-to-depth ratio. Considering that the data around the landing site might be affected by the impact of the lander during the landing process resulting in inaccurate DEM results, we selected the PCAM images of the third lunar day of Yutu-2.

3.2. The Mosaic of the Descent Images

We registered and mosaicked the descent images using the method in Section 2.1. For the acquired LCAM mosaic image, we selected 100 pairs of homonymous points with a uniform distribution between the LCAM mosaic image and the georeferenced LROC NAC images and 50 pairs of corresponding impact craters with diameters larger than 50× pixels to calculate the registration error of the LCAM mosaic image and the diameter deviation of the corresponding impact craters, respectively. The mean and root mean square error (RMSE) of the homonymous points deviation, as well as the RMSE and mean value of the diameter difference of the corresponding impact craters were calculated separately. The results show that the mean value of the homonymous point deviation is 2 pixels and the RMSE is 1 pixel. Moreover, the mean value of impact crater diameter deviation is 2 pixels and the RMSE is 1 pixel, which satisfies the need for subsequent degradation statistical analysis of the crater population.

Based on the obtained LCAM mosaic image, the study area centered on the landing point was defined, covering an area of 146,687 m^2 (408 m from east to west and 386 m from north to south), as shown in Figure 5b. The resolution is 0.023 m/pixel in the central area and 0.46 m/pixel on the edge. The descent images used in this paper were captured along the optical axis during landing; thus, the size of the extractable impact craters gradually decreased as the ground resolution of the descent images increased, especially in the central landing area where the extractable size could reach the decimeter level. Due to the improved spatial resolution, impact craters as small as 15 cm in diameter could be identified over the central landing area. Therefore, the descent images provided a reliable data source for analyzing impact craters around the landing site.

3.3. Impact Crater Distribution Results

In total, 6316 impact craters were identified and extracted manually. The crater distribution map and the boundaries of craters with different sizes are circled in Figure 6. As shown in Figure 6, influenced by the high central resolution and low edge resolution of the LCAM mosaic image, the impact craters are very dense near the center of the landing site and gradually become sparse as they approach the boundary (the pink line of Figure 6) of the study area. At the same time, smaller size (0.15–0.5 m) impact craters are also distributed around the landing site, while the impact craters become larger at the boundary of the study area. This is because the resolution of the mosaic image obtained from LCAM images is highest around the landing site at 0.023 m/pixel, where the largest number of small impact craters can be identified and decreases to 0.46 m/pixel closer to the boundary of the study area, where only relatively large (>2 m) impact craters can be identified due to resolution limitations. The study data are a mosaic of LCAM images from higher to lower resolution, and the difference in resolution leads to different minimum sizes of impact craters that can be identified in various parts of the study area. As shown in Figure 6, the study area is divided into four parts, where the area within the red line, the area within the yellow line, the area within the aquamarine line, and the entire mosaic image range corresponds to the minimum diameter of identifiable impact craters of 0.15, 0.5, 1, and 2 m, accordingly. The minimum diameter of identifiable impact craters corresponding to each of the study areas is divided in equal proportion to the total number of impact crater populations, making about 2000 impact craters in each of the four areas. Therefore, the red outlined area, yellow outlined area, aquamarine outlined area, and pink outlined area (the entire area) contain 2341, 2160, 2926, and 2018 craters with diameters of 0.15 m or more, 0.5 m or more, 1 m or more, and 2 m or more, respectively.

The completeness of the impact crater population statistics is mainly limited by the resolution of the images, with impact crater diameters greater than 10× pixels having a high confidence level [40]. Therefore, to make the statistical analysis of the crater degradation significant, we calculated histograms of the size distribution of the craters above the minimum identified size for the above four study areas and considered the middle 95% of the craters in each area to have a high confidence level. For the four study areas, the minimum statistical diameter of the impact crater is 0.268 m, which is greater than 10× pixels. As shown in Figure 7, for the red, yellow, aquamarine, and pink outlined areas obtained above, the corresponding diameter intervals of 95% craters in each area are 0.268–1.725 m (contains 2223 impact craters), 0.512–2.612 m (contains 2050 impact craters), 1.026–8.963 m (contains 2781 impact craters), and 2.040–15.449 m, (contains 1916 impact craters). This allows us to perform a degradation analysis of the crater population with crater diameters in the range of 0.268–15.449 m.

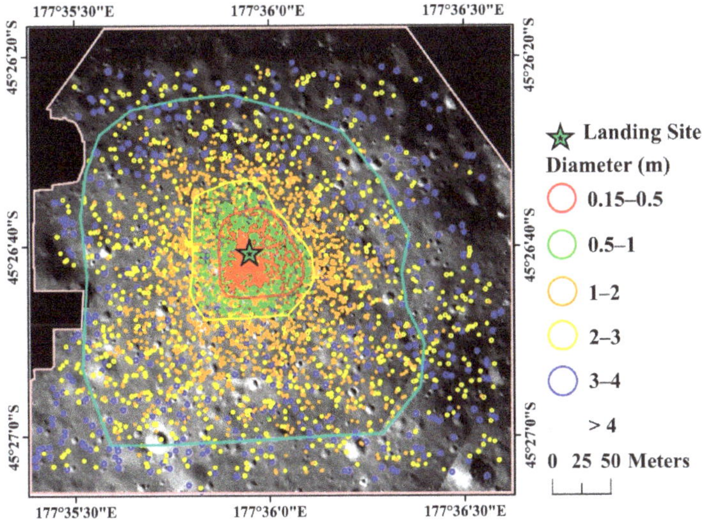

Figure 6. The mosaicked LCAM image with boundaries of areas defined by different sizes of visible craters. The red, the yellow, the aquamarine, and the pink outlined areas correspond to impact craters with a minimum diameter of 0.15, 0.5, 1, and 2 m that can be identified, respectively, as constrained by the limitations of the resolution of descent images.

3.4. Impact Crater Degradation and Crater Size

All craters in the landing area were classified into degradation levels according to the morphology of the impact craters. The study areas with the classified craters are visualized in Figure 8. Detailed degradation results for the crater population are provided in Table 1.

As shown in Figure 8, the C1 craters (red, fresh craters) in all four study areas are rare, and the distribution is sparse. C2 impact craters (aquamarine, slightly degraded craters) cover more area than C1 craters. C3 craters (yellow, moderately degraded craters) and C4 craters (green, severely degraded craters) extend over a large range of the landing area. Therefore, the craters in the landing area are dominated by C3 and C4 craters, which occupy more than 90% of the craters in the landing area.

In the above, we obtained a 95% confidence interval for the diameter range of impact craters in each study area, and the following statistical analysis of the degradation of the impact crater population was performed using the craters within the diameter range of each study area as a statistical sample. We then calculated the proportion of C1, C2, C3, and C4 in each impact crater statistical diameter interval. The percentage of each degradation level was also calculated for the total number of impact craters in each study area. Additionally, considering that the C3 and C4 types are more likely to be confused and that misclassification may still occur despite the multiple voting method used in the previous section, we also used the C3 + C4 type of impact crater count to represent the more severely degraded craters.

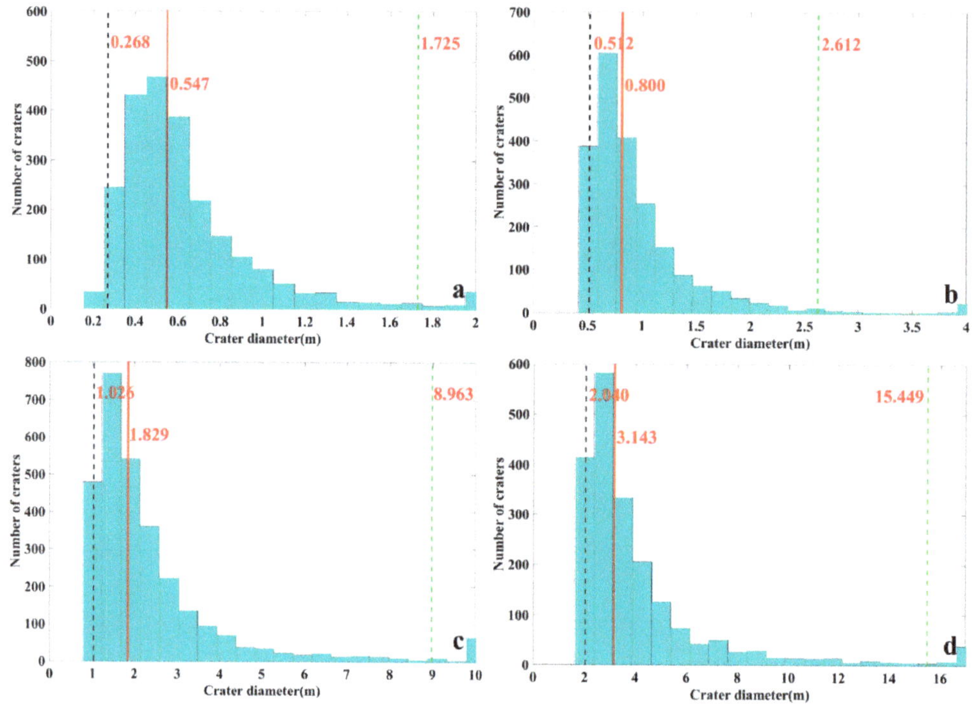

Figure 7. Histogram of the crater diameter distribution in each study region, where the red line is the median of the crater diameter distribution, and the black and the green dashed lines are the diameters corresponding to 2.5% and 97.5% of the number of impact craters, respectively. The interval between the two dashed lines corresponds to the diameter distribution of 95% of the impact craters. (**a**–**d**) show histograms of the crater diameter distribution of the red, yellow, aquamarine, and pink outlined areas, respectively.

As shown in Table 1, for the total number of impact craters counted in each study area, C1 craters are the lowest proportion (less than 1%) and C4 craters are the highest (greater than 64%), with C3 + C4 craters accounting for more than 93% of the craters, indicating that the study area is heavily degraded. Furthermore, in those areas, the smaller the impact crater diameter is, the fewer C1–C3 impact craters and the more C4 impact craters there are. In addition, as the crater diameter increases, the proportion of C1–C3 impact craters increases, while the proportion of C4 impact craters decreases (the decrease is 79.46% to 63.67% for the red outlined area and 79.53% to 57.62% for the yellow outlined area). In addition, the proportion of C3 + C4 type impact craters, i.e., more heavily degraded impact craters, decreases as the diameter of the impact craters increases. This finding suggests that the smaller the craters are, the easier they degrade.

For the aquamarine outlined area and the full study area-wide crater population, the 95% crater diameter distribution ranges from 1.026 to 8.963 m and 2.040 to 15.449 m. Overall, the number of craters within the aquamarine outlined area shows an increasing trend for C1, C2, and C3 craters and a decreasing trend for C4 craters as the diameter increases. The percentage of C4 craters decreases from 72.41% to 57.18% for the aquamarine outlined area and from 72.65% to 56.37% for the overall study area. For both areas, overall, the proportion of C3 + C4 craters tends to decrease as the crater diameter increases.

For the population of impact craters within the four study areas in the diameter range of 0.268–15.449 m, the proportion of craters in the C3 + C4 degradation class decreases with increasing diameter.

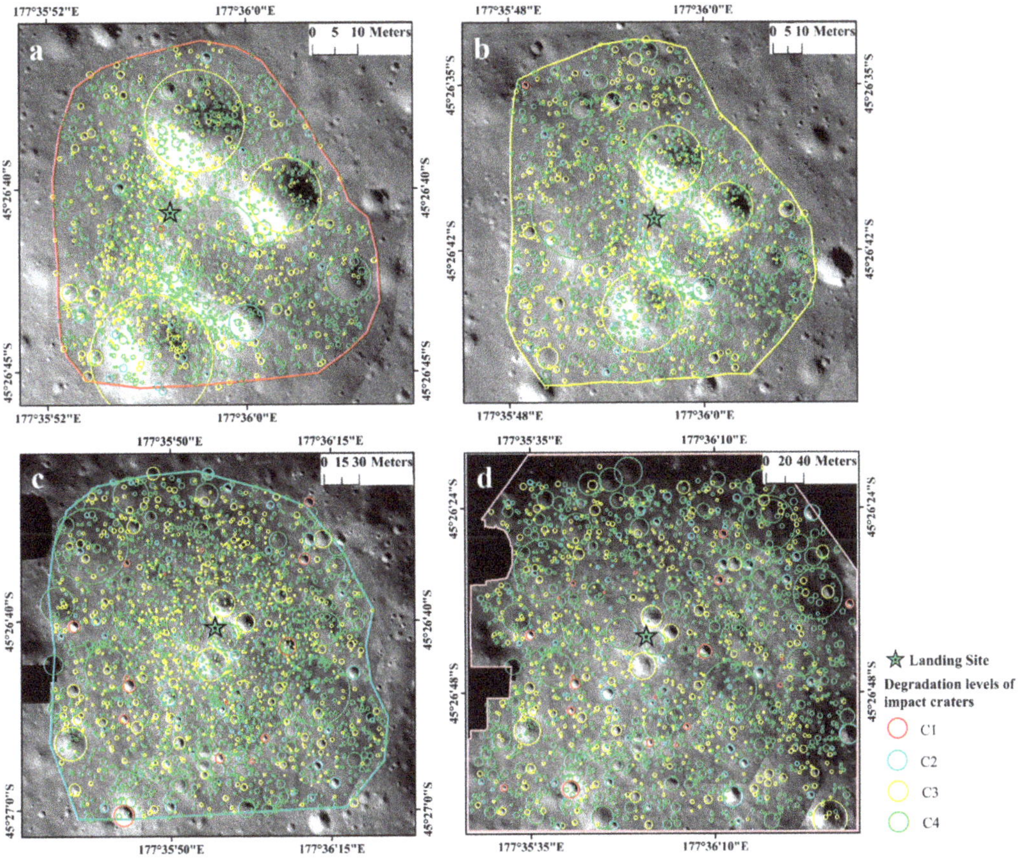

Figure 8. Crater distribution map with different degradation levels in different study areas. (**a**–**d**) show the red, yellow, aquamarine, and pink outlined study areas. Red, aquamarine, yellow, and green circles are C1, C2, C3, and C4 impact craters.

In the above, we considered 95% of the impact craters in each study area with a high level of confidence, and then we performed statistical analysis on those parts of the impact crater population. The analysis shows a trend of degradation of impact craters on the centimeter to meter scale (0.268–15.449 m). The overall population of impact craters in this size range also shows a trend toward more severe degradation with smaller crater sizes. The specific degradation of these small impact crater populations is summarized as follows:

(1) For each study area, the proportion of C4 type impact craters is the highest, exceeding 50% for most areas, while the proportion of fresh impact craters is less than 10%, indicating that impact crater degradation in the Chang'e-4 landing area is severe. The sizes of the craters in the landing area were relatively small (the maximum diameter was 67 m).

(2) As the size of the impact craters increased, the proportions of C1, C2, and C3 craters all gradually increased, while the proportion of C4 craters gradually decreased. The C3 + C4 impact craters show the same degradation pattern, which is consistent in all four study areas (for the effective statistical interval, i.e., diameter range corresponding to 95% of the impact crater samples in the study area).

Table 1. Impact crater population diameter and degradation level statistics for the four study areas.

	Data Range	**0–25%**	**25–50%**	**50–75%**	**75–100%**	**Total**
	Diameter (m)	**0.268–0.430**	**0.430–0.547**	**0.547–0.714**	**0.714–1.725**	**0.268–1.725**
Red outlined area (4334 m^2)	C1	0.00%	0.00%	0.00%	0.18%	0.84%
	C2	1.26%	1.08%	1.26%	4.68%	6.00%
	C3	19.28%	20.68%	23.38%	31.47%	28.91%
	C4	79.46%	78.24%	75.36%	63.67%	64.25%
	C3 + C4	98.74%	98.92%	98.74%	95.14%	93.16%
	Data range	0–25%	25–50%	50–75%	75–100%	Total
	Diameter (m)	0.512–0.627	0.627–0.800	0.800–1.059	1.059–2.612	0.512–2.612
Yellow outlined area (9569 m^2)	C1	0.00%	0.00%	0.00%	0.39%	0.40%
	C2	0.97%	1.56%	1.95%	5.86%	4.06%
	C3	19.49%	25.00%	29.63%	36.13%	28.30%
	C4	79.53%	73.44%	68.42%	57.62%	67.24%
	C3 + C4	99.03%	98.44%	98.05%	93.75%	95.54%
	Data range	0–25%	25–50%	50–75%	75–100%	Total
	Diameter (m)	1.026–1.381	1.381–1.829	1.829–2.597	2.597–8.963	1.026–8.963
Aquamarine outlined area (71,763 m^2)	C1	0.14%	0.00%	0.14%	1.29%	0.10%
	C2	1.44%	3.31%	2.88%	8.62%	2.59%
	C3	26.01%	25.90%	28.39%	32.90%	27.56%
	C4	72.41%	70.79%	68.59%	57.18%	69.76%
	C3 + C4	98.42%	96.69%	96.97%	90.09%	97.32%
	Data range	0–25%	25–50%	50–75%	75–100%	Total
	Diameter (m)	2.040–2.492	2.492–3.141	3.141–4.408	4.408–15.449	2.040–15.449
Pink outlined area (146,687 m^2)	C1	0.21%	0.21%	0.63%	2.30%	0.04%
	C2	2.51%	4.80%	5.43%	11.27%	2.07%
	C3	24.63%	31.11%	29.85%	30.06%	23.71%
	C4	72.65%	63.88%	64.09%	56.37%	74.18%
	C3 + C4	97.29%	94.99%	93.95%	86.43%	97.89%

3.5. *In Situ Qualitative and Quantitative Validation*

We used TCAM images from the Chang'e-4 lander and panoramic camera (PCAM) images from the Yutu-2 Rover for the in-situ cross-check; the results are shown in Figures 9 and 10. The TCAM images were used to qualitatively analyze impact craters of different degradation levels from a side-view perspective. The PCAM images are used to obtain the DOM and DEM, where the DEM can be used to calculate the crater depth-to-diameter ratios of impact craters and to quantitatively analyze impact craters of different degradation levels.

For TCAM, we performed a degradation analysis of the impact crater within part of the TCAM panoramic surround image (the region bounded by two white circles in Figure 9b, corresponding to the area between the two white circles of the LCAM mosaic image in Figure 9c), which was used to support our classification of the degradation of impact craters. The circular area contains 2371 impact craters in the LCAM mosaic image, representing approximately 36% of all impact craters. Considering the large distortion of the terrain camera data, we compared the morphologies of 100 pairs of more discernible craters and input the results into the classification of the crater degradation levels. In Figure 9b, we show 12 crater pairs that have been labeled in the image to provide a side-view morphology of the impact craters at different degradation levels from the TCAM.

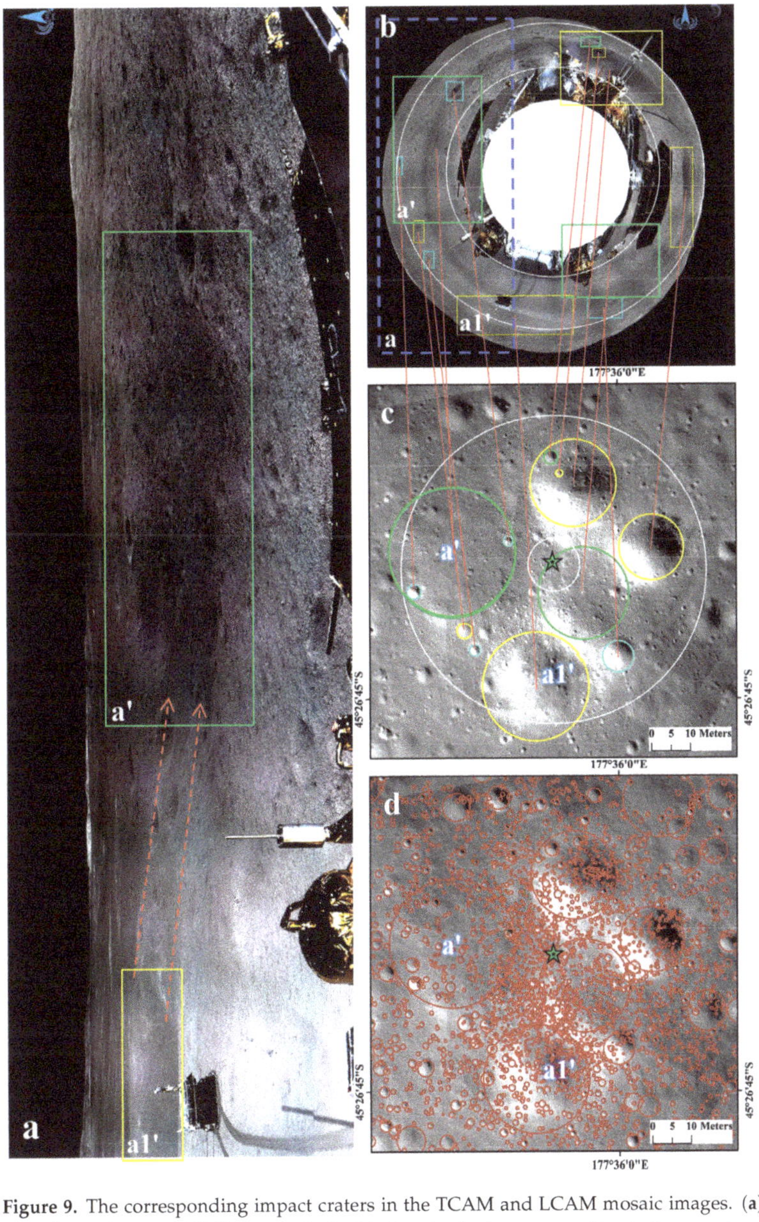

Figure 9. The corresponding impact craters in the TCAM and LCAM mosaic images. (**a**) Part of the enhanced ground TCAM image (cylindrical projection), corresponding to the area shown in the blue dashed box in (**b**). The red dashed arrow shows the direction of ejection from crater a1′ to crater a′. (**b**) Ground TCAM image (azimuthal projection). The red solid lines show 12 corresponding craters with different degradation levels. Aquamarine, yellow, and green outlined areas in (**b**,**c**) are C2, C3, and C4 craters. The areas within the white circles in (**b**,**c**) are the study areas where TCAM images were used to verify the degradation level of impact craters. (**c**) The LCAM mosaic image. (**d**) Distribution of LCAM impact craters corresponding to the TCAM image. The green pentagram is the Chang′e-4 landing site.

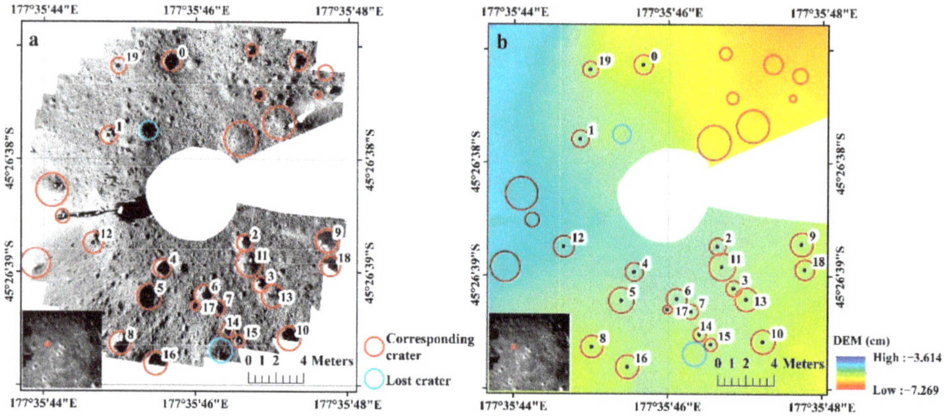

Figure 10. DOM (**a**) and DEM (**b**) of the third lunar day PCAM of Yutu-2, where the red impact craters are the corresponding craters of the PCAM and LCAM, and the blue represents impact craters not identified by the LCAM. The areas shown in (**a**,**b**) correspond to the red area of the LCAM mosaic image thumbnail in the lower left corner of the figure.

As shown in Figure 9b,c, 12 pairs of corresponding craters (four C2, five C3, and three C4 craters) on the ground TCAM image and LCAM mosaic image are linked by red lines. C2 craters (aquamarine) have clear circular shapes, and the inner contrast between bright and dark is clear on the ground TCAM image. C3 craters (yellow) have circular shapes but no discernible rim. C4 craters (green) have hidden circular shapes, the bottoms are infilled, and the crater rims disappear. The classified degradation levels on the LCAM mosaic image are consistent with the TCAM camera images.

Moreover, there are many impact craters of different sizes around the landing point, with varying degrees of degradation. Most craters on the surface are small craters, and their distribution is dispersed across the surface and consistent with their distribution in the descent images. However, only a few small craters have a clear circular shape. Most small impact craters are severely degraded, which is why smaller craters have a higher C4 ratio in the distribution analysis results (Table 1).

For quantitative validation of impact crater degradation levels, we used the PCAM image of the third lunar day of Yutu-2 and obtained the georeferenced DOM (0.0014 m/pixel) and DEM (0.01 m/pixel) of the region (Figure 10). The correspondence between the impact crater degradation level and the crater depth-to-diameter ratio in this area was verified. This study area is within the aquamarine outlined area of the LCAM mosaic image, an area where we statistically analyzed craters from 1.026 to 8.963 m in the above section. For the impact craters in this range, a total of 25 craters can be established to correspond to the LCAM mosaic data and the PCAM obtained DOM and DEM (red impact craters in Figure 10), while compared to the PCAM, two craters are not identified on the LCAM mosaic image in this diameter range. These two impact craters are relatively small, highly degraded, and not identified due to the resolution limitations of this area in the LCAM mosaic image. Thus, the total number of impact craters obtained from the above content impact crater count may be underestimated, but our method identifies at least 92% of the impact craters in the study area and still has a high degree of reliability.

For the relationship between the level of crater degradation and the crater depth-to-diameter ratio, we selected the craters numbered 0–19 in Figure 10b for the analysis because these craters are in a region with low elevation change and the topography has less influence. However, for the craters numbered 15 and 17, we did not use these two craters for the calculation of the depth-to-diameter ratio because the difference between their sizes and the size of the largest crater in the region exceeded 1 m, considering that the depth-to-diameter ratio was calculated in the case of similar crater sizes. Then, we sorted the diameter-to-depth ratios of the above 18 craters from lowest to highest diameter-to-depth ratio and show the results in Table 2. Since there are only craters with degradation levels of C3 and C4 in the study area, for these 18 craters, the diameter-to-depth ratios of C4 craters tend to be smaller, while those of C3 craters tend to be larger (Table 2). Overall, both C3 and C4 impact craters are very shallow, with maximum depths of less than 0.2 cm and impact crater diameter-to-depth ratios of less than 1.121×10^{-3}. Consequently, the DEM-calculated crater depth-to-diameter ratios validate our classification of impact crater degradation levels C3 and C4 as relatively accurate.

Table 2. Diameter-to-depth ratio of the impact crater in the third lunar day experiment area from the Yutu-2 PCAM.

Depth-to-Diameter Ratio ($\times 10^{-3}$)	Crater Number	Diameter (m)	Depth (cm)	Crater Degradation Levels
0.023	19	2.015	0.005	C4
0.043	5	1.907	0.008	C4
0.288	14	1.017	0.029	C4
0.321	12	1.592	0.051	C4
0.363	13	1.744	0.063	C4
0.385	7	1.123	0.043	C4
0.390	0	1.466	0.057	C3
0.432	10	1.798	0.078	C3
0.450	6	1.531	0.069	C3
0.465	18	1.499	0.070	C3
0.495	9	1.734	0.086	C3
0.518	8	1.673	0.087	C3
0.557	3	1.176	0.065	C3
0.601	4	1.374	0.083	C3
0.700	11	2.040	0.143	C3
0.794	2	1.256	0.100	C3
0.804	1	1.295	0.104	C3
1.121	16	1.752	0.196	C3

4. Discussion

With continued bombardment, the impact crater density of any diameter will eventually reach an upper limit, especially for smaller impact craters [41,42]. A crater population smaller than a given diameter is considered to be in equilibrium when the production rate of craters is balanced by the removal rate [15]. Impact-crater-based lunar surface age determinations should always be performed on impact crater populations larger than the equilibrium onset diameter (D_{eq}, for a given crater population, the diameter of the largest crater at which the population is in equilibrium), otherwise, misleading dating results will be obtained [15,43].

The methods of Gault [43] and Hartmann [44] are the widely used methods for evaluating the equilibrium state of impact crater populations, both of which regard certain empirical crater densities as the equilibrium level. Therefore, cumulative size–frequency distribution (SFD) slopes around -2 are often used as empirical equilibrium levels for different impact crater populations. However, reference to the method of Gault and Hartmann can lead to different interpretations of impact crater populations, especially for small-sized impact crater populations in older geological units [36]. In particular, the age of the geological units in our study area is approximately 3.6–3.7 Gyr old [36], while the impact craters are mostly below 2 m in diameter and there may also be many secondary impact craters. We used the Craterstats2 [45] program to plot Hartmann's equilibrium lines [44] (black lines in Figure 11) for the craters in the four study areas and found that the empirical equilibrium levels of Hartmann were not appropriate for our four study areas [44]. Alternatively, we used a power-law index to obtain the cumulative SFD slope according to Xiao and Werner [15]. The details are as follows.

In the above, we divided the LCAM mosaic image into four study areas (Figure 6). In the statistical analysis of impact craters at each degradation level, the small number of impact craters outside the 95% impact crater population, limited by the resolution and size of the study area, will lead to low confidence in the results of the obtained impact crater degradation statistics. Therefore, we performed degradation statistics for 95% of the impact craters in each study area in the above section. However, when analyzing the crater accumulation process, we analyzed the full population of craters to obtain a localized geographical evolution of the four study areas.

For each of the four study areas, the equilibrium state of their impact crater populations was evaluated. We fitted power laws to craters in each study area using the technique by Clausi et al. [46] to obtain the cumulative SFD slope (the power law exponent is less than the cumulative SFD slope by 1) and the equilibrium onset diameter D_{eq} (as shown in Table 3) [10,15,16,47]. The results show that the crater populations in equilibrium have a cumulative SFD slope of <−2 for both the red and yellow outlined areas (−2.696 and −2.392 for the red and yellow regions, respectively), while the crater populations in equilibrium have a cumulative SFD slope of about −2 for both the aquamarine and pink outlined areas (−1.952 and −1.917 for the red and yellow regions, respectively).

Table 3. Summary of the equilibrium states of the crater populations on the counting areas.

Counting Areas	Cumulative SFD Slopes of Crater Population in Equilibrium	D_{eq} (m)
Red outlined area	−2.696	0.779
Yellow outlined area	−2.392	0.810
Aquamarine outlined area	−1.952	2.090
Pink outlined area	−1.917	3.037

In previous studies, the cumulative SFD slope of decimeters to hundreds of meters in equilibrium for the crater population was −2, while the production SFD slope was −3 [15,43,49]. In contrast, only the cumulative SFD slopes of crater populations in the equilibrium of the pink outline area and aquamarine outline area were approximately −2, a parameter that has been considered to represent equilibrium in previous studies, while the equilibrium onset diameters of the craters in those areas are 3 and 2 m. For impact crater populations with higher resolution and smaller diameter ranges, the red outlined area and the yellow outlined area are found to have the highest resolution areas with steeper cumulative SFD slopes (of −2.696 and −2.392, respectively). In previous studies, smaller diameter impact crater populations were more likely to reach equilibrium [15]; thus, the red outlined area should be the most likely of the study areas to be in equilibrium. However, this area has the highest cumulative SFD slopes; it seems that the statistical area is far from equilibrium, while the large areas easily reach equilibrium.

Regarding the inconsistency of our findings relative to previous impact crater population equilibrium studies, we also suggest that it is possible that the Chang'e-4 landing area is old with a thick regolith layer [24], and thus may have a large distribution of secondary impact craters. Such secondary impact craters may be more widely distributed in the centimeter diameter range. A large distribution of secondary impact craters would result in a steeper cumulative SFD slope [50], resulting in a higher resolution area with a steeper cumulative SFD slope.

Surface resurfacing may also be causing the red and the yellow outlined areas to not be in equilibrium. Figure 11 shows the cumulative SFDs for four statistical areas. Figure 11a,b show inflection points, it seems that some magma resurfacing events occurred in the landing area, resulting in some small craters being infilled and thus disappearing in the crater counting results. However, recent research [51] shows that the landing area is 70 m higher than the surrounding lunar basalt and that the surface materials were mainly ejecta from the Finsen and Alder craters. Therefore, the surface may have not been resurfaced by lava flows.

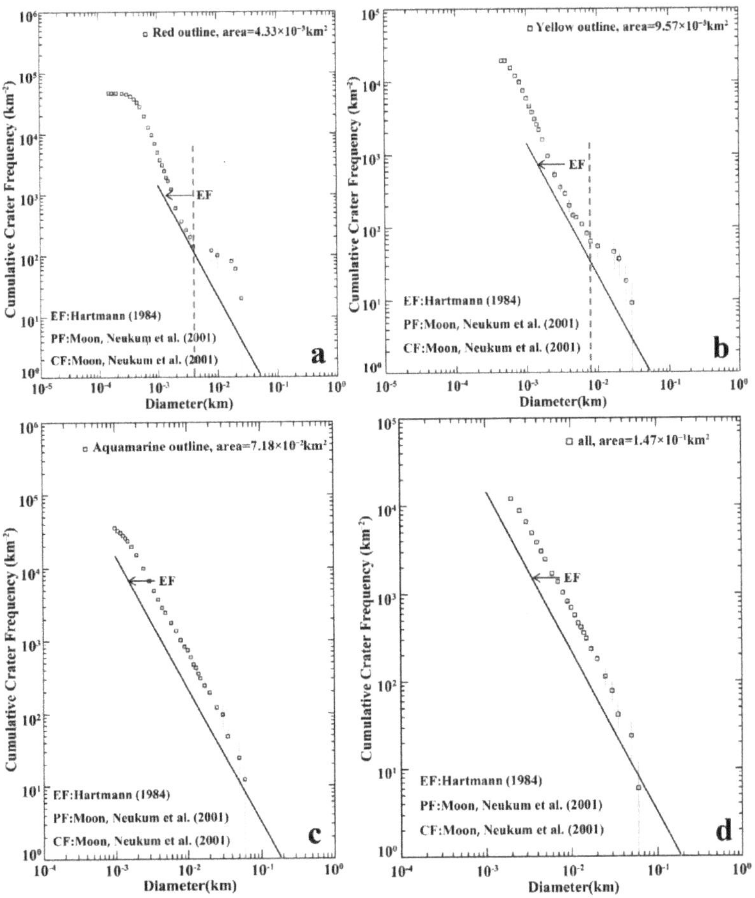

Figure 11. Craterstats2 program calculated cumulative SFD and equilibrium lines in the red, yellow, aquamarine, pink outlined areas. EF refer to the equilibrium function of Hartmann [44]. PF and CF refer to the production function and the chronology function of Neukum et al. [48], respectively. Red dashed lines show the inflection points. The equilibrium lines (criterions from Hartmann [44]) are displayed as black lines in (**a**–**d**). (**a**) Red outlined area. (**b**) Yellow outlined area. (**c**) Aquamarine outlined area. (**d**) Pink outlined area.

As shown in Figure 9a, the ejecta from C3 type crater a1′ infilled and partly resurfaced the neighboring C4 type crater a′ and caused a lower crater density in Figure 9d. The corresponding crater diameters at the inflection points of Figure 11a,b are 3 and 8 m, respectively, and the crater excavation depths can be scaled by diameters (0.084:1 for a simple lunar crater [41,52]); thus, the initial depth for craters with diameters of 3 and 8 m are 25.3 and 67.2 cm. The depth results show that the smaller the craters, the easier they are infilled by ejecta, while the larger craters show bigger depths to resist this influence.

Considering the existence of five craters with diameters of 15–22 m in the red statistical area and their different degradation levels, these five impact craters formed at different times due to the different degradation statuses. During the impact formation process, the materials were ejected into the surrounding area, thereby covering the neighboring tiny craters and producing an effect like magma resurfacing events.

This ejecta reshaping process may have caused the reduction in the number of craters smaller than 3 m in diameter in the red statistical area. Then, the cumulative SFD of the crater population was reset, and the process of crater accumulation restarted. As the statistical area gradually becomes larger, the crater size corresponding to the inflection points gradually becomes larger (8 m shown in Figure 11b) and nearly disappears in Figure 11c,d. This shows that the surface is increasingly difficult to resurfaced by ejecta materials as the craters become larger, and this is likely the reason that the inflection points of the statistical curve of impact craters gradually increase until it disappears.

Unlike previous works [15,42,43], small craters (diameter smaller than 3 m in this work) show an unusual cumulative SFD slope in equilibrium that smaller craters could be resurfaced by the ejecta from neighboring craters and uneasily reach equilibrium [16]. Considering that, the smaller the impact crater mentioned above is, the more severe the degradation is, the smaller the size of the impact crater is, and the more difficult it is for them to achieve crater equilibrium due to secondary craters and the local resurfacing events caused by ejecta from neighboring craters.

5. Future Research Considerations

Considering that Chang'e-4 is the first exploration mission to the far side of the Moon, its research data are of high scientific value. Compared with the lunar orbiter images, the Chang'e-4 lander and Yutu-2 rover provide a large number of high-resolution images in situ, such as the lander's LCAM images, Yutu-2 PCAM images, navigation camera images, etc. Combining these data and obtaining possible DEMs will help subsequent detailed population degradation analyses of centimeter- to meter-scale impact craters around the landing site and rover route. Furthermore, statistical analysis of the small-sized impact crater population degradation can also be extended to other landing areas for future lunar landing exploration and even for other celestial bodies, providing more detailed information for analyzing local geological evolution (e.g., potential surface resurfacing).

6. Conclusions

Small-sized (centimeter to meter scale) craters are the focus of our study. In contrast to the existing studies, which have mainly concentrated on large-sized craters (hundreds of meters), our study considers the degradation statistics of these small-sized crater populations to provide detailed information on the analysis of the surface resurfacing of local areas. Based on this, 6316 craters with diameters from 0.15 to 67 m were identified and classified into different degradation levels using their morphological differences on the high-resolution Chang'e-4 LCAM mosaic images. The ground TCAM image and the crater diameter-to-depth ratio obtained from the DEM of the Yutu-2 PCAM validate the classification of crater degradation levels from a qualitative and quantitative perspective, respectively. The results of the study indicated that:

(1) The crater population is severely degraded, with over 90% of the craters being moderately degraded (C3) and severely degraded (C4). Moreover, like the degradation pattern of the large crater population, the smaller the crater size, the more severe the degradation.

(2) The red and yellow outlined areas cover more small-sized craters (with minimum crater diameters of 0.15 and 0.5 m, respectively) than the other two study areas in this paper; unlike previous findings, they have more difficulty reaching equilibrium (cumulative SFD slopes of −2.696 and −2.392 for these two areas). This may be due to secondary craters and/or resurfacing events caused by ejecta from adjacent craters.

Author Contributions: Z.K. and T.H. conceived the idea. T.H. and Z.Y. wrote the manuscript. T.H., Z.Y. and Z.K. discussed and revised the manuscript. H.L., J.Z., D.Z., Y.C. and H.G. collected and processed data. All authors have read and agreed to the published version of the manuscript.

Funding: This work was supported by the National Key R&D Program of China (Grant No. 2019YFE0123300), the National Natural Science Foundation of China (Grant Nos. 41602215 and 41872207), the Pre-research Project on Civil Aerospace Technologies Funded by China National Space Administration (CNSA) (No. D020204), the National Key R&D Program of China (Grant No.

2020YFE0202100), and the 2021 Graduate Innovation Fund Project of China University of Geosciences, Beijing (ZY2021YC008).

Data Availability Statement: The mosaic of Chang'e-4 descent images and all 6316 classified craters can be found in "Hu, T. & Kang, Z. (2020). The mosaic of Chang\'E-4 descent images and the classified craters [Data set]. Zenodo. https://doi.org/10.5281/zenodo.3600427." For the specific crater degradation levels classification results can be downloaded from the Science Data Bank.

Acknowledgments: The Chang'e-4 mission was carried out by the Chinese Lunar Exploration Program, and the data can be downloaded from the Data Release and Information Service System of China's Lunar Exploration Program (https://moon.bao.ac.cn/ce5web/searchOrder_dataSearchData.search, accessed on 4 January 2022).

Conflicts of Interest: The authors declare no conflict of interest.

References

1. Li, C.; Zuo, W.; Wen, W.; Zeng, X.; Gao, X.; Liu, Y.; Fu, Q.; Zhang, Z.; Su, Y.; Ren, X. Overview of the Chang'e-4 mission: Opening the frontier of scientific exploration of the lunar far side. *Space Sci. Rev.* **2021**, *217*, 1–32. [CrossRef]
2. Liu, Z.; Di, K.; Li, J.; Xie, J.; Cui, X.; Xi, L.; Wan, W.; Peng, M.; Liu, B.; Wang, Y. Landing site topographic mapping and rover localization for Chang'e-4 mission. *Sci. China Inf. Sci.* **2020**, *63*, 1–12. [CrossRef]
3. Yang, C.; Zhao, H.; Bruzzone, L.; Benediktsson, J.A.; Liang, Y.; Liu, B.; Zeng, X.; Guan, R.; Li, C.; Ouyang, Z. Lunar impact crater identification and age estimation with Chang'E data by deep and transfer learning. *Nat. Commun.* **2020**, *11*, 1–15. [CrossRef] [PubMed]
4. Guo, D.; Fa, W.; Wu, B.; Li, Y.; Liu, Y. Millimeter-to Decimeter-Scale Surface Roughness of the Moon at the Chang'e-4 Exploration Region. *Geophys. Res. Lett.* **2021**, *48*, e2021GL094931. [CrossRef]
5. Wu, B.; Li, Y.; Liu, W.C.; Wang, Y.; Li, F.; Zhao, Y.; Zhang, H. Centimeter-resolution topographic modeling and fine-scale analysis of craters and rocks at the Chang'E-4 landing site. *Earth Planet. Sci. Lett.* **2021**, *553*, 116666. [CrossRef]
6. Jia, M.; Di, K.; Yue, Z.; Liu, B.; Wan, W.; Niu, S.; Liu, J.; Cheng, W.; Lin, Y. Multi-scale morphologic investigation of craters in the Chang'e-4 landing area. *Icarus* **2021**, *355*, 114164. [CrossRef]
7. Hiesinger, H.V.; Van Der Bogert, C.; Pasckert, J.; Funcke, L.; Giacomini, L.; Ostrach, L.; Robinson, M. How old are young lunar craters? *J. Geophys. Res. Planets* **2012**, *117*, E00H10. [CrossRef]
8. Geiss, J.; Rossi, A.P. On the chronology of lunar origin and evolution. *Astron. Astrophys. Rev.* **2013**, *21*, 1–54. [CrossRef]
9. Massironi, M.; Cremonese, G.; Marchi, S.; Martellato, E.; Mottola, S.; Wagner, R.J. Mercury's geochronology revised by applying Model Production Function to Mariner 10 data: Geological implications. *Geophys. Res. Lett.* **2009**, *36*, L21204. [CrossRef]
10. Basilevsky, A.T. On the evolution rate of small lunar craters. In Proceedings of the 7th Lunar Planetary Science Conference, Houston, TX, USA, 15–19 March 1976; Volume 1, pp. 1005–1019.
11. Craddock, R.A.; Howard, A.D. Simulated degradation of lunar impact craters and a new method for age dating farside mare deposits. *J. Geophys. Res. Planets* **2000**, *105*, 20387–20401. [CrossRef]
12. Molaro, J.; Byrne, S. Rates of temperature change of airless landscapes and implications for thermal stress weathering. *J. Geophys. Res. Planets* **2012**, *117*, E10011. [CrossRef]
13. Fassett, C.I.; Thomson, B.J. Crater degradation on the lunar maria: Topographic diffusion and the rate of erosion on the Moon. *J. Geophys. Res. Planets* **2014**, *119*, 2255–2271. [CrossRef]
14. Wilhelms, D.E.; John, F.; Trask, N.J. *The Geologic History of the Moon*; USGS Professional Paper 1348; Geological Survey: Reston, VA, USA, 1987; pp. 2330–7102.
15. Xiao, Z.; Werner, S.C. Size-frequency distribution of crater populations in equilibrium on the Moon. *J. Geophys. Res. Planets* **2015**, *120*, 2277–2292. [CrossRef]
16. Xie, M.; Xiao, Z.; Xu, A. Time-dependent production functions of lunar simple craters on layered targets with consideration of topographic degradation. *Geophys. Res. Lett.* **2019**, *46*, 10987–10996. [CrossRef]
17. Xie, M.; Zhu, M.H.; Xiao, Z.; Wu, Y.; Xu, A. Effect of topography degradation on crater size-frequency distributions: Implications for populations of small craters and age dating. *Geophys. Res. Lett.* **2017**, *44*, 10171–10179. [CrossRef]
18. Wood, C.; Head, J.; Cintala, M. Crater degradation on Mercury and the moon-Clues to surface evolution. In Proceedings of the 8th Lunar and Planetary Science Conference Proceedings, Houston, TX, USA, 14–18 March 1977; Volume 3, pp. 3503–3520.
19. Ivanov, M.; Basilevsky, A. Morphology and size–frequency distribution of kilometer-scale impact craters on Callisto and Ganymede derived from Galileo data. *Sol. Syst. Res.* **2002**, *36*, 447–457. [CrossRef]
20. Li, K. *Study on Small-Scale Lunar Craters' Morphology and Degradation*; Wuhan University: Wuhan, China, 2013.
21. Howard, A.D. Simulating the development of Martian highland landscapes through the interaction of impact cratering, fluvial erosion, and variable hydrologic forcing. *Geomorphology* **2007**, *91*, 332–363. [CrossRef]
22. Kreslavsky, M.A.; Head, J.W.; Neumann, G.A.; Rosenburg, M.A.; Aharonson, O.; Smith, D.E.; Zuber, M.T. Lunar topographic roughness maps from Lunar Orbiter Laser Altimeter (LOLA) data: Scale dependence and correlation with geologic features and units. *Icarus* **2013**, *226*, 52–66. [CrossRef]

23. Xie, M.; Liu, T.; Xu, A. Ballistic Sedimentation of Impact Crater Ejecta: Implications for the Provenance of Lunar Samples and the Resurfacing Effect of Ejecta on the Lunar Surface. *J. Geophys. Res. Planets* **2020**, *125*, e06113. [CrossRef]
24. Qiao, L.; Ling, Z.; Fu, X.; Li, B. Geological characterization of the Chang'e-4 landing area on the lunar farside. *Icarus* **2019**, *333*, 37–51. [CrossRef]
25. Chapman, C. The system of lunar craters. *Quadr. I Comm.* **1963**, *2*, 71–78.
26. Hirabayashi, M.; Minton, D.A.; Fassett, C.I. An analytical model of crater count equilibrium. *Icarus* **2017**, *289*, 134–143. [CrossRef]
27. Povilaitis, R.; Robinson, M.; Van der Bogert, C.; Hiesinger, H.; Meyer, H.; Ostrach, L. Crater density differences: Exploring regional resurfacing, secondary crater populations, and crater saturation equilibrium on the moon. *Planet. Space Sci.* **2018**, *162*, 41–51. [CrossRef]
28. Minton, D.A.; Fassett, C.I.; Hirabayashi, M.; Howl, B.A.; Richardson, J.E. The equilibrium size-frequency distribution of small craters reveals the effects of distal ejecta on lunar landscape morphology. *Icarus* **2019**, *326*, 63–87. [CrossRef]
29. Salaün, Y.; Marlet, R.; Monasse, P. Line-based robust SfM with little image overlap. In Proceedings of the 2017 International Conference on 3D Vision (3DV), Qingdao, China, 10–12 October 2017; pp. 195–204.
30. Li, R.; Fei, M.; Xu, F.; Matthies, L.H.; Olson, C.F.; Arvidson, R.E. Localization of Mars rovers using descent and surface-based image data. *J. Geophys. Res. Planets* **2002**, *107*, FIDO 4-1–FIDO 4-8. [CrossRef]
31. Kang, Z.Z.; Zhang, Z.X.; Yang, F.L. Relative Orientation and Epipolar Arrangement based on Forward Moving Image Pairs along the Optical Axis. *Acta Geod. Cartogr. Sin.* **2007**, *36*, 56–61.
32. Barlow, N.G. Constraining geologic properties and processes through the use of impact craters. *Geomorphology* **2015**, *240*, 18–33. [CrossRef]
33. Wang, J.; Ming, D.; Cheng, W.; Sensing, R. Identification of Multiscale Spatial Structure of Lunar Impact Crater: A Semivariogram Approach. *IEEE J. Sel. Top. Appl. Earth Obs.* **2021**, *14*, 2731–2739. [CrossRef]
34. Huang, Y.H.; Minton, D.A.; Hirabayashi, M.; Elliott, J.R.; Richardson, J.E.; Fassett, C.I.; Zellner, N.E. Heterogeneous impact transport on the Moon. *J. Geophys. Res. Planets* **2017**, *122*, 1158–1180. [CrossRef]
35. Kinczyk, M.J.; Prockter, L.M.; Byrne, P.K.; Susorney, H.C.; Chapman, C.R. A morphological evaluation of crater degradation on Mercury: Revisiting crater classification with MESSENGER data. *Icarus* **2020**, *341*, 113637. [CrossRef]
36. Ling, Z.; Qiao, L.; Liu, C.; Cao, H.; Bi, X.; Lu, X.; Zhang, J.; Fu, X.; Li, B.; Liu, J. Composition, mineralogy and chronology of mare basalts and non-mare materials in Von Kármán crater: Landing site of the Chang'E−4 mission. *Planet. Space Sci.* **2019**, *179*, 104741. [CrossRef]
37. Di, K.; Zhu, M.H.; Yue, Z.; Lin, Y.; Wan, W.; Liu, Z.; Gou, S.; Liu, B.; Peng, M.; Wang, Y. Topographic evolution of Von Kármán crater revealed by the lunar rover Yutu-2. *Geophys. Res. Lett.* **2019**, *46*, 12764–12770. [CrossRef]
38. Garrick-Bethell, I.; Zuber, M.T. Elliptical structure of the lunar South Pole-Aitken basin. *Icarus* **2009**, *204*, 399–408. [CrossRef]
39. Wu, B.; Li, F.; Hu, H.; Zhao, Y.; Wang, Y.; Xiao, P.; Li, Y.; Liu, W.C.; Chen, L.; Ge, X. Topographic and Geomorphological Mapping and Analysis of the Chang'E-4 Landing Site on the Far Side of the Moon. *Photogramm. Eng. Remote Sens.* **2020**, *86*, 247–258. [CrossRef]
40. Wang, Y.; Xie, M.; Xiao, Z.; Cui, J. The minimum confidence limit for diameters in crater counts. *Icarus* **2020**, *341*, 113645. [CrossRef]
41. Melosh, H.J. *Impact Cratering: A Geologic Process*; Oxford University Press: New York, NY, USA; Clarendon Press: Oxford, UK, 1989; pp. 1–255.
42. Richardson, J.E. Cratering saturation and equilibrium: A new model looks at an old problem. *Icarus* **2009**, *204*, 697–715. [CrossRef]
43. Gault, D.E. Saturation and equilibrium conditions for impact cratering on the lunar surface: Criteria and implications. *Radio Sci.* **1970**, *5*, 273–291. [CrossRef]
44. Hartmann, W.K. Does crater 'saturation equilibrium' occur in the solar system? *Icarus* **1984**, *60*, 56–74. [CrossRef]
45. Michael, G.; Neukum, G. Planetary surface dating from crater size–frequency distribution measurements: Partial resurfacing events and statistical age uncertainty. *Earth Planet. Sci. Lett.* **2010**, *294*, 223–229. [CrossRef]
46. Clauset, A.; Shalizi, C.R.; Newman, M.E. Power-law distributions in empirical data. *SIAM Rev.* **2009**, *51*, 661–703. [CrossRef]
47. Crater Analysis Techniques Working Group. Standard techniques for presentation and analysis of crater size-frequency data. *Icarus* **1979**, *37*, 467–474. [CrossRef]
48. Neukum, G.; Ivanov, B.; Hartmann, W.K. Cratering records in the Inner solar system in relation to the lunar reference system. *Space Sci. Rev.* **2001**, *96*, 55–86. [CrossRef]
49. Moore, H.J. Density of small craters on the lunar surface. In *United States Geological Survey Astrogeology Studies*; Annual Program Report: Washington, DC, USA, 1964; pp. 34–51.
50. Xiao, Z.; Strom, R.G. Problems determining relative and absolute ages using the small crater population. *Icarus* **2012**, *220*, 254–267. [CrossRef]
51. Di, K.; Liu, Z.; Liu, B.; Wan, W.; Peng, M.; Wang, Y.; Gou, S.; Yue, Z.; Xin, X.; Jia, M. Chang'e-4 lander localization based on multi-source data. *J. Remote Sens.* **2019**, *23*, 177–184.
52. Thomson, B.J.; Grosfils, E.B.; Bussey, D.B.J.; Spudis, P.D. A new technique for estimating the thickness of mare basalts in Imbrium Basin. *Geophys. Res. Lett.* **2009**, *36*, L12201. [CrossRef]

remote sensing

MDPI

Article

Quantitative Inversion of Lunar Surface Chemistry Based on Hyperspectral Feature Bands and Extremely Randomized Trees Algorithm

Shuangshuang Wu [1,2], Jianping Chen [1,2,*], Li Li [3], Cheng Zhang [1,2], Rujin Huang [1,2] and Quanping Zhang [1,2]

1 School of Earth Sciences and Resources, China University of Geosciences, Beijing 100083, China
2 Beijing Key Laboratory of Development and Research for Land Resources Information, Beijing 100083, China
3 College of Geomatics, Xi'an University of Science and Technology, Xi'an 710054, China
* Correspondence: 3s@cugb.edu.cn

Abstract: In situ resource utilization (ISRU) is required for the operation of both medium and long-term exploration missions to provide metallic materials for the construction of lunar base infrastructure and H_2O and O_2 for life support. The study of the distribution of the lunar surface elements (Fe, Ti, Al, and Si) is the basis for the in situ utilization of mineral resources. With the arrival of the era of big data, the application of big data concepts and technical methods to lunar surface chemistry inversion has become an inevitable trend. This paper is guided by big data theory, and the Apollo 17 region and the area near the Copernicus crater are selected for analysis. The dimensionality of the first-order differential spectral features of lunar soil samples is reduced based on Pearson correlation analysis and the successive projections algorithm (SPA), and the extremely randomized trees (Extra-Trees) algorithm is applied to Chang'E-1 Interference Imaging Spectrometer (IIM) data to establish a prediction model for the lunar surface chemistry and generate FeO, TiO_2, Al_2O_3, and SiO_2 distribution maps. The results show that the optimum number of variables for FeO, TiO_2, Al_2O_3, and SiO_2 is 17, 5, 8, and 30, respectively. The accuracy of the Extra-Trees model using the best variables was improved over that of the original band model, with determination coefficients (R^2) of 0.962, 0.944, 0.964, and 0.860 for FeO, TiO_2, Al_2O_3, and SiO_2, and root mean square errors (RMSEs) of 1.028, 0.672, 0.942, and 0.897, respectively. The modeling feature variables and model preference methods in this study can improve the inversion accuracy of chemical abundance to some extent, demonstrating the potential of IIM data in predicting chemical abundance and providing a good data basis for lunar geological evolution studies and ISRU.

Keywords: moon; machine learning; successive projections algorithm; extremely randomized trees; lunar chemistry

1. Introduction

From ancient times to the present day, humankind has never stopped exploring the vastness of the universe. The Moon, the closest celestial body to the Earth, has always been a prime target for human astronomical activities and is naturally the first step for humankind to step out of the cradle of the Earth and into the vastness of the universe [1]. As humankind is currently facing a shortage of resources, the exploitation and utilization of the rich lunar reserves of mineral resources can effectively alleviate the problem of resource shortages for humankind in the future. Given the high cost of Earth–Moon transportation, in situ resource utilization (ISRU) has become a fundamental technological guarantee for the establishment of a lunar base [2–4]. Currently, the most important ISRU project is to obtain H_2O and O_2 for life support through ilmenite. However, increasing human understanding and knowledge of the moon has revealed that any oxide can also be reduced to its constituent elements through reduction processes and produce H_2O and O_2 as byproducts, which is a key issue worthy of study [5]. Therefore, the elements of greatest interest to us are the lunar surface elements Fe, Ti, Al, and Si. The study of the distribution of Fe, Ti, Al, and Si is the basis for assessing the prospects for the exploitation of lunar mineral resources and the rational exploitation of resources. Fe, Ti, Al, and Si can all be used as sources of building materials for lunar base construction, with Fe mainly existing in lunar mafic minerals, Ti accounting for a large proportion of the high-titanium mare basalt [6,7], almost all Al on the lunar surface found in highland plagioclase [8], and Si is always present in the form of silicates in nearly every rock and mineral grain on the Moon. Therefore, this paper focuses on the inversion of Fe, Ti, Al, and Si on the lunar surface to study the distribution and abundance of mineral resources and to lay the foundation for a better assessment of the lunar mineral resources.

Currently, multisource remote sensing techniques based on γ-ray, X-ray, and multi/hypers pectral data are widely used for lunar surface chemistry inversion [9–16]. Compared with high-energy techniques, multi/hyperspectral data have the advantage of high resolution; therefore, many studies invert chemistry abundances based on multi/hyperspectral data. The outcomes of these studies include FeO and TiO_2 distribution maps with a resolution of 100 m obtained using Clementine data [17]; FeO, TiO_2, Al_2O_3, SiO_2, CaO, and MgO distribution maps with a resolution of 200 m obtained based on IIM data [18–21]; FeO distribution maps with a resolution of 20 m obtained based on Moon Mineralogy Mapper (M^3) data [10]; a TiO_2 distribution map with a resolution of 400 m obtained from the Lunar Reconnaissance Orbiter Camera (LROC) Wide Angle Camera (WAC) [16]; and FeO, TiO_2, Al_2O_3, CaO, and MgO distribution maps with a resolution of 59 m obtained based on Kaguya Multiband Imager (MI) data [22,23].

Among the four elements found on the lunar surface, namely, Fe, Ti, Al, and Si, Fe and Ti are transition metals that can undergo ligand field transitions (d-d transitions). Accompanying this electronic transition from low to high energy levels, the transition metals Fe^{2+} and Ti^{3+} produce absorption features in the visible to near-infrared range [18,24–26]; thus, the spectral features of the lunar surface can be directly interpreted based on mineralogical principles [7,24,26]. Early quantitative spectroscopic analysis of lunar samples demonstrates the potential for inversion of lunar surface chemistry and mineral components using band ratio methods [27]. Currently, this approach has been widely used for FeO and TiO_2 inversion [16,17,19,28–32]. However, elemental abundance inversion is not limited to the chromophore elements Fe and Ti. Although nonchromophore elements do not have absorption characteristics in the visible NIR, they affect reflectance [18]. Therefore, many studies have used regression models to establish statistical relationships between elemental abundance and spectral characteristics to predict chemical abundance [20,21,33,34]. With continuous research, a variety of machine learning models have been applied for oxide inversion. Regardless of whether the input variables are full or feature bands, machine learning methods have displayed good performance in inversion [7,22,35–37]. For example, Korokhin et al. [7] proposed a nonlinear method based on an artificial neural network (ANN) inversion of TiO_2 and avoided problems related to the limited number of bands and the subjective selection of band combinations compared with the traditional linear regression method. Zhang et al. [37] used principal component analysis (PCA) combined with a support vector machine (SVM) to estimate the abundance of chemical compositions (SiO_2, Al_2O_3, FeO, MgO, and TiO_2) and their maturity indicators (Is/FeO), where PCA was used to downscale the reflectance spectra of lunar soil samples for screening and the SVM was used to build a predictive oxide content model. Wang et al. [22] used a particle swarm optimization-support vector machine (PSO-SVM) algorithm based on KAGUYA MI data to generate oxide abundance maps with a spatial resolution of up to 59 m/pixel, and the map was relatively free of topographic shadows.

The selection of appropriate modeling feature variables and machine learning models is particularly important for the accuracy of oxide inversion. Feature band selection methods can improve the efficiency and accuracy of machine learning models to some extent. However, methods that combine machine learning and multi/hyperspectral data for the inversion of the lunar surface oxide content do not fully consider the preferential selection of model feature variables. For example, Shkuratov [33] used multiple linear regression models for three bands of 750, 915, and 965 nm from SMART-1 data. Wang &

Niu [35] used 19 (571 nm–865 nm) bands of IIM data to construct a model after removing anomalous bands. Wu [18] selected 22 bands in the spectral range of 561–918 nm for modeling based on the magnitude of the signal-to-noise ratio. Sun et al. [20] screened four single bands and three band ratios with high correlations in the 513 nm–891 nm range based on Pearson correlation coefficients for modeling. Most of the above methods do not consider the interrelationships between pairs of spectral features or between oxides and spectral features, and adding all available bands to a model can have a negative impact on the accuracy and generalization ability of the model. In response to the above research deficiencies, Pearson correlation coefficients are adopted in this paper for the initial screening of spectral features. Then, input parameters are secondarily screened based on clustering analysis combined with the successive projections algorithm (SPA) to reduce the covariance interference among spectral features while ensuring a high correlation between input spectral features and oxides.

The extremely randomized trees (Extra-Trees) machine learning algorithm is selected to build the oxide inversion model and improve the oxide inversion accuracy. The Extra-Trees algorithm can effectively describe the small-sample, high-dimensional, and complex nonlinear relationship between the oxide content and spectral reflectance by integrating several weak learners to obtain a strong learner [38]. This approach shows a clear advantage in the inversion of the oxide content. For model sample set partitioning, Jin et al. [39], Li [27,40], and Zhang et al. [37] divided the training and test sets according to particle size. Zhou et al. [41] randomly selected 2/3 of the samples as the training set for modeling and 1/3 of the samples as the validation set. These segmentation methods are often random. In this paper, we use sample set partitioning based on the joint x-y distance (SPXY) algorithm to partition the samples in a way that maximizes the characterization of the sample distribution and improves the stability of the model.

The inversion of lunar surface chemistry based on the concept and analysis methods of big data is an inevitable trend in future development. The selection of suitable modeling feature variables and machine learning models is crucial for achieving high-accuracy oxide inversion. Guided by big data theory, big data concepts and technical methods are applied to lunar chemistry inversion, and a prediction method for feature band selection combined with an Extra-Trees model is proposed. First, the first-order difference bands are initially screened according to the Pearson correlation coefficient, and they are then downscaled using the bisecting K-means (BKM) algorithm combined with the SPA to determine the best band combination. Second, the Lunar Soil Characterization Consortium (LSCC) training sample is used to build an Extra-Trees prediction model. Finally, the model is applied to predict the oxide abundance in the Apollo 17 region of the lunar surface and the region near the Copernicus crater. The experimental results are compared with those of representative models [15,17,30,31], and it is found that the model shows good agreement with them, providing a new idea and method for the inversion of oxide abundances.

2. Materials and Methods

This paper proposes a method to invert the distribution of lunar surface chemistry, targeting the need for ISRU. Figure 1 shows the technological flow chart of this paper, which includes the following main steps:

(1) Feature band selection: The sensitive regions of each chemistry were initially screened according to Pearson correlation coefficients, and then clustering analysis combined with SPA was used for secondary screening to determine the best combination of bands.
(2) Construction of an Extra-Trees model: Seventy-six LSCC samples with reflectance and oxide content data were used as model inputs for training and testing.
(3) Prediction of chemical abundance: The IIM reflectance data were put into the model to estimate the lunar surface chemical abundance.

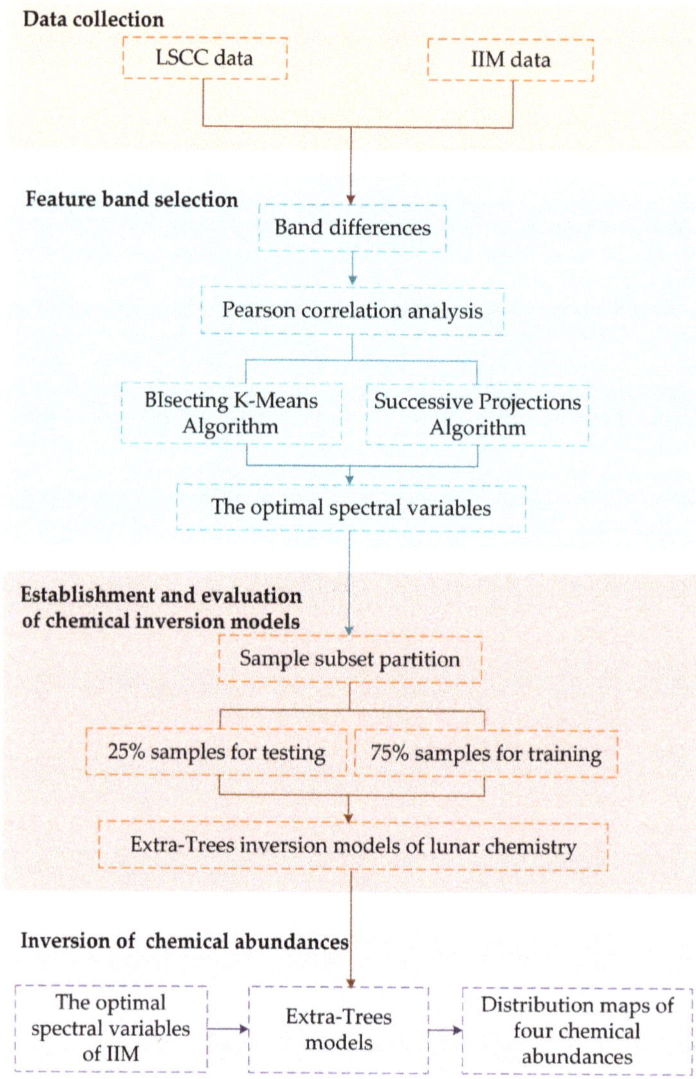

Figure 1. Flow chart of lunar surface chemical inversion.

2.1. Data Description

2.1.1. LSCC Data

The actual measured samples are important for studying the material components of the lunar surface and are the only standard for testing the effectiveness of the inversion. The LSCC soil sample data are stored in a spectral library created through the spectral analysis of samples collected from the six Apollo lunar landings. The samples include 10 samples of highland lunar soil from Apollo 14 and Apollo 16 and 9 samples of mare lunar soil from Apollo 11, 12, 15, and 17 [42]. Since the lunar surface spectral characteristics are influenced by the particle size of the material, the LSCC screened the measured samples for particle size and divided them into four groups, namely, <10 μm, 10–20 μm, 20–45 μm, and <45 μm, for a total of 76 subsamples [27,42–44]. All measured spectra of lunar surface samples were obtained at the RELAB laboratory at Brown University using a phase angle of 30° at a sampling interval of 5 nm in the spectral range of 0.3–2.6 μm [40–42]. Composition data for the LSCC soils used in this paper are shown in Table 1.

Table 1. Composition data for LSCC soils used in this paper.

Region	Mission	Number	Size (μm)	FeO	TiO_2	Al_2O_3	SiO_2
mare	Apollo 11	10,084	<10	12.00	7.25	15.90	42.10
			10–20	14.70	7.94	13.20	41.20
			20–45	15.50	8.30	12.00	41.30
	Apollo 12	12,001	<10	12.50	2.78	14.90	46.00
			10–20	15.90	2.96	12.30	45.00
			20–45	16.90	3.20	11.00	45.30
		12,030	<10	14.30	3.01	13.90	46.20
			10–20	17.20	3.32	10.70	46.30
			20–45	17.60	3.74	10.50	46.10
	Apollo 15	15,041	<10	11.00	1.79	16.40	46.60
			10–20	14.40	1.88	13.50	46.20
			20–45	15.20	2.03	12.50	46.10
		15,071	<10	9.59	1.57	17.10	46.90
			10–20	15.40	1.88	12.90	45.70
			20–45	15.60	2.33	12.40	45.80
	Apollo 17	70,181	<10	12.70	6.54	15.40	41.50
			10–20	15.50	7.88	12.70	40.40
			20–45	16.00	8.11	11.50	40.70
		71,061	<10	14.80	7.89	13.80	40.20
			10–20	17.50	8.94	10.80	39.50
			20–45	18.50	9.48	9.33	39.20
		71,501	<10	13.50	8.27	14.50	40.40
			10–20	16.40	9.83	11.60	39.00
			20–45	17.80	10.70	9.94	38.40
		79,221	<10	11.30	5.83	15.90	42.30
			10–20	15.00	7.21	12.90	40.90
			20–45	15.80	7.38	11.60	40.50
highland	Apollo 14	14,141	<10	7.66	1.51	19.20	49.20
			10–20	9.46	1.71	17.20	48.40
			20–45	11.60	1.96	15.00	47.20
		14,163	<10	8.83	2.07	18.90	47.20
			10–20	10.10	1.88	17.00	47.40
			20–45	11.50	2.00	15.40	47.10
		14,259	<10	7.82	2.02	19.30	47.90
			10–20	9.71	1.96	17.40	47.50
			20–45	11.00	1.99	15.80	47.10
		14,260	<10	8.10	1.94	19.10	47.80
			10–20	9.84	1.98	17.30	47.50
			20–45	10.70	1.86	16.30	47.40
	Apollo 16	61,221	<10	3.64	0.50	28.50	44.50
			10–20	4.40	0.54	27.50	44.50
			20–45	4.62	0.56	27.20	44.50
		61,141	<10	3.66	0.59	27.40	44.90
			10–20	5.14	0.64	25.60	44.60
			20–45	5.15	0.58	26.10	44.50
		62,231	<10	3.63	0.58	27.40	45.00
			10–20	4.86	0.61	26.30	44.70
			20–45	5.31	0.58	25.70	44.50
		64,801	<10	3.84	0.61	27.70	44.80
			10–20	4.78	0.68	26.30	44.50
			20–45	4.82	0.63	26.50	44.60
		67,461	<10	3.35	0.34	29.40	44.50
			10–20	4.64	0.39	27.80	44.10
			20–45	4.93	0.44	27.30	44.40
		67,481	<10	3.61	0.42	29.10	44.50
			10–20	4.04	0.40	28.40	44.40
			20–45	5.19	0.49	26.70	44.70

2.1.2. IIM Data

The IIM, one of the payloads of Chang'E-1, has achieved the first international application of interferometric imaging spectroscopy in the field of planetary exploration.

The wavelength range of IIM data is 480–960 nm, with a spectral resolution of 335 cm^{-1}, a total of 32 bands, and a spatial resolution of 200 m/pixel [18,45,46]. Critical to establishing a statistical relationship between the hyperspectral band characteristics of the samples and the mineral element content is the requirement that the laboratory-measured LSCC samples and the IIM image absolute calibrations have the same photometric calibration [7,21,33,47,48]. The IIM 2C data downloaded from the Ground Application System of the Lunar Exploration Project have been preprocessed with radiometric correction, geometric correction, and photometric correction [46,49]. The IIM data used in this paper are based on the processing of 2C data with further recalibration processing, which includes reflectance correction using Apollo samples. Therefore, the photometric conditions between the IIM image and the LSCC sample data are the same after preprocessing, and the laboratory-measured LSCC samples can be used together with the IIM data.

IIM data quality evaluation is also an important prerequisite for performing scientific inversion. As indicated by signal-to-noise ratio analysis, the first four bands of IIM 2C data contain considerable noise, and the 32nd band has the highest noise level (lowest signal-to-noise ratio observed). Therefore, 26 bands of data in the range between 513 and 891 nm (bands 5–30) are used in this paper.

2.2. Feature Band Selection

The reflectance spectra used in this paper are between 513 and 891 nm, and band difference calculation results in 325 spectral bands. If all these spectral bands are input into the model for prediction, the number of bands will be high, as will the correlations between adjacent bands, which will inevitably lead to an increase in the redundancy of spectral information and adversely affect the accuracy and generalization of the model [37]. Therefore, it is necessary to filter the sensitive bands that play a key role in the model.

Pearson correlation analysis is used to correlate the differentially transformed spectral features and the oxide content to find the band differences with correlation coefficients that pass the significance test at the 0.01 level. The results are used to identify the sensitive feature regions of the differentially transformed spectra.

Due to the large number and continuous distribution of selected wavelengths, secondary screening is required within the sensitive regions identified with Pearson correlation analysis. Considering the random nature of the initial bands of the SPA, to reduce the possibility of the selection of invalid initial bands, the range of random initial bands is restricted to the corresponding clusters by combining the above algorithm with a clustering algorithm. Fine screening is performed separately in each cluster to eliminate the presence of covariance bands. The feature wavelengths selected by the SPA are those that are most representative.

2.2.1. Bisecting K-Means Algorithm

The bisecting K-means algorithm is an improvement on and expansion of the K-means algorithm, and it essentially bifurcates the selected clusters until the specified number of clusters is reached while satisfying a criterion based on the sum of the squared error (SSE) [50]. The SSE is defined as shown in Equation (1). The greatest advantage of this algorithm over the traditional K-means method is that it is simple and fast to implement and finds the globally optimal solution. Therefore, in this paper, the dichotomous K-means algorithm is used for the clustering analysis of sensitive feature regions with differences, and the initial K value is set to 3.

$$SSE = \sum_{i=1}^{k} \sum_{x \in C_i} dist(c_i, x)^2 \tag{1}$$

where k is the number of clusters, c_i is the cluster center of cluster C_i, and x is a sample in that cluster.

2.2.2. Successive Projections Algorithm

The successive projections algorithm (SPA) is a forward iterative feature variable selection method that minimizes the covariance in vector space, minimizes the redundant information in the original spectral matrix, eliminates the effect of covariance, and can be used for spectral feature wavelength screening [51,52]. The SPA starts with one wavelength and introduces the wavelength with the largest projection vector into the set of wavelength variables in each iteration by analyzing the projection of the vector until a specified number of wavelengths is reached [53–55]. The exact procedure of the algorithm is as follows.

1. In the 1st iteration ($n = 1$), any column of wavelength $k(0)$ is chosen and denoted as x_j, where $j = 1, \ldots, J$.
2. The wavelengths that are not included in the set are identified as $s = \{j, 1 \leq j \leq J, j \notin \{k(0), \ldots, k(n-1)\}\}$.
3. The projection of the initialized band x_j with an unselected wavelength in orthogonal space is calculated as

$$P_{x_j} = x_j - \left(x_j^T x_{k(n-1)}\right) x_{k(n-1)} \left(x_{k(n-1)}^T x_{k(n-1)}\right)^{-1} \quad (2)$$

4. The maximum wavelength of the projection vector is calculated:

$$k(n) = \arg\left(\max\left(\|P_{x_j}\|\right)\right) \quad (3)$$

5. $n = n + 1$; if $n < N$, return to step 2.
6. The final combination of wavelength variables is determined.

2.3. Sample Subset Partition

The division of the sample data set will affect the accuracy of the model estimates to some extent. The SPXY algorithm was proposed by Galvao et al. [56]. The objective is to use the physical-chemical variable y and the spectral variable x to calculate the intersample distance, fully characterize the sample distribution, effectively cover the multidimensional vector space, increase intersample variability and representativeness, and improve model stability [57,58]. The distance equation is as follows [56]:

$$d_x(p,q) = \sqrt{\sum_{j=1}^{J} \left[x_p(j) - x_q(j)\right]^2} \; p, q \in [1, N] \quad (4)$$

$$d_y(p,q) = \sqrt{\left(y_p - y_q\right)^2} = \left|y_p - y_q\right| \; p, q \in [1, N] \quad (5)$$

To ensure that the variables x and y give the same weight to a sample, $d_x(p,q)$ and $d_y(p,q)$ are divided by the maximum value in the data set, respectively. Thus, the formula after normalization is

$$d_{xy}(p,q) = \frac{d_x(p,q)}{max_{p,q\in[1,N]} d_x(p,q)} + \frac{d_y(p,q)}{max_{p,q\in[1,N]} d_y(p,q)} \; p, q \in [1, N] \quad (6)$$

In this study, the SPXY algorithm was used to divide the training and test sets of 76 LSCC samples based on the four oxide contents as y variables and the spectral features as x variables, and the specific division results are shown in Table 2. The training set accounts for 75% of all samples and contains 57 samples, and the test set accounts for 25% of all samples and contains 19 samples. The statistical parameters of the oxide content in the test set are generally within the same ranges as those of the training set, and the sample set is divided reasonably and effectively, which helps improve the stability and reliability of the model.

Table 2. Statistical parameters of sample set partitioning.

Elements	Training Set					Test Set				
	Count	Max	Min	Mean	Std	Count	Max	Min	Mean	Std
FeO	57	18.50	3.35	10.50	5.01	19	16.40	3.66	10.83	4.40
TiO_2	57	10.70	0.34	3.49	3.21	19	7.94	0.35	2.38	2.41
Al_2O_3	57	29.40	9.33	18.90	6.83	19	27.40	11.60	16.52	4.96
SiO_2	57	49.20	38.40	44.35	2.75	19	47.60	40.40	44.90	2.31

2.4. Extra-Trees Regression

Extra-Trees is an integrated algorithm based on decision trees with good generalization and robustness and was proposed by Geurts et al. [38]. The Extra-Trees algorithm is trained using all training samples, and different decision trees are constructed according to different features. The score of each random classification node for K random features is calculated with the Score function (Equation (7)), and the node with the highest score is selected as the splitting node [59,60]. When a new sample is input, multiple decision trees in the integrated model score it, and the final prediction is based on the average of all decision tree predictions [61].

$$Score_R(s,S) = \frac{var\{y|S\} - \frac{|S_l|}{|S|}var\{y|S_l\} - \frac{|S_r|}{|S|}var\{y|S_r\}}{var\{y|S\}} \quad (7)$$

where $var\{y|S\}$ is the variance of the output y in sample S and l and r denote the left and right divergence values of the nodes, respectively.

Unlike random forests that build each tree by sampling with feedback, Extra-Trees uses the entire training sample to establish each tree, which can effectively reduce the effects of the bias and variance of the sample set [62]. Moreover, unlike random forests that obtain the best bifurcation within a random subset, Extra-Trees obtains the bifurcation of each decision tree completely randomly [38].

In this paper, the Extra-Trees algorithm is used as the lunar surface oxide content prediction model, and the flow chart of this model is shown in Figure 2. The best combination of bands after secondary screening is input into the model, and each oxide content is output to construct the Extra-Trees oxide content prediction model. The implementation of the Extra-Trees model in this study is based on the ExtraTreesRegressor algorithm provided in the sklearn package of the Python language. The algorithm has two main parameters, including the minimum size of the samples for splitting the nodes, n_{min}, and the size of the randomly selected attributes for each node, k [38]. According to Geurts's experimental results, the default parameter settings are generally satisfactory in terms of accuracy and computational efficiency; therefore, the default parameters are used.

2.5. Evaluation Indicators

The accuracy and predictive ability of the model are mainly evaluated in terms of the coefficient of determination (R^2) and root mean square error (RMSE). R^2 is used to evaluate the correlation between the predicted and true values of the sample, and the closer R^2 is to 1, the higher the correlation between the predicted and true values (Equation (8). The RMSE is used to evaluate the predictive ability of the model for a given data set, and the smaller the RMSE is, the better the predictive ability of the model [37,52].

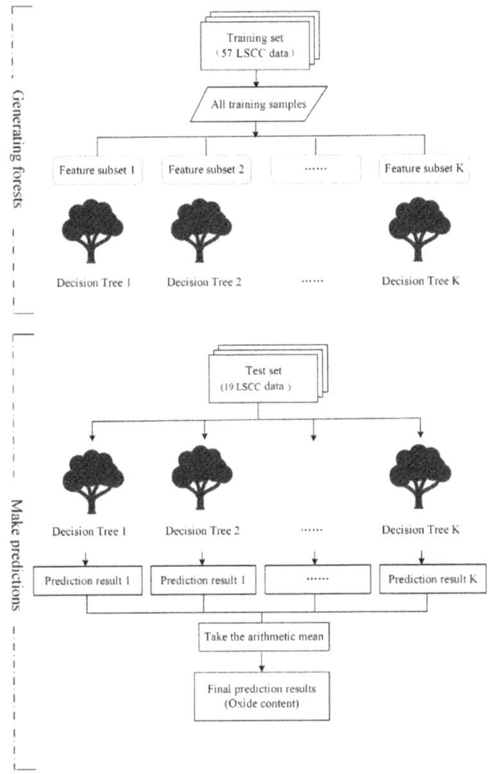

Figure 2. Flow chart of the Extra-Trees model.

$$R^2 = 1 - \frac{\sum_{i=1}^{m} (\check{c}_i - c_i)^2}{\sum_{i=1}^{m} (c_i - \overline{c_i})^2} \tag{8}$$

$$RMSE = \sqrt{\frac{\sum_{i=1}^{m} (\check{c}_i - c_i)^2}{n}} \tag{9}$$

where $\check{c}_i$, c_i and $\overline{c_i}$ denote the true, predicted, and mean values of the ith sample, respectively, and n denotes the number of samples.

3. Results

3.1. Correlation Coefficients between Elements and Reflectance

In this paper, IIM reflectivity data in the range of 513–891 nm were used as the basis for the estimation and analysis of the lunar surface oxide content. First, the standard bidirectional emissivity data for 76 LSCC samples (300–2600 nm) were resampled into the wavelength range of IIM data (513–891 nm) using a linear interpolation method. The spectral data after resampling were differentially transformed to obtain the reflectivity spectral band difference.

The linear correlation coefficients between LSCC reflectance and the content were calculated (Figure 3). The results showed that the correlation coefficients did not vary significantly within 513–891 nm, and the correlation coefficients between FeO, TiO_2, Al_2O_3, and SiO_2 and reflectance remained at approximately 0.75, 0.58, 0.79, and 0.25, respectively. Additionally, positive correlations were observed between the Al_2O_3 and SiO_2 contents and spectral reflectance, and negative correlations were observed between the FeO and TiO_2 contents and spectral reflectance.

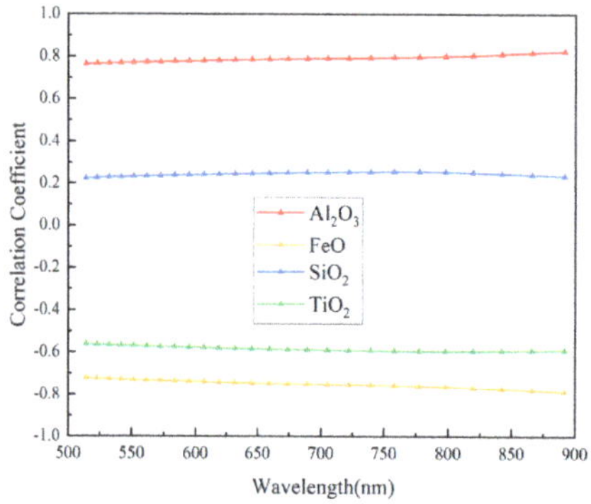

Figure 3. Linear correlation coefficients between the contents of the four oxides and LSCC reflectance data.

The correlations between the elemental contents and LSCC reflectance are mainly concentrated in the grain size range of 10–20 μm, as the optical information at these sizes is most similar to that of the bulk soil [18,63]. Figure 4 demonstrates the relationship between the 891 nm reflectance and the content of each oxide (FeO and TiO_2) in the LSCC samples of size 10–20 μm. The FeO and TiO_2 contents decrease with increasing reflectance, highlighting the negative correlations between the FeO and TiO_2 contents and spectral reflectance. However, the relationship between elements and IIM reflectance is not necessarily linear. For some oxides, it is difficult to describe the complex relationship between them and the spectra based on conventional linear regression [36], and suitable nonlinear models are needed to invert the major oxide contents.

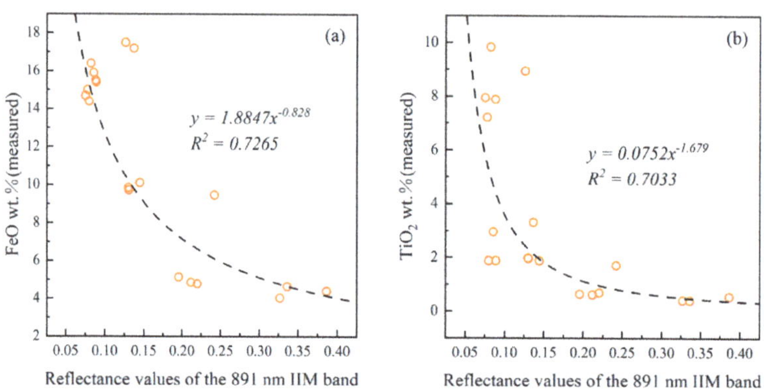

Figure 4. Relationship between the 891 nm reflectance and the (**a**) FeO and (**b**) TiO_2 contents in LSCC samples with 10–20 μm grain sizes.

Moreover, there are also correlations among elements. As shown in Figure 5, there is a significant inverse correlation between the FeO and Al_2O_3 contents and a significant inverse correlation between TiO_2 and SiO_2. These findings are consistent with the conclusion that there are positive correlations between the Al_2O_3 and SiO_2 contents and spectral reflectance and negative correlations between the FeO and TiO_2 contents and spectral reflectance, as shown in Figure 3.

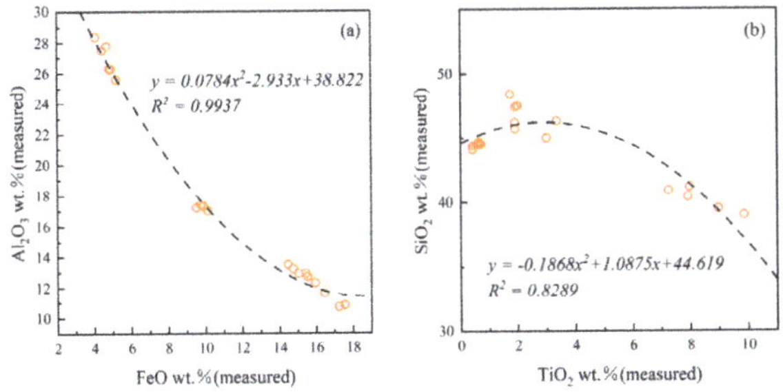

Figure 5. Correlations between (**a**) FeO and Al_2O_3 and (**b**) TiO_2 and SiO_2 in the LSCC samples with 10–20 μm grain sizes.

Figure 6 shows the correlations between the reflectance band differences and the major oxides, and the color of each point in the figure indicates the magnitude of the correlation coefficient between the band differences and the oxide contents. The results show that the band differences display the highest correlations with FeO and Al_2O_3 and the lowest correlations with TiO_2 and SiO_2. The maximum value of the correlation coefficient between FeO and the band difference is located at 541 nm–532 nm (0.861); most of the correlation coefficients between Al_2O_3 and the band difference of IIM are approximately 0.80, and the maximum value is 0.870. Additionally, the correlation between TiO_2 and the band difference of IIM decreases from the lower left to the upper right corner, and the maximum value appears at 532 nm–522 nm (0.710). The correlation between SiO_2 and the band difference is poor, with a maximum value of 0.427, and most of the correlation coefficients are approximately 0.40. In general, for the four oxides, the high correlations are mainly concentrated in the lower left corner, and the band differences among the longer bands (upper right corner in the figure) are poorly correlated. Correlation calculations for the original bands and all band differences in the IIM spectral range separately show that band differences can be used to effectively enhance the correlations between reflectance and the oxide contents to determine the best combination of bands for subsequent modeling.

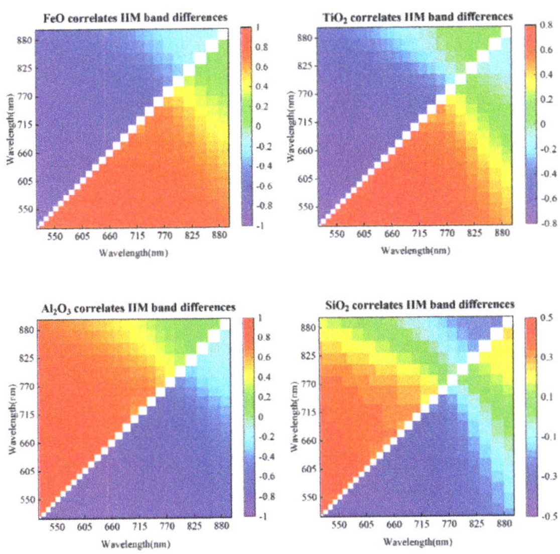

Figure 6. Correlation coefficients between spectral differences and chemical abundances. Each point represents the band difference between the x− and y−axes, and the intensity of each point is the correlation between the band difference and chemical abundance.

3.2. Screening of Feature Bands

The band differences that passed the 0.01 significance test were screened twice, and the sensitive bands identified through BKM clustering analysis combined with SPA are shown in Table 3. These bands are used as the inputs to the model.

Table 3. Final screening results for feature bands.

Oxide	Clustering Categories	Number of Bands	Selected Band (nm)
FeO	Category 1	5/91	522–583, 631–777, 645–721, and 659–842, and 659–891
	Category 2	9/123	513–757, 513–819, 522–659, 551–721, 551–777, 561–842, 606–777, 606–891, and 618–819
	Category 3	3/99	532–583, 739–819, and 739–891
TiO_2	Category 1	1/81	513–631
	Category 2	1/119	522–673
	Category 3	3/87	606–659, 721–819, and 739–797
Al_2O_3	Category 1	1/91	522–631
	Category 2	4/123	513–891, 561–797, 606–757, and 645–842
	Category 3	3/98	513–561, 739–891, and 739–819
SiO_2	Category 1	10/125	513–618, 522–594, 532–583, 541–572, 551–645, 561–673, 572–689, 583–631, 606–659, and 631–757
	Category 2	12/46	513–631, 513–673, 522–659, 522–689, 532–645, 551–721, 561–739, 572–721, 572–757, 583–777, 606–739, and 618–777
	Category 3	8/32	513–721, 513–777, 522–705, 522–819, 532–739, 551–757, 561–777, and 583–797

3.3. Establishment and Evaluation of the Extra-Trees Model

The spectral features obtained based on the band reduction screening were used as the independent variables for the Extra-Trees modeling analysis. Additionally, the lunar surface oxide content was used as the dependent variable, and the SPXY algorithm was applied to divide the sample set and establish the Extra-Trees model. To effectively analyze the results of variable screening, the modeling results for the original bands were also applied for comparison. The prediction results (Figures 7 and 8) show that the coefficients of determination R^2 of the FeO, TiO_2, Al_2O_3, and SiO_2 prediction models after differential transformation and variable filtering were 0.962, 0.944, 0.964, and 0.860, respectively, which are all greater than 0.850, indicating good prediction ability for the four oxide contents. The model accuracy is improved compared with that obtained based on the original bands, further verifying the importance of the differential transformation of the original bands and reduced-dimension filtering of the variable information. The modeling accuracy of feature band selection was compared with the accuracy reported by Wu [18] (Table 4), and the accuracy for all four oxide contents after differential transformation and variable screening was improved. Therefore, in this paper, the differential variables obtained based on the SPA were adopted as the inputs to the model for prediction.

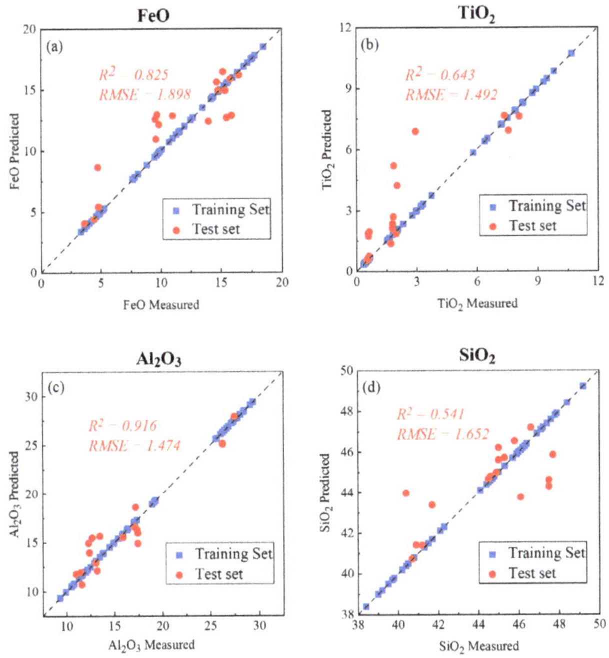

Figure 7. Training and testing accuracy of the Extra-Trees model with chemical abundance based on the original waveband: (**a**) FeO, (**b**) TiO_2, (**c**) Al_2O_3, and (**d**) SiO_2.

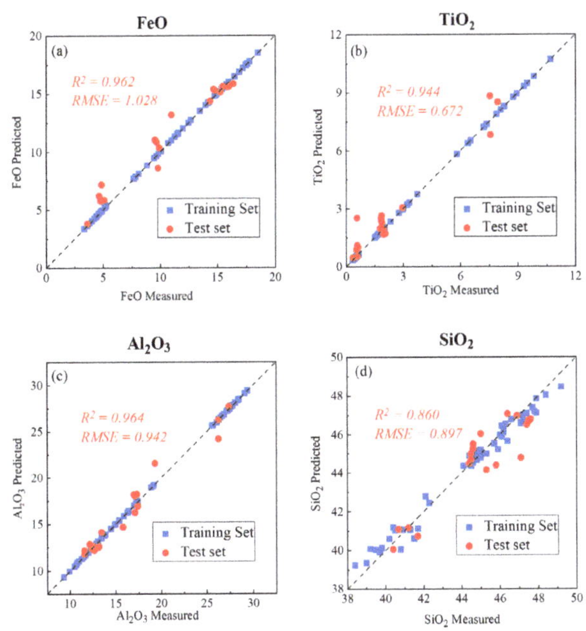

Figure 8. Training and testing accuracy of the Extra-Trees model with chemical abundance based on feature band selection: (**a**) FeO, (**b**) TiO_2, (**c**) Al_2O_3, and (**d**) SiO_2.

Table 4. Comparison of the accuracy of the four chemical abundance tests with the results of Wu [18].

	FeO (wt.%)	TiO_2 (wt.%)	Al_2O_3 (wt.%)	SiO_2 (wt.%)
Calibration in this work (Extra-Trees model based on feature band selection)				
R^2	1	1	1	0.974
RMSE	0.012	0	0.032	0.438
Validation in this work (Extra-Trees model based on feature band selection)				
R^2	0.962	0.944	0.964	0.860
RMSE	1.028	0.672	0.942	0.897
Calibration by Wu [18]				
R^2	0.90	0.69	0.92	0.76
RMSE	1.58	2.00	1.68	0.75
Validation by Wu [18]				
R^2	0.88	0.59	0.90	0.67
RMSE	1.76	2.26	1.92	0.91

3.4. Extra-Trees Modelling in the Apollo 17 Area

The Apollo 17 area is close to the junction of Serenitatis and Tranquillitatis, and the coordinates of the landing site are (30°44′58.3″E, 20°9′50.5″N). The geological environment of the area is complex; according to the analysis of Apollo17 samples, the rock types in the area are mainly overlying mare basalts and orange-grey breccia [64].

3.4.1. Extra-Trees Modelling of FeO

Fe mainly exists in lunar mafic minerals, and understanding the abundance and distribution of Fe can aid in understanding the nature and origin of the Moon and provide an indicator of lithology to distinguish between mare and highland [65]. In this paper, Extra-Trees modeling is performed with the effective bands obtained by the Pearson correlation coefficients and SPA. Due to the large number of bands of IIM data that undergo band-difference transformation, the coverages overlap. The sensitive bands that play a key role in the model are first screened, and the model is then trained and tested; however, the regression results are not the only reference used to evaluate the model. To further verify the accuracy of the model, it was applied to the Apollo 17 area (landing site coordinates: 30°44′58.3″ E, 20°9′50.5″ N) to evaluate the accuracy of FeO inversion.

The FeO training and testing accuracies for the Extra-Trees model after variable screening are shown in Figure 8a, where the blue dots represent the training set, and the red dots represent the test set. The coefficient of determination (R^2) for the test set is 0.962, and the RMSE is approximately 1.028. The FeO inversion results (Figure 9) show that the predicted FeO abundance in the region near the Apollo 17 landing site ranges from 7.64 to 15.72 wt.%, and the FeO content of the mare area is significantly higher than that of the highland area, with good consistency regarding the distribution of FeO content values in comparison with the results of Lucey et al. [17].

3.4.2. Extra-Trees Modelling of TiO_2

The distribution of the TiO_2 content varies greatly among different types of mare basalts; thus, it is the key to classifying mare basalt types and is important for the exploitation of lunar ilmenite resources [6]. The optimal band selection for TiO_2 is similar to that in the FeO model. First, the 287 bands that passed the $p < 0.01$ significance test were used to establish the sensitive region for TiO_2, and the best band combination was selected by clustering analysis and applying the SPA. The results are shown in Table 4.

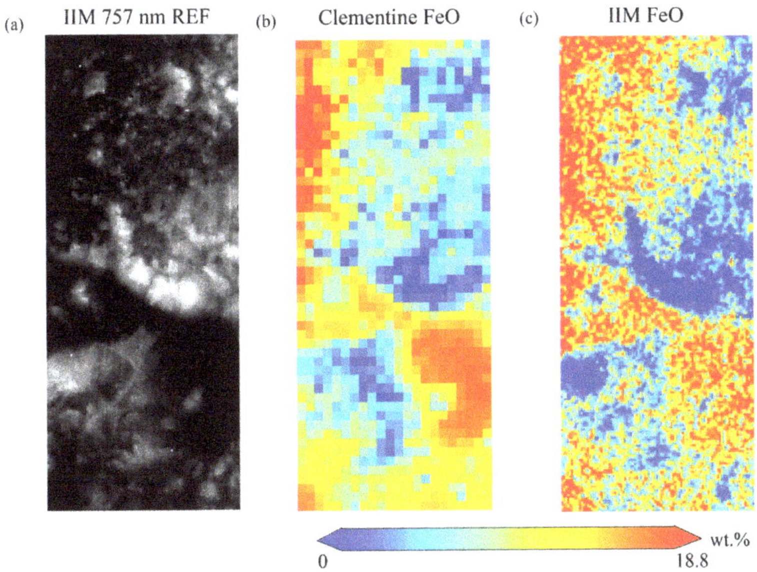

Figure 9. Apollo 17 area: (**a**) IIM 757 nm reflectance image, (**b**) FeO contents from Lucey et al. [17], and (**c**) FeO prediction results of the Extra-Trees model based on feature band selection.

The TiO_2 training and testing accuracies of the Extra-Trees model after variable screening are shown in Figure 8b, where the coefficient of determination (R^2) for the test set is 0.944 and the RMSE is approximately 0.672. The model was applied for TiO_2 inversion in the test area (Figure 10), and the results show that the TiO_2 abundance ranges from 2.21 to 9.37 wt.%; the high-value areas are mainly located in the mare, and the low-value areas are mainly in the highland, which is generally similar to the results reported by Lucey et al. [17]. The inversion results display a similar trend, but the lower limit value of TiO_2 is higher, and the contents are slightly higher in the lower left corner of the figure, which may be due to the effect of topographic shadowing from the light angle.

3.4.3. Extra-Trees Modelling of Al_2O_3

Almost all Al on the lunar surface is present in plagioclase, and the distribution of the Al_2O_3 content is related to the formation and evolution of the lunar crust; notably, levels are highest in highland calcarenite rocks and intermediate in mare basalts and can be used to assist in distinguishing mare basalts from highland plagioclase. To assess the accuracy of its inversion, an initial screening based on Pearson correlation coefficients was carried out, followed by BKM clustering analysis combined with the SPA algorithm to find the best band combinations for Al_2O_3. From Table 3, it can be seen that the eight best band differences were finally screened and used as input to the model for modeling.

The training and testing accuracies of the Al_2O_3 Extra-Trees model after variable screening are shown in Figure 8c. The coefficient of determination (R^2) for the test set with this method is 0.964, and the RMSE is approximately 0.942. The model was applied to the Apollo 17 region for Al_2O_3 inversion, and the results (Figure 11) show that the Al_2O_3 abundance obtained with this method ranges from 11.75–24.43 wt.%, with a mean value of 14.70 wt.%. The abundance of Al_2O_3 in the highlands region is lower than that in the mare region, and this correlates negatively with FeO.

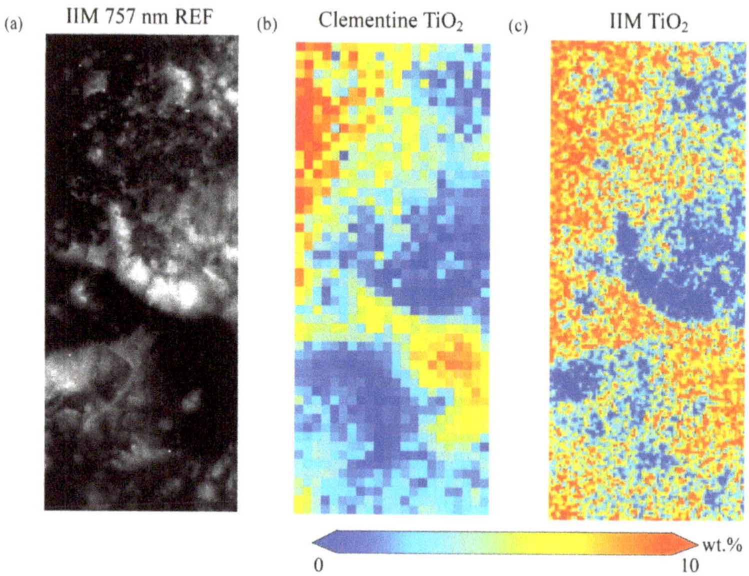

Figure 10. Apollo 17 area: (**a**) IIM 757 nm reflectance image, (**b**) TiO_2 contents from Lucey et al. [17], and (**c**) TiO_2 prediction results of the Extra-Trees model based on feature band selection.

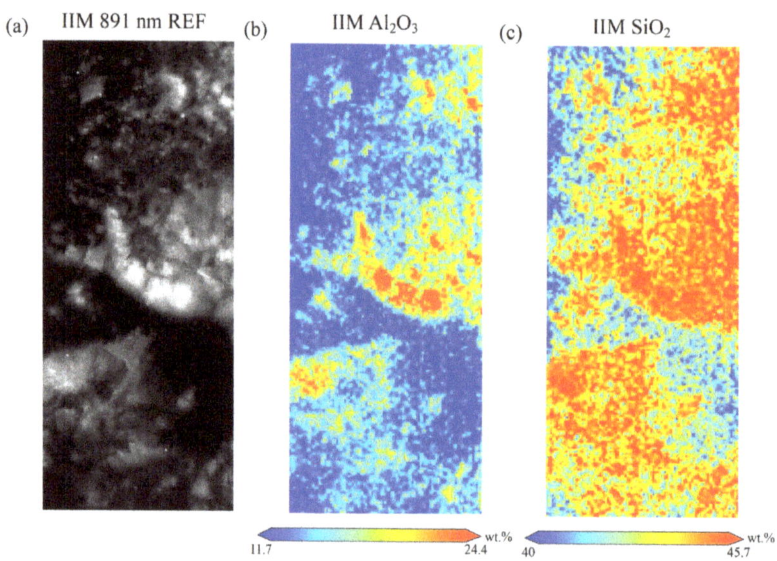

Figure 11. Apollo 17 area: (**a**) IIM 891 nm reflectance image, (**b**) Al_2O_3, (**c**) SiO_2 prediction results of the Extra-Trees model based on feature band selection.

3.4.4. Extra-Trees Modeling of SiO_2

SiO_2 is in nearly every rock and mineral grain on the Moon as the fundamental component of silicate minerals. The optimal waveband selection for SiO_2 is similar to that for FeO, and the optimum number of bands for SiO_2 is shown in Table 3, with a total of 30 band differences.

The training and testing accuracies of the SiO_2 Extra-Trees model after variable screening are shown in Figure 8d. The R^2 value for the test set obtained with this method is 0.860, and the RMSE is approximately 0.897. The model was applied for SiO_2 inversion in the Apollo 17 region (Figure 11c), and the results show that the SiO_2 abundance obtained with this method ranges from 41.08 to 45.67 wt.%, with a mean value of 43.84 wt.%. SiO_2 is commonly found in lunar rocks, with higher levels in the highland than in the mare, which is negatively correlated with TiO_2.

3.5. Oxide Content Mapping for the Copernicus Crater Region

3.5.1. Regional Distribution of Lunar Surface Chemistry

Using ArcGIS as the platform, a 1:2,500,000 lunar geological map of the study area was mapped (Figure 12), which covers a geographical area of approximately 125×10^4 km^2 at longitudes of $-52°$ to 6.6° W and latitudes of $-2.8°$ to 28° N. The study area is mainly occupied by the Copernicus crater, Kepler crater, and Aristarchus crater, located in the transition zone between the mare and the lunar highlands in the south of the Oceanus Procellarum. The study area is rich in geomorphic features and material types, and the rocks are mainly mare basalts, followed by KREEP rocks and ferroan anorthositic suite. Linear structures are widely developed and distributed and are of great importance to studies of lunar diagenesis and geological evolution.

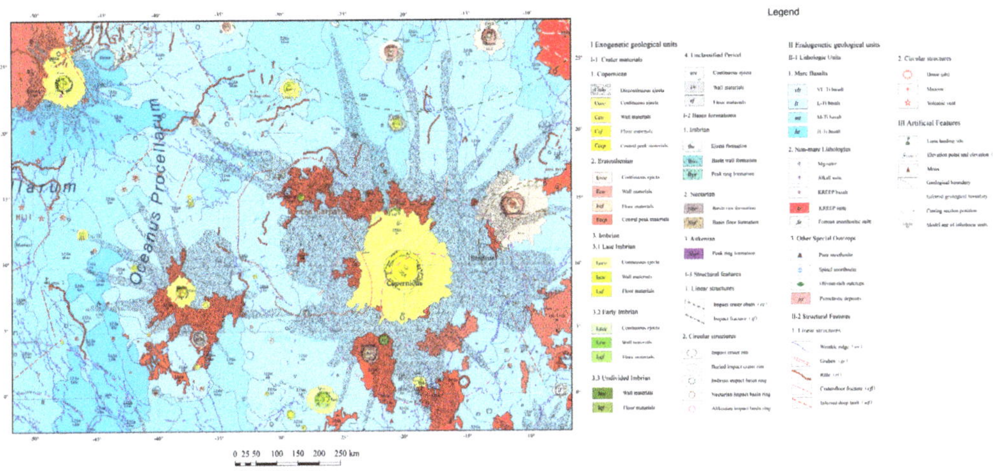

Figure 12. A 1:2,500,000 geological map [66] of the study area.

In the study area, the mare basalt is widely distributed in the lower and flat region, including south of Oceanus Procellarum and the impact basin floor, basically occupying the entire study area. These mare basalts formed between 19 Ga and 37 Ga ago according to the isotopic age data of lunar samples and the dating results of crater size-frequency distribution (CSFD) for mare basalt units. Meanwhile, a large number of tectonic developments are present, and the main linear formations are rills, wrinkle ridges, and crater-floor fractures, with wrinkle ridges being the most numerous, widespread, and characteristic linear structures, mainly within the mare basalt. Rills are also widespread and distributed around the wrinkle ridges; the circular structures are mainly domes, volcanic vents, craters, and impact basins.

The main geological evolution in the study area includes endodynamic geological evolution processes, such as differentiation of magma ocean, plutonic magmatism, and volcanism, and exodynamic geological evolution processes such as impact. During the Magma-Oceanian period, the primary ferroan anorthositic crust mainly formed, followed by the formation of the Copernicus-H basin in the Aitkenian, the Imbrain basin in the Imbrian, the

Timocharis crater in the Eratothenian, and the Copernicus, Kepler, and Aristarchus craters in the Copernican.

An Extra-Trees model based on variable screening was used to invert the oxide contents in the region near the Copernicus crater using IIM data, and frequency histograms were plotted based on the oxide contents (Figures 13 and 14).

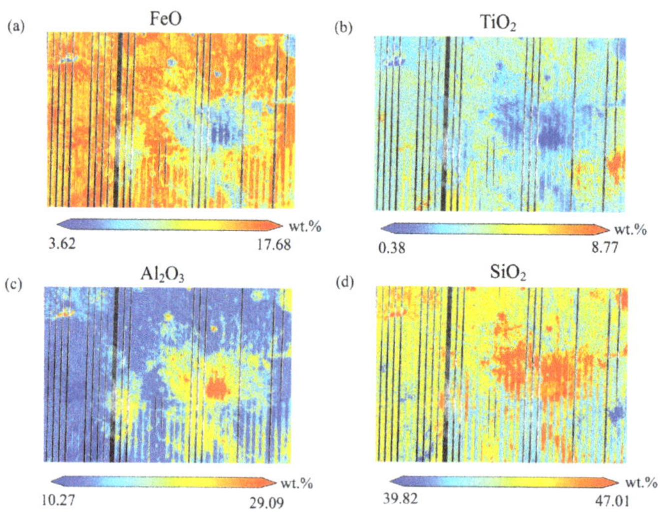

Figure 13. Predicted results of oxide contents in the study area: (**a**) FeO, (**b**) TiO_2, (**c**) Al_2O_3 and (**d**) SiO_2.

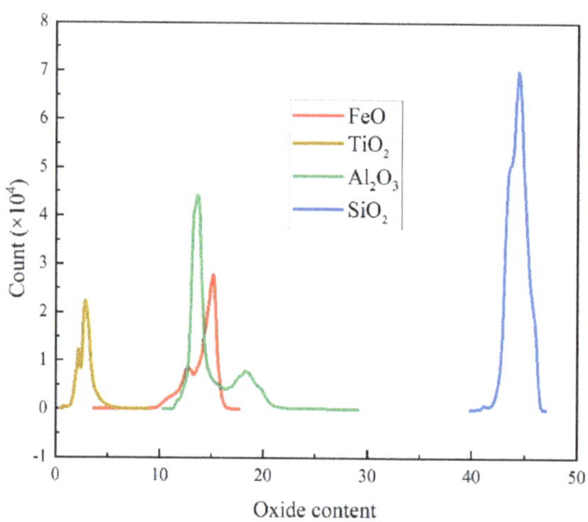

Figure 14. Smoothed histograms of four oxide contents in the study area.

Figure 13 shows the predicted results of oxide contents (FeO, TiO_2, Al_2O_3, and SiO_2) in the Copernicus crater region. As shown in Figure 13, the FeO content ranges from 3.61 to 17.68 wt.%, with a mean value of 12.51 wt.%. The FeO content in the mare is significantly higher than that in the highland, with the lower FeO content mainly located in the Copernicus crater, Aristarchus crater, and Kepler crater floor, and the high FeO content area mainly located on the southwest side of Kepler crater. The frequency histogram in Figure 14 also shows a clear bimodal distribution, with the first peak at approximately 12.80 wt.% and the second peak at approximately 15.1 wt.%. The TiO_2 content ranges from

0.38 to 8.88 wt.%, with a mean value of 2.60 wt.%. Its content is significantly higher in the mare than in the highland, and there is a positive correlation between the FeO content in areas with high TiO_2 content in the mare basalt area. The low TiO_2 content areas are mainly located around the Copernicus, Aristarchus, and Kepler craters, and the high TiO_2 content areas are mainly located on the southwestern side of the Kepler crater and the northeastern side of the Copernicus crater, while the rest of the mare has moderate TiO_2 content. The frequency histogram distribution shows a clear bimodal distribution, with the first peak at approximately 2.16 wt.% and the second peak at approximately 2.82 wt.%. The Al_2O_3 content ranges from 10.27–29.09 wt.%, with a mean value of 13.56 wt.%. It is clearly higher in the highland than in the mare, with an inverse correlation with FeO content. The high-value areas are mainly located in the vicinity of craters, with lower levels in the mare, especially on the southwestern and western sides of the Kepler crater. The frequency histogram of Al_2O_3 shows a clear bimodal distribution, with the first peak of approximately 13.57 wt.% and the second peak of approximately 18.25 wt.%. The SiO_2 content ranges from 39.82 to 47.01 wt.%, with an average value of 39.53 wt.%. The SiO_2 content is high throughout the study area, ranging from 39.82 to 47.01 wt.%, with a mean value of 39.53 wt.%. The content is significantly higher in the highland than in the mare, which shows an inverse correlation with TiO_2. The high-value areas are mainly located within the crater and its sputtering, with lower levels in the rest of the area. The SiO_2 first peak of the frequency histogram is approximately 41.11 wt.%, and the second peak is approximately 44.25 wt.%.

3.5.2. Comparison with Previous Works

Figures 15 and 16 show the FeO and TiO_2 content distributions generated based on the different data/methods. First, the correlation analysis was carried out with the FeO content distribution maps generated in this paper (Table 5). The correlations between the FeO inversion results in this paper and the inversion results of Clementine, IIM (band ratio), and LP GRS data were 0.91, 0.89, and 0.90, respectively, and the correlation was basically stable at approximately 0.90. The correlations between the TiO_2 inversion and the Clementine and IIM (band ratio) inversions (Table 6) were 0.61 and 0.65, respectively. Due to the low resolution of the GRS TiO_2 data, this correlation was not analyzed. As seen in Figure 15, the FeO and TiO_2 contents of the inversions in this paper are lower than the mean values of the inversions of other models. To better compare the discrepancy between the inversion results of this paper and the Clementine FeO results. the IIM FeO map is compared with the Clementine FeO product resampled to the same resolution (0.5°), and the correlation between the two products is 91%, with good agreement, as shown in Figure 17. The difference between the results of this paper and Clementine may be due to the influence of topographic shadows caused by the camera angle, post-image processing, and differences in the oxide inversion algorithm. Additionally, using IIM data, the difference between the results of this paper and the band ratio-based method is mainly due to differences in the oxide inversion algorithm and post-image processing. In contrast to the IIM optical image data, which can only capture the material component of the lunar surface, the LP GRS data are not affected by topographic shadows, light angles, etc., and can be detected to a depth of 20–30 cm below the lunar surface [15]. Therefore, the difference between the results of this paper and the LP GRS is mainly because the GRS data are not affected by topographic shadows and because the GRS data are deeper than the IIM data. The results from different data and methods indicate that, despite the different absolute abundance ranges, their relative abundances are similar to each other [19,33].

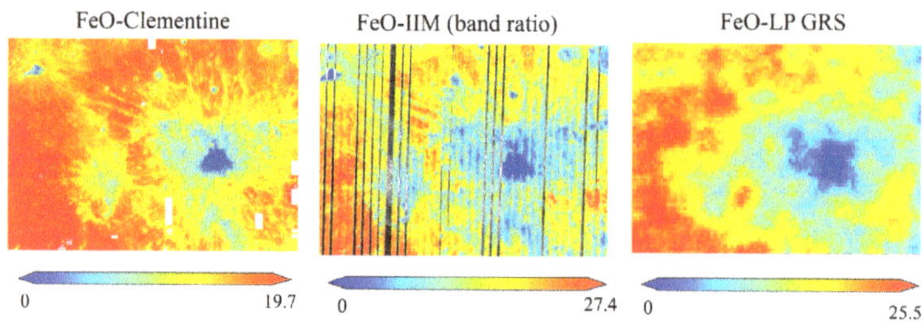

Figure 15. Distribution of FeO based on different data/methods.

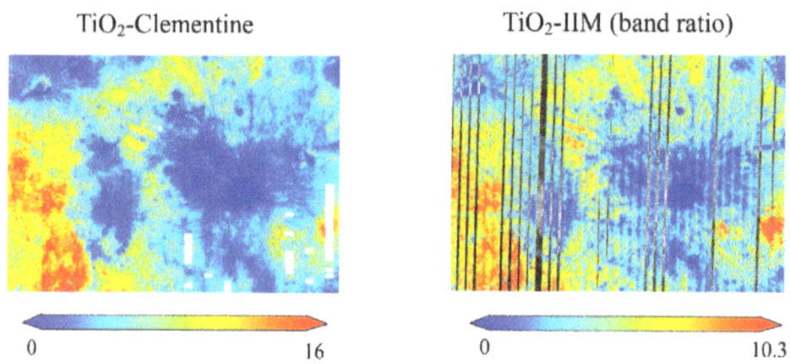

Figure 16. Distribution of TiO_2 based on different data/methods.

Table 5. Correlation coefficients based on different data/methods for FeO inversion.

Parameters	Statistical Values of Predicted FeO			
	This Work	**Clementine [17]**	**IIM (Band Ratio) [30]**	**LP GRS [15]**
Average	12.51	13.98	13.8	15.82
Standard Deviation	4.7	5.61	6.19	6.26
Correlation Coefficients	1	0.91	0.89	0.9

Table 6. Correlation coefficients based on different data/methods for TiO_2 inversion.

Parameters	Statistical Values of Predicted TiO_2			
	This Work	**Clementine [17]**	**IIM (Band Ratio) [31]**	**LP GRS [15]**
Average	2.6	4.1	2.68	2.87
Standard Deviation	1.36	2.94	1.64	1.67
Correlation Coefficients	1	0.61	0.65	0.48

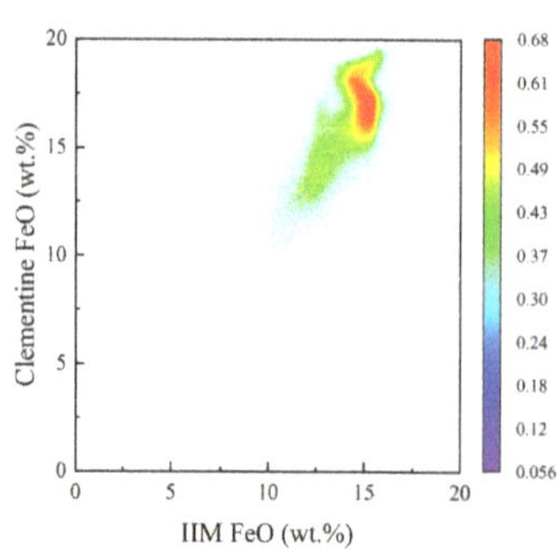

Figure 17. Two-dimensional scatter plot of the difference between the Clementine FeO and IIM FeO abundances.

The existence of negative correlations between FeO and Al_2O_3 and TiO_2 and SiO_2 was verified. Specifically, two-dimensional density plots of FeO and Al_2O_3 and two-dimensional density plots of TiO_2 and SiO_2 were established (Figure 18). The results show that the correlation between FeO and Al_2O_3 is 75%, and the correlation between TiO_2 and SiO_2 is 76%. These results are consistent with the previous conclusion that Fe is negatively correlated with Al content and that a decreasing Si phenomenon tends to occur along with the enrichment of Ti elements [64].

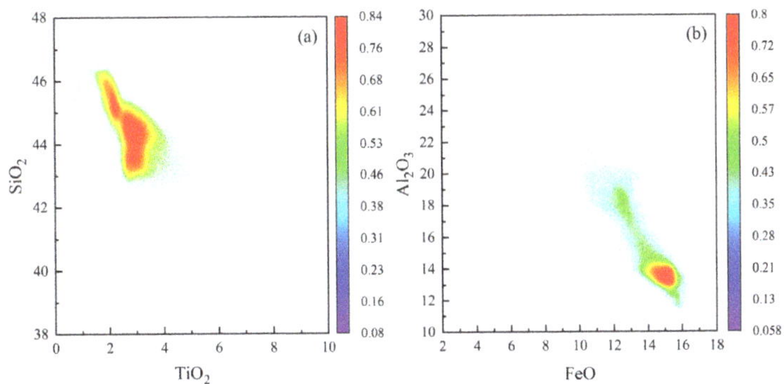

Figure 18. (**a**) Two-dimensional density plot between IIM FeO and IIM Al_2O_3 and (**b**) Two-dimensional density plot between IIM TiO_2 and IIM SiO_2.

4. Discussion

In this paper, a new method of feature band selection combined with Extra-Trees is presented to predict oxide contents on the lunar surface. The results show that the proposed method has good potential for assessing the distribution of oxide contents using spectral features.

4.1. Comparison with Other Similar Studies

Previous studies mainly used two bands to retrieve oxide contents according to the absorption characteristics of mineral elements in the visible near-infrared (VNIR) band. For example, 750 nm and 950 nm [17], 757 nm and 891 nm [21], and 757 nm and 918 nm [19] were used to retrieve the FeO content. Additionally, wavelengths of 415 nm and 750 nm [17,67], 531 nm and 757 nm [19], and 321 nm and 415 nm [16] were used to retrieve TiO_2 content. Due to the complexity of the lunar soil composition, the ability to describe the nonlinear relationships between spectral features and oxide contents from only two bands is very limited [36], and the accuracy of composition inference is highly dependent on the selected inversion models [68]. In addition, nonchromophore elements do not display absorption characteristics in the VNIR region, which makes it difficult to use most methods in nonchromophore element (Al and Si) inversion. Currently, some scholars [18,20,54] have established the relationships between spectral features and oxide contents based on regression models to invert these oxide contents. Therefore, the inversion of oxide contents on the lunar surface using regression models has become a popular research topic.

Spectral pretreatment methods can improve the correlations between spectral features and oxide contents, as has been well established in previous studies. For example, Li [40] attempted to improve the performance of a model by using a log(1/reflectance) transformation approach. Pieters [26] and Sun et al. [20] applied a band-ratio approach to improve the above correlations. Wu [18] experimentally demonstrated that the correlations obtained considering band differences were superior to those obtained with the band-ratio technique. Therefore, the first-order band-difference transformation method was chosen in this paper to improve the correlations between spectral features and oxide contents. After the preprocessing of spectral data, feature band selection is crucial. However, previous studies paid little attention to the optimization of feature variables in models. In this study, on the basis of earlier methods [20,37], the Pearson correlation coefficient and SPA are used to select characteristic bands, thus ensuring that high correlations between input spectral features and oxides are retained and reducing the multicollinearity among spectral features. With expanding research, a variety of machine-learning models have been applied to the inversion of lunar surface oxides. The Extra-Trees algorithm selected in this paper can effectively describe the nonlinear characteristics of the studied features and provides strong generalization ability and robustness. This feature band selection approach combined with the Extra-Trees prediction method for predicting oxide contents on the lunar surface yielded reasonable accuracy for the sample test set (R^2 values of 0.962, 0.944, 0.964, and 0.86 for FeO, TiO_2, Al_2O_3, and SiO_2, respectively, and RMSEs of 1.028, 0.672, 0.942, and 0.897, respectively) and the regional scale (Apollo17 area; the area near the Copernicus crater).

4.2. Future Prospects

(1) The LSCC sample data used in this paper are mainly distributed in the low-latitude area of the lunar nearside, and the number of samples is small; consequently, the distribution characteristics of the oxide contents on the lunar surface cannot be comprehensively reflected, and there are limitations in both quantity and region. This leads to some uncertainty in the prediction results. In the future, more samples should be obtained to supplement the sample data limitations in certain regions of the Moon, such as at middle and high latitudes, and increase the number of samples, which is expected to improve the accuracy of chemical abundance distribution characteristics.

(2) There are spectral anomalies at the edges of different orbit images of IIM hyperspectral data, and factors such as solar azimuth and topographic relief will lead to shadows in the images, which will inevitably increase errors in the inversion results. In future research, we should consider how to mitigate the spectral anomalies and topographic shadows in IIM data or explore the potential of using higher-quality and higher-resolution remote sensing data for inversion, such as the M^3 data obtained with the Indian satellite Chandrayaan-1 and the MI data obtained with Kaguya in Japan, to improve the prediction accuracy of the model.

(3) The continuity of hyperspectral data greatly enriches the amount of information available in remote sensing data, but it can also lead to issues such as information redundancy and high correlations between bands. Therefore, determining how to obtain the best combination of sensitive bands is an important step in the application of hyperspectral data. The sensitive band screening method applied in this paper provides a reference for other chemistry inversion research. In the future, more band screening algorithms can be applied for feature selection with lunar hyperspectral data to reasonably select the best number and combination of bands and improve accuracy.

(4) In the era of big data, big data theory and technology are important tools for solving practical problems. The application of machine learning and deep learning algorithms for lunar chemistry inversion is still in its infancy. Although the Extra-Trees model developed in this paper provides good prediction ability, there is still room for further improvement. Thus, a future development direction is to complete lunar surface oxide inversion through better machine learning and deep learning methods.

5. Conclusions

The plan to return to the Moon in the new era is to develop and utilize lunar resources, establish a lunar base, and use the Moon as a springboard to gradually carry out deep space exploration. Through comprehensive remote sensing exploration of the Moon to analyze and assess the distribution of lunar mineral resources, it can provide a long-term stable resource reserve for the sustainable development of human society. In this paper, using IIM data, the original reflectance as the initial input and the oxide contents of LSCC samples as the desired outputs, we use a combination of the mathematical transformation of spectral data (first-order difference method), a feature band selection technique (with the SPA) and a machine learning modeling method (Extra-Trees). The correlations between reflectance and oxide contents are improved to identify the difference bands most sensitive to oxide levels and construct a prediction model with excellent inversion ability for lunar surface oxide (FeO, TiO_2, Al_2O_3, and SiO_2) contents. This approach can solve the problems of low accuracy and poor generalizability encountered with traditional modeling methods based on raw spectral data. The main conclusions of this paper are as follows.

(1) The correlation calculations for the original bands and all band differences in the IIM spectral range separately show that considering the band differences can enhance the correlations between reflectance and oxide contents. Thus, the best combination of spectral bands can be used in subsequent modeling, and the accuracy of the model can be improved. Moreover, the calculated interelement correlations suggest that Fe is negatively correlated with Al and that Si depletion is often accompanied by the enrichment of Ti.

(2) In total, 325 band differences were initially screened using Pearson correlation coefficients, and then secondary downscaling screening was performed according to BKM combined with SPA. Consequently, 17, 5, 8, and 30 feature bands were retained for the four oxides (FeO, TiO_2, Al_2O_3, and SiO_2, respectively) for modeling. The SPA encompasses most of the spectral information associated with samples, effectively reduces the complexity of modeling and reduces the covariance interference among spectral features.

(3) Machine learning algorithms are increasingly integrated with lunar surface chemistry inversion in the big data era. Big data undoubtedly provide new data-driven research methods and can effectively overcome the relatively limited numbers of lunar samples. Moreover, the high dimensional and complex nonlinear relationships between oxide contents and spectral reflectance can be better described. In this paper, we apply the Extra-Trees algorithm for chemistry inversion to predict the distribution of chemistry on the lunar surface. The results show that the R^2 values of the test sets for FeO, TiO_2, Al_2O_3, and SiO_2 are 0.962, 0.944, 0.964, and 0.860, respectively, and the RMSE values are 1.028, 0.672, 0.942, and 0.897, respectively, improving the modeling accuracy over the original bands.

(4) The average contents of the four oxides (FeO, TiO_2, Al_2O_3, and SiO_2) in the region near the Copernicus crater are 12.51 wt.%, 2.60 wt.%, 13.56 wt.%, and 39.53 wt.%, respectively. The oxide content distributions display obvious variations, and the frequency histograms show a clear bimodal distribution. The comparison of the present result with representative models shows that the model in this paper provides good agreement in the inversion of oxide abundance, thus providing a new idea and method for the inversion of oxide abundance.

Author Contributions: Conceptualization, S.W. and J.C.; methodology, S.W.; software, S.W.; validation, S.W., J.C., and L.L.; formal analysis, S.W.; resources, C.Z. and R.H.; data curation, S.W., J.C., and C.Z.; writing—original draft preparation, S.W.; writing—review and editing, S.W., J.C., and L.L.; visualization, S.W., C.Z., and Q.Z.; supervision, S.W., L.L., and R.H.; funding acquisition, J.C. All authors have read and agreed to the published version of the manuscript.

Funding: This research was funded by the Geological Survey Project of China Geological Survey, grant No. 42932022001, and the Natural Science Basis Research Plan in Shaanxi province of China, grant No. 2020JQ-742.

Data Availability Statement: Chang'E-1 IIM data are provided by "The Science and Application Center for Moon and Deep Space Exploration" at http://Moon.bao.ac.cn/.

Acknowledgments: The authors gratefully acknowledge Zongcheng Ling for providing the improved Chang'E-1 IIM imagery and Chang Liu and Yiwen Pan for their guidance in mineralogy. The authors also thank the cartographers who compiled the geological map of the Moon.

Conflicts of Interest: The authors declare no conflictd of interest.

References

1. Rossi, A.P.; van Gasselt, S. (Eds.) *Planetary Geology*; Springer International Publishing: Cham, Switzerland, 2018; ISBN 978-3-319-65177-4.
2. Anderson, S.L.; Sansom, E.K.; Shober, P.M.; Hartig, B.A.D.; Devillepoix, H.A.R.; Towner, M.C. The Proposed Silicate-Sulfuric Acid Process: Mineral Processing for In Situ Resource Utilization (ISRU). *Acta Astronaut.* **2021**, *188*, 57–63. [CrossRef]
3. Mills, J.N.; Katzarova, M.; Wagner, N.J. Comparison of Lunar and Martian Regolith Simulant-Based Geopolymer Cements Formed by Alkali-Activation for in-Situ Resource Utilization. *Adv. Space Res.* **2022**, *69*, 761–777. [CrossRef]
4. O'Leary, B. Mining the Apollo and Amor Asteroids. *Science* **1977**, *197*, 363–366. [CrossRef] [PubMed]
5. Anand, M.; Crawford, L.A.; Balat-Pichelin, M.; Abanades, S.; Westrenen, W.V.; Peraudeau, G.; Jaumann, R.; Seboldt, W. A Brief Review of Chemical and Mineralogical Resources on the Moon and Likely Initial in Situ Resource Utilization (ISRU) Applications. *Planet. Space Sci.* **2012**, *74*, 42–48. [CrossRef]
6. Melendrez, D.E.; Johnson, J.R.; Larson, S.M.; Singer, R.B. Remote Sensing of Potential Lunar Resources: 2. High Spatial Resolution Mapping of Spectral Reflectance Ratios and Implications for Nearside Mare TiO_2 Content. *J. Geophys. Res.* **1994**, *99*, 5601. [CrossRef]
7. Korokhin, V.V.; Kaydash, V.G.; Shkuratov, Y.G.; Stankevich, D.G.; Mall, U. Prognosis of TiO_2 Abundance in Lunar Soil Using a Non-Linear Analysis of Clementine and LSCC Data. *Planet. Space Sci.* **2008**, *56*, 1063–1078. [CrossRef]
8. Taylor, S.R. The Unique Lunar Composition and Its Bearing on the Origin of the Moon. *Geochim. Cosmochim. Acta* **1987**, *51*, 1297–1309. [CrossRef]
9. Athiray, P.S.; Narendranath, S.; Sreekumar, P.; Dash, S.K.; Babu, B.R.S. Validation of Methodology to Derive Elemental Abundances from X-Ray Observations on Chandrayaan-1. *Planet. Space Sci.* **2013**, *75*, 188–194. [CrossRef]
10. Bhatt, M.; Mall, U.; Wöhler, C.; Grumpe, A.; Bugiolacchi, R. A Comparative Study of Iron Abundance Estimation Methods: Application to the Western Nearside of the Moon. *Icarus* **2015**, *248*, 72–88. [CrossRef]
11. Chevrel, S.D.; Pinet, P.C.; Daydou, Y.; Maurice, S.; Lawrence, D.J.; Feldman, W.C.; Lucey, P.G. Integration of the Clementine UV-VIS Spectral Reflectance Data and the Lunar Prospector Gamma-Ray Spectrometer Data: A Global-Scale Multielement Analysis of the Lunar Surface Using Iron, Titanium, and Thorium Abundances: Lunar Surface Multielement Analysis. *J. Geophys. Res.* **2002**, *107*, 15-1–15-14. [CrossRef]
12. Lawrence, D.J.; Feldman, W.C.; Barraclough, B.L.; Binder, A.B.; Elphic, R.C.; Maurice, S.; Thomsen, D.R. Global Elemental Maps of the Moon: The Lunar Prospector Gamma-Ray Spectrometer. *Science* **1998**, *281*, 1484–1489. [CrossRef] [PubMed]
13. Lemelin, M.; Lucey, P.G.; Song, E.; Taylor, G.J. Lunar Central Peak Mineralogy and Iron Content Using the Kaguya Multiband Imager: Reassessment of the Compositional Structure of the Lunar Crust: Lunar Central Peak Mineralogy and Iron. *J. Geophys. Res. Planets* **2015**, *120*, 869–887. [CrossRef]
14. Narendranath, S.; Athiray, P.S.; Sreekumar, P.; Kellett, B.J.; Alha, L.; Howe, C.J.; Joy, K.H.; Grande, M.; Huovelin, J.; Crawford, I.A.; et al. Lunar X-Ray Fluorescence Observations by the Chandrayaan-1 X-Ray Spectrometer (C1XS): Results from the Nearside Southern Highlands. *Icarus* **2011**, *214*, 53–66. [CrossRef]
15. Prettyman, T.H.; Hagerty, J.J.; Elphic, R.C.; Feldman, W.C.; Lawrence, D.J.; McKinney, G.W.; Vaniman, D.T. Elemental Composition of the Lunar Surface: Analysis of Gamma Ray Spectroscopy Data from Lunar Prospector: Lunar Elemental Composition. *J. Geophys. Res.* **2006**, *111*, E12. [CrossRef]
16. Sato, H.; Robinson, M.S.; Lawrence, S.J.; Denevi, B.W.; Hapke, B.; Jolliff, B.L.; Hiesinger, H. Lunar Mare TiO_2 Abundances Estimated from UV/Vis Reflectance. *Icarus* **2017**, *296*, 216–238. [CrossRef]
17. Lucey, P.G.; Blewett, D.T.; Jolliff, B.L. Lunar Iron and Titanium Abundance Algorithms Based on Final Processing of Clementine Ultraviolet-Visible Images. *J. Geophys. Res.* **2000**, *105*, 20297–20305. [CrossRef]
18. Wu, Y. Major Elements and Mg# of the Moon: Results from Chang'E-1 Interference Imaging Spectrometer (IIM) Data. *Geochim. Cosmochim. Acta* **2012**, *93*, 214–234. [CrossRef]
19. Yan, B.; Xiong, S.Q.; Wu, Y.; Wang, Z.; Dong, L.; Gan, F.; Yang, S.; Wang, R. Mapping Lunar Global Chemical Composition from Chang'E-1 IIM Data. *Planet. Space Sci.* **2012**, *67*, 119–129. [CrossRef]
20. Sun, L.; Ling, Z.; Zhang, J.; Li, B.; Chen, J.; Wu, Z.; Liu, J. Lunar Iron and Optical Maturity Mapping: Results from Partial Least Squares Modeling of Chang'E-1 IIM Data. *Icarus* **2016**, *280*, 183–198. [CrossRef]
21. Xiang, W.; Jianping, C.; Yanbo, X.; Yongchun, Z.; Bokun, Y.; Yunzhao, W. Inversion of the Main Mineral Compositions and Subdivision of Tectonic Units on Lunar LQ-4 Based on Chang'e Data. *Acta Geol. Sin. Engl. Ed.* **2015**, *89*, 1882–1894. [CrossRef]
22. Wang, X.; Zhang, J.; Ren, H. Lunar Surface Chemistry Observed by the KAGUYA Multiband Imager. *Planet. Space Sci.* **2021**, *209*, 105360. [CrossRef]
23. Lemelin, M.; Lucey, P.G.; Miljković, K.; Gaddis, L.R.; Hare, T.; Ohtake, M. The Compositions of the Lunar Crust and Upper Mantle: Spectral Analysis of the Inner Rings of Lunar Impact Basins. *Planet. Space Sci.* **2019**, *165*, 230–243. [CrossRef]
24. Burns, R.G. *Mineralogical Applications of Crystal Field Theory*, 2nd ed.; Cambridge University Press: Cambridge, UK, 1993; ISBN 978-0-521-43077-7.
25. Jaumann, R. Spectral-Chemical Analysis of Lunar Surface Materials. *J. Geophys. Res.* **1991**, *96*, 22793. [CrossRef]
26. Pieters, C. Statistical Analysis of the Links among Lunar Mare Soil Mineralogy, Chemistry, and Reflectance Spectra. *Icarus* **2002**, *155*, 285–298. [CrossRef]

27. Li, L. Partial Least Squares Modeling to Quantify Lunar Soil Composition with Hyperspectral Reflectance Measurements. *J. Geophys. Res.* **2006**, *111*, E04002. [CrossRef]
28. Blewett, D.T.; Lucey, P.G.; Hawke, B.R.; Jolliff, B.L. Clementine Images of the Lunar Sample-Return Stations: Refinement of FeO and TiO_2 Mapping Techniques. *J. Geophys. Res.* **1997**, *102*, 16319–16325. [CrossRef]
29. Gillis-Davis, J.J.; Lucey, P.G.; Hawke, B.R. Testing the Relation between UV–Vis Color and TiO_2 Content of the Lunar Maria. *Geochim. Cosmochim. Acta* **2006**, *70*, 6079–6102. [CrossRef]
30. Ling, Z.; Zhang, J.; Liu, J.; Zhang, W.; Bian, W.; Ren, X.; Mu, L.; Liu, J.; Li, C. Preliminary Results of FeO Mapping Using Imaging Interferometer Data from Chang'E-1. *Chin. Sci. Bull.* **2011**, *56*, 376–379. [CrossRef]
31. Ling, Z.; Zhang, J.; Liu, J.; Zhang, W.; Zhang, G.; Liu, B.; Ren, X.; Mu, L.; Liu, J.; Li, C. Preliminary Results of TiO_2 Mapping Using Imaging Interferometer Data from Chang'E-1. *Chin. Sci. Bull.* **2011**, *56*, 2082–2087. [CrossRef]
32. Lucey, P.G.; Taylor, G.J.; Malaret, E. Abundance and Distribution of Iron on the Moon. *Science* **1995**, *268*, 1150–1153. [CrossRef]
33. Shkuratov, Y.G. Composition of the Lunar Surface as Will Be Seen from SMART-1: A Simulation Using Clementine Data. *J. Geophys. Res.* **2003**, *108*, 5020. [CrossRef]
34. Sun, L.; Ling, Z. Partial Least Squares Modeling of Lunar Surface FeO Content with Clementine Ultraviolet-Visible Images. In *Planetary Exploration and Science: Recent Results and Advances*; Jin, S., Haghighipour, N., Ip, W.-H., Eds.; Springer: Berlin/Heidelberg, Germany, 2015; pp. 1–20. ISBN 978-3-662-45051-2.
35. Wang, X.; Niu, R. Lunar Titanium Abundance Characterization Using Chang'E-1 IIM Data. *Sci. China Phys. Mech. Astron.* **2012**, *55*, 170–178. [CrossRef]
36. Xia, W.; Wang, X.; Zhao, S.; Jin, H.; Chen, X.; Yang, M.; Wu, X.; Hu, C.; Zhang, Y.; Shi, Y.; et al. New Maps of Lunar Surface Chemistry. *Icarus* **2019**, *321*, 200–215. [CrossRef]
37. Zhang, X.; Li, C.; Lü, C. Quantification of the Chemical Composition of Lunar Soil in Terms of Its Reflectance Spectra by PCA and SVM. *Chin. J. Geochem.* **2009**, *28*, 204–211. [CrossRef]
38. Geurts, P.; Ernst, D.; Wehenkel, L. Extremely Randomized Trees. *Mach. Learn.* **2006**, *63*, 3–42. [CrossRef]
39. Jin, M.; Ding, X.; Han, H.; Pang, J.; Wang, Y. An Improved Method Combining Fisher Transformation and Multiple Endmember Spectral Mixture Analysis for Lunar Mineral Abundance Quantification Using Spectral Data. *Icarus* **2022**, *380*, 115008. [CrossRef]
40. Li, L. Quantifying Lunar Soil Composition with Partial Least Squares Modeling of Reflectance. *Adv. Space Res.* **2008**, *42*, 267–274. [CrossRef]
41. Zhou, P.; Zhao, Z.; Wei, G.; Huo, H. Mineral Content Estimation of Lunar Soil of Lunar Highland and Lunar Mare Based on Diagnostic Spectral Characteristic and Partial Least Squares Method. *Appl. Sci.* **2022**, *12*, 1197. [CrossRef]
42. Taylor, L.A.; Pieters, C.M.; Morris, R.V.; Keller, L.P.; Wentworth, S.J. Integration of the Chemical and Mineralogical Characteristics of Lunar Soils with Reflectance Spectroscopy. In Proceedings of the Lunar and Planetary Science Conference, Houston, TX, USA, 15–19 March 1999.
43. Lawrence, D.J.; Feldman, W.C.; Elphic, R.C.; Little, R.C.; Prettyman, T.H.; Maurice, S.; Lucey, P.G.; Binder, A.B. Iron Abundances on the Lunar Surface as Measured by the Lunar Prospector Gamma-Ray and Neutron Spectrometers: Iron Abundances on the Lunar Surface. *J. Geophys. Res.* **2002**, *107*, 13-1–13-26. [CrossRef]
44. Taylor, L.A.; Pieters, C.; Patchen, A.; Taylor, D.H.; Morris, R.V.; Keller, L.P.; Mckay, D.S. Mineralogical Characterization of Lunar Highland Soils. *Lunar Planet. Sci.* **2003**, *34*, 1774.
45. Tingyan, X.; Fujiang, L.; Li, L.; Chan, L.; Ying, Z.; Le, Q.; Rong, Y.; Xiaopo, Z.; Meng, Z.; Qi, Z. Prognosis of Ti Abundance in Sinus Iridum Using a Nonlinear Analysis of Chang' E-1 Interference Imaging Spectrometer Imagery. *Earth Space Sci.* **2015**, *2*, 187–193. [CrossRef]
46. Zhu, F.; Liu, J.H.; Ren, X.; Chen, T.Q.; Liu, J. Spectrum Reconstruction for Chang'e-1 Imaging Interferometer Data Using Modified Periodogram Method. In Proceedings of the International Conference on Electronics, Palanga, Lithuania, 13–15 June 2016.
47. Pieters, C.M. The Moon as a Spectral Calibration Standard Enabled by Lunar Samples: The Clementine Example. *New Views Moon 2 Underst. Moon Through Integr. Divers. Datasets* **1999**, 47–49.
48. Pieters, C.M. Bidirectional Spectroscopy of Returned Lunar Soils: Detailed "Ground Truth" for Planetary Remote Sensors. In Proceedings of the Lunar and Planetary Science Conference, Houston, TX, USA, 18–22 March 1991.
49. Wu, Y.; Zheng, Y.; Zou, Y.; Chen, J.; Xu, X.; Tang, Z.; Xu, A.; Yan, B.; Gan, F.; Zhang, X. A Preliminary Experience in the Use of Chang'E-1 IIM Data. *Planet. Space Sci.* **2010**, *58*, 1922–1931. [CrossRef]
50. Steinbach, M.; Karypis, G.; Kumar, V. *A Comparison of Document Clustering Techniques*; Technical Report; University of Minnesota: Minneapolis, MN, USA, 2000.
51. Araújo, M.C.U.; Saldanha, T.C.B.; Galvão, R.K.H.; Yoneyama, T.; Chame, H.C.; Visani, V. The Successive Projections Algorithm for Variable Selection in Spectroscopic Multicomponent Analysis. *Chemom. Intell. Lab. Syst.* **2001**, *57*, 65–73. [CrossRef]
52. Ji, S.; Gu, C.; Xi, X.; Zhang, Z.; Hong, Q.; Huo, Z.; Zhao, H.; Zhang, R.; Li, B.; Tan, C. Quantitative Monitoring of Leaf Area Index in Rice Based on Hyperspectral Feature Bands and Ridge Regression Algorithm. *Remote Sens.* **2022**, *14*, 2777. [CrossRef]
53. de Araújo Gomes, A.; Galvão, R.K.H.; de Araújo, M.C.U.; Véras, G.; da Silva, E.C. The Successive Projections Algorithm for Interval Selection in PLS. *Microchem. J.* **2013**, *110*, 202–208. [CrossRef]
54. Xu, Y.; Wang, J.; Xia, A.; Zhang, K.; Dong, X.; Wu, K.; Wu, G. Continuous Wavelet Analysis of Leaf Reflectance Improves Classification Accuracy of Mangrove Species. *Remote Sens.* **2019**, *11*, 254. [CrossRef]

55. Guo, P.; Li, T.; Gao, H.; Chen, X.; Cui, Y.; Huang, Y. Evaluating Calibration and Spectral Variable Selection Methods for Predicting Three Soil Nutrients Using Vis-NIR Spectroscopy. *Remote Sens.* **2021**, *13*, 4000. [CrossRef]
56. Galvao, R.; Araujo, M.; Jose, G.; Pontes, M.; Silva, E.; Saldanha, T. A Method for Calibration and Validation Subset Partitioning. *Talanta* **2005**, *67*, 736–740. [CrossRef]
57. Yang, Z.; Xiao, H.; Zhang, L.; Feng, D.; Zhang, F.; Jiang, M.; Sui, Q.; Jia, L. Fast Determination of Oxide Content in Cement Raw Meal Using NIR Spectroscopy with the SPXY Algorithm. *Anal. Methods* **2019**, *11*, 3936–3942. [CrossRef]
58. Chen, W.; Chen, H.; Feng, Q.; Mo, L.; Hong, S. A Hybrid Optimization Method for Sample Partitioning in Near-Infrared Analysis. *Spectrochim. Acta Part A Mol. Biomol. Spectrosc.* **2021**, *248*, 119182. [CrossRef] [PubMed]
59. Galelli, S.; Castelletti, A. Assessing the Predictive Capability of Randomized Tree-Based Ensembles in Streamflow Modelling. *Hydrol. Earth Syst. Sci.* **2013**, *17*, 2669–2684. [CrossRef]
60. Calhoun, P.; Hallett, M.J.; Su, X.; Cafri, G.; Levine, R.A.; Fan, J. Random Forest with Acceptance–Rejection Trees. *Comput. Stat.* **2020**, *35*, 983–999. [CrossRef]
61. Simm, J.; Magrans De Abril, I.; Sugiyama, M. Tree-Based Ensemble Multi-Task Learning Method for Classification and Regression. *IEICE Trans. Inf. Syst.* **2014**, *97*, 1677–1681. [CrossRef]
62. Liang, T.; Liang, S.; Zou, L.; Sun, L.; Li, B.; Lin, H.; He, T.; Tian, F. Estimation of Aerosol Optical Depth at 30 m Resolution Using Landsat Imagery and Machine Learning. *Remote Sens.* **2022**, *14*, 1053. [CrossRef]
63. Taylor, L.A.; Pieters, C.M.; Keller, L.P.; Morris, R.V.; McKay, D.S. Lunar Mare Soils: Space Weathering and the Major Effects of Surface-Correlated Nanophase Fe. *J. Geophys. Res.* **2001**, *106*, 27985–27999. [CrossRef]
64. Hörz, F.; Grieve, R.; Heiken, G.; Spudis, P.; Binder, A. *Lunar Source Book*; Cambridge University Press: Cambridge, UK, 1991.
65. Johnson, J.R.; Larson, S.M.; Singer, R.B. Remote Sensing of Potential Lunar Resources: 1. Near-side Compositional Properties. *J. Geophys. Res. Planets* **1991**, *96*, 18861–18882. [CrossRef]
66. Ji, J.; Guo, D.; Liu, J.; Chen, S.; Ling, Z.; Ding, X.; Han, K.; Chen, J.; Cheng, W.; Zhu, K.; et al. The 1:2,500,000-Scale Geologic Map of the Global Moon. *Sci. Bull.* **2022**, S2095927322002316. [CrossRef]
67. Otake, H.; Ohtake, M.; Hirata, N. Lunar Iron and Titanium Abundance Algorithms Based on SELENE (KAGUYA) Multiband Imager Data. In Proceedings of the Annual Lunar and Planetary Science Conference, Woodlands, TX, USA, 19–23 March 2012.
68. Ma, M.; Li, B.; Chen, S.; Lu, T.; Lu, P.; Lu, Y.; Jin, Q. Global Estimates of Lunar Surface Chemistry Derived from LRO Diviner Data. *Icarus* **2022**, *371*, 114697. [CrossRef]

remote sensing

MDPI

Article

Study of the Buried Basin C-H, Based on the Multi-Source Remote Sensing Data

Xiaojian Xu [1,2,3], Teng Hu [1,2,3,*], Zhizhong Kang [1,2,3], Xing Du [1,2,3] and Lin Zhao [1,2,3]

1 School of Land Science and Technology, China University of Geosciences (Beijing), Beijing 100083, China
2 Research Center of Lunar and Planetary Remote Sensing Exploration, China University of Geosciences (Beijing), Beijing 100083, China
3 Subcenter of International Cooperation and Research on Lunar and Planetary Exploration, Center of Space Exploration, Ministry of Education of the People's Republic of China, Beijing 100083, China
* Correspondence: huteng@cugb.edu.cn

Abstract: We use multi-source remote sensing data to identify the details of a mascon south-east of the lunar Copernicus crater. Studies of the topography, gravity, geochronology and mineral are combined to prove that the mascon is a buried peak-ring basin with diameters of about 130 km and 260 km. The underground structure is covered by 890 m thick mare basalts, as determined by analyzing the spectral features of the impact crater, Copernicus H. The determination of the crater size–frequency distribution (CSFD) suggests that the impact that created the C-H basin occurred earlier than 3.9 Ga. Then, a Hawaiian-style eruption in the late Imbrian period formed the Sinus Aestuum-I dark mantling deposit (DMD). Soon, mare basalts covered the basin several times from 3.8 Ga. Finally, the ejecta from the Copernicus impact event at about 800 Ma, and the weathering processes caused the disappearance of the C-H basin rim from the lunar surface.

Keywords: Copernicus; buried basin; mascons; multi-source remote sensing data; planetary geology; planetary topography; planetary geomorphology

1. Introduction

Over the last half-century of lunar exploration, vast amounts of images and spectral data were collected, as well as data describing the topography, gravity and magnetism. Combining these multi-source data allowed scientists to learn the general process of planetary geology evolution. Craters are one of the most noticeable geomorphological features on the lunar surface. Most recent studies focus on their morphological characteristics and spatial distributions [1–4]. The use of multi-source data allowed to scientists reconstruct the geological history of the lunar craters, among the most noticeable geomorphological features on the Moon. For example, the repetitive lava filling events in Sinus Iridum Bay were identified by combining images, topography, spectral data and dating results [5]. In addition to craters and basins exposed on the lunar surface, there are also some basins buried by basalt. These buried basins have positive gravity anomalies and are called mascons. Mascons were discovered in 1968, in association with the giant circular basins [6]. Numerous theories of the relationship between mascons and the mechanisms for creation of the circular basins have been suggested, such as: mantle plugs upwelling into giant impact basins followed by volcanic filling [7,8], the puddling of high-density meteorite material [9], crater in a low-density lunar crust filled by normal-density basaltic lava [10], and the uplift of the basin center [11].

Prior to the gravity model derived from the doppler tracking of the Lunar Prospector (LP) spacecraft, mascons were distributed at equatorial latitudes on the lunar nearside, filled with mare. The LP model helped to identify seven new mascons on both sides of the Moon: three of them have no visible mare fill, supporting the mantle plug contribution [12].

Gravity Recovery and Interior Laboratory (GRAIL), launched in 2011, returned high resolution gravity data leading to the discovery of more mascons. The discoveries were not limited to the impact basins, but included apparently buried basins. Scientists proposed a new term for these small sizes mascons: quasi-circular mass anomalies (QCMAs) [13] and positive Bouguer gravity anomalies (PBGAs) [14]. With new 900 or 1200 degree gravity models, the additional small mascons have been identified. Evans et al. revealed more than 100 QCMAs on the lunar nearside and inferred a 1.5 km lower bound on the average thickness for the nearside lunar mare [13]. Two buried basins (200 km and 160 km in diameter), located in the north-western rim of Lacus Somniorum and Maria Tranquillitatis, validated the forward modeling approach [15]. Neumann et al. doubled the GRAIL inventory of the lunar basins and determined the main ring diameters of the previously known degraded basins [16]. In Oceanus Procellarum, four PBGAs were recognized which also support the inference of Evans et al. [14]. The GRAIL data also played an important role in clarifying the theory that mascons are caused by an isostatic adjustment, cooling and contraction of a voluminous melt pool [11,17–19]. Klokočník et al. used gravity aspects (the gravity disturbance, the Marussi tensor, two gravity invariants, dimensionality ratio, the strike angles, and the virtual deformations) to study the Copernicus area and amended the definition of 'mascon' into: "An object that has dilatation regime of the virtual deformations in the crater center while the rims are in compressional regime [20]". However, most of the above studies were based on the gravitational and topographical data and hardly brought in remote sensing images to try to prove their inference.

Copernicus is one of the most prominent craters on the Moon. It constitutes a classic example of a relatively young, well-preserved lunar impact crater. Copernicus is found in the Oceanus Procellarum, located at 10°N, 20°W, near the southern rim of the Imbrium Basin (Mare Imbrium) impact structure (Figure 1). Copernicus measures 93 km (58 miles) in diameter. The abundant geomorphic features are developed in the impact crater, such as the central peak, the ladder-like crater wall, the catena, and the extensive fusion coating [21]. Copernicus has a young geological age and a well-defined ray system and is classified as a highland crater. As the geological map shows (Figure 1c), the crater is located on Montes Carpatus. This is a highly diverse morphological area and is characterized by peaks and elevated terrains (Nbm in Figure 1c). In addition, it is punctuated by valleys and low-elevation corridors that are mostly flooded with mare-like materials (Em and Im2 in Figure 1c) [22]. The occurrence of the Copernicus impact resulted in the formation of an ejecta blanket (The largest Cc part in Figure 1c). What's more, there are many Imbrian and Eratosthenian craters in the area (Ic and Ec in Figure 1c). Thus, the impact occurred on a highly stratigraphically complex and heterogeneous region of the lunar surface, mostly composed of feldspathic highland materials rather than mare [23].

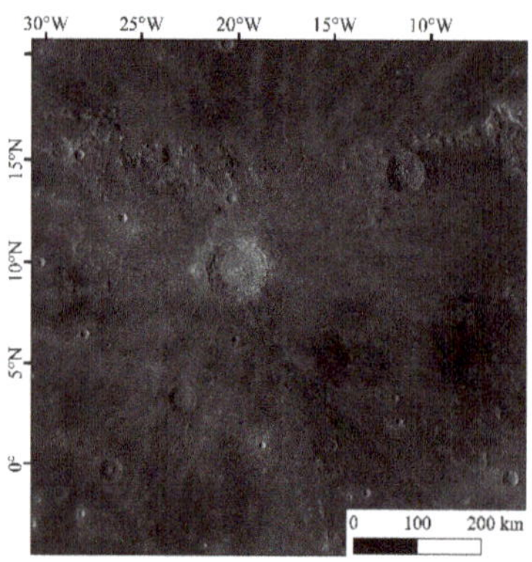

(a)

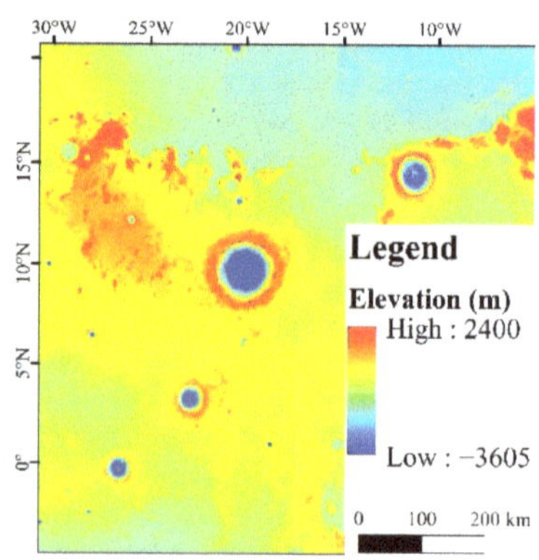

(b)

Figure 1. *Cont.*

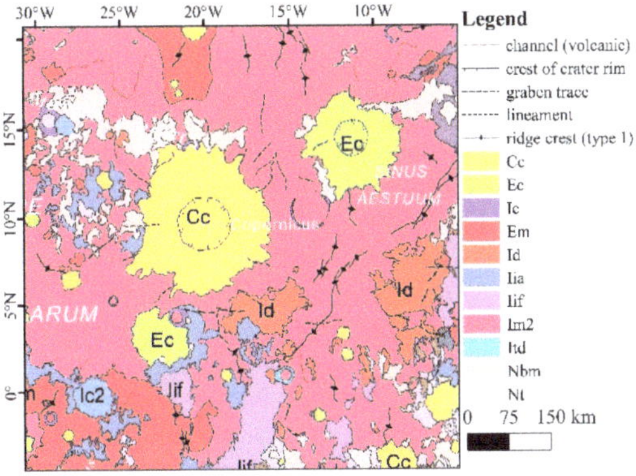

(c)

Figure 1. Copernicus crater and the surrounding area. (**a**) Copernicus area on a LROC WAC Global Morphology Mosaic; (**b**) Copernicus area on a LROC WAC DTM GLD100; (**c**) Copernicus area on a "Unified Geologic Map of the Moon, 1:5M, 2020".

In the south-eastern area of the Copernicus crater, a buried basin is present, called C-H [16]. The classification of the basin's type and the diameter is uncertain [13,16,20,24]. Such cases suggest that the current basin catalogs need further correction. As a start, in this paper, gravity, images, topography and spectral data are used to survey the morphology features and discuss the evolution of the C-H basin.

2. Data and Methods

2.1. Datasets

In this study, we make use of a high-resolution gravity field model of the degree and order 1200 in spherical harmonics, obtained by NASA's Gravity Recovery and Interior Laboratory mission (GRGM1200A) [25]. A digital elevation model (DEM), created by Lunar Orbiter Laser Altimeter (LOLA) and SELenological and Engineering Explorer (SELENE) Kaguya Team called Moon LRO LOLA-SELENE Kaguya TC DEM Merge(59 m/pixel) [26] and Chang'e-2 digital orthophoto map (CE2 DOM, 7 m/pixel) show the high-resolution topography. The Kaguya TC Ortho Mosaic Global image (7.4 m/pixel) helps to analyze the morphology and chronology of the C-H basin. Olivine and optical maturity (OMAT) parameter images (59 m/pixel) from Kaguya MI, are also used [27]. Some olivine-bearing places, such as the Copernicus central peaks, represent an olivine lithology occurring at depth [28] and high regional olivine contents might show the existence of basalts in the western mare [29]. The OMAT is the degree to which a particular lunar soil possesses the quantitative characteristics consistent with that exposure [30]. It shows the time that the lunar surface was exposed to the space environment to distinguish the type of surface. In addition, the "Unified Geologic Map of the Moon, 1:5M, 2020" from NASA, is used to learn the geology background of the research area [31]. The Moon LRO LROC WAC Global Morphology Mosaic (100 m/pixel) and Moon LROC WAC DTM GLD100 (118 m/pixel) are also used to display the study area. Both of them are available through Astropedia (https://astrogeology.usgs.gov/site/annex (accessed on 30 May 2022).

2.2. Gravity

In this study, the field of Bouguer gravity data was truncated to the degree and order 600 and derived from the GRAIL GRGM900C gravity model, after subtracting the gravity resulting from the topography, that assumed a density of 2500 kg/m^3 [13,25,32,33]. The stretched renderer is adopted to display the continuous pixel values across a gradual ramp of colors and to improve the display. Thus, the mascons can be directly observed on the Bouguer gravity data. The proposed basins may be recognized by their distinctive circular outlines in the Bouguer gravity map view [16]. For the peak-ring basins, such as the C-H basin, several features are important: (1) the roughly circular rings, which are sometimes discontinuous. (2) A central positive Bouguer anomaly that is confined within the peak-ring. (3) A negative Bouguer anomaly annulus that extends from the edge of the positive anomaly outward to the rim crest. (4) A number of degraded basins, lacking interior peak-rings, have diameters and gravity patterns similar to those of the well-preserved peak-ring basins [19].

2.3. Topography

In this paper, the terrain analysis and morphologic analysis are introduced for the recognition of the C-H basin and to determine their diameters. The terrain analysis is based on the DEM. On the edge of an impact basin, there is a circle of a raised rim which is usually the highest part of the basin. What's more, a peak-ring of mountains first develops within the crater, at about 140 km in diameter and the diameter of this inner "peak-ring" is roughly half the crater rim diameter [34]. The rim and ring are the most visible characteristics of the peak-ring basins on the mosaic images. However, craters on the Moon suffer from various mass-wasting processes and meteoroid bombardment. As a result, the rim height should decrease, due to the downslope movement of the rim and the ejecta materials, over time [35]. Furthermore, most of the basin rim and peak-ring may disappear from the surface in the event of severe geological processes, such as an impact and volcanism.

As previous research suggests, some mascons surrounded by linear rilles or mare ridges are impact craters/basins filled with basalts [34]. Therefore, the geological map is used to check if there are rilles or ridges around the mascon beside Copernicus.

2.4. Geochronology

The date of the C-H basin was measured using the crater size–frequency distribution (CSFD) [36]. This approach requires an accurate measurement of the crater diameters within and the well-defined geologic units [5]. The craters were mapped with "CraterTools", an ArcGIS plug-in [37]. With three evenly distributed points identified on the crater rims, the diameters would be accurate. The secondary craters, which were distributed in chains, were excluded from the dating process [38]. The Copernicus ejecta is from the "Unified Geologic Map of the Moon, 1:5M, 2020" and used to eliminate the influence of the division of the geological units. The resurfacing events may occur several times in the area and erase the relatively small craters. As a result, based on the craters of different sizes, the CSFD shows the ages that correspond to the different resurfacing events [39]. For example, a once buried basin has resurfaced, the smallest group stands for the age of latest resurfacing event; the middle group represents the time of the basalts fill, the largest group denotes the date of the impact event. The dates of the different events were fitted by a differential fit using the Craterstats 2.0 tool, with the lunar chronology function (CF) and the production function (PF) from Neukum (2001) [36].

2.5. Olivine Contents and OMAT

We searched for a large enough fresh crater in the mascon to establish the outcrop and to study the underground layer. Olivine and OMAT, from spectrum data, suggest the nature of fill material through the outcrop. This may show the original surface of the C-H basin. Then, the depth of the C-H basin can be estimated, when combined with the DEM.

Lucey et al. provided algorithms to inverse the OMAT based on the Clementine UV/VIS images [30].

$$OMAT = \left[(R_{750} - x_0)^2 + \left(\frac{R_{950}}{R_{750}} - y_0 \right)^2 \right]^2 \quad (1)$$

where x_0 is the reflectance of the origin, y_0 is the ratio value of the origin, R_{750} is the reflectance at 750 nm of a pixel or spectrum, and R_{950} is the reflectance at 950 nm of a pixel or spectrum. For the Clementine data, the constants are $x_0 = 0.08$ and $y_0 = 1.1$.

In this study, the OMAT were derived from the Kaguya MI data. Thus, the regression coefficients, described by Lemelin et al. [27], should be applied to the Multiband Imager data at 750 (R_{750}) and 950 (R_{950}) nm, in order to match the reflectance of the Clementine data at these bands ($R750_{corr}$, $R950_{corr}$).

$$\begin{aligned} R750_{corr} &= (R_{750} * 1.51) + 0.020 \\ R950_{corr} &= (R_{950} * 1.38) + 0.020 \end{aligned} \quad (2)$$

The method, developed by Lemelin et al. use Hapke's radiative transfer equations and the reflectance data from the Kaguya Multiband Imager are used to obtain the abundances of olivine [27]. This method computes a spectrum library of the mineral mixtures of plagioclase, orthopyroxene, clinopyroxene, and olivine. Then, we compared this with the central peak spectra with the library, to arrive at an estimated composition. To assign a mineralogical composition to a MI pixel spectrum, Lemelin et al. use an evenly weighted average of the correlation between the library and the unknown spectrum and the sum of the absolute difference in the reflectance between the library and the given pixel spectrum, to find the closest spectral match. They compute the FeO abundance, consistent with the stoichiometric abundance of each mineral mixture, in the spectral library. Once the closest spectral match in the library is obtained, a gradient descent algorithm is used to refine the abundance of plagioclase, orthopyroxene, clinopyroxene, and olivine [40].

3. Results

3.1. Bouguer Gravity Anomaly

The Bouguer gravity anomaly is calculated by removing the separately measured topography. A typical peak-ring basin usually has a large positive Bouguer gravity anomaly within the peak-ring and a small or negative anomaly between the peak-ring and the inward edge of the rim [19]. This pattern is helpful to recognize the degraded basins that lack apparent rings.

There is a Bouguer gravity anomaly which is located to the south-east of Copernicus and fits the feature pattern. It has a typical positive Bouguer gravity anomaly and a negative Bouguer anomaly annulus that extends from the edge of the positive anomaly of the mascons, as shown in Figure 2. The Bouguer gravity anomaly has a center at 7.42°N 17.87°W and 123 km in diameter. In the mascons, the maximum positive Bouguer anomaly is 373.04 mGal, the minimum value is 263.85 mGal, and the mean value is 338.72 mGal.

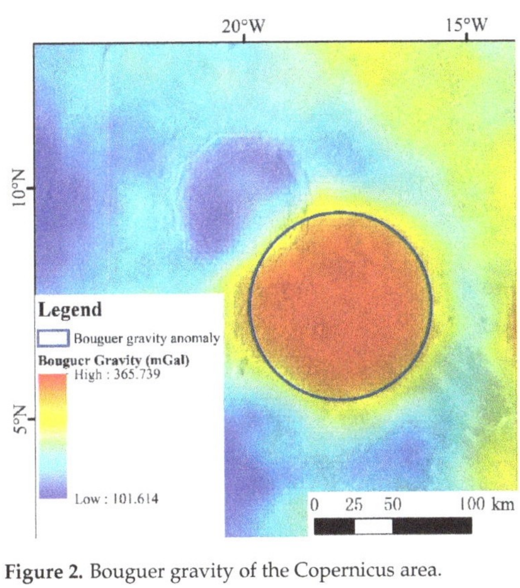

Figure 2. Bouguer gravity of the Copernicus area.

3.2. Morphology and Topography

Some previous studies gave out a diverse scale of the C-H basin. Neumann et al. measured the mascon as a buried peak-ring basin which is located at 7.2°N 18.2°W, with a rim diameter = 260 km and a peak-ring diameter = 130 km [16] (blue circle in Figure 3). The results of Liu et al. confirm the basin type and the location of the central part of the basin, but they provide different values for the rim-to-rim diameter (483 km) and for the ring diameter (221 km) [24] (pink circle in Figure 3). Evans et al. measured the mascon as a single ring basin, located at 7.28°N 18.26°W and 175 km in diameter (orange circles in Figure 3) [13]. The study of Klokočník et al. did not focus on the features except the gravity and magnetism. They only pointed out that the diameter of the C-H basin had ahigh gravity anomaly that is about 150–160 km and they did not indicate the center position of the C-H basin [20].

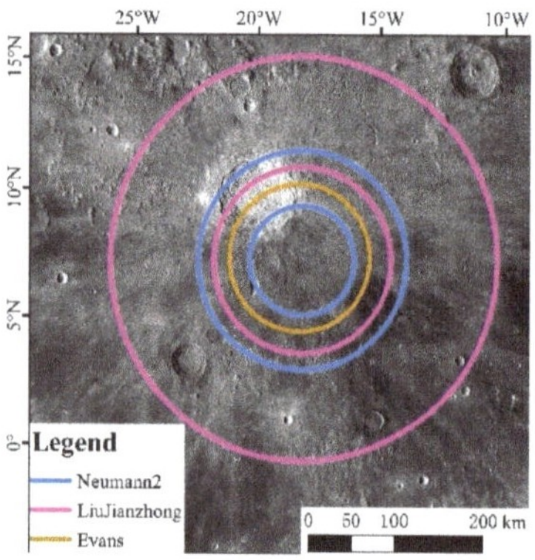

Figure 3. Previous studies of the C-H basin, as shown on the LROC WAC Global Morphology Mosaic. Blue circles from Neumann et al. [16]; pink circles from Evans et al. [13]; orange circle from Liu et al. [24].

Most of the basin rim and the peak-ring disappeared from the lunar surface. However, at the north-eastern part of Neumann's basin rim, two rims are still visible, as can be observed in Figure 4. Both surviving rims are about −1140 m high.

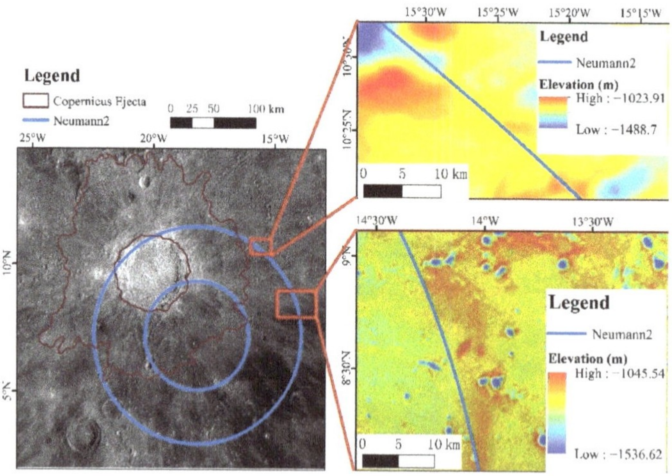

Figure 4. Surviving rim on the basin rim, as given by Neumann. The elevations are from the LOLA-SELENE Kaguya TC DEM.

Solomon and Head [41] discovered that all of the mascon basins display an associated tectonic feature: the linear rilles or mare ridges tend to be concentric to the mascon. The C-H basin might match this feature. There are rilles to the south-west and ridges to the south-east, as exhibited by the red arrows in Figure 5.

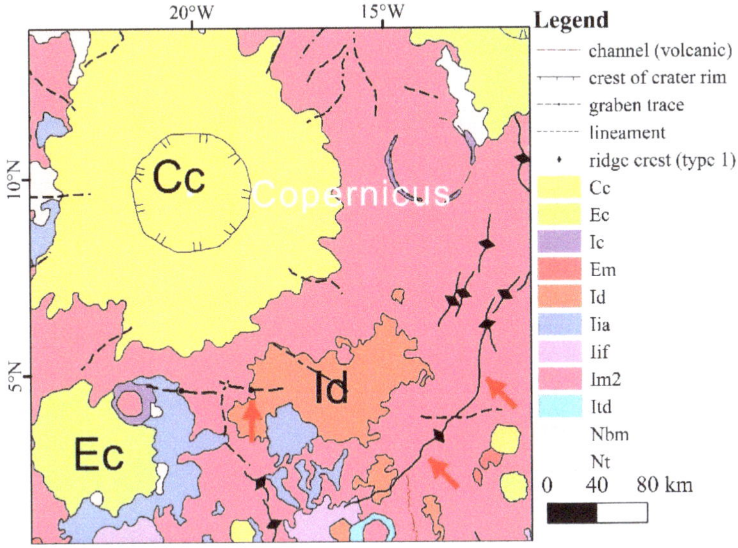

Figure 5. Geology map of the Copernicus area showing the linear rilles and mare ridges.

3.3. *Age*

We pick the intersection of the Copernicus ejecta and the Bouguer gravity anomaly, which is 123 km in diameter and inside all of the three previous basins, as the dating area, to avoid different previous opinions about the scale of the C-H basin and the different geological units, as well as the abundant secondary craters' influence on the dating result (Figure 6a). The 8848 small craters located in the C-H basin were signed with a diameter range from 100 m to 4500 m. Two-hundred and two of them are distributed in chains and were recognized as secondary craters (Figure 6a yellow circle) and they were rejected, in order to eliminate any errors. The small sized craters suggest the age of the surface ejecta and the large sized craters show age of the unit below the ejecta [42]. In our experiment, the craters between 170–1400 m gave 790 Ma, which represent the age of Copernicus (Figure 6b). Those between 3000–4500 m, separately proved an Imbrian age of 3.9 Ga (Figure 6b). In order to check the accuracy of the dating result, we also investigated the craters on the Copernicus ejecta (Figure 7a). A similar age of 800 Ma supports the age of Copernicus. Both Copernicus ejecta dating results approach to the 800 Ma isotope dating result [43–45], suggesting an accurate experiment result. The unit below the ejecta is 3.8 Ga (Figure 7b), and belongs to the Imbrian period. The Imbrian ages of Copernicus show that the C-H basin was refreshed by Imbrian lava. To allow for the Imbrian upper Mare unit (Im2), located to the south-east of Copernicus, in Figure 5, the units below the ejecta might be Im2.

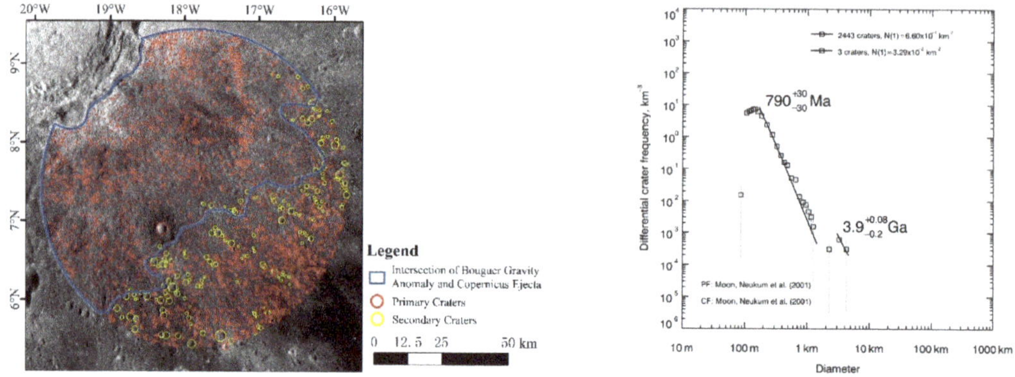

(**a**) (**b**)

Figure 6. Bouguer gravity anomaly CSFD dating. (**a**) Primary craters (red) and secondary craters (yellow) in the intersection of the Copernicus ejecta and the Bouguer gravity anomaly (blue); (**b**) Differential crater frequency and geologic time of the Bouguer gravity anomaly.

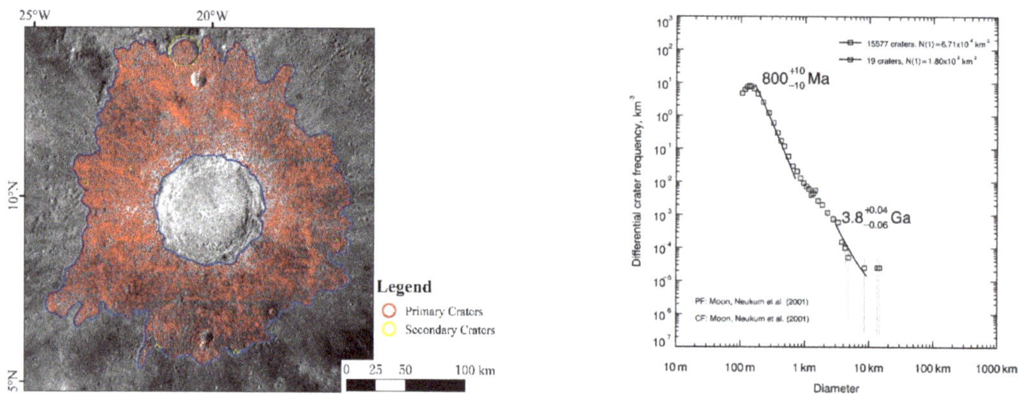

(**a**) (**b**)

Figure 7. Copernicus ejecta CSFD dating. (**a**) Primary craters (red) and secondary craters (yellow) in the intersection of the Copernicus ejecta; (**b**) Differential crater frequency and geologic time of the Copernicus ejecta.

3.4. Mineral

The Copernicus Crater H (Figure 8), provided a partial clue, that gave us access to learn more about the C-H basin. The fresh crater, located on the ejecta blanket of the Copernicus crater (6.89°N 18.29°W) and 4.4 km in diameter. Copernicus H is a classic dark-haloed crater which is surrounded by dark deposits extending almost two crater diameters from the rim crest [46]. The depth is 890 m, as measured on the LOLA-SELENE Kaguya TC DEM. The profile indicates that Copernicus H is a flat-bottomed crater (Figure 9). Because the mass wasting and the downslope movement have disrupted the regolith maturation process, the fresh basalts, relatively free of the weathering effects, are often exposed on the steep inner walls of Copernicus H [46]. The mineral maps from the Kaguya MI images are used to study this. The OMAT map in Figure 10a demonstrates that the crater rim wall has a high OMAT value, indicating that the crater wall is weakly weathered. Hence, Copernicus H is believed to accurately show the material composition of the lower stratum. The geology map and the CSFD dating help to identify the unit on the rim wall of Copernicus H as Im. This conforms to the high olivine content on the crater rim wall in Figure 10b.

Bell et al. also believed that Copernicus H has excavated mare basalt from beneath the ejecta material [46].

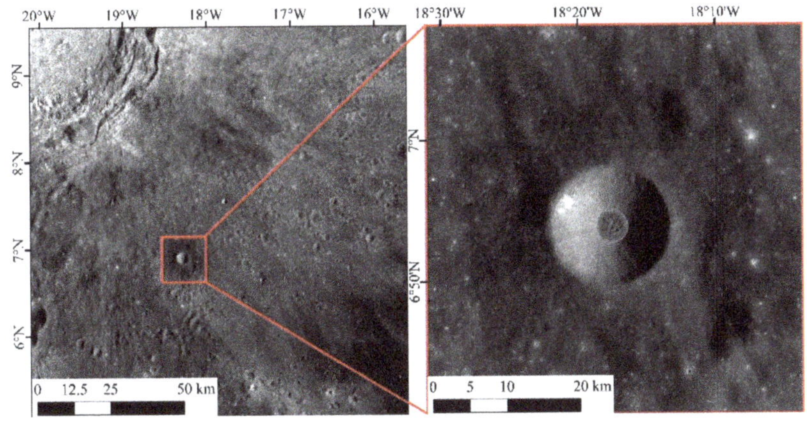

Figure 8. Position of Copernicus H crater on the Kaguya TCortho Mosaic.

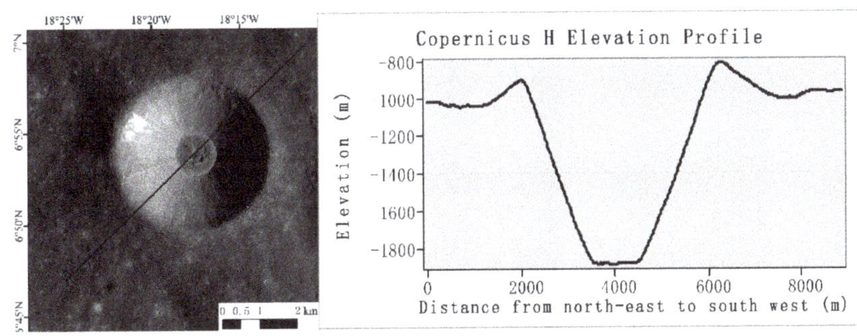

Figure 9. Elevation Profile of Copernicus H from the LOLA-SELENE Kaguya TC DEM.

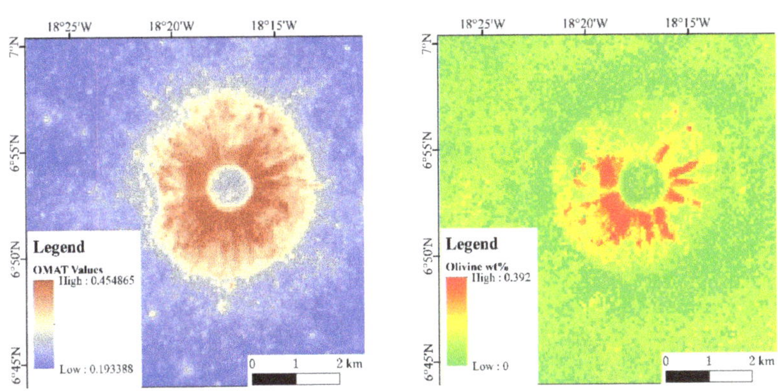

(**a**) (**b**)

Figure 10. Mineral maps of the OMAT and olivine of Copernicus H. (**a**) OMAT in Copernicus H; (**b**) Olivine values on Copernicus H.

Some regions of the Moon exhibit dark mantling deposits that were formed by fire-fountain style eruptions, similar to some Hawaiian eruptions [47–51]. These regions are called dark mantle deposits (DMDs) which are diffuse deposits with a very low albedo [51]. DMDs are the remnants of pyroclastic eruptions. Unlike the effusively emplaced mare basalts, the pyroclastic eruptions were more energetic because the erupted material contained more volatiles and formed volcanic glass beads. In some cases, pyroclastics are found in small, localized areas, surrounding a vent. The largest of the DMDs are near the edges of the lunar mare [51]. They cover many thousands of square kilometers and include a range of knobs and other highland features [51]. Thus, because lava only flows downhill, these units cannot be lava flows. Instead, they seem to mark areas where a thin layer has been draped over an older surface. They may have resembled Hawaiian fire fountains, but on a much larger scale. There are also many smaller dark mantling units on the Moon. Most of these features are only a few kilometers in diameter. They are almost always located near the mare or in the large crater floors. Many also lie along clear fault lines [51]. Since most have a small central pit or crater, they are likely sites for small volcanic explosions. Some of these small eruptions may have released gases from shallow lunar intrusions.

The DMD area Sinus Aestuum-I [48] is located on the south-east part of the mascon (Figure 11). It is 90 km across and situated on small highland hills. The OMAT map on Figure 11 shows that the Sinus Aestuum-I DMD has a lower OMAT value than the surrounding mare and has been heavily disrupted by the ejecta and the secondary impacts. This evidence suggests that the Sinus Aestuum-I DMD formed earlier than the surrounding mare, as well as the Copernicus crater.

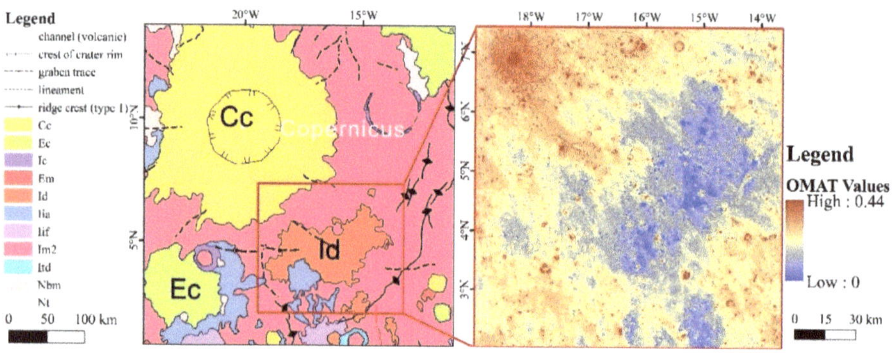

Figure 11. Position and the OMAT map of Sinus Aestuum-I DMD.

4. Discussion

4.1. Confirmation and Characteristics of the C-H Basin

Because Neumann et al., Liu et al. and Evans et al. have contrasting conclusions about the scale and the type of the C-H basin, confirming the accurate scale and type becomes an important work. The mascon fits all of the features of a peak-ring basin (in Section 2.2). It should be classified as a peak-ring crater. Liu et al. (pink circles in Figure 12) gave a much larger scale than Neumann et al. (blue circles in Figure 12). However, the rim, given by Liu et al., overlays the rims of the Imbrium basin, as given by Neumann et al. (white circles in Figure 12). The topographic feature suggests that the highland, north of Copernicus, should belong to the rim of the Imbrium basin. In addition, the morphology and topography results in this paper show a surviving rim which fits the rim given by Neumann. From our results, the C-H basin has a rim diameter of 260 km and a peak-ring diameter of 130 km. The center of the basin is 7.2°N 18.2°W.

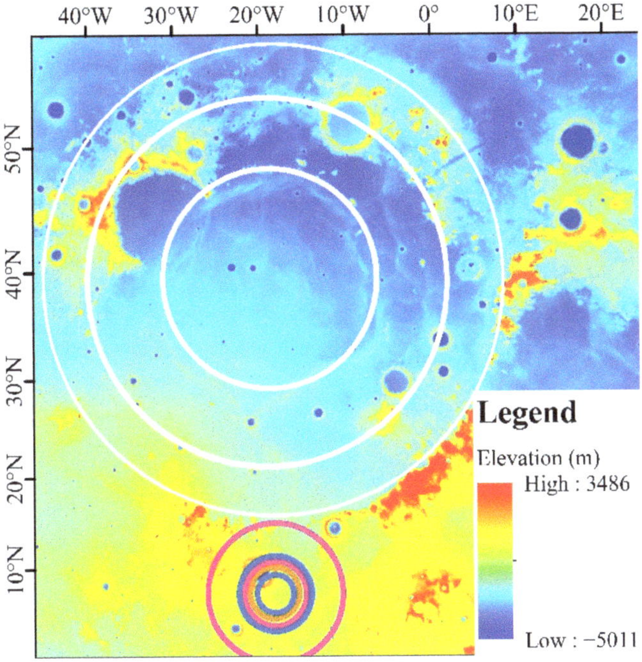

Figure 12. Previous study results and the Imbrium basin, shown on the LROC WAC DTM GLD100. Blue circles from Neumann et al. [16]; pink circles from Evans et al. [13]; orange circle from Liu et al. [24]. White circles from Neumann et al. are rims of the Imbrium basin [16].

4.2. Age of the Sinus Aestuum-I DMD

The existence of the Sinus Aestuum-I DMD suggests that there are a variety of volcanoes in the C-H basin area. Clarifying the ages and sequences of these volcanic activities are meaningful to reconstruct the history of the region and to understand the evolution of the lunar geological activities. Although, the Sinus Aestuum-I DMD has fewer superposed craters than the nearby Im2 units, analyses of the Apollo 17 samples from Taurus–Littrow indicated that the DMDs could not be unequivocally associated with youth on the lunar landscape. Due to the abnormally rapid degradation of the small craters in the unconsolidated pyroclastic material, the exact ages of the DMDs are not at all clear without samples [49,51]. Therefore, we can only judge the sequence of the DMD, Im2 and Copernicus ejecta according to their intersection relations and spectral characteristics. As result of the minerals, the Sinus Aestuum-I DMD has a lower OMAT value than Im2 and Copernicus (Figure 11). Considering the slope of the Sinus Aestuum-I DMD is not obviously lower than Im2 and the Copernicus ejecta (Figure 13d), such a difference is mainly caused by the different formation ages. On the LROC WAC Global Morphology Mosaic and the OMAT map, the radial texture from Copernicus can be seen in the falls on the Sinus Aestuum-I DMD (Figure 13a). This phenomenon, combined with the result of the minerals, show that the Copernicus ejecta is younger than the Sinus Aestuum-I DMD.

Weitz et al. showed that the Sinus Aestuum-I DMD shares a centralized vent with the Sinus Aestuum-II DMD (Figure 14). The central vent would be at 4°N 12°W [48]. However, the DMD at this location and other lowlands became buried by Im. Such a case and the OMAT map together, indicate that the Hawaiian-style eruption which formed the Sinus Aestuum-I DMD and the Sinus Aestuum-II DMD is earlier than Im.

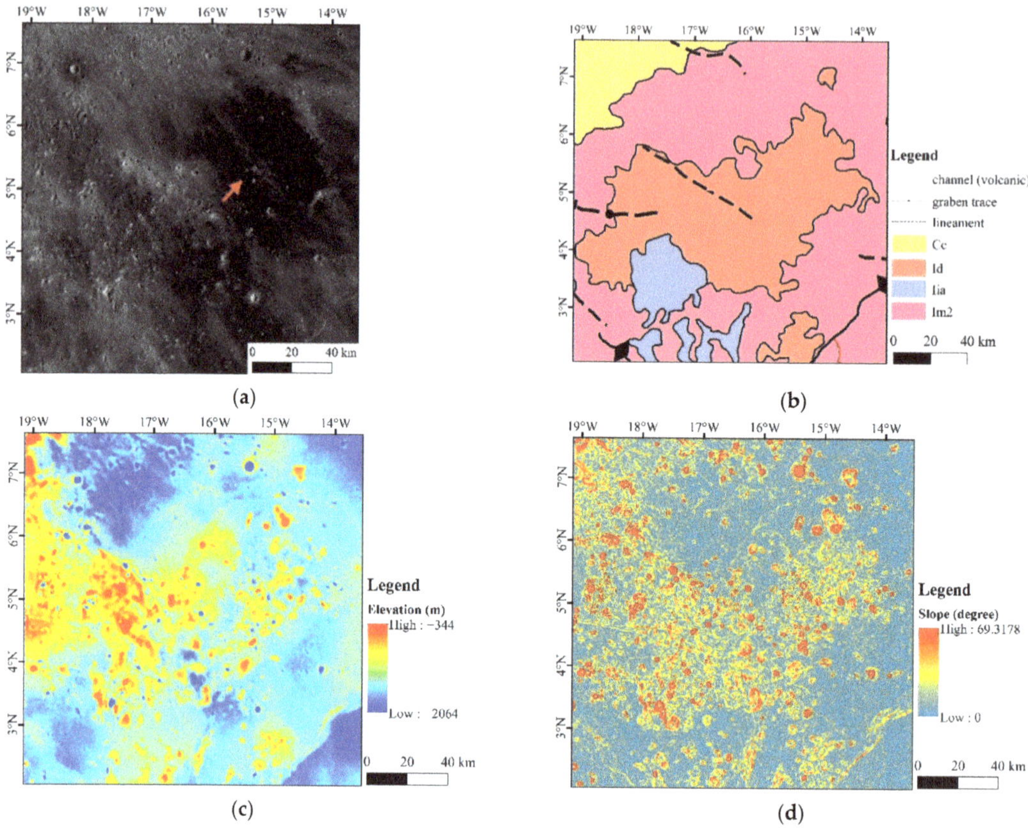

Figure 13. Sinus Aestuum-I DMD. (**a**) Sinus Aestuum-I DMD on the Kaguya TCortho Mosaic, the red arrow points to the radial texture from Copernicus; (**b**) Geological map of the Sinus Aestuum-I DMD; (**c**) Elevation of the Sinus Aestuum-I DMD; (**d**) Slope of the Sinus Aestuum-I DMD.

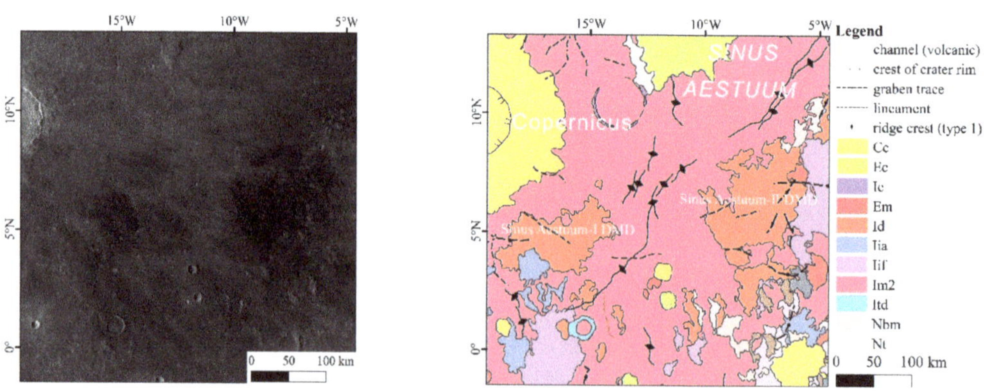

Figure 14. Aestuum I DMD and the Sinus Aestuum-II DMD. (**a**) Aestuum I DMD and the Sinus Aestuum-II DMD on the LROC WAC Global Morphology Mosaic; (**b**) Geology of the Aestuum I DMD and the Sinus Aestuum-II DMD.

4.3. Olivine Rich Strata in Copernicus H

An early study by Bell et al., suggested that the Copernicus H crater exposes the excavation of mare basalt from beneath the Copernicus ejecta deposits, based on the basalt excavation theory of the dark-haloed crater [46]. According to the theory, the rich olivine in the Copernicus H wall comes from Im. Following years of research, a variety of olivine origin theories have been presented and indicated that the olivine-bearing lithologies of Copernicus are diverse and did not all form in the same way [22,28]. Therefore, it is necessary to study the differences between the rich olivine area in Copernicus H and the other regions, and to clarify the type of olivine in the Copernicus H wall. There is an olivine-bearing crystallizing melt pond in the northern wall of Copernicus (Figure 15). The crystallizing melt pond breached and flowed down the crater wall because of the geological process of the impact melt [28]. However, the prominent line nature of the melt pond in the northern wall of Copernicus does not exit from the wall of Copernicus H (Figures 9 and 10).

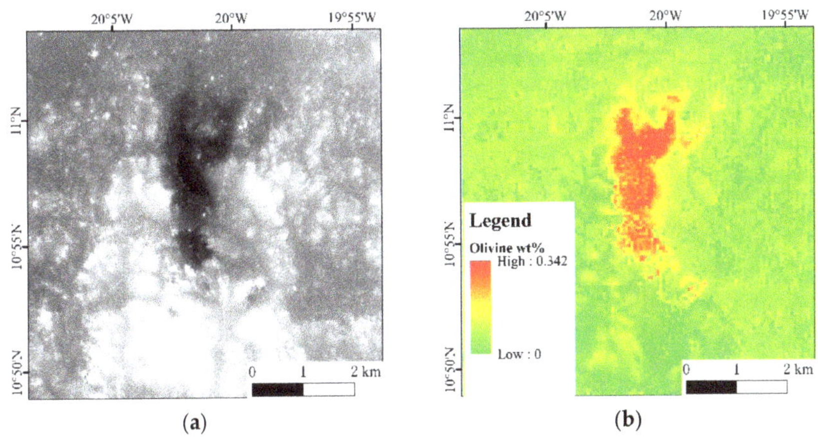

Figure 15. Olivine-bearing exposure in the northern wall of Copernicus. (**a**) Olivine-bearing exposure shown on the CE2 DOM G122; (**b**) Olivine map of the exposure.

What's more, the Sinus Aestuum-I DMD is another olivine-bearing occurrence near the C-H basin. We found two craters in the Sinus Aestuum-I DMD from which we can learn more about the olivine rich strata. As Figure 16 shows, compared with the Sinus Aestuum-I DMD, olivine inside both craters is poorer. Considering that the depths of these two craters are 190 m (Figure 16b,c) and 350 m (Figure 16d,e), respectively, and much shallower than Copernicus H (Figure 9). Thus, olivine in Copernicus H is also different from the olivine in the Sinus Aestuum-I DMD. Above all, the opinion of Bell et al. that Copernicus H exposes Im2 under the Copernicus ejecta is still reliable [46].

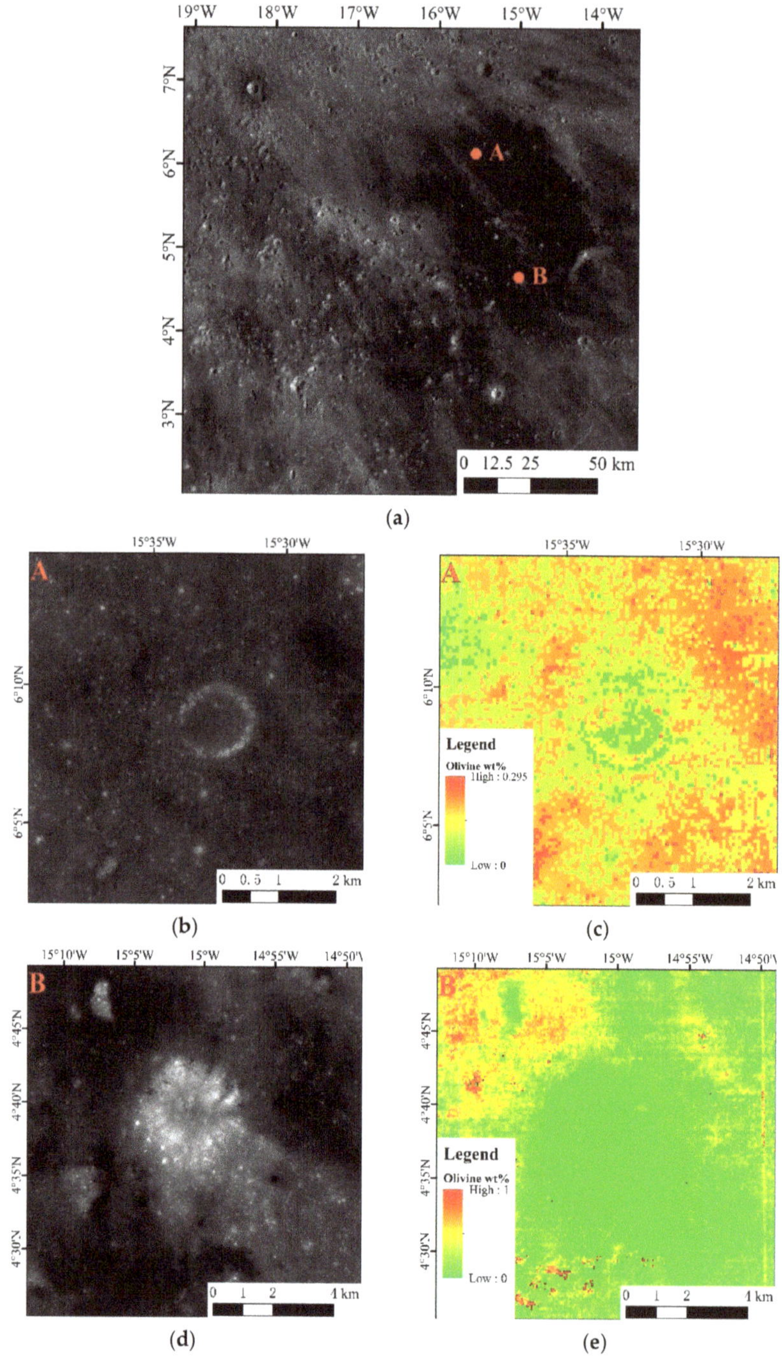

Figure 16. Two craters in the Sinus Aestuum-I DMD. (**a**) Position of the craters on the Sinus Aestuum-I DMD shown on the Kaguya TCortho Mosaic; (**b**) Crater A shown on the CE2 DOM G222; (**c**) Olivine map of Crater A; (**d**) Crater B shown on the CE2 DOM G222 and G223; (**e**) Olivine map of Crater B.

4.4. Evolution of the C-H Basines

The olivine value in the flat bottom of Copernicus H is similar to the ejecta from it. Figure 10a,b suggests that the flat bottom has a similar composition to the crater wall and has not exposed the original surface of the C-H basin. Thus, the depth of the C-H basin surface should be deeper than that of Copernicus H (890 m, Figure 9). Meanwhile, In Figure 10a, the rich olivine rim wall of Copernicus suggests that it is basalt which is located above the original surface of the C-H basin. Thus, 3.8 Ga should correspond to the Imbrian basalt and the age of the C-H basin should older than 3.9 Ga. This age range covers the result of Liu et al. [24]. Considering the Stadius crater (Figure 17), which is located 186 km east of Copernicus, was obviously buried by Im2, the assumption that the C-H basin was buried by the Imbrian basalt, and the rim of the C-H basin is eroded by the ejecta from the Copernicus impact event and the weathering processes, is reasonable. Evans et al. divided the QCMAs with diameters greater than 80 km into high/low-relative-Bouguer-anomaly groups, interpreted as craters formed prior to and after the onset of the local mare [13]. However, they did not specify this characteristic of the C-H basin in their study. According to our assumption, the C-H basin formed prior to Im2 and should be divided into the high-relative-Bouguer-anomaly groups.

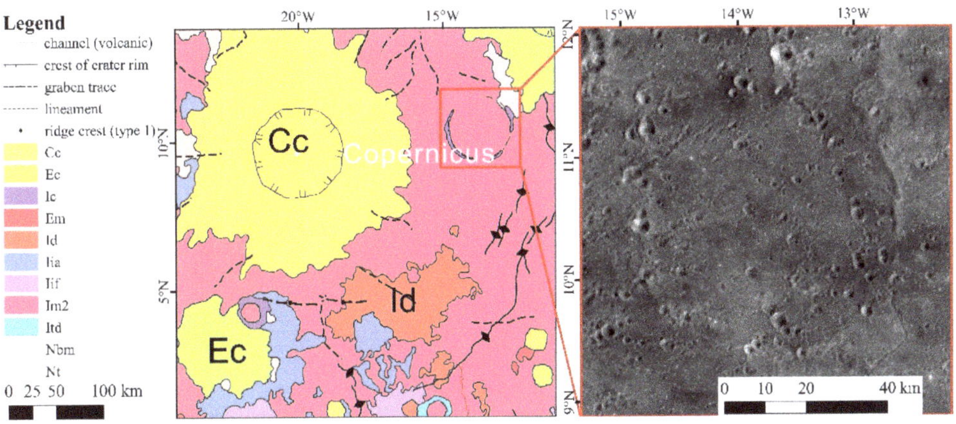

Figure 17. Stadius crater on the geological map and the Kaguya TCortho Mosaic.

The ages of the Sinus Aestuum-I DMD (the Id part in Figure 17) cannot be obtained from the CSFD because their unconsolidated nature is thought to have artificially reduced the number of small craters observed on their surfaces [49,51]. Considering that the Sinus Aestuum-I DMD has a lower OMAT value than the mare, the corresponding Hawaiian-style eruption event should be earlier than the formation of the last Imbrian lava event.

Based on the age of Copernicus, the evolution of the C-H basin can be expressed as: an impact event occurred more than 3.7 Ga. From 3.7 Ga to 3.4 Ga, the mantle plugs upwelling into giant impact basins, followed by the volcanic filling in Copernicus and the nearby area, thus forming the positive Bouguer gravity anomaly buried basin and the Im2 units. During that time, a Hawaiian-style eruption formed the whole Sinus Aestuum DMD. Some DMDs on high places, such as Aestuum-I and II survived from the last Imbrian lava event. Then ejecta from the Copernicus impact event in about 820 Ma, and the weathering processes cause the rim of the C-H basin to almost disappear from the lunar surface. Finally, Copernicus H is exposed to the original surface.

5. Conclusions

We use multi-source remote sensing data including: images, topography, spectrum to analyze the positive Bouguer gravity anomaly to the south-east of Copernicus. Our results show that the mascon (QCMA) is a buried basin. The details of the C-H basin lead to the following conclusions:

1. The C-H basin's basic parameters are updated and located in 7.2°N 18.2°W, with a rim 260 km in diameter and a peak-ring 130 km in diameter. It should be classified as peak-ring basin and divided into the high-relative-Bouguer-anomaly groups.
2. The DEM and the spectral data on Copernicus H suggest that the thickness of Im2 in the C-H basin is 890 m. What's more, the surface of the C-H basin is buried deeper than 890 m.
3. The C-H basin impact event occurred earlier than 3.9 Ga, maybe during the Aitkennian Epoch [24]. Then, a Hawaiian-style eruption formed the Sinus Aestuum-I DMD. Soon, Im2 filled it in, from 3.8 Ga. In addition to the intrusive basalts, the ejecta from the Copernicus impact event in about 800 Ma and the weathering processes also cause the disappearance of the C-H basin's rim from the lunar surface. The Bouguer gravity anomaly was formed by the mantle upwelling [20]. However, the lunar dynamo likely ceased sometime between 1.92 Ga–0.80 Ga [52], earlier than the formation of Copernicus. There was not enough heat for the mantle upwelling and formation. Therefore, the Copernicus crater does not have a "mascon" characteristic.

Author Contributions: Conceptualization, X.X.; methodology, X.X. and T.H.; formal analysis, X.X. and T.H.; data curation, X.X., X.D. and L.Z.; writing—original draft preparation, X.X.; supervision, Z.K.; project administration, Z.K.; funding acquisition, Z.K. All authors have read and agreed to the published version of the manuscript.

Funding: This work was supported by National Key R&D Program of China (Grant Nos. 2020YFE0202100 and 2019YFE0123300), the National Natural Science Foundation of China (Grant No. 41872207), the Pre-research Project on Civil Aerospace Technologies Funded by China National Space Ad-ministration (CNSA) (No. D020204).

Data Availability Statement: GRGM1200A bouguer data gggrx_1200l_boug_l660 are obtained from NASA's Planetary Data System (PDS): https://ode.rsl.wustl.edu/moon/indexDatasets.aspx (accessed on 30 May 2022). Moon LRO LOLA-SELENE Kaguya TC DEM Merge, olivine and optical maturity (OMAT) parameter images from Kaguya MI and Kaguya TCortho Mosaic Global image are available at https://astrogeology.usgs.gov/site/annex (accessed on 30 May 2022). Unified Geologic Map of the Moon, 1:5M, 2020 from NASA can be downloaded at https://astrogeology.usgs.gov/search/map/Moon/Geology/Unified_Geologic_Map_of_the_Moon (accessed on 30 May 2022). Moon LRO LROC WAC Global Morphology Mosaic can be downloaded at https://astrogeology.usgs.gov/search/map/Moon/LRO/LROC_WAC/Lunar_LRO_LROC-WAC_Mosaic_global_100m_June2013.cub (accessed on 30 May 2022). Moon LROC WAC DTM GLD100 can be downloaded at https://astrogeology.usgs.gov/search/map/Moon/LRO/LROC_WAC/Lunar_LROC_WAC_GLD100_79s79n_118m_v1_1.cub (accessed on 30 May 2022). CE2 DOM can be downloaded at https://moon.bao.ac.cn/ce5web/moonGisMap.search (accessed on 30 May 2022).

Conflicts of Interest: The authors declare no conflict of interest.

References

1. Kang, Z.; Wang, X.; Hu, T.; Yang, J. Coarse-to-Fine Extraction of Small-Scale Lunar Impact Craters From the CCD Images of the Chang'E Lunar Orbiters. *IEEE Trans. Geosci. Remote Sens.* **2019**, *57*, 181–193. [CrossRef]
2. Jia, M.N.; Di, K.C.; Yue, Z.Y.; Liu, B.; Wan, W.H.; Niu, S.L.; Liu, J.Z.; Cheng, W.M.; Lin, Y.T. Multi-scale morphologic investigation of craters in the Chang' e-4 landing area. *Icarus* **2021**, *355*, 114164. [CrossRef]
3. Robbins, S.J. A New Global Database of Lunar Impact Craters >1–2 km: 1. Crater Locations and Sizes, Comparisons With Published Databases, and Global Analysis. *J. Geophys. Res. Planets* **2019**, *124*, 871–892. [CrossRef]
4. Wang, Y.; Wu, B.; Xue, H.; Li, X.; Ma, J. An Improved Global Catalog of Lunar Impact Craters (≥1 km) With 3D Morphometric Information and Updates on Global Crater Analysis. *J. Geophys. Res. Planets* **2021**, *126*, e2020JE006728. [CrossRef]
5. Hu, T.; Kang, Z.; Massironi, M.; Hiesinger, H.; van der Bogert, C.H.; Gamba, P.; Brunetti, M.T.; Melis, M.T. Geological evolution of the Sinus Iridum basin. *Planet. Space Sci.* **2020**, *194*, 105134. [CrossRef]

6. Muller, P.M.; Sjogren, W.L. Mascons: Lunar Mass Concentrations. *Science* **1968**, *161*, 680–684. [CrossRef] [PubMed]
7. Wise, D.U.; Yates, M.T. Mascons as structural relief on a lunar 'moho'. *J. Geophys. Res.* **1970**, *75*, 261–268. [CrossRef]
8. Phillips, R.J.; Dvorak, J. The origin of lunar mascons: Analysis of the Bouguer gravity associated with Grimaldi. In Proceedings of the Conference on Multi-ring Basins: Formation and evolution, Houston, TX, USA, 10–12 November 1980; p. 91.
9. Urey, H.C. Mascons and the History of the Moon. *Science* **1968**, *162*, 1408–1410. [CrossRef]
10. Conel, J.E.; Holstrom, G.B. Lunar Mascons: A Near-Surface Interpretation. *Science* **1968**, *162*, 1403–1405. [CrossRef]
11. Andrews-Hanna, J.C. The origin of the non-mare mascon gravity anomalies in lunar basins. *Icarus* **2013**, *222*, 159–168. [CrossRef]
12. Konopliv, A.S.; Binder, A.B.; Hood, L.L.; Kucinskas, A.B.; Sjogren, W.L.; Williams, J.G. Improved gravity field of the moon from lunar prospector. *Science* **1998**, *281*, 1476–1480. [CrossRef] [PubMed]
13. Evans, A.J.; Soderblom, J.M.; Andrews-Hanna, J.C.; Solomon, S.C.; Zuber, M.T. Identification of buried lunar impact craters from GRAIL data and implications for the nearside maria. *Geophys. Res. Lett.* **2016**, *43*, 2445–2455. [CrossRef]
14. Deutsch, A.N.; Neumann, G.A.; Head, J.W.; Wilson, L. GRAIL-identified gravity anomalies in Oceanus Procellarum: Insight into subsurface impact and magmatic structures on the Moon. *Icarus* **2019**, *331*, 192–208. [CrossRef] [PubMed]
15. Sood, R.; Chappaz, L.; Melosh, H.J.; Howell, K.C.; Milbury, C.; Blair, D.M.; Zuber, M.T. Detection and characterization of buried lunar craters with GRAIL data. *Icarus* **2017**, *289*, 157–172. [CrossRef]
16. Neumann, G.A.; Zuber, M.T.; Wieczorek, M.A.; Head, J.W.; Baker, D.M.H.; Solomon, S.C.; Smith, D.E.; Lemoine, F.G.; Mazarico, E.; Sabaka, T.J.; et al. Lunar impact basins revealed by Gravity Recovery and Interior Laboratory measurements. *Sci. Adv.* **2015**, *1*, e1500852. [CrossRef] [PubMed]
17. Melosh, H.J.; Freed, A.M.; Johnson, B.C.; Blair, D.M.; Andrews-Hanna, J.C.; Neumann, G.A.; Phillips, R.J.; Smith, D.E.; Solomon, S.C.; Wieczorek, M.A.; et al. The Origin of Lunar Mascon Basins. *Science* **2013**, *340*, 1552–1555. [CrossRef]
18. Montesi, L.G.J. Solving the Mascon Mystery. *Science* **2013**, *340*, 1535–1536. [CrossRef]
19. Baker, D.M.H.; Head, J.W.; Phillips, R.J.; Neumann, G.A.; Bierson, C.J.; Smith, D.E.; Zuber, M.T. GRAIL gravity observations of the transition from complex crater to peak-ring basin on the Moon: Implications for crustal structure and impact basin formation. *Icarus* **2017**, *292*, 54–73. [CrossRef]
20. Klokočník, J.; Kostelecký, J.; Cílek, V.; Kletetschka, G.; Bezděk, A. Gravity aspects from recent gravity field model GRGM1200A of the Moon and analysis of magnetic data. *Icarus* **2022**, *384*, 115086. [CrossRef]
21. Zhou, C.; Zhang, Y.; Chen, S.; Zhu, B. Analyzing the Magnesium (Mg) Number of Olivine on the Lunar Surface and Its Geological Significance. *Remote Sens.* **2019**, *11*, 1544. [CrossRef]
22. Bugiolacchi, R.; Mall, U.; Bhatt, M.; McKenna-Lawlor, S.; Banaszkiewicz, M.; Brønstad, K.; Nathues, A.; Søraas, F.; Ullaland, K.; Pedersen, R.B. An in-depth look at the lunar crater Copernicus: Exposed mineralogy by high-resolution near-infrared spectroscopy. *Icarus* **2011**, *213*, 43–63. [CrossRef]
23. McCord, T.B.; Charette, M.P.; Johnson, T.V.; Lebofsky, L.A.; Pieters, C.; Adams, J.B. Lunar spectral types. *J. Geophys. Res.* **1972**, *77*, 1349–1359. [CrossRef]
24. Liu, J.; Liu, J.; Yue, Z.; Zhang, L.; Wang, J.; Zhu, K. Characterization and interpretation of the global lunar impact basins based on remote sensing. *Icarus* **2022**, *378*, 114952. [CrossRef]
25. Goossens, S.; Sabaka, T.J.; Wieczorek, M.A.; Neumann, G.A.; Mazarico, E.; Lemoine, F.G.; Nicholas, J.B.; Smith, D.E.; Zuber, M.T. High-Resolution Gravity Field Models from GRAIL Data and Implications for Models of the Density Structure of the Moon's Crust. *J. Geophys. Res. Planets* **2020**, *125*, e2019JE006086. [CrossRef]
26. Barker, M.K.; Mazarico, E.; Neumann, G.A.; Zuber, M.T.; Haruyama, J.; Smith, D.E. A new lunar digital elevation model from the Lunar Orbiter Laser Altimeter and SELENE Terrain Camera. *Icarus* **2016**, *273*, 346–355. [CrossRef]
27. Lemelin, M.; Lucey, P.G.; Gaddis, L.R.; Hare, T.; Ohtake, M. Global Map Products from the Kaguya Multiband Imager at 512 ppd: Minerals, FeO, and OMAT. In Proceedings of the 47th Annual Lunar and Planetary Science Conference, The Woodlands, TX, USA, 21–25 March 2016; p. 2994.
28. Dhingra, D.; Pieters, C.M.; Head, J.W. Multiple origins for olivine at Copernicus crater. *Earth Planet. Sci. Lett.* **2015**, *420*, 95–101. [CrossRef]
29. Lucey, P.G. Mineral maps of the Moon. *Geophys. Res. Lett.* **2004**, *31*, 1–4. [CrossRef]
30. Lucey, P.G.; Blewett, D.T.; Taylor, G.J.; Hawke, B.R. Imaging of lunar surface maturity. *J. Geophys. Res.* **2000**, *105*, 20377–20386. [CrossRef]
31. Fortezzo, C.M.; Spudis, P.D.; Harrel, S.L. Release of the Digital Unified Global Geologic Map of the Moon at 1:5,000,000-Scale. In Proceedings of the Lunar and Planetary Science Conference, The Woodlands, TX, USA, 16–20 March 2020; p. 2760.
32. Konopliv, A.S.; Park, R.S.; Yuan, D.-N.; Asmar, S.W.; Watkins, M.M.; Williams, J.G.; Fahnestock, E.; Kruizinga, G.; Paik, M.; Strekalov, D.; et al. High-resolution lunar gravity fields from the GRAIL Primary and Extended Missions. *Geophys. Res. Lett.* **2014**, *41*, 1452–1458. [CrossRef]
33. Lemoine, F.G.; Goossens, S.; Sabaka, T.J.; Nicholas, J.B.; Mazarico, E.; Rowlands, D.D.; Loomis, B.D.; Chinn, D.S.; Neumann, G.A.; Smith, D.E.; et al. GRGM900C: A degree 900 lunar gravity model from GRAIL primary and extended mission data. *Geophys. Res. Lett.* **2014**, *41*, 3382–3389. [CrossRef]
34. Melosh, H.J. *Impact Cratering: A Geologic Process*; Oxford University Press: New York, NY, USA, 1989.
35. Du, J.; Fa, W.; Wieczorek, M.A.; Xie, M.; Cai, Y.; Zhu, M.-H. Thickness of Lunar Mare Basalts: New Results Based on Modeling the Degradation of Partially Buried Craters. *J. Geophys. Res. Planets* **2019**, *124*, 2430–2459. [CrossRef]

36. Neukum, G.; Ivanov, B.A.; Hartmann, W.K. Cratering Records in the Inner Solar System in Relation to the Lunar Reference System. *Space Sci. Rev.* **2001**, *96*, 55–86. [CrossRef]
37. Kneissl, T.; van Gasselt, S.; Neukum, G. Map-projection-independent crater size-frequency determination in GIS environments—New software tool for ArcGIS. *Planet. Space Sci.* **2011**, *59*, 1243–1254. [CrossRef]
38. Michael, G.G.; Platz, T.; Kneissl, T.; Schmedemann, N. Planetary surface dating from crater size–frequency distribution measurements: Spatial randomness and clustering. *Icarus* **2012**, *218*, 169–177. [CrossRef]
39. Michael, G.; Neukum, G. Planetary Surface Dating from Crater Size-Frequency Distribution Measurements: Differential Forms of Production Function Polynomials. In Proceedings of the Lunar and Planetary Science Conference, The Woodlands, TX, USA, 1–5 March 2010.
40. Lemelin, M.; Lucey, P.G.; Song, E.; Taylor, G.J. Lunar central peak mineralogy and iron content using the Kaguya Multiband Imager: Reassessment of the compositional structure of the lunar crust. *J. Geophys. Res. Planets* **2015**, *120*, 869–887. [CrossRef]
41. Solomon, S.C.; Head, J.W. Lunar Mascon Basins: Lava filling, tectonics, and evolution of the lithosphere. *Rev. Geophys.* **1980**, *18*, 107–141. [CrossRef]
42. Yue, Z.; Yang, M.; Jia, M.; Michael, G.; Di, K.; Gou, S.; Liu, J. Refined model age for Orientale Basin derived from zonal crater dating of its ejecta. *Icarus* **2020**, *346*, 113804. [CrossRef]
43. Eberhardt, P.; Geiss, J.; Grögler, N.; Stettler, A. How old is the crater copernicus? *Moon* **1973**, *8*, 104–114. [CrossRef]
44. Alexander, E.C., Jr.; Bates, A.; Coscio, M.R., Jr.; Dragon, J.C.; Murthy, V.R.; Pepin, R.O.; Venkatesan, T.R. K/Ar dating of lunar soils II. *Lunar Planet. Sci. Conf. Proc.* **1976**, *1*, 625–648.
45. Hiesinger, H.; van der Bogert, C.H.; Pasckert, J.H.; Funcke, L.; Giacomini, L.; Ostrach, L.R.; Robinson, M.S. How old are young lunar craters? *J. Geophys. Res. Planets* **2012**, *117*, 1–15. [CrossRef]
46. Bell, J.F.; Hawke, B.R. Lunar dark-haloed impact craters: Origin and implications for Early Mare volcanism. *J. Geophys. Res. Solid Earth* **1984**, *89*, 6899–6910. [CrossRef]
47. Chen, Y.; Wang, X.; Liu, J.; Ren, X.; Huang, H.; Zhang, H.; Li, C. The Identification and Analysis of Gas-Related Volcanic Features within Chang'e-5 Landing Region. *Remote Sens.* **2021**, *13*, 3879. [CrossRef]
48. Weitz, C.M.; Head, J.W.; Pieters, C.M. Lunar regional dark mantle deposits: Geologic, multispectral, and modeling studies. *J. Geophys. Res. Planets* **1998**, *103*, 22725–22759. [CrossRef]
49. Gaddis, L.R.; Staid, M.I.; Tyburczy, J.A.; Hawke, B.R.; Petro, N.E. Compositional analyses of lunar pyroclastic deposits. *Icarus* **2003**, *161*, 262–280. [CrossRef]
50. Jawin, E.R.; Besse, S.; Gaddis, L.R.; Sunshine, J.M.; Head, J.W.; Mazrouei, S. Examining spectral variations in localized lunar dark mantle deposits. *J. Geophys. Res. Planets* **2015**, *120*, 1310–1331. [CrossRef]
51. Gustafson, J.O.; Bell, J.F.; Gaddis, L.R.; Hawke, B.R.; Giguere, T.A. Characterization of previously unidentified lunar pyroclastic deposits using Lunar Reconnaissance Orbiter Camera data. *J. Geophys. Res. Planets* **2012**, *117*, 1–21. [CrossRef]
52. Mighani, S.; Wang, H.P.; Shuster, D.L.; Borlina, C.S.; Nichols, C.I.O.; Weiss, B.P. The end of the lunar dynamo. *Sci. Adv.* **2020**, *6*. [CrossRef]

remote sensing

MDPI

Article

A Generative Adversarial Network for Pixel-Scale Lunar DEM Generation from High-Resolution Monocular Imagery and Low-Resolution DEM

Yang Liu [1,2], Yexin Wang [1,*], Kaichang Di [1,3], Man Peng [1], Wenhui Wan [1] and Zhaoqin Liu [1]

[1] State Key Laboratory of Remote Sensing Science, Aerospace Information Research Institute, Chinese Academy of Sciences, Beijing 100101, China
[2] University of Chinese Academy of Sciences, Beijing 100049, China
[3] CAS Center for Excellence in Comparative Planetology, Hefei 230026, China
* Correspondence: wangyx@aircas.ac.cn; Tel.: +86-10-64807987

Abstract: Digital elevation models (DEMs) provide fundamental data for scientific and engineering applications in lunar exploration missions. Lunar DEMs have been mainly generated by laser altimetry and stereophotogrammetry. Complementarity to stereo photogrammetry, reflection-based surface reconstruction methods, such as shape from shading (SFS), have been studied and applied in lunar DEM reconstruction from a single image. However, this method often suffers from solution ambiguity and instability. In this paper, we propose a generative adversarial network (GAN)-based method that is able to generate high-resolution pixel-scale DEMs from a single image aided by a low-resolution DEM. We have evaluated the accuracy of the reconstructed high-resolution DEMs from 25 LROC NAC images of four regions using LROC NAC DEMs (2 m/pixel) as ground truth. The experimental results demonstrate good accuracy and adaptability to changes in illumination conditions. The root mean square error (RMSE) can reach about 2 m in areas where the degree of elevation variation is less than 100 m, and the RMSE value ranges from around 3 m to 10 m without considering the degree of the elevation variation in large-area reconstruction. As high-resolution monocular images and low-resolution DEMs are available for the entire lunar surface, the proposed GAN-based method has great potential in high-resolution lunar DEM reconstruction for lunar mapping applications.

Keywords: lunar exploration; digital elevation model; high-resolution imagery; generative adversarial network

1. Introduction

Digital elevation models (DEM) of the Moon provide the fundamental data in lunar topographic, morphological, and geological studies. They are also essential in engineering applications in lunar exploration missions, e.g., in landing site evaluation, safe landing, rover navigation, etc. [1–6]. Lunar global DEMs have been produced using laser altimetry or optical images. Representative examples include the lunar orbiter laser altimeter (LOLA) LDEM1024 with a resolution of 1024 pixels per degree (ppd) (~30 m at the equator) [7], 1024 ppd DEM derived from 10 m-resolution SELENE terrain camera (TC) stereo images [8], and 20 m grid spacing DEM derived from 7 m-resolution Chang'e-2 stereo images [9]. SLDEM2015, which was produced by coregistration and a combination of SELENE TC DEM and LOLA data, is also widely used. It covers latitudes within $\pm 60^{\circ}$ at a horizontal resolution of 512 ppd (~60 m at the equator) [10].

Lunar local DEMs with higher resolution (e.g., meter-level resolution) have also been produced, mostly by stereo photogrammetry, which requires target area covering by stereo images. As a rule of thumb, the resolution (grid spacing) of a DEM generated by stereophotogrammetry is about three times that of the stereo imagery [11]. Currently, the highest-resolution lunar orbital images (up to 0.5 m/pixel) are acquired by lunar reconnaissance orbiter camera (LROC) narrow-angle camera (NAC), and the local DEMs generated from LROC NAC stereo images are usually of 2–5 m resolution [12], which are the highest-resolution DEMs derived from orbital images. The LROC was not designed with built-in stereo observation capability, and NAC stereo images can only be acquired from neighboring orbits using off-nadir slew. As a result, NAC stereo observations are very limited, though NAC monocular images have covered almost the entire lunar surface. Therefore, it is highly desirable and will be widely applicable to reconstruct pixel-scale DEMs from monocular high-resolution orbital images, particularly LROC NAC images.

In fact, complementary to stereo photogrammetry, shape from shading (SFS) (also called photoclinometry) techniques have been studied and applied in the planetary mapping field [11,13–15], which reconstruct a DEM from a single image or multiple nonstereo images. SFS techniques have also been widely researched in the field of computer vision and artificial intelligence [16,17]. In general, the multiple-image SFS method is limited by a lack of multiple images with different illumination conditions covering the same area. According to [18], single-image SFS is underconstrained and faces serious issues. First, the assumption of a uniform or known albedo is needed, but determining the albedo from a single image is underconstrained. Second, as shading actually reflects the depth of a point relative to its neighbors, it is difficult to get clues to restore shapes from coarse-scale depth variations that produce small changes throughout the entire image. A number of approaches have been developed to overcome the underconstrained issues, including learning albedos from training data [18], separation of albedos, and shape optimization with the aid of coarse DEMs, etc. [15,19].

With the development of deep learning, a neural network can learn the "hidden" mapping relationship between two types of data and realize the "end-to-end" transformation of data under the condition of sufficient training samples. Deep neural network has been applied to DEM reconstruction using a single image and has shown some advantages in establishing reliable mapping and eliminating noise and image uncertainty. In order to improve the performance of DEM reconstruction, strategies of deepening the depth of the network and increasing the amount of training samples can be used [20]. There is no atmospheric influence on the lunar surface, and the change of image intensities is mainly caused by the topography, supposing that the albedo does not change severely in a local area. Low-resolution DEMs are available for the entire lunar surface and can be used as a constraint to enable the network to obtain low-frequency terrain information, while high-frequency terrain information can be obtained from high-resolution image. This makes it possible to use a network to generate a high-resolution DEM with higher accuracy from high-resolution monocular imagery and existing low-resolution DEM.

In this paper, a high-resolution lunar DEM reconstruction method is proposed based on generative adversarial network (GAN), which generates pixel-scale lunar DEM from monocular LROC NAC imagery aided by existing low-resolution DEM, e.g., SLDEM. For convenience, we call the lunar DEM reconstruction network LDEMGAN. It adopts the U-Net structure as the generator network, uses the convolution layer instead of the pooling layer for feature extraction, takes the discrimination result of the discriminator network instead of the image similarity as the loss, adds the gradient loss item, and uses a variety of training data to train the network. In this study, four regions are selected for reconstruction testing using 25 LROC NAC image (2 m/pixel) and SLDEM (~60 m/pixel). The reconstruction results are verified with DEMs (2 m/pixel) produced by LROC NAC images, and the effects of different lunar surface conditions and data quality on the reconstruction results are analyzed. The effects of adding low-resolution DEM and influence of illumination conditions of the images are also discussed.

Section 2 of this paper reviews the relevant research. Section 3 presents the proposed method in detail. Section 4 gives the experimental results and discussions. Conclusions are given in Section 5.

2. Related Work

With rapid development of deep learning technology, monocular depth estimation approaches based on deep neural networks have been studied and applied in various fields of computer vision, such as robotics, path planning, autonomous navigation, etc. Due to difficulties of absolute depth recovery, as well as the demand of comparable accuracy to that of traditional triangulation methods, monocular depth estimation is still an active and challenging topic. Currently, the commonly used deep neural network models for monocular depth recovery are the convolutional neural network (CNN), recurrent neural network (RNN), and generative adversarial network (GAN) [21]. As RNN is usually employed in the field of video sequence processing, this section only focuses on reviewing and discussing monocular depth restoration that uses CNN and GAN models.

CNN models have been well developed and applied in monocular depth estimation. Eigen et al. [22] adopted a two-scale convolutional network to first estimate the depth of the scene on a global scale and then add details on a local scale. The work also raised and explained the basic problem of solving monocular depth estimation in terms of supervised learning [22]. On this basis, Eigen et al. subsequently proposed an improved framework constructed by three-scale network structure for monocular depth estimation [23]. Thereafter, Laina et al. [24] proposed a fully convolutional network structure to capture monocular features at multiple scales with fewer parameters, and solved the problem of high training costs for fully connected layers. An up-convolution decoding module was designed to improve the processing speed of the network [24]. Ma et al. proposed a CNN-residual network, and the comparison to the network with the hybrid convolutional fully connected network and to the fully convolutional network demonstrated the effectiveness and optimality of the proposed network [25]. Fu et al. added a spacing-increasing discretization strategy that splits the depth according to different intervals and recast the network learning as an ordinal regression problem to reduce the training loss weight in regions with large depth values [26]. Facil et al. used the U-Net to encode images and then add sensor parameters for decoding, which enhances the generalization ability of the network [27]. In order to shorten the computing time of the network, Wolk et al. proposed a lightweight U-Net to improve the operation speed, which can realize real-time depth estimation [28].

The GAN model was proposed by Goodfellow in 2014 and has been proved applicable for monocular depth estimation tasks [29]. Jung et al. used GAN for the first time to implement monocular depth estimation by supervised learning [30]. The generative network consists of a global-scale U-Net [31] and a fully convolutional fine-scale network. The discriminative network is designed according to Li's work [32]. Instead of using a multiscale generator, Lore et al. used two GANs for depth estimation at both global and local scales [33].

In the field of planetary remote sensing, since the orbital images of the Moon's surface, especially high-resolution stereo images, are very scarce compared to those of the Earth, DEM generation by monocular high-resolution images is needed more than traditional photogrammetry methods. Therefore, in recent years, several studies have used depth estimation networks to reconstruct relative heights from planetary orbital images and to restore relative heights to absolute heights using control points or existing coarse DEM. Several works have discussed combining monocular relative height estimation and scale information recovery for planetary surface reconstruction. Chen et al. proposed a CNN-based digital terrain model (DTM) reconstruction method for Mars using Mars Reconnaissance Orbiter Context Camera (CTX) images [20]. Tao et al. proposed a GAN-based network called MADNet for fast reconstruction of DTM from a single image of Mars [34]. Subsequently, MADNet 2.0 was developed, which used the optimized single-scale inference and multiscale reconstruction to improve the resolution of reconstructed DTM [35]. These methods all used a two-step strategy to reconstruct the terrain, i.e., firstly, an image is input into the deep network to restore relative depth, and then absolute height is recovered by stretching the relative depth using a scale value calculated by existing low-resolution data.

In these previous works, low-resolution data are only used for scale recovery such that the reconstruction result is affected greatly by scale value. To make full use of the coarse DEM, Hao et al. proposed using the low-resolution DEM and the image as the input of the CNN, and the CNN prediction results and the image obtain a high-resolution DEM by SFS [36]. However, the reconstruction results generated by SFS are affected by lighting conditions. Intending to make full use of low-resolution DEM in terrain reconstruction, improve adaptability to light conditions and combine the advantages of GAN, this paper proposes a GAN-based reconstruction network for lunar DEM generation—LDEMGAN. The main features of this work are:

(1) The proposed approach takes both monocular image and a coarse DEM as input of the GAN.
(2) The coarse DEM is fully used to provide control information in the proposed LDEMGAN so as to improve the stability and accuracy of high-resolution DEM reconstruction.
(3) In order to generate DEM at the same high resolution as the input image, the reconstruction process can be executed iteratively until the needed DEM resolution is obtained.

3. Method

3.1. Overview of the Proposed Approach

In this study, LROC NAC imagery is used to reconstruct a pixel-scale DEM through LDEMGAN with the help of control information provided by SLDEM. Figure 1 shows the overall process of the proposed approach, which consists of training and reconstruction phases. In the beginning, the registered LROC NAC imagery and SLDEM data are preprocessed. (1) The centers of the craters with diameters of about 200 m are selected as control points, and the LROC NAC images and SLDEM are registered by affine transformation. The registration accuracy needed to reach the subpixel level of SLDEM (the registration error needs to be less than the grid spacing of SLDEM). (2) The NAC image performs downsampling if needed and the SLDEM is upsampled to the same pixel size. (3) They are both clipped to the size of 256×256 due to the requirement of the network with an overlap of 100 pixels between adjacent image blocks. (4) The clipped data are normalized to [0, 1]. (5) The clipped image and SLDEM data are set as channel 1 and channel 2, respectively, to synthesize the input data of LDEMGAN. Note that in the training phase, the corresponding NAC DEM is also preprocessed in the same way so as to take the role as ground truth.

In the training phase, the preprocessed high-resolution image and low-resolution DEM of the same area with the same grid are taken as input of LDEMGAN, which outputs depth reconstruction results. The ground truth is preprocessed DEM with identical resolution of the input imagery, and supervised training is carried out to obtain the reconstruction network.

In the reconstruction phase, the preprocessed high-resolution image and low-resolution DEM covering the same area are input into LDEMGAN to generate DEM that has identical resolution to the input image. However, in practice, the resolutions of the input image and the DEM that provide control information differ too much, such as NAC and SLDEM in our case. Therefore, a high-resolution DEM generation process can be carried out iteratively, which is shown in Figure 1. In our case, first the low-resolution SLDEM is upsampled to 10 m grid spacing and an input high-resolution NAC image is downsampled to the same pixel size of the upsampled SLDEM. Next, they are simultaneously input into the network to generate DEM with identical resolution to the downsampled input image which is 10 m/pixel. As the resolution of the generated DEM is lower than the NAC image, the previous steps should be implemented again. This time, the generated DEM from the previous step is upsampled to the same pixel size of the NAC image. Then, they are both input into the reconstruction network to obtain a DEM with the same resolution as the NAC image. Afterwards, scale restoration will be carried out to convert the generated DEM from relative height into absolute height.

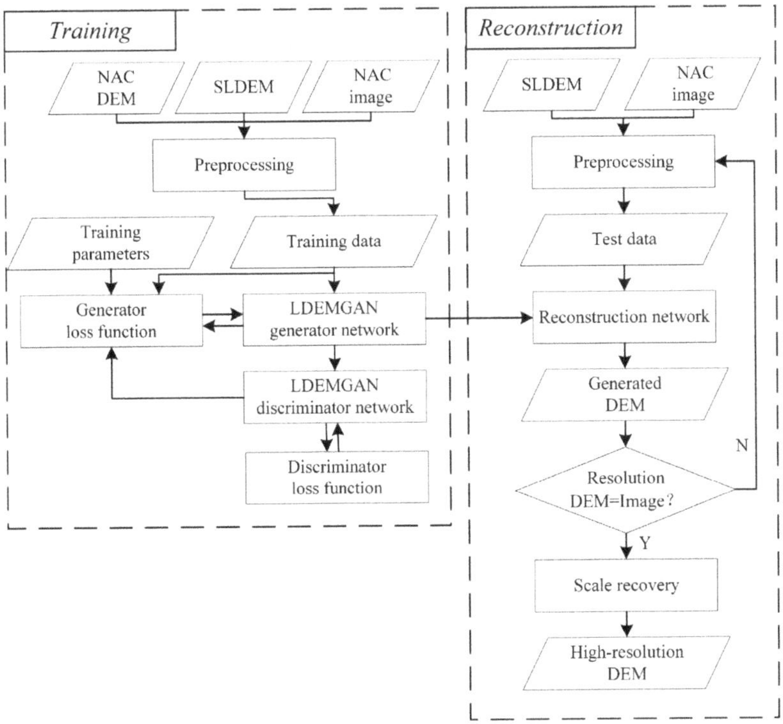

Figure 1. Overall process of the proposed method.

It is worth mentioning that there are often geometric inconsistencies between data captured from different sensors; therefore, geometric registration of the NAC image and SLDEM is required beforehand in this study. In fact, the accuracy of geometric registration can directly affect the final reconstruction result.

3.2. Network Architecture

Figure 2 shows the LDEMGAN network architecture used for high-resolution DEM generation. According to the principle of the GAN network, it can be divided into a generator network G and a discriminator network D. The generator network adopts a U-Net architecture, adding skip connections between the encoding part and the decoding part. The CNN structured encoder consists of 7 BCDL modules, which encode a 256 × 256 image into a feature vector with a size of [2, 2, 1024]. The BCDL module consists of a batch normalization layer, a convolution layer (1 × 1 kernel, and with increasing number of feature maps of 64, 128, 256, 512, 1024, 1024, 1024, stride 1), a downsampling convolution layer (4 × 4 kernel, stride 2, padding 1), and a LeakyReLU activation function. LeakyReLU is able to solve the zero gradient problem for negative values, as shown in Equation (1):

$$\text{LeakyReLU}(x) = \begin{cases} x & x > 0 \\ 0.2x & x \leq 0 \end{cases} \tag{1}$$

where x is the input tensor. The decoder employs a BUR model and adds skip connections, enabling direct transfer of low-frequency information. The BUR module consists of a batch normalization layer, an upsampling convolution layer (4 × 4 kernel, and with decreasing number of feature maps of 1024, 1024, 512, 256, 128, 64, 1, stride 2, padding 1), and an activation function. The activation function used for the BUR-1–BUR-6 is ReLU:

$$\mathrm{ReLU}(x) = \begin{cases} x & x > 0 \\ 0 & x \leq 0 \end{cases} \quad (2)$$

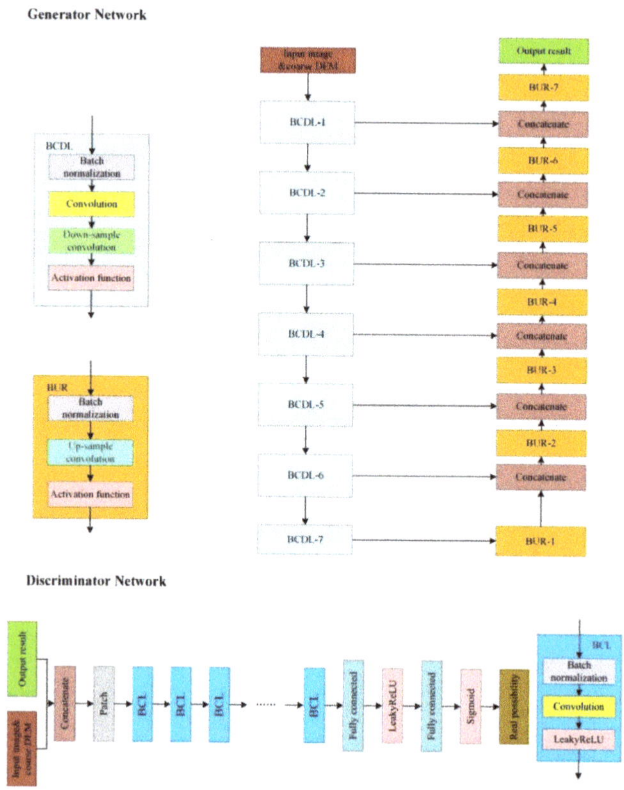

Figure 2. Schematic diagram of the LDEMGAN structure.

The TanhL activation function is used in the BUR-7, so that the range of the output result can be constrained within [0, 1], as shown below:

$$\mathrm{TanhL} = \frac{1}{2}\left(\frac{e^x - e^{-x}}{e^x + e^{-x}} + 1\right) \quad (3)$$

The optimization objective function of training the generator network parameters is as follows:

$$\hat{\theta}_G = \underset{\theta_G}{\mathrm{argmin}} \frac{1}{N} \sum_{n=1}^{N} \mathrm{L_G}\left(H^n_{pred}, H^n_{\mathrm{truth}}\right) \quad (4)$$

Among them, N is the total number of images, $\mathrm{L_G}(\cdot)$ is the loss function of the generator network G (see Section 3.3), θ_G is the parameter of the generator network. H^n_{truth} is the nth ground truth DEM data of the corresponding input image, and H^n_{pred} is the nth reconstruction results derived by the generator network. H^n_{pred} is calculated as:

$$H^n_{pred} = \mathrm{G}_{\theta_G}(I_n, H^n_{coarse}) \quad (5)$$

where G_{θ_G} is the generator network with network layer parameters θ_G, I_n is the nth input high-resolution image, and H^n_{coarse} is the corresponding low-resolution DEM that provides control information in the same area.

According to the structure of the GAN network, the output of the discriminator network D is a probability value that denotes whether its input sample is real or fake. The discriminator network used in this paper adopts the principle of PatchGAN to capture the high-frequency information. PatchGAN divides the output of the last layer into M blocks, and discriminates each block individually:

$$P_H = \frac{1}{M}\sum_{m=1}^{M} \mathrm{D}((I^m, H^m_{coarse}), H^m) \tag{6}$$

where H^m is the H^n_{pred} calculated by the generator network or the ground truth surface data H^n_{truth}. Therefore, the optimization objective for the adversarial network parameters is:

$$\hat{\theta}_D = \underset{\theta_D}{\mathrm{argmin}} \frac{1}{N}\sum_{n=1}^{N} \mathrm{L_D}\left(I^n, H^n_{pred}, H^n_{\mathrm{truth}}\right) \tag{7}$$

where $\mathrm{L_D}(\cdot)$ is the loss function of the discriminator network D (see Section 3.3).

3.3. Loss Function

The loss function of the LDEMGAN consists of generator network loss and discriminator network loss. The losses of the generator network include the loss representing the difference between the prediction and the ground truth, the loss of local structure, and the loss of the discriminator. Usually, the L_2 loss function is used for the generator network, which reflects the cumulative squared difference between the value predicted by the model and the ground truth value. However, it can bring over smooth predictions. Hence, the Huber loss is used, as shown in Equation (8):

$$\mathrm{L}_{Huber}(H_{pred}, H_{truth}) = \begin{cases} (H_{pred} - H_{truth})^2/2 & \left|H_{pred} - H_{truth}\right| < \delta \\ \delta\left|H_{pred} - H_{truth}\right| - \frac{1}{2}\delta^2 & otherwise \end{cases} \tag{8}$$

where δ is one fifth of the maximum of the difference of prediction and the ground truth. When the difference is less than δ, the Huber loss is consistent with L_2. Otherwise, the Huber loss equals L_1.

In order to maintain the relationship between each pixel and its adjacent pixels, a local structure loss is added, as shown below:

$$\mathrm{L}_{grad} = \frac{1}{MN}\sum_{i=1}^{M}\sum_{j=1}^{N}\left[(\nabla_x(H_{pred}, H_{truth})^2 + \nabla_y\left(H_{pred}, H_{truth}\right)^2\right] \tag{9}$$

where M is the number of image rows and N is the number of image columns. By considering ∇_x and ∇_y, the x directional and y directional gradient differences between the predicted value and the ground truth calculated in local windows, respectively, the consistencies of the predicted value and the ground truth can be preserved on a local scale.

The loss of discriminator refers to the discrimination of the generated prediction data by the discriminator network in LDEMGAN, indicating the similarity between the generated prediction data and the ground truth, as shown below:

$$\mathrm{L}_{GAN} = \log(1 - \mathrm{D}((I, H_{coarse}), \mathrm{G}(I, H_{coarse}))) \tag{10}$$

where I is the input image and H_{coarse} is the input DEM (low-resolution DEM).

Therefore, the overall loss of the generator network can be represented by:

$$L_G = \alpha L_{Huber} + \beta L_{grad} + \gamma L_{GAN} \quad (11)$$

where α, β, γ are the coefficients to balance the losses.

The loss function of the discriminator network is shown as:

$$L_D = \frac{1}{2}[\log(D((I, H_{coarse}), G(I, H_{coarse}))) + \log(1 - D((I, H_{coarse}), H_{truth}))] \quad (12)$$

3.4. Scale Recovery

As normalization of the input data is needed before entering the generator network for better convergence, the values of the network output are within [0, 1]. Thus, postprocessing is required to recover the reconstruction results to the correct scale as elevation information. This study uses the normalized low-resolution DEM as a constraint, as shown in Equation (10). In addition, in the scale recovery process, statistics of this low-resolution DEM are used to restore the absolute elevation. First, mean and standard deviation of the low-resolution DEM are calculated. Then the reconstruction results are stretched to obtain the absolute heights, as shown in Equation (13). By adding geographic information to the restored absolute heights, scale recovery is finally completed.

$$DEM_{abs} = m + sd \cdot DEM_{pred} \quad (13)$$

In the above equation, DEM_{abs} is the absolute elevation value, m is the mean value of the low-resolution DEM, and sd is the standard deviation of the low-resolution DEM.

3.5. Network Training

As mentioned in Section 3.1, practically the DEM restoration is carried out iteratively. In our case, two reconstruction networks are employed and the structures of them are identical. In fact, the restoration processes are similar in each iteration, so theoretically one set of trained network parameters should work for both networks. However, in order to obtain high-precision results, there is no doubt that the two networks should be trained separately with training data the same as the actual situation.

Meanwhile, in order to enhance the universality of the network, the number of training samples should be sufficient so that the network can avoid fuzzy solution caused by the changes of illumination conditions or various landform characteristics. For adapting illumination changes, the training data can be expanded from two aspects. First, the existing training data can be rotated and flipped to simulate different sun azimuths. Second, images captured at different sun altitude angles of the same area that pair with the same corresponding coarse DEM should be added to enhance the network's adaptation to shadow properties. For adapting to changes of landform characteristics in various topographic regions (e.g., lunar mare or highlands), the network performance can always be improved if the training data are adequate. Generally, if the feature characteristics of the area to be processed are hugely different from the areas that used for training the networks, the recommended practice is to employ transfer training by using a small amount of data from the destination area.

4. Experimental Results

4.1. Reconstruction Network Training

As described in the above sections, two networks with identical structure and trained by different data are used to implement DEM reconstruction in a resolution-progressive manner. The first network aims at generating DEM at a resolution of 10 m/pixel, which is denoted as LDEMGAN-10 m. To train this network, the input data are downsampled LROC NAC images of 10 m/pixel and normalized SLDEM of 60 m/pixel (upsampled to the same grid spacing). Ground truth is downsampled LROC NAC DEMs at grid spacing of 10 m/pixel. The second network intends to generate DEM of 2 m/pixel and is denoted as LDEMGAN-2 m. To train this network, the input data are LROC NAC image with resolution of 2 m/pixel, together with the LROC NAC DEM downsampled to 10 m/pixel, yet with similar pixel size to the NAC image. The ground truth is the LROC NAC DEM of its original resolution. For clearer demonstration, the 10 m reconstruction network and the 2 m reconstruction network are denoted as LDEMGAN-10 m and LDEMGAN-2 m, respectively. From this, it can be seen that the areas that satisfy the conditions for producing training data should include LROC NAC images, LROC NAC DEMs, and SLDEM at the same time.

According to the availability of data by the above requests, eight regions were selected for training the reconstruction networks. Some image samples of training data are shown in Figure 3: samples with different lunar surface features (Figure 3a), samples with different lighting conditions in the same area (Figure 3b), and samples after rotation or flipping (Figure 3c).

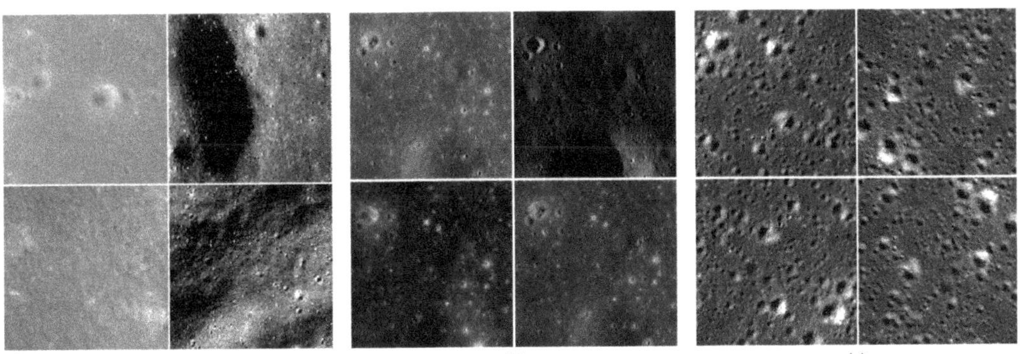

(a) (b) (c)

Figure 3. Examples of LROC NAC images from our training data: (**a**) images with different surface topography and features; (**b**) images with different lighting conditions in the same area; (**c**) images after rotation or flipping to simulate different sun azimuths.

A total of 89 scenes of LROC NAC images (2 m/pixel) and their corresponding LROC NAC DEM (2 m/pixel), together with SLDEM (60 m/pixel) covering these areas were selected to make training samples. A total of 6428 sets of training samples were made for training the first 10 m/pixel reconstruction network, and 4024 sets of training samples were made for training the second 2 m/pixel reconstruction network.

For LDEMGAN-10 m, the number of training epochs is 200, and the initial learning rate is 2×10^{-4}, which is linearly reduced to 10^{-6} after 100 epochs. The coefficients in Equation (11) are set to $\alpha = 0.5$, $\beta = 0.5$, and $\gamma = 0.8$, respectively. Next, LDEMGAN-2 m is obtained through migration learning based on LDEMGAN-10 m. In addition, during the LDEMGAN-2 m training, γ is reduced to 0.3 in Equation (11), and training epochs is set to 100. This experiment used the pytorch1.10 platform to build the network and was carried out on two RTX 3090 graphics cards. The training took a total of 5.7 h.

4.2. Testing

4.2.1. Testing Dataset

In order to test the reconstruction network, another four areas (including 25 LROC NAC images) different from training data areas containing LROC NAC DEM that aims at evaluating the reconstruction results are selected as test areas. The four areas are Apollo 11, Imbrium, Lichtenberg, and Hponds. The LROC NAC images of these areas are superimposed on SLDEM (Figure 4). The experimental areas are named according to what the LROC NAC DEMs contain.

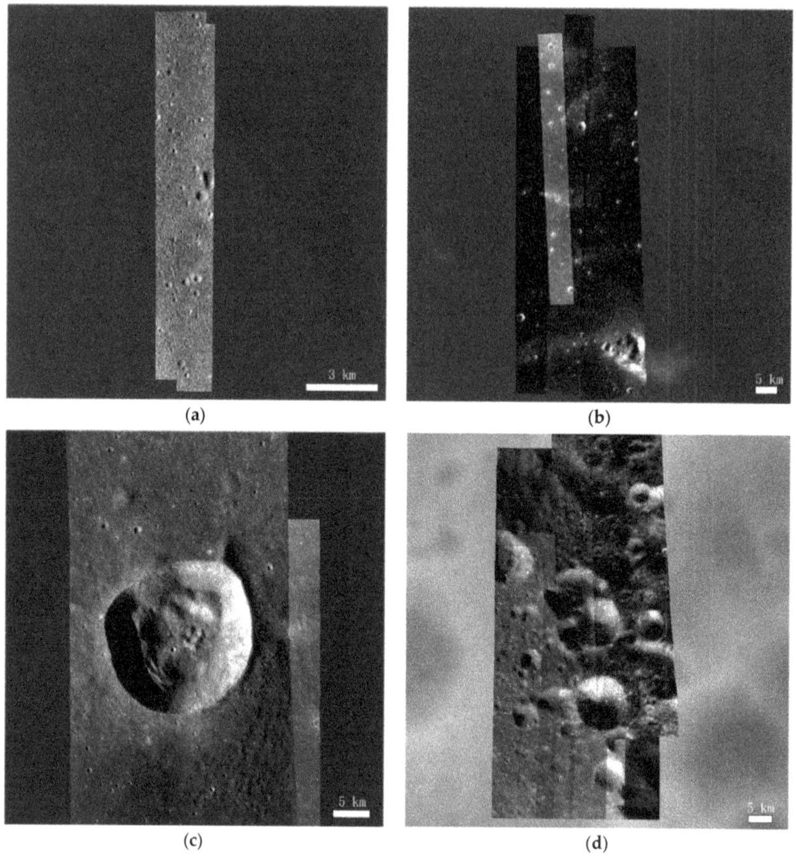

Figure 4. The LROC NAC images of the four test areas displayed on SLDEM: (**a**) Apollo 11; (**b**) Imbrium; (**c**) Lichtenberg; (**d**) Hponds.

Details of the four areas are shown in Table 1. The images used in each area are as follows: (1) Apollo 11: m150361817le, m150361817re, m150368601re; (2) Imbrium: m183697099le, m183704246re, m1106080949le, m1106080949re, m1175588941le, m1226179201re, m1265076277re; (3) Lichtenberg: m104955555le, m1096922142re, m1096929289le, m1096929289re, m1096936436le, m1096943583le, m1096936436re; (4) Hponds: m1100101542le, m1100101542re, m1102444668le, m1102444668re, m1110719713le, m1110719713re, m1113084806le, m1113084806re. The above data are downloaded from https://ode.rsl.wustl.edu/moon/index.aspx (accessed on 10 July 2022).

Table 1. Information of the four test areas.

ID	Test Area	Centre Latitude, Longitude (N, E/°)	Number of LROC NAC Images	SLDEM Resolution (m/pixel)
1	Apollo 11	0.77, 23.44	3	55.22
2	Imbrium	28.66, 333.88	7	55.51
3	Lichtenberg	31.68, 292.43	7	59.22
4	Hponds	41.79, 167.36	8	59.20

4.2.2. Reconstruction Results and Analysis

The reconstruction results of the four areas are shown in Figure 5. We use the LROC NAC DEMs of the experimental areas to evaluate the elevation accuracy of the reconstruction results. We choose two metrics, mean absolute error (MAE) and root mean square error (RMSE), to evaluate the reconstruction results. The two metrics can reflect the overall reconstruction performance of the statistics area, and the analysis of the local reconstruction effect is further analyzed by the profile graph and the absolute difference plot.

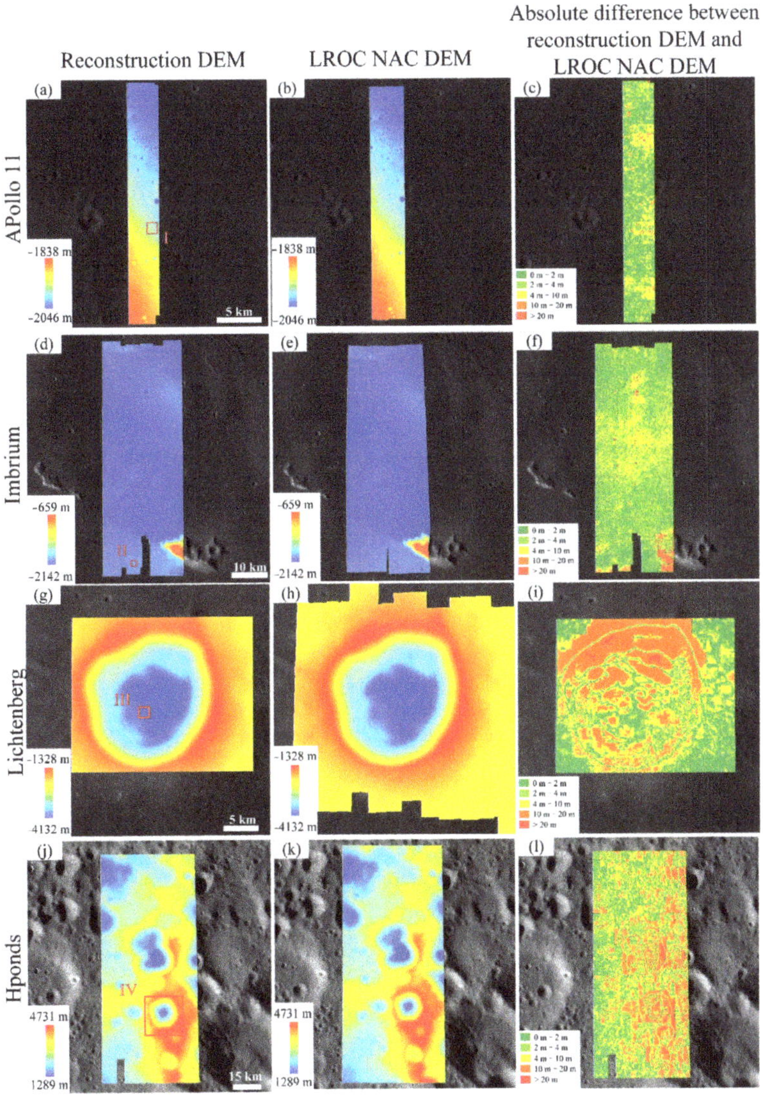

Figure 5. Reconstruction results and comparisons with LROC NAC DEMs. Each row shows one of the four experimental areas. The first column shows reconstruction results; the second column shows LROC NAC DEM as ground truth; the third column shows differences between the reconstruction results and LROC NAC DEMs.

The reconstruction effect of Apollo 11 (Figure 5a–c) is better than others in most parts, because the training data contain a considerable number of areas that have similar geomorphic characteristics and this area is relatively flat with less elevation variation. Approximately 50% of pixels have elevation differences less than 2 m, and more than 90% of pixels have elevation differences less than 10 m. The MAE of the whole experimental area is 2.76 m and the RMSE is 3.02 m, revealing high-precision reconstruction performance.

The reconstruction effect of Imbrium (Figure 5d–f) is similar to that of Apollo 11, and high reconstruction accuracy has been achieved in most parts of the area, the MAE and the RMSE is less than 4 m for the whole area, and more than 85% of pixels have elevation differences less than 10 m. The region with relatively large error in the Figure 5f bottom right corner is Mons La hire, which is a hillside with large topographic relief. In this region, the variation in image brightness (change in gray scale) is less than the other regions, yet the elevation change is relatively large; therefore, the reconstruction network may not be able to extract much effective information, leading to a comparatively large reconstruction error.

Experimental area Lichtenberg (Figure 5g–i) contains a large impact crater with diameter of 19 km and depth of 2.2 km. As illustrated in Figure 5i, large reconstruction errors mainly occur in the area with large topographic relief and less gray scale variation, such as near the upper edge of the impact crater.

The Hponds experimental area (Figure 5j–l) is located in the Chandler area on the far side of the Moon. Impact craters are densely and overlappingly distributed in this area. There are a large number of impact craters in this area with diameter more than 500 m and depth more than 100 m, which brings the existence of shading regions with brightness close to 0 in the impact crater. Both the large variation in elevation and shadows caused larger absolute reconstruction errors than the other test areas.

According to the above analysis of the reconstruction results, there are no obvious artifacts in the reconstruction results. The reconstructed elevation accuracy is related to the elevation and brightness ranges of the input area. The statistics of the reconstruction errors in the four areas are listed in Table 2. As can be found in Table 2, experimental areas Apollo 11 and Imbrium, which have moderate terrain fluctuation, have lower proportions of elevation errors (i.e., the differences between reconstructed results and LROC NAC DEMs) than the other two, which have relatively drastic topographic reliefs. The MAE and RMSE in these four areas also reflect this trend.

Table 2. The overall statistics of the reconstruction errors in the four areas.

Area	Reconstruction Error <2 m (%)	Reconstruction Error <4 m (%)	Reconstruction Error <10 m (%)	MAE (m)	RMSE (m)
Apollo 11	48.32	73.06	91.54	2.76	3.02
Imbrium	42.27	70.47	85.67	3.03	3.64
Lichtenberg	27.21	45.11	68.89	9.26	10.52
Hponds	25.52	45.95	62.51	9.43	10.65

4.2.3. Detailed Analysis of the Reconstruction Results in Subareas

In order to obtain a more detailed analysis of the reconstruction results, four typical subareas (numbered I–IV in Figure 5a,d,g,j) are selected to illustrate more elaborate analysis. Figure 6 shows the zoomed subareas, the corresponding SLDEM, the reconstruction results obtained by LDEMGAN, LROC NAC DEMs as ground truth for comparisons, and sectional views of the reconstruction results and gray scale profiles of the LROC NAC image, respectively. From Figure 6, it can be seen that the reconstructed DEMs from the monocular LROC NAC images have good consistency with LROC NAC DEMs, showing much finer elevation information than SLDEM, which provides control information.

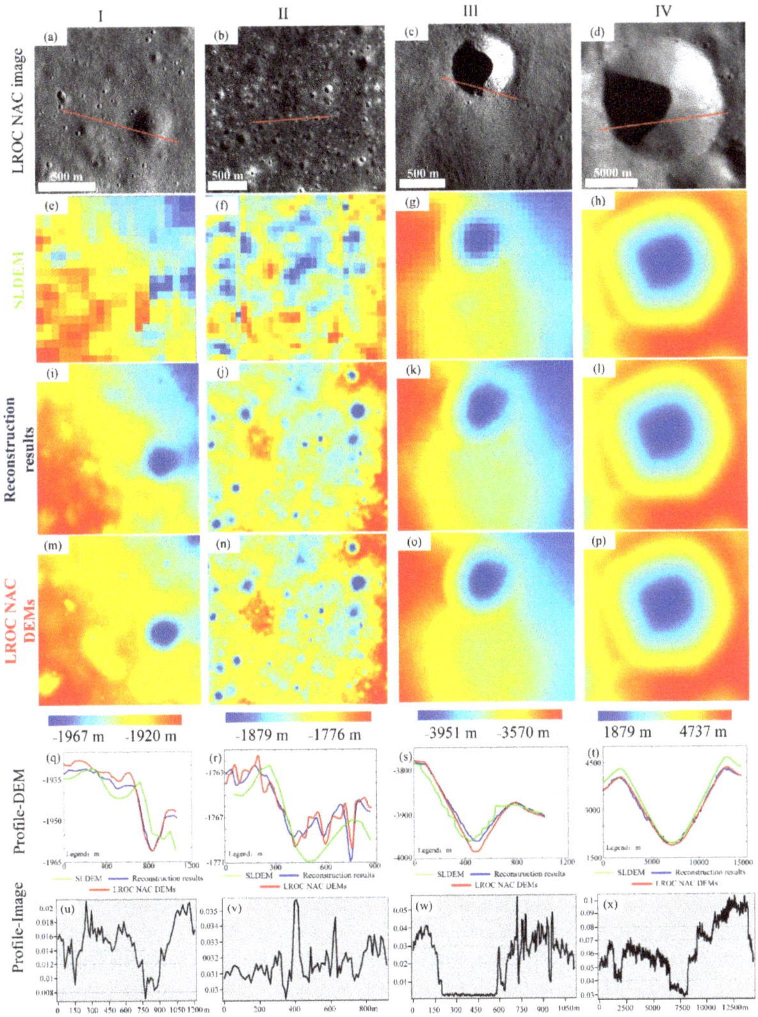

Figure 6. More detailed analysis of the results in the four subareas I–IV annotated in Figure 5. Each column shows one subarea. Row 1: LROC NAC image of 2 m/pixel showing the zoomed subareas. Row 2: SLDEM of 60 m/pixel. Row 3: network reconstruction results of 2 m/pixel. Row 4: LROC NAC DEM of 2 m/pixel as ground truth for comparison purposes. Row 5: DEM profiles of the four subareas with their profile locations shown in the first row. The green curves represent SLDEM, the blue curves represent network reconstruction results, and the red curves represent LROC NAC DEMs. Row 6: gray scale profiles of the LROC NAC image.

In subarea I (Figure 6a,e,i,m), the MAE of the reconstruction result is 1.26 m and the RMSE is 1.84 m, as demonstrated in Table 3. The maximum elevation error is 6.43 m, which is relatively small compared to the other subareas. From the elevation profiles (Figure 6d), it is shown that LDEMGAN can recover detailed DEM that coincides with the ground truth. Even the control information provided by SLDEM is with coarse grid and has shift in the bottoms of the curves, which demonstrate that the details in the image have more impact on the result.

Table 3. Reconstruction error of subareas.

Subarea	Degree of Elevation Variation (m)	Reconstruction Error <2 m (%)	MAE (m)	RMSE (m)	Max Error (m)
I	47.68	88.53	1.26	1.84	6.43
II	103.55	80.10	1.93	2.26	8.20
III	381.20	65.51	7.28	8.63	19.70
IV	2858.62	38.72	9.43	10.33	28.82

In subarea II (Figure 6b,f,j,n), when SLDEM is greatly different from the LROC NAC DEMs, as shown in Figure 6f,n, the reconstruction result is still highly consistent with the ground truth, from which it can be deduced that the input LROC NAC image takes more credit on recovering depth values. The MAE of the reconstruction result in this subarea is 1.93 m, the RMSE is 2.26 m, and the maximum elevation error is 8.20 m.

In subarea III (Figure 6c,g,k,o), the reconstruction result shows consistency with SLDEM (Figure 6g), yet with more details, with a reconstruction MAE of 7.28 m, RMSE of 8.63 m, and a maximum error of 19.70 m compared to the ground truth LROC NAC DEM (Figure 6k). It can be seen from the profile of this subarea (Figure 6s) that the center location of the crater in the reconstruction result coincides more with the ground truth than that of SLDEM, but there is an elevation drift of the reconstruction result from the real bottom value in the shadow region of the crater as the gray scale profile shown in Figure 6w. We found that when image brightness of the region is close to 0, the reconstruction network would be unable to translate much effective information from the image and may rely more on SLDEM to predict the elevation.

As can be seen from Figure 6a–c, the scales of subareas I, II and III are similar. The reconstruction errors rise with the increase of the degree of the elevation variation in these subareas.

In subarea IV (Figure 6d,h,p), the reconstruction result of a large impact crater with diameter of approximately 13.2 km is shown. From the profile (Figure 6t), the reconstruction results are better than the SLDEM (Figure 6h), especially on the edges of the craters. However, the RMSE is relatively larger than other subareas, which is 10.33 m, and the maximum error is 28.82 m. This is because the data are normalized at an imagery size of 256×256 to relative scale values before inputting into LDEMGAN and are recovered to absolute values according to the elevation range of this area. In consequence, deep craters or terrain with large elevation change would result in a relatively large absolute error.

In summary, based on the reconstruction results of these four subareas, the LDEMGAN can produce DEM with rich detail comparable to the NAC DEMs with the assistance of control information supplied by SLDEM. The accuracy of the reconstruction results is related to both the brightness variation of the input image and the degree of regional topographic relief. When the input images have less shading (the gray scale value is close to 0) and the degree of elevation variation is less than 100 m, the percentage of the reconstruction error less than 2 m (less than the input image grid spacing) can be more than 80%, the MAE can be less than 2 m, and the RMSE can reach below 2 m.

4.3. Discussion

4.3.1. Effect of Control Information

As demonstrated in the previous sections, the reconstruction results may rely more on the input image. In order to verify the necessity of control information instead of considering the process as just a style transfer operation from image to elevation, the reconstruction results with and without control information are compared in this subsection. The Apollo 11 landing area is selected as the test area, the input images are LROC NAC images (m150361817re, m150361817le), and the input low-resolution DEM for providing control information is downsampled LROC NAC DEMs (nac_dtm_apollo11) with grid spacing of 10 m/pixel. Here, using downsampled LROC NAC DEM instead of SLDEM can verify the improvement of the reconstruction effect under ideal conditions such that the low-resolution data are completely correct. At the same time, considering that the degree of elevation change mentioned in Section 4.2 will affect the absolute error, the result of accuracy evaluation in this subsection applies relative elevation without scale recovery (i.e., the value range is [0, 1]) so that the error magnitude of each input datum is the same.

Figure 7 shows the comparison results using different inputs: using only LROC NAC image as input or using both LROC NAC image and coarse DEM as input. For better visualization, the relative elevation values are stretched to [0, 255] for display in Figure 7. Figure 7c,h,m shows the reconstruction result taking only LROC NAC images as input. Although the trend of the elevation changes of these reconstruction results are consistent with the real lunar surface, which can indicate the locations of big craters, the extent of elevation change is inaccurate, especially in the areas that are relatively flat. In this case, the MAE of predicated results using only image is 0.11, the RMSE is 0.18, and the maximum mean error is 0.21. After adding the downsampled low-resolution DEM (coarse DEM) as input (Figure 7d,i,n), the reconstruction of elevation change extent is greatly improved while maintaining the detail. The error of each pixel of the reconstruction results is less than 0.14, the MAE is less than 2×10^{-3} and the RMSE of the pixels of all predicated results is 4.2×10^{-3}, which indicates great improvement in the reconstruction. This experiment verifies the necessity of using coarse DEM as control information, indicating that by adding coarse DEM in the reconstruction network, the low-frequency part can be preserved and noise can be reduced through the reconstruction process.

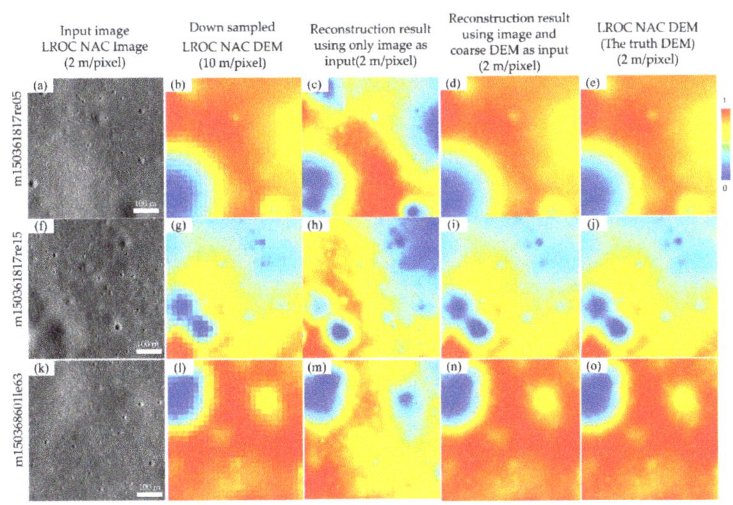

Figure 7. Comparisons of image reconstruction results with and without the help of coarse DEM. Each row shows results of one area. Column 1: The input image of LROC NAC image of 2 m/pixel. Column 2: The coarse DEM obtained by downsampling LROC NAC DEM from 2 m/pixel to 10 m/pixel. Column 3: The reconstructed result taking only image as input. Column 4: The reconstructed result using both image and coarse DEM as input. Column 5: The ground truth DEM of LROC NAC DEM (2 m/pixel).

4.3.2. Influence of Light Conditions

Similar to the traditional SFS method, the foundation of recovering elevation information from images through LDEMGAN is image photometric content changes, such as diverse intensity values due to reflection of lights from the topographic relief of the lunar surface. Under different lighting conditions, such as solar altitude angle, the imaged landforms of the lunar surface vary a lot, especially in the regions with severe elevation changes. Although during the training process, images taken under multiple solar altitude angles are considered to build the training dataset, relatively extreme cases, such as images captured at too high or too low solar altitude angles, are excluded to ensure network convergence. Therefore, in this subsection, elevation reconstructions are carried out on the images captured under different lighting conditions. The tested images also include some with solar altitude angles that are not included during reconstruction network training. In this way, how different illumination conditions affect reconstruction results can be analyzed.

We selected 29 images captured at different solar altitude angles (see Table 4 for details), combined their SLDEM covering the same area as input, and reconstructed DEMs through LDEMGAN established in Section 4.1, and part of the test results are shown in Figure 8. On the whole, the reconstruction results (Figure 8e–h) of each image captured at different solar altitude angles (Figure 8a–d) are visually consistent, and the maximum RMSE of all reconstructed results is 3.27 m, which shows the network can adapt to the change in lighting conditions to maintain the consistency of reconstruction results.

Table 4. Solar altitude angle information of the experimental images used for testing influence of light conditions.

ID	Image	Solar Altitude Angle (°)	ID	Image	Solar Altitude Angle (°)
1	m1171733209le	0.22	16	m1159956344re	43.01
2	m1277607697le	5.04	17	m1129325970re	44.44
3	m102014464re	8.24	18	m1114014396re	46.40
4	m102000149re	10.25	19	m1114007294re	47.39
5	m1126986303re	15.50	20	m1121074527re	48.91
6	m1157600009re	16.36	21	m1121081627re	49.94
7	m1123426954re	21.27	22	m1197661005le	58.69
8	m1169345323re	23.45	23	m1282310415le	60.55
9	m1123448288le	24.38	24	m106719774le	62.77
10	m1350328511le	28.86	25	m1192915976re	71.81
11	m113799518re	35.79	26	m1118716779re	76.14
12	m104362199re	36.03	27	m1116358477le	76.28
13	m129133239re	36.87	28	m1149338748re	76.90
14	m1129347305re	41.49	29	m1254093845le	80.50
15	m1129340193re	42.47			

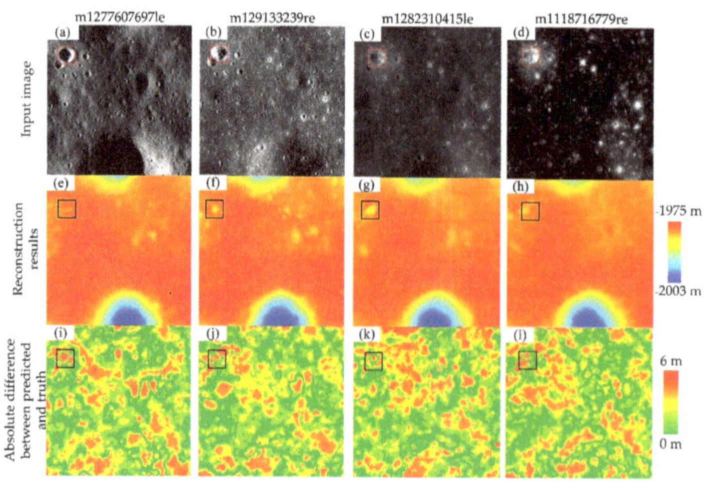

Figure 8. Comparison of reconstruction results using images captured at different solar altitude angles. Each column shows one result of one image. Row 1: the input image of LROC NAC image, solar altitude angle: m1277607697le, 5.04°; m129133239re, 36.87°; m1282310415le, 60.55°; m1118716779re, 76.14°. Row 2: the reconstruction results. Row 3: the difference between the reconstruction results (2 m/pixel) and the LROC NAC DEM (ground truth, 2 m/pixel).

On the local scale, when the SLDEM due to the resolution limitation is also unable to provide effective information of the region to network, too high or too low solar altitude angle can lead to a decline in reconstruction accuracy. Especially, impact craters that cannot be shown in SLDEM (the region marked in red in image and in black in reconstruction results) have a noticeable difference. When the solar altitude angle is lower than 25° (as shown in Figure 8a, m1277607671e, solar altitude angle 5.04°), the RMSE of the marked region is 5.88 m. When the image brightness (gray scale value) of the region is close to 0, the reconstruction network could be unable to translate much effective information from the image. This results in a large reconstruction error in this region (the RMSE is 5.88 m). When the solar altitude angle is higher than 70° (as shown in Figure 8d, m1118716779re, solar altitude angle 76.14°), the RMSE of the marked region is 6.23 m, and the RMSE of the whole image is 4.48 m. The high solar altitude angle causes overexposure of the image. In this kind of image, small impact craters in the image are bright entirely or mostly scattered on the edge without distinguishable intensity changes, which leads to a certain degree of increase in reconstruction error.

According to the experiment results shown and analyzed in this subsection, it is proved that the network can be applied to different lighting conditions through the training method described in Section 4.1, and the different elevation restoration results caused by diverse lighting conditions have limited impact on the reconstruction results with acceptable errors.

4.3.3. Discussion of Experimental Results

In terms of network structure, the generator network of our proposed LDEMGAN adopts the U-Net structure, which takes both image and coarse DEM as input. The coarse DEM, providing low-frequency information in the network, is taken as control information for depth reconstruction (this is confirmed by the experiments in Section 4.3.1). However, this is based on the assumption that the input image and DEM are accurate and their spatial positions are consistent, which makes the accuracy of the reconstruction result affected by the quality and spatial consistency of the input data. In other words, the offset and systematic error in the spatial position of the input data will lead to unsatisfactory reconstruction results. In this work, as registration error is less than one grid spacing of the coarse DEM, the network is able to handle this error level, yet more research is needed to be done concerning precise registration and systematic error elimination in future work.

The input of LDEMGAN is a single image and a low-resolution DEM, which is not limited by stereo image coverage. The LDEMGAN is less affected by illumination changes (see Section 4.3.2 for details). In Section 4.2, the LROC NAC DEM as the ground truth was generated by stereophotogrammetry, and the RMSE and MAE in Table 2 are able to demonstrate that the DEM reconstructed by LDEMGAN is very close to the results reconstructed by stereophotogrammetry. Therefore, LDEMGAN is able to be used for DEM reconstruction of a wider area of the Moon compared to stereophotogrammetry. In addition, it is worth noting that the evaluation indicators MAE and RMSE used in this work are aimed at accuracy in a statistical sense, which can reflect overall reconstruction accuracy, but cannot reflect the spatial distribution of the error. In this paper, the spatial distribution of errors is mainly analyzed by the visual effect of the absolute difference plot (third column of Figure 5) and the profile of the subregion (fifth row of Figure 6). Because no obvious regularities affecting the reconstruction accuracy were found from the absolute difference maps or profile maps, the in-depth analysis was not continued in this paper. If the subsequent application requires an in-depth study of the anisotropy or spatial distribution law of the reconstruction effect, more relevant indicators will be needed for experiments and evaluation.

From Figure 6, it can be seen that the reconstruction results of the network in the area with large changes in elevation are not as accurate as that with small changes. We suppose that this may be related to the following reasons. (1) The elevation variation of the region where the training data was made is small, resulting in less ability of the network to learn to extract such informative features. (2) The scale recovery is carried out according to Equation (13). When two areas with identical mean value but different standard deviations (*sd*), e.g., greater degree of elevation change results in larger *sd* value, the one with larger *sd* value probably leads to larger errors according to this scale recovery method. (3) Scale recovery only depends on low-resolution DEM. The low-resolution DEM may have systematic and random errors that are biased from the real lunar surface. These errors will reflect on the reconstruction results through scale recovery. From the above discussion, in future work, we need to firstly improve the normalization and scale recovery mechanisms, and then add various training data so that the network may identify and learn the different degree of elevation change. In this case, systematic errors caused by data quality may be detected and corrected for better DEM reconstruction performance.

In subarea IV shown in Figure 6t, it seems that the reconstruction results are not affected by the errors of SLDEM, and the reconstruction results are close to the ground truth. The mean value of SLDEM in this subarea is close to the ground truth, but its standard deviation is large. We deduce that gradient loss L_{grad} included in the loss function in Equation (11) of LDEMGAN may take this credit of eliminating errors. When the network finds that the gradient information extracted from the image is inconsistent with the elevation change of SLDEM, it will trust the high-frequency information obtained from the image more.

5. Conclusions

In this paper, according to the current situation that there are high-resolution images and low-resolution DEMs covering the whole lunar surface, but few high-resolution stereo pairs, we propose a high-resolution lunar DEM reconstruction method based on GAN. This method can reconstruct a high-resolution DEM using a single high-resolution image and a much lower-resolution DEM in the corresponding area. The grid spacing of the reconstructed DEM is as high as that of the high-resolution input image.

The experimental results verify the feasibility and effectiveness of the proposed method and verify the necessity of using much coarser DEM taken as control information in the network, instead of only using it to recover absolute depth. The RMSE of the reconstruction results can reach about 2 m in the area where the degree of elevation variation is less than 100 m, and the RMSE value ranges from around 3 m to 10 m without considering the degree of the elevation variation in large area reconstruction. In addition, the reconstruction accuracy is proved stable under nonextreme illumination conditions (the solar altitude angle is between 25° and 75°). In the future, we will continue to conduct research on anisotropy of the reconstruction results, so as to further optimize and improve the network.

Due to the current and near-future data situation on the lunar surface, the proposed approach has great potential for building global high-resolution DEMs for the lunar surface to support more scientific and engineering applications, such as topographic characterization, landing site selection, navigation, positioning, path planning, decision support, etc.

Author Contributions: Conceptualization, K.D., Y.W. and Y.L.; methodology, Y.L. and Y.W.; software, Y.L. and Y.W.; validation, M.P., W.W. and Z.L.; formal analysis, Y.W.; investigation, K.D., Y.W. and M.P.; resources, W.W. and Z.L.; writing—original draft preparation, Y.L.; writing—review and editing, K.D. and Y.W; project administration, Y.W.; funding acquisition, K.D., Y.W. and M.P. All authors have read and agreed to the published version of the manuscript.

Funding: This work was supported by the National Key Research and Development Program of China (grant 2018YFB1305004) and the National Natural Science Foundation of China (grants 41941003 and 41771488).

Data Availability Statement: Publicly available datasets were analyzed in this study. The datasets can be found here: https://ode.rsl.wustl.edu/moon/.

Acknowledgments: The authors thank all those who worked on the NASA Planetary Data System archive to make the LROC imagery and SLDEM available.

Conflicts of Interest: The authors declare no conflict of interest.

References

1. Kokhanov, A.A.; Karachevtseva, I.P.; Zubarev, A.; Patraty, V.; Rodionova, Z.; Oberst, J. Mapping of potential lunar landing areas using LRO and SELENE data. *Planet. Space Sci.* **2018**, *162*, 179–189. [CrossRef]
2. Liu, Z.Q.; Di, K.C.; Peng, M.; Wan, W.; Liu, B.; Li, L.; Yu, T.; Wang, B.; Zhou, J.; Chen, H. High precision landing site mapping and rover localization for Chang'e-3 mission. *Sci. China-Phys. Mech. Astron.* **2015**, *58*, 1–11. [CrossRef]
3. Liu, Z.Q.; Di, K.C.; Li, J.; Xie, J.; Cui, X.; Xi, L.; Wan, W.; Peng, M.; Liu, B.; Wang, Y.; et al. Landing site topographic mapping and rover localization for Chang'e-4 mission. *Sci. China Inf. Sci.* **2020**, *63*, 140901. [CrossRef]
4. Wu, B.; Li, F.; Ye, L.; Qiao, S.; Huang, J.; Wu, X.; Zhang, H. Topographic modeling and analysis of the landing site of Chang'E-3 on the Moon. *Earth Planet. Sci. Lett.* **2014**, *405*, 257–273. [CrossRef]
5. Wu, B.; Li, F.; Hu, H.; Zhao, Y.; Wang, Y.; Xiao, P.; Li, Y.; Liu, W.C.; Chen, L.; Ge, X.; et al. Topographic and Geomorphological Mapping and Analysis of the Chang'E-4 Landing Site on the Far Side of the Moon. *Photogramm. Eng. Remote Sens.* **2020**, *86*, 247–258. [CrossRef]
6. Di, K.; Oberst, J.; Karachevtseva, I.; Wu, B. Topographic Mapping of The Moon in the 21st Century: From Hectometer to Millimeter Scales. *ISPRS J. Photogramm. Remote Sens.* **2020**, *XLIII-B3-2020*, 1117–1124. [CrossRef]
7. Smith, D.E.; Zuber, M.T.; Jackson, G.B.; Cavanaugh, J.F.; Neumann, G.; Riris, H.; Sun, X.; Zellar, R.S.; Coltharp, C.; Connelly, J.; et al. The Lunar Orbiter Laser Altimeter Investigation on the Lunar Reconnaissance Orbiter Mission. *Space Sci. Rev.* **2010**, *150*, 209–241. [CrossRef]
8. Haruyama, J.; Hara, S.; Hioki, K.; Iwasaki, A.; Morota, T.; Ohtake, M.; Matsunaga, T.; Araki, H.; Matsumoto, K.; Ishihara, Y.; et al. Lunar Global Digital Terrain Model Dataset Produced from SELENE (Kaguya) Terrain Camera Stereo Observations. In Proceedings of the Lunar & Planetary Science Conference, Woodlands, TX, USA, 19–23 March 2012.
9. Li, C.L.; Ren, X.; Liu, J.J.; Wang, F.F.; Wang, W.R.; Yan, W.; Zhang, G.H. A New Global and High Resolution Topographic Map Product of the Moon from Chang'e-2 Image Data. In Proceedings of the Lunar & Planetary Science Conference, Woodlands, TX, USA, 16–20 March 2015.
10. Barker, M.K.; Mazarico, E.M.; Neumann, G.; Zuber, M.; Haruyama, J.; Smith, D. A new lunar digital elevation model from the Lunar Orbiter Laser Altimeter and SELENE Terrain Camera. *Icarus* **2016**, *273*, 346–355. [CrossRef]
11. Wu, B.; Liu, W.C.; Grumpe, A.; Wöhler, C. Construction of pixel-level resolution DEMs from monocular images by shape and albedo from shading constrained with low-resolution DEM. *J. Photogramm. Remote Sens.* **2018**, *140*, 3–19. [CrossRef]
12. Henriksen, M.R.; Manheim, M.R.; Burns, K.; Seymour, P.; Speyerer, E.; Deran, A.; Boyd, A.; Howington-Kraus, E.; Rosiek, M.; Archinal, B.; et al. Extracting accurate and precise topography from LROC narrow angle camera stereo observations. *Icarus* **2017**, *283*, 122–137. [CrossRef]
13. Kirk, R.L.; Barrett, J.M.; Soderblom, L.A. Commission IV, Working Group IV/9. Photoclinometry Made Simple ... ? In Proceedings of the "Advances in Planetary Mapping 2003" Workshop, Houston, TX, USA, 22 March 2003.
14. Lohse, V.; Heipke, C. Multi-image shape-from-shading: Derivation of planetary digital terrain models using clementine images. *Remote Sens.* **2004**, *35*, 828–833.
15. Grumpe, A.; Belkhir, F.; Wöhler, C. Construction of lunar DEMs based on reflectance modelling. *Adv. Space Res.* **2014**, *53*, 1735–1767. [CrossRef]
16. Horn, B.K. Understanding Image Intensities. *Artif. Intell. Rev.* **1977**, *8*, 201–231. [CrossRef]
17. Zhang, R.; Tsai, P.S.; Cryer, J.; Shah, M. Shape from shading: A survey. *IEEE Trans. Pattern Anal. Mach. Intell.* **1999**, *21*, 690–706. [CrossRef]
18. Barron, J.T.; Malik, J. High-Frequency Shape and Albedo from Shading using Natural Image Statistics. In Proceedings of the IEEE Conference on Computer Vision and Pattern Recognition (CVPR), Colorado Springs, CO, USA, 20–25 June 2011.
19. Danzl, R.; Scherer, S. Integrating Shape from Shading and Shape from Stereo for Variable Reflectance Surface Reconstruction from SEM Images. In Proceedings of the 26th Workshop of the Austrian Association for Pattern Recognition, Graz, Austria, 10 September 2002; pp. 281–288.
20. Chen, Z.Y.; Wu, B.; Liu, W. Mars3DNet: CNN-Based High-Resolution 3D Reconstruction of the Martian Surface from Single Images. *Remote Sens.* **2021**, *13*, 839. [CrossRef]
21. Khan, F.; Salahuddin, S.; Javidnia, H. Deep Learning-Based Monocular Depth Estimation Methods-A State-of-the-Art Review. *Sensors* **2020**, *20*, 2272. [CrossRef] [PubMed]
22. Eigen, D.; Puhrsch, C.; Fergus, R. Depth Map Prediction from a Single Image using a Multi-Scale Deep Network. In Proceedings of the 28th Conference on Neural Information Processing Systems (NIPS), Montreal, QC, Canada, 8–13 December 2014.
23. Eigen, D.; Fergus, R. Predicting Depth, Surface Normals and Semantic Labels with a Common Multi-Scale Convolutional Architecture. In Proceedings of the 2015 IEEE International Conference on Computer Vision, Santiago, Chile, 7–13 December 2015; pp. 2650–2658.
24. Laina, I.; Rupprecht, C.; Belagiannis, V.; Tombari, F.; Navab, N. Deeper Depth Prediction with Fully Convolutional Residual Networks. In Proceedings of the 2016 Fourth International Conference on 3D Vision (3DV), Stanford, CA, USA, 25–28 October 2016; pp. 239–248.

25. Ma, X.; Geng, Z.; Bie, Z. Depth Estimation from Single Image Using CNN-Residual Network. Semantic Scholar. Available online: http://cs231n.stanford.edu/reports/2017/pdfs/203.pdf (accessed on 21 July 2021).
26. Fu, H.; Gong, M.M.; Wang, C.; Batmanghelich, K.; Tao, D. Deep Ordinal Regression Network for Monocular Depth Estimation. In Proceedings of the IEEE Conference on Computer Vision and Pattern Recognition, Salt Lake City, UT, USA, 18–23 June 2018; pp. 2002–2011.
27. Facil, J.M.; Ummenhofer, B.; Zhou, H.; Montesano, L.; Brox, T.; Civera, J. CAM-Convs: Camera-Aware Multi-Scale Convolutions for Single-View Depth. In Proceedings of the IEEE/CVF Conference on Computer Vision and Pattern Recognition, Long Beach, CA, USA, 15–20 June 2019; pp. 11826–11835.
28. Wolk, D.; Ma, F.C.; Yang, T.-J.; Karaman, S.; Sze, V. FastDepth: Fast Monocular Depth Estimation on Embedded Systems. In Proceedings of the 2019 International Conference on Robotics and Automation (ICRA), Montreal, QC, Canada, 20–24 May 2019; pp. 6101–6108.
29. Goodfellow, I.; Pouget-Abadie, J.; Mirza, M.; Xu, B.; Warde-Farley, D.; Ozair, S.; Courville, A.; Bengio, Y. Generative Adversarial Networks. *Commun. Acm.* **2020**, *63*, 139–144. [CrossRef]
30. Jung, H.; Kim, Y.; Min, D.; Oh, C.; Sohn, K. Depth prediction from a single image with conditional adversarial networks. In Proceedings of the 2017 IEEE International Conference on Image Processing (ICIP), Beijing, China, 17–20 September 2017; pp. 1717–1721.
31. Ronneberger, O.; Fischer, P.; Brox, T. U-Net: Convolutional Networks for Biomedical Image Segmentation. In *Medical Image Computing and Computer-Assisted Intervention—MICCAI 2015*; Springer: Cham, Switzerland, 2015; Volume 9351, pp. 234–241.
32. Li, J.; Jia, J.J.; Xu, D. Unsupervised Representation Learning of Image-Based Plant Disease with Deep Convolutional Generative Adversarial Networks. In Proceedings of the Chinese Control Conference (Ccc), Wuhan, China, 25–27 July 2018; pp. 9159–9163.
33. Lore, K.G.; Reddy, K.; Giering, M.; Bernal, E.A. Generative adversarial networks for depth map estimation from RGB video. In Proceedings of the IEEE/CVF Conference on Computer Vision and Pattern Recognition (CVPR), Salt Lake City, UT, USA, 18–22 June 2018.
34. Tao, Y.; Muller, J.P.; Conway, S.J.; Xiong, S. Large Area High-Resolution 3D Mapping of Oxia Planum: The Landing Site for the ExoMars Rosalind Franklin Rover. *Remote Sens.* **2021**, *13*, 3270. [CrossRef]
35. Tao, Y.; Muller, J.P.; Xiong, S.; Conway, S.J. MADNet 2.0: Pixel-Scale Topography Retrieval from Single-View Orbital Imagery of Mars Using Deep Learning. *Remote Sens.* **2021**, *13*, 4220. [CrossRef]
36. Hao, C.; Xuanyu, H.; Oberst, J. Pixel-Resolution DTM Generation for The Lunar Surface Based on a Combined Deep Learning and Shape-From-Shading (SFS) Approach. In Proceedings of the ISPRS Annals of the Photogrammetry, Remote Sensing and Spatial Information Sciences, Nice, France, 6–11 June 2022; Volume V-3-2022, pp. 511–516.

remote sensing

MDPI

Article

Diverse Geological Evolution of Impact Basins on the Moon

Jiayin Deng [1,2], Weiming Cheng [1,2,3,4,*] and Yimeng Jiao [1,2]

1 State Key Laboratory of Resources and Environmental Information System, Institute of Geographic Sciences and Natural Resources Research, Chinese Academy of Sciences, Beijing 100101, China
2 University of Chinese Academy of Sciences, Beijing 100049, China
3 Jiangsu Centre for Collaborative Innovation in Geographical Information Resource Development and Application, Nanjing 210023, China
4 Chinese Academy of Sciences, Center for Excellence in Comparative Planetology, Hefei 230052, China
* Correspondence: chengwm@lreis.ac.cn

Abstract: Impact basins are the dominant landforms on the lunar surface, and their geological evolution varies. This research studied the diversity in the geological evolution of three impact basins: the Dirichlet–Jackson Basin, the *Nectaris* Basin, and the *Orientale* Basin. First, the regional topography and geomorphology of the three basins were studied using the SLDEM2015 digital elevation model (DEM). Clementine ultraviolet–visible (UVVIS) data and Moon Mineralogy Mapper (M3) data were used to study the chemical composition and mineralogical composition of the three basins. Additionally, the lunar crust thickness data have been used to study the subsurface structure of the three basins. The topographical analogies of the three basins indicate that the shapes of the basins are cavity-like. However, the shape of the Dirichlet–Jackson basin is not an obvious cavity compared with the other basins. The positions with minimum and maximum crustal thickness of the three basins are located at the center and the rim. The uplift of the crust-mantle interface of the *Nectaris* Basin and *Orientale* Basin is relatively larger than in the Dirichlet–Jackson Basin. Below the center of the maria of the *Nectaris* Basin and *Orientale* Basin, collapses occurred at the crust–mantle interface. The concentrations of FeO and TiO_2 in the non-mare formation of the basin and maria show expected bimodal distributions. Moreover, we found exposures of olivine-rich materials in the *Nectaris* Basin and *Orientale* Basin which are located in the Rosse and Maunder craters, respectively. These exposures of olivine may be explained by the fact that the formation of the large impact basin, which might penetrate and blast away the upper lunar crust, excavating deep-seated material.

Keywords: moon; impact basins; geological features; chemical composition; mineralogical composition

1. Introduction

Basin-forming impact events are one of the major geologic processes that occurred during the evolution of the Moon, and they record the history of meteorite bombardment in the early Solar System. Impact basins are the dominating topographic and morphological features [1–3] that were formed by large-scale impact events, which represent major exogenous resurfacing episodes on the Moon. During the late stage of lunar history, many impact basins were filled with volcanic materials (mostly basalts) to form the maria, which reflects the evolution representing the major endogenic geological processes of moon resurfacing. The formation of impact basins was a key process during the initial periods of geological evolution on the Moon, which affects both the lunar mantle and crust and is responsible for the formation of megaregolith [4,5]. Recently, NASA, JAXA, and ISRO missions, Lunar Reconnaissance Orbiter (LRO), Kaguya, Chandrayaan-1, and the Gravity Recovery and Interior Laboratory (GRAIL) have provided high-resolution images and topographic, hyperspectral, and gravity data, considerably enhancing our ability to understand the geological evolution of the Moon.

Lunar impact basins have been identified, catalogued, and studied for a long time [2,6–9]. For example, Spudis [6] studied the geology and composition of important basin deposits, including the *Orientale* and *Nectaris* basins, and systematically summarized the geological characteristics of another 45 lunar impact basins. Thus far, studies of the lunar basins have mostly focused on the structure of individual basins or considered the compositional characteristics of the basins [10–12]. In our research, we conducted a comparative analysis of three lunar basins, Dirichlet–Jackson, *Nectaris*, and *Orientale*, which were formed at different moments of early lunar history. The Dirichlet–Jackson Basin was formed earlier in the pre-Nectarian period, and its rim has been considerably degraded since its formation. The formation of the *Nectaris* Basin marked the beginning of the Nectarian period [1], and the *Orientale* Basin is the youngest multi-ring impact basin on the Moon [6].

Our first goal was to study the topographical and morphological characteristics of these basins. Our second goal was to analyze the chemical composition and mineralogy of a variety of associated materials. The mineralogy of the lunar surface is relatively simple, and the absorbance properties of lunar minerals and associated glasses are dominated by only a few elements. Thus, in our study, we focused on the contents of FeO and TiO_2 in the basin regions. Our third goal was to highlight analogies and differences among the basins of different ages and locations. Based on our results, we analyzed the diversity among the three basins and propose a possible explanation for the diversity.

2. Geological Overview of the Studied Basins

The lunar stratigraphy consists of five major time units, periods, from oldest to youngest: pre-Nectarian, Nectarian, Imbrian, which includes the Early Imbrian and Upper Imbrian epochs, Eratosthenian, and Copernican [1]. Stöffler et al. [13] derived a revised and time-calibrated lunar stratigraphy. In this study, we adhered to this terminology (Figure 1a). The basins which we studied (Figure 1b) characterize the pre-Nectarian period (Dirichlet–Jackson Basin) period and mark the beginnings of the Nectarian (*Nectaris* Basin) and Imbrian (*Orientale* Basin) periods.

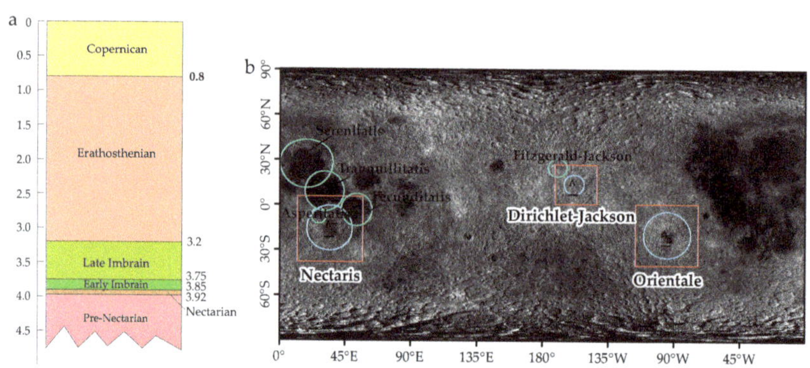

Figure 1. (**a**) Geologic time scale of the Moon revised by Stöffler et al. [13]. (**b**) LRO WAC Global Mosaic (resolution 100 m/pixel) showing the locations of three study areas surrounding the selected basins. The map is displayed in simple cylindrical projection.

The Dirichlet–Jackson Basin (Figure 2a) is located on the far side of the Moon. Using the Bouguer anomaly data of the GRAIL mission, Neumann et al. [9] validated the existence of the Dirichlet–Jackson Basin. More recently, Satya Kumar et al. [14] analyzed the gravity anomalies generated by the GRAIL mission in detail and studied the sub-surface structure of the Dirichlet–Jackson Basin. The basin has no mare filling, and its surface is covered by ejecta from the Fitzgerald–Jackson basin and subsequent impact craters.

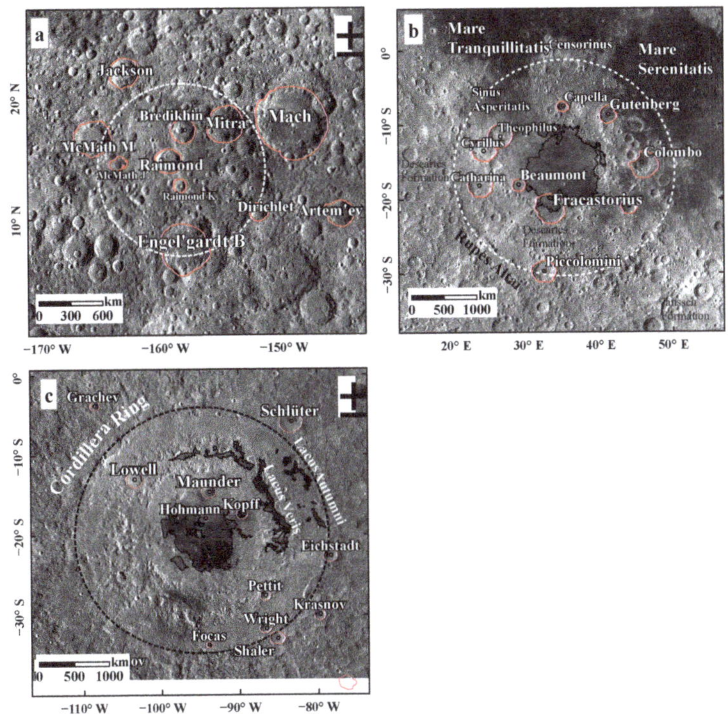

Figure 2. WAC images of the three basins in Mercator projection. (**a**) Map of Dirichlet–Jackson Basin centered at 158.2°W, 13.4°N. (**b**) Map of Nectaris Basin centered at 35.1°E, 15.6°S. (**c**) Map of Orientale Basin centered at 94.8°W, 20.1°S. The red lines show the craters labeled in the maps.

The *Nectaris* Basin (Figure 2b) is on the nearside of the Moon and is located at 35.1°E, 15.6°S, south of *Mare Tranquillitatis*, and west of *Mare Fecunditatis*. Pike and Spudis [15] classified the *Nectaris* Basin as a multi-ring basin. The western portion of the rim of the basin is marked by a scarp of Rupes Altai, and Janssen Formation [6,16] and Descartes Formation characterize the southeastern and western regions adjacent to the basin, respectively. The Janssen Formation is exposed southeast of the basin, which extends outward from the basin rim (Figure 2b). The Janssen Formation is similar to the Fra Mauro Formation of the Imbrium Basin and the Hevelius Formation of the *Orientale* Basin, all of which consist of continuous deposits [17]. The Descartes Formation is exposed to the west of the basin, the knobby deposits of which resemble materials of the Apenninus Formation of the Imbrium Basin.

The *Orientale* Basin (Figure 2c) is centered at 94.8°W, 20.1°S, southwest of *Oceanus Procellarum*, and represents the lunar youngest and best-preserved multiring basin [10,18,19]. The basin is located in the transitional region between the highlands and the *Maria*, which is known as the boundary between thin crust and thick crust [20]. There are three ejecta units associated with the basin: the Maunder Formation, the Montes Rook Formation, and the Hevelius Formation. The Maunder Formation consists of light smooth plains and represents the melt sheet produced by the basin-forming impact [18]. The Montes Rook formation is a knobby unit mainly located between the Outer Rook Mountains and the Cordillera Mountains [21]. The Hevelius Formation is the main ejecta blanket of *Orientale*, which represents the materials ejected outside the basin. Recently, Yue et al. [22] redefined the absolute model age of the *Orientale* Basin and showed that the *Orientale* Basin was likely formed approximately 3.80 Ga ago.

3. Datasets and Methodology

We conducted a study of the three basins on the new datasets provided by the following instruments: the Clementine UVVIS spectrometer, Lunar Reconnaissance Orbiter Wide Angle Camera (LROC-WAC) (http://pds.lroc.asu.edu/data/LRO-L-LROC-5-RDR-V1.0/LROLRC_2001/DATA/BDR/WAC_GLOBAL/, accessed on 30 August 2020), Lunar Orbiter Laser Altimeter (LOLA), SELenological and Engineering Explorer Terrain Camera (SELENE TC), and Moon Mineralogy Mapper (M3). In this study, we used LROC-WAC (resolution 100 m/pixel) [23] to study the geomorphology of the three basins. The maps of topography obtained by LOLA and SELENE TC merged Digital Elevation Model (SLDEM2015) were used to analyze the morphology of the three basins. The spatial resolution of SLDEM data is about 60 m/pixel and the vertical accuracy of SLDEM data is ~3–4 m [24]. The SLDEM can be downloaded from http://darts.isas.jaxa.jp/planet/pdap/selene/, accessed on 18 February 2021. Spectral data taken by the Clementine mission [25] and hyperspectral data of M^3 onboard Chandrayaan-1 spacecraft were used to study the FeO and TiO_2 distributions and mineralogy of the three basins. The UVVIS Clementine data can be found at NASA PDS and Derived Products (https://astrogeology.usgs.gov/search/map/Moon/Clementine/UVVIS/Lunar_Clementine_UVVIS_WarpMosaic_5Bands_200m, accessed on 30 August 2020). The M3 data are available at PDS Geosciences Node (https://pdsimage2.wr.usgs.gov/Missions/Chandrayaan_1/M3/CH1M3_0004/, accessed on 4 February 2021). The Clementine UVVIS camera has five spectral channels with a resolution of 115 m/pixel. The wavelength ranges from 415 to 1000 nm [25]. The M3 data have 86 spectral bands, from visible to near-infrared, with wavelengths ranging from 0.42 to 3.0 μm, and its spectral resolution is 20 or 40 nm [26,27]. The global modes of M3 data exhibit two optical periods (OP) and provide data with five different sets of observations, as discussed by Boardman et al. [26] and Besse et al. [28]. The Level 2 data from the instrument's Optical Period 2c were used to analyze the mineralogy of the three basins, and the reflectance data were thermally, geometrically, and photometrically corrected [27,29]. The resolution of the Level 2 data from the instrument's Optical Period 2c is approximately 280 m/pixel [30]. In addition, to estimate the major features of the lunar interiors in regions surrounding the study basins, we used data from the updated version of the crust thickness map, which is based on the GRAIL extended mission gravity model JGGRAIL_900C11A [31].

3.1. Topographic Analyses

We constructed the DEMs for the Dirichlet–Jackson, *Nectaris*, and *Orientale* basins using the SLDEM digital elevation data and conducted quantitative topographic analyses for the basins. We extracted the elevation profile of the three basins from A to A′ and determined the diameters of the rings of the basins using the CraterTools extension of ArcGIS [32]. As complementary information to the topographic characteristics of the basins, we calculated the surface slopes in regions surrounding the basins at a 118 m baseline based on the SLDEM data. The slope variations may help to emphasize processes acting over geologic time [33].

3.2. Composition Analysis

Several algorithms have been proposed to derive the content of FeO and TiO_2 [34–37]. The FeO and TiO_2 abundances in the regions of the studied basins were calculated according to the algorithms proposed by Lucey et al. [36]. The algorithms are defined as:

$$\mathrm{FeO(wt.\%)} = -17.427 \times arctan(\frac{\frac{R_{950}}{R_{750}} - 1.19}{R_{750} - 0.08}) - 7.565 \quad (1)$$

$$\mathrm{TiO_2(wt.\%)} = 3.708 \times (arctan(\frac{\frac{R_{415}}{R_{750}} - 0.42}{R_{750}}))^{5.979} \quad (2)$$

where R_λ is the reflectance value for a given wavelength λ.

To analyze the compositions of the different landforms of the Dirichlet–Jackson, *Nectaris*, and *Orientale* basins, sets of randomly distributed points (pixels) were sampled according to the area of three basins. In order to determine representative samples, the number of points was proportional to the areas of each landform. In total, 200,000 sample points (pixels) were randomly extracted from the *Nectaris* Basin and *Orientale* Basin, and 60,000 sample points (pixels) were randomly extracted from Dirichlet–Jackson Basin. This method was successfully applied to the study of the northeast part of the South Pole-Aitken (SPA) basin [38], and the number of measurement points was sufficiently large to provide a representative sample of either FeO or TiO_2 content. Using the results of a previous study, we mapped the sub-formations of the three basins and analyzed the FeO and TiO_2 contents in the different sub-formation units.

3.3. Mineralogy Analysis

Different minerals have various spectral absorption characteristics (Figure 3). Here, we used the hyperspectral data from M3 to study variations in the mineralogical composition of the surface in the Dirichlet–Jackson, *Nectaris*, and *Orientale* basin regions. There is no spatial information of the OP2C data of M3. Therefore, in the data processing, the spatial information of M3 1B level data was coupled with level 2 data so that level 2 data had spatial information. Twelve OP2C data strips covering the Dirichlet–Jackson Basin were mosaiced into a single image. We repeated the same data processing for the *Nectaris* and *Orientale* basins.

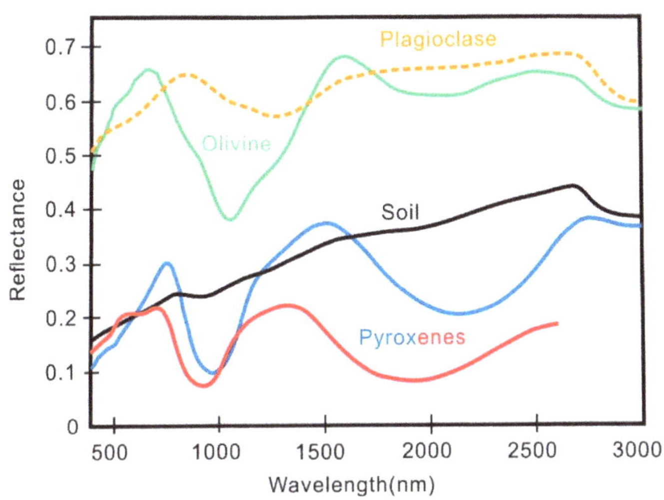

Figure 3. Laboratory reflectance spectra of representative lunar minerals and regolith of M3 data (modified from Pieters et al. [30]).

To reduce the signal-to-noise ratio of spectral data and comprehensively describe the basic mineralogical characteristics (e.g., the abundance of mafic silicates and soil maturity of lunar regolith, members of the M3 science team have developed several mineral indicator parameters. Here, we used the integrated band depth (IBD) to analyze and capture the fundamental mineralogical properties of the surfaces in the basin regions because the IBD is very sensitive to the chemical characteristics of minerals, which can more intuitively highlight the absorption characteristics of minerals [39]. The IBD is defined as the integral of the band depth within the spectral range of the absorption spectrum [40,41]. Here, we used the crystal field absorptions at 1000 nm, and 2000 nm, and a reflectance of 1580 nm, to synthesize RGB false-color images, where RGB is the integrated band depth around

2000 nm (red), 1000 nm (green), and reflectance at 1580 nm (blue), respectively. Using this method, the mineral absorption characteristic parameter images of three basins were obtained to characterize the mineralogical diversity in the basin regions [19]. The definitions of the integrated band depths for 1000 nm and 2000 nm are as follows [41]:

$$\mathrm{IBD}(1000\ \mathrm{nm}) = \sum_{\mathrm{n}=0}^{\mathrm{n}=26} \left(1 - \frac{R(789 + 20n)}{R_c(789 + 20n)}\right) \tag{3}$$

$$\mathrm{IBD}(2000\ \mathrm{nm}) = \sum_{\mathrm{n}=0}^{\mathrm{n}=21} \left(1 - \frac{R(1658 + 40n)}{R_c(1658 + 40n)}\right) \tag{4}$$

where R is the reflectance value for a given wavelength and R_c is the reflectance value after removing the continuum [19,41]. The wavelengths at which the calculation starts are 789, and 1658, and 20 and 40 are the calculation intervals.

4. Results

4.1. Photogeological Analysis

In this study, the SLDEM topography maps were drawn in the Mercator projection, which causes little distortion in low-latitude zones where all three studied basins are situated. We constructed topographic profiles based on the SLDEM to study the configuration of the Dirichlet–Jackson, *Nectaris*, and *Orientale* basins (Figure 4). The SLDEM topography map (Figure 4) indicates distinct differences in morphological characteristics among the three basins.

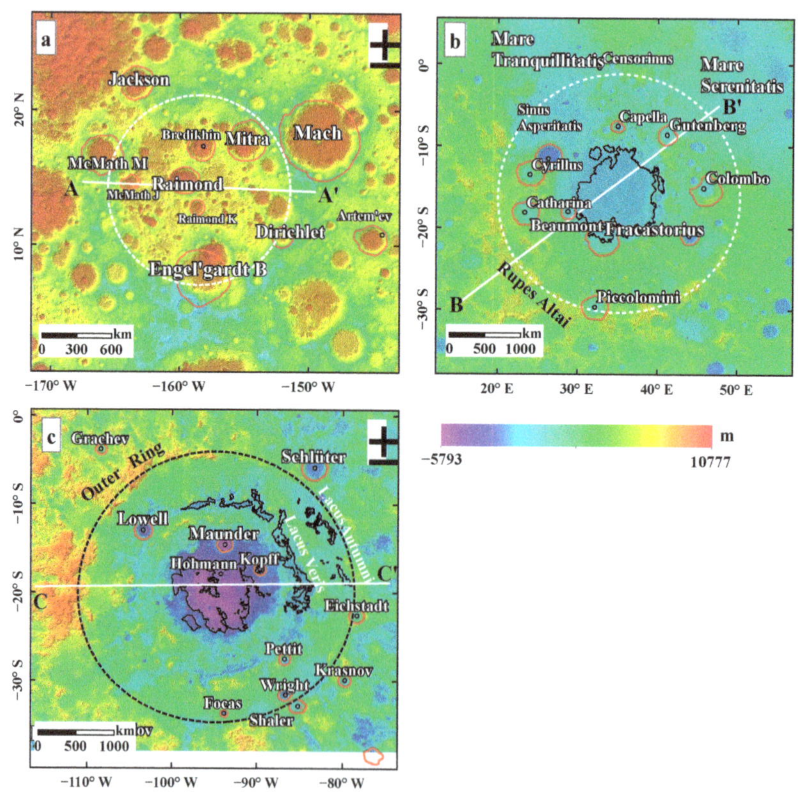

Figure 4. *Cont.*

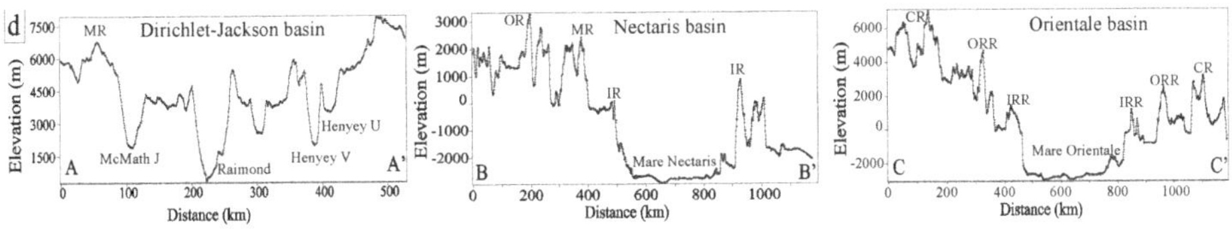

Figure 4. Topography map of Dirichlet–Jackson Basin, *Nectaris* Basin, and *Orientale* Basin and their surrounding areas (based on SLDEM data). Major rings of the basins are labeled by dashed lines. The red lines show the craters labeled in the maps. (**a**) Dirichlet–Jackson Basin. (**b**) *Nectaris* Basin. (**c**) *Orientale* Basin. (**d**) Topographic profiles across the three basins derived from SLDEM elevation data.

The Dirichlet–Jackson Basin (Figure 4a) is located in the lunar farside highlands. The topographic profile (AA′, Figure 4d) shows that the basin floor is not flat because it is distorted by numerous large craters. The diameter of the Dirichlet–Jackson Basin is about 430 km (Table 1), roughly half the diameter of the *Nectaris* Basin. The average elevations for the floor and rim crests are approximately 4.1 km and 6.3 km from the 12 radical profiles 30° apart, respectively. Thus, the mean depth of the basin is approximately 3.8 km. The most prominent topographic features within the Dirichlet–Jackson Basin are impact craters, some of which are 2–4.5 km deep. The depth of the Raimond crater is larger than the Dirichlet–Jackson Basin, possibly because the Raimond crater formed after the Dirichlet–Jackson Basin, and the basin underwent later modification and degradation after its formation.

Table 1. Mean FeO contents in the sub-formations of the three basins.

Basin Sub-Formations	Mean FeO Content (STD) [1] of Dirichlet–Jackson	Mean FeO Content (STD) of *Nectaris*	Mean FeO Content (STD) of *Orientale*
Mare	–	12.95 (1.80)	11.75 (2.81)
Floor formation	8.64 (1.44)	7.30 (2.55)	4.57 (2.19)
Peak-ring formation	8.76 (1.19)	6.28 (2.41)	4.22 (1.99)
Wall formation	8.19 (1.42)	6.49 (2.05)	4.26 (1.54)
Rim formation	9.12 (1.31)	6.42 (2.15)	4.92 (1.28)

[1] STD is the value of standard deviation.

The *Nectaris* Basin (Figure 4b) was partly flooded with mare basalts from the north. Both the SLDEM topography map and the BB′ topographic profile (Figure 4d) shows that the surface of the *Mare Nectaris* is at a relatively low elevation and exhibits a flatter topography. The B–B′ profile clearly shows the positions of the inner and outer rings of the basin. The average elevation of the floor and the outer rim crest are approximately 2.8 km and 0.3 km from the 12 radical profiles 30° apart, respectively, and the depth of the *Nectaris* basin is about 3.1 km. The topographic map and the profile across the *Nectaris* Basin (B–B′) show that the northwestern portion of the basin, which is near the Serenitatis basin, is significantly lower, and the bottom of the basin is relatively flat. The outer ring of *Nectaris* Basin is relatively well preserved but appears breached to the northwest of the *Sinus Asperitatis* (Figure 4b). The northwest part of the inner ring is missing probably due to the filling of basalt from the North. In places where the inner ring is prominent, the height reaches ~3–3.5 km above the mare surface and the diameter of the inner ring is approximately 440 km. The outer ring forms Rupes Altai in its southwestern segment, with a diameter of ~880 km (Table 1).

The *Orientale* Basin (Figure 4c) is the youngest multi-ring basin. The CC′ profile in Figure 4d shows the position of the Inner Rook Ring, the Outer Montes Rook, and the Cordillera Ring. The topography of the *Orientale* Basin is asymmetrical, as shown by the elevation profile in Figure 4d. The topography in the west of the *Orientale* Basin is significantly higher than that in the east, which is likely because the *Orientale* Basin is located in the transitional region between the highlands and the mare. The average elevations for the floor and the rim crest in the Cordillera Ring are approximately −2.8 km and 2.3 km, respectively; the depth of the northeastern portion of the basin and its surroundings is significantly higher than the terrains at the eastern section of the basin. The diameters of the Inner Rook Ring (IRR) and Outer Rook ring are approximately ~490 km and ~690 km, respectively. The Cordillera Ring (CR), the outermost topographical rim of the *Orientale* basin, is approximately 920 km in diameter.

4.2. Local Crustal Characteristics

Wieczorek et al. [31] used the LOLA topographic and the GRAIL gravity data to construct models of crust thickness and relief of the crust–mantle boundary ("Moho") on the Moon. These models provide important data for study of the subsurface structure of impact basins.

Figure 5 shows that the crustal thickness in the region of the Dirichlet–Jackson Basin varies from 28 km to 69 km, with an average of 56 km. The crust–mantle interface profile shows that the minimum and maximum thickness of the crust in the Dirichlet–Jackson Basin region is located near the craters Raimond and Henyey U, respectively (Figure 5). According to the crust–mantle interface profile across the center of the basin, the crust–mantle interface in the basin is only slightly uplifted, approximately 20 km.

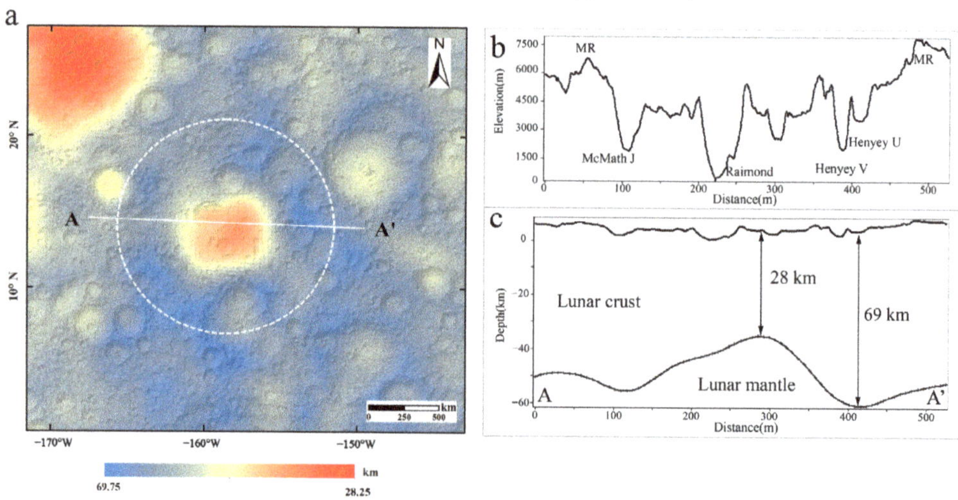

Figure 5. (**a**) Crustal thickness map of Dirichlet–Jackson Basin with shaded relief topography; (**b**) topographic profile from A to A′ based on SLDEM, where MR means the Main Rim of the basin; (**c**) subsurface characteristics of the Dirichlet–Jackson Basin along the AA′ profile.

Figure 6 shows that the crustal thickness in the region surrounding the *Nectaris* Basin varies from 5 km to 56 km, with an average of 32 km. The area of minimum thickness occurs below the mare fill in the basin, and the areas of maximum thickness correspond to the basin middle ring (Figure 6). The crust–mantle interface profile across the center of the basin shows that the mantle uplift is relatively large, approximately 35 km. Below the center of the mare (red arrow), the crust–mantle interface exhibits large-scale collapse, and the subsidence depth is approximately 3 km.

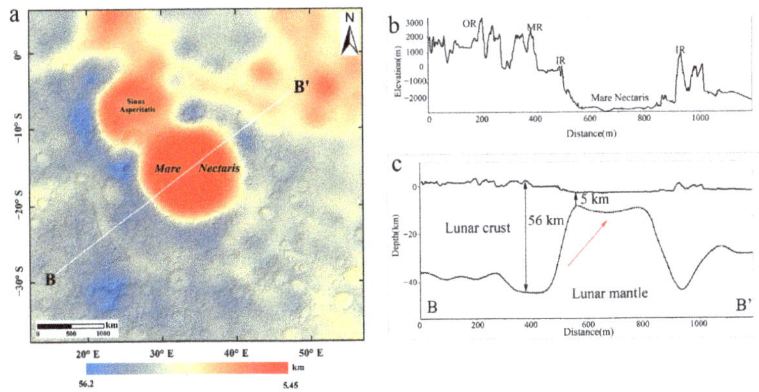

Figure 6. (a) Crustal thickness map of the *Nectaris* Basin with shaded relief topography; (b) topographic profile from B to B′ based on SLDEM, IR means Inner Ring, MR means Middle Ring, OR means Outer Ring; (c) subsurface characteristics of the *Nectaris* Basin along the AA′ profile.

The crustal thickness in the *Orientale* Basin region ranges from 6 km to 61 km, with an average of 42 km. As shown in Figure 7, the minimum thickness is located at the center of the basin floor. The thickest crust in the *Orientale* Basin region is in the western section of the Outer Montes Rook, consistent with the results of Ishihara et al., 2009. The crustal thickness is marked by a black double arrow in Figure 7, and notably decreases from the east to the west. The crust–mantle interface in the *Orientale* Basin is columnar, similar to that in the *Nectaris* basin. However, the scale of the mantle uplift is approximately 45 km, larger than in the *Nectaris* basin.

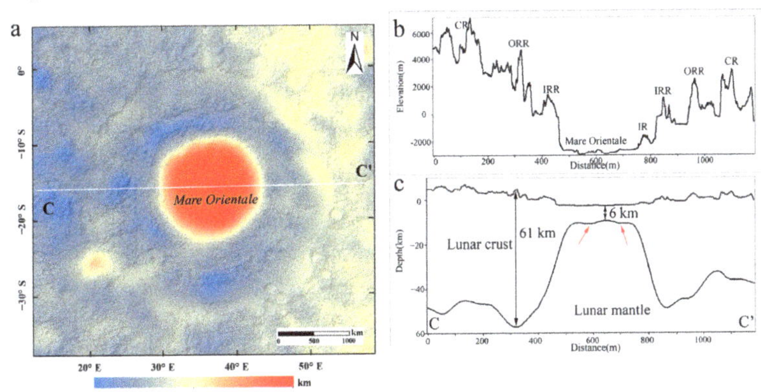

Figure 7. (a) Crustal thickness map of *Orientale* Basin with shaded relief topography; (b) topographic profile from C to C′ based on LOLA, IR means Inner Ring, IRR means Inner Rook Ring, ORR means Outer Rook Ring, and CR means Cordillera Ring; (c) subsurface characteristics of *Orientale* Basin along with the AA′ profile. Red arrows indicate subsidence of the crust and mantle interface.

4.3. Geochemical Results

4.3.1. FeO and TiO_2 Contents of the Basins

FeO and TiO_2 contents are commonly used in lunar geology. The FeO content in the region of the Dirichlet–Jackson Basin varies from 0 to 17.8 wt.% (Figure 8a), with a mean content of 8.4 wt.%. The TiO_2 content for the same region varies from 0 to 1.6 wt.%, with a mean of 0.5 wt.% (Figure 9a). The iron–poor ejecta from the Jackson crater overlays a significant portion of the Dirichlet–Jackson Basin. However, there are regions with higher concentrations of FeO inside and outside of the basin, e.g., at Mitra and Mach craters in the northeastern part of the basin (Figure 8a). The TiO_2 content within the basin floor is higher than the regions outside the basin. Figure 9a shows that the floor of the Jackson and Raimond craters is low content in Titanium. The ejecta from the Jackson crater have higher concentrations of TiO_2 and overlay the floor of the Dirichlet–Jackson Basin.

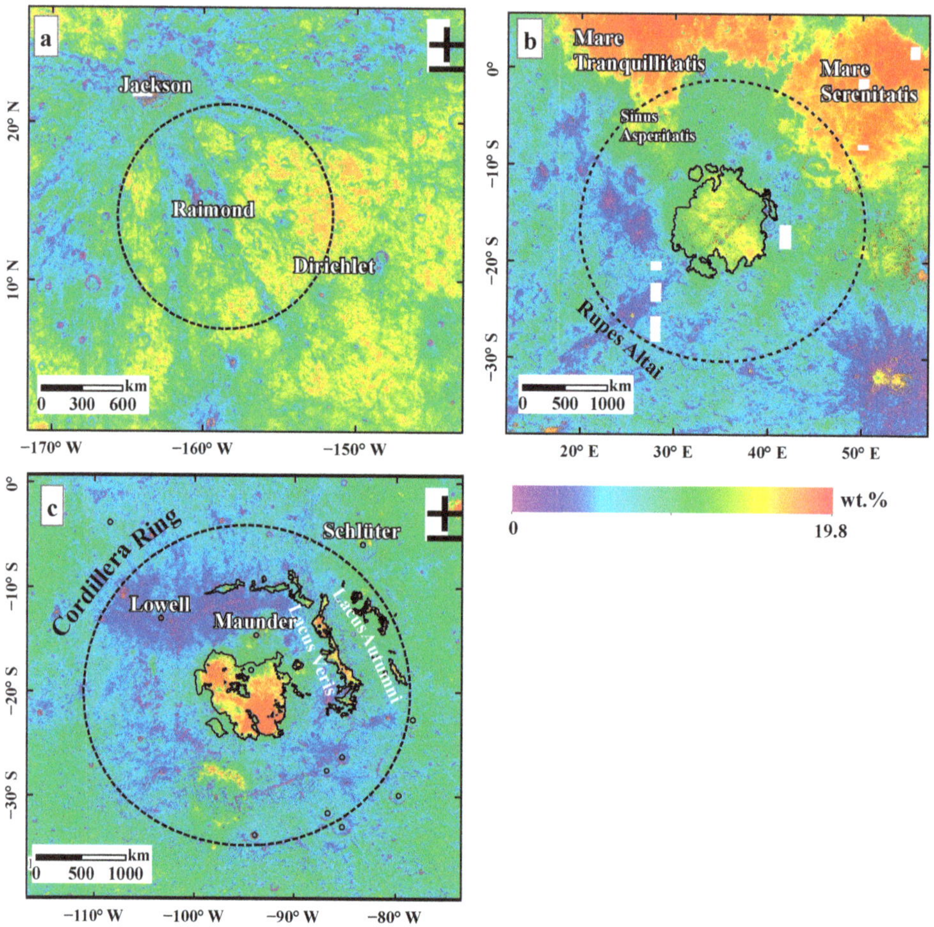

Figure 8. FeO contents of Dirichlet–Jackson Basin (**a**), *Nectaris* Basin (**b**), and *Orientale* Basin (**c**). The black lines outline *Maria* in the *Nectaris* and *Orientale* basins. The maps are in Mercator projection.

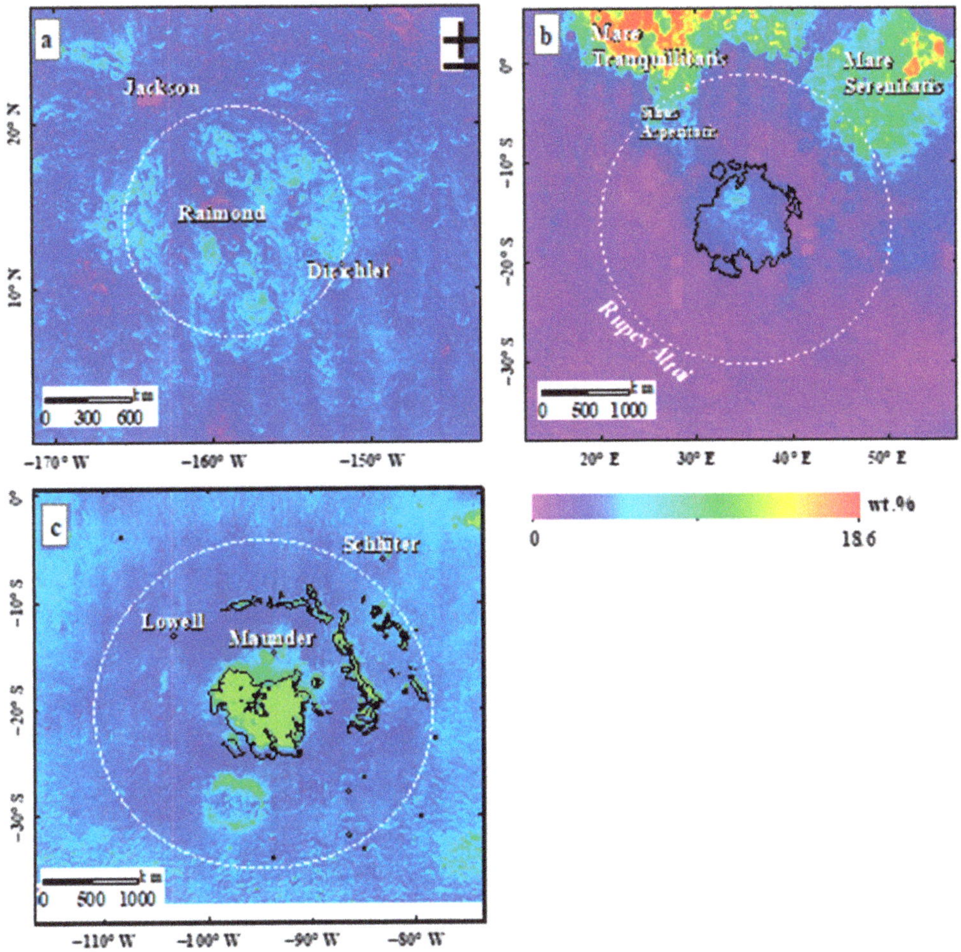

Figure 9. TiO_2 contents of Dirichlet–Jackson Basin (**a**), the *Nectaris* Basin (**b**), and the *Orientale* Basin (**c**). The black lines outline *Maria* in the *Nectaris* and *Orientale* basins. The map is in Mercator projection.

Figure 8b shows that the FeO content of the *Nectaris* Basin and the surrounding area varies from 0 to ~19.8 wt.%, with a mean value of 14.9 wt.%. The abundance of TiO_2 in the basin region varies from 0 to ~18.6 wt.% (Figure 9b), with a mean value of 1.4 wt.%. The FeO and TiO_2 concentrations in *Mare* Tranquillitatis and *Mare* Fecunditatis are much higher than in *Mare Nectaris*, with mean contents of 1.8 wt.% of TiO_2 and 14.9 wt.% of FeO. The highest titanium content in the *Mare Nectaris* area is around 6 wt.%. The values of FeO and TiO_2 contents in *Mare Nectaris* indicate that the mare materials belong to the low-titanium basalts.

As shown in Figure 8c, the FeO content in *Orientale* Basin region varies from 0 wt.% to 19.6 wt.% with the mean of ~9.2 wt.%. The TiO_2 content in the same region varies from 0 wt.% to 9.6 wt.% (Figure 9c) with the mean of 2.7 wt.% (Table 2). Similar to the *Nectaris* Basin, the areas with the higher iron and titanium content correspond to the mare material within the *Orientale* Basin.

Table 2. Mean TiO_2 content in the sub-formations of the three basins.

Basin Sub-Formations	Mean TiO_2 Content (STD) [1] of Dirichlet–Jackson	Mean TiO_2 Content (STD) of *Nectaris*	Mean TiO_2 Content (STD) of *Orientale*
Mare	–	2.54 (0.67)	2.48 (1.35)
Floor formation	0.55 (0.16)	1.04 (0.46)	0.56 (0.38)
Peak-ring formation	0.55 (0.13)	0.91 (0.54)	0.52 (0.29)
Wall formation	0.57 (0.15)	0.87 (0.49)	0.42 (0.15)
Rim formation	0.46 (0.11)	0.74 (0.55)	0.50 (0.16)

[1] STD is the value of standard deviation.

4.3.2. Geochemical Characteristics of the Key Basins

In the completed Chinese 1:2,500,000 global Lunar geological Mapping project [42,43], basin landforms are classified as peak-ring sub-formation, basin floor sub-formation, basin wall sub-formation, and basin rim sub-formation. They are identified within Dirichlet–Jackson, *Nectaris* and *Orientale* basins based on the WAC image (Figure 10). Here, we calculated and compared the FeO and TiO_2 contents of different sub-formations in the Dirichlet–Jackson, *Nectaris*, and *Orientale* basins. To analyze the chemical compositions of the basins, Tables 1 and 2 show the results of the mean contents of FeO and TiO_2 in the different basin sub-formations.

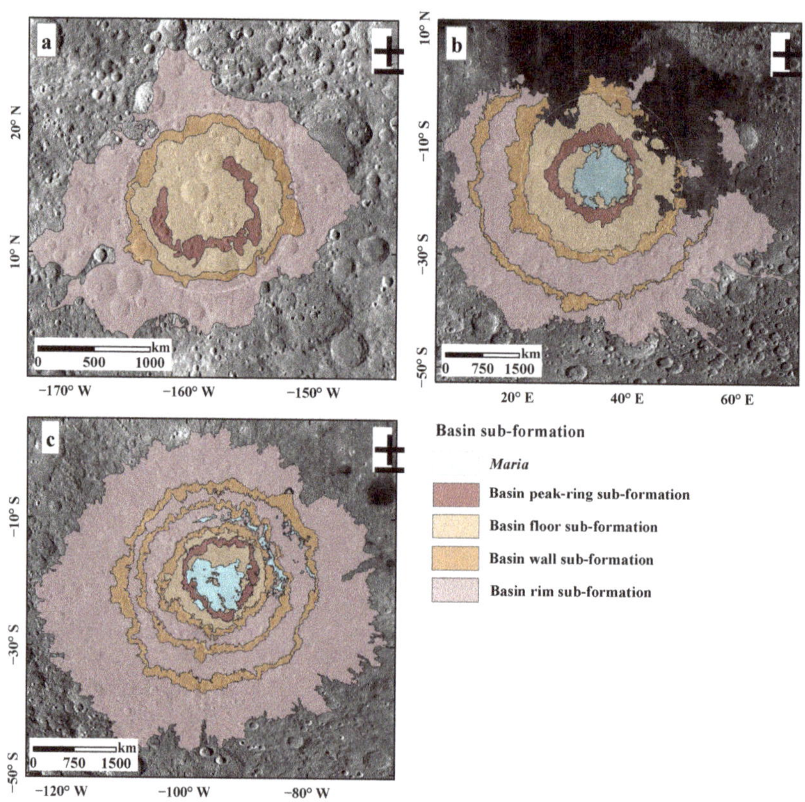

Figure 10. The landforms of the Dirichlet–Jackson Basin (a), the *Nectaris* Basin (b) and the *Orientale* Basin (c) (modified from Liu et al. [44]).

Figure 11 shows the model concentrations of FeO and TiO_2 of different landforms. The concentrations of FeO mainly range between ~6 and ~12 wt.%, whereas the TiO_2 abundance varies from 0 to ~1 wt.%. The concentrations of FeO and TiO_2 in the *Nectaris* Basin and the *Orientale* Basin are similar, concentrated between ~3 and ~9 wt.% and 0 and ~5 wt.%, respectively. The concentrations of FeO and TiO_2 in the Dirichlet–Jackson Basin peak-ring, floor, and wall sub-formation are mainly concentrated at ~9.5 wt.% and ~0.5 wt.%, respectively, whereas the concentrations of FeO and TiO_2 in the rim sub-formation are mainly concentrated at ~7.5 wt.% and 0.4 wt.%, respectively (Figure 11a,b). The concentrations of FeO in *Mare Nectaris* and *Orientale* vary from ~12 to ~15 wt.%, whereas in the *Nectaris* Basin and the *Orientale* Basin, the concentrations in the peak-ring, floor, wall, and rim sub-formations range ~3 to ~9 wt.% (Figure 11c,e). The trend in the concentrations of TiO_2 is similar to FeO (Figure 11d). The concentrations of TiO_2 in *Mare Nectaris* vary from ~2 to ~4 wt.%, whereas in the *Nectaris* Basin peak-ring, floor, wall, and rim sub-formations from ~0 to ~2 wt% (Figure 11d). The concentrations of TiO_2 in *Mare Nectaris* vary from ~1 to ~3 wt.%, while in the *Nectaris* Basin, peak-ring, floor, wall, and rim sub-formation is from ~0 to ~1 wt.% (Figure 11f). The concentrations of FeO and TiO_2 in the non-mare sub-formation of the basin are similar, showing expected bimodal distributions together with the concentrations of FeO and TiO_2 in the *Mare* area.

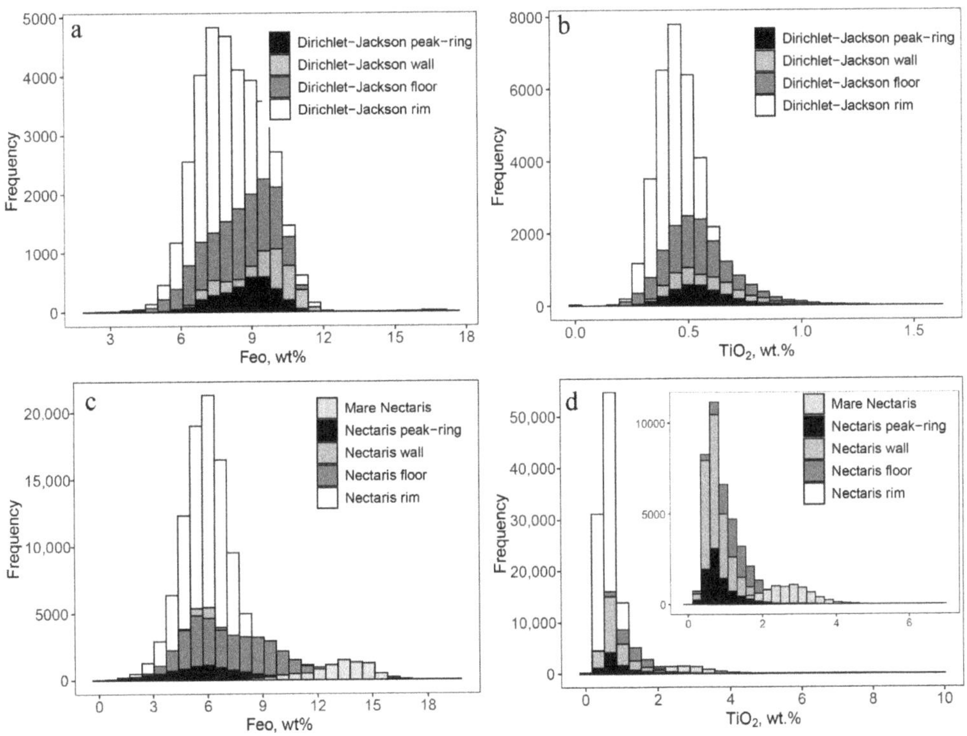

Figure 11. *Cont.*

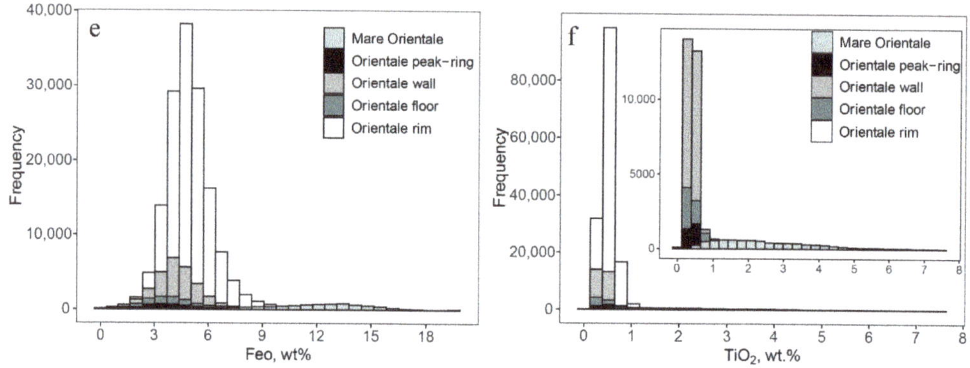

Figure 11. The frequency distribution of FeO and TiO_2 contents in the landforms of Dirichlet–Jackson Basin (**a**,**b**), *Nectaris* Basin (**c**,**d**) and *Orientale* Basin (**e**,**f**).

4.4. Mineralogical Results

The lunar regolith is dominated by phases such as plagioclase, pyroxene, olivine, ilmenite, and impact-generated glasses. Pyroxene spectra exhibit two major absorption bands near 1000 nm and 2000 nm both due to Fe^{2+} crystal field transitions. Reflectance spectra of olivine exhibited a major absorption peak near 1000 nm. In our paper, the M3 data were used to superimpose a certain range near the main mineral absorption peaks based on the IBD method, and IBD maps for the three basins were derived (Figure 12).

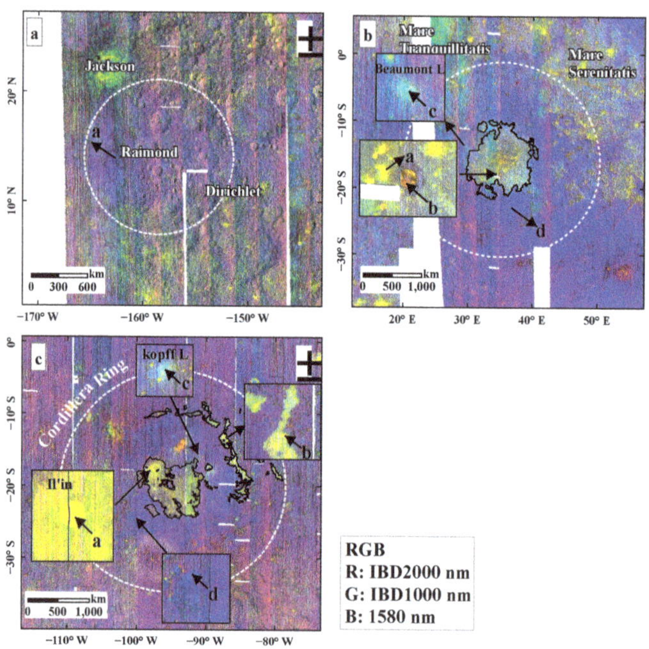

Figure 12. Composite map of the spectral parameters of the Dirichlet–Jackson Basin (**a**), *Nectaris* Basin (**b**), and *Orientale* Basin (**c**) calculated from the M3 data. The white dotted lines show the main rings of the basins.

The blue color of the Dirichlet–Jackson Basin region in the composite spectral parameter map (e.g., point a in Figure 12a) indicates that the region is dominated by plagioclase, which likely corresponds to the lunar crust composition before the formation of the basin.

The *Nectaris* mare regions and the surrounding highlands exhibit clear differences on the IBD map. The materials in *Mare Nectaris* near the Rosse crater (e.g., point a in Figure 12b) display strong absorptions at both 1000 nm and 2250 nm (Figure 13a), indicating that the regolith in this area contains fresh pyroxene materials. At the wall of the Rosse crater (e.g., point b in Figure 12b), the materials show relatively weak absorption characteristics at 1000 nm and 2250 nm (Figure 13b). The *Mare Nectaris* appears yellow-orange, displaying absorption characteristics at 1000 nm and 2000 nm indicating a mainly pyroxene composition. The materials in the southern wall of the Beaumont L crater (e.g., point c in Figure 12b) display an absorption band at 1000 nm but no absorption band at 2250 nm (Figure 13c), indicating that the minerals are olivine-rich. Additionally, the materials in the surrounding highlands at the composite map (point d in Figure 12b) are blue, indicating feldspathic materials (Figure 13d).

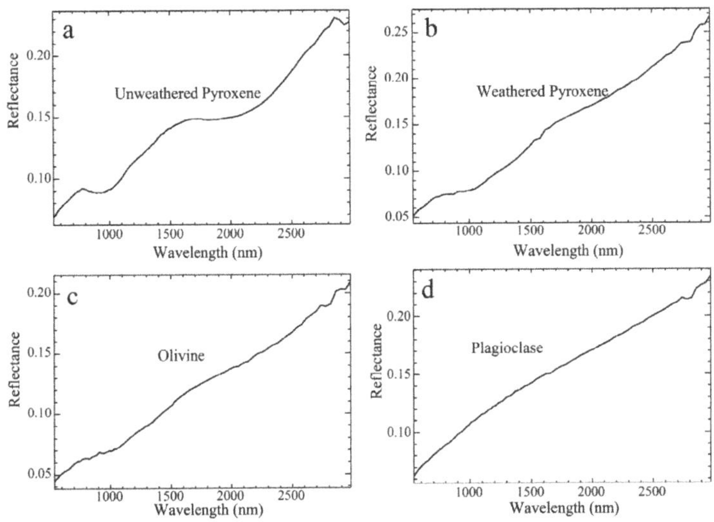

Figure 13. The spectral reflectance curve of the materials in *Nectaris* Basin: (**a**) unweathered pyroxene-rich basalt, (**b**) weathered pyroxene-rich materials at the Rosse crater wall, (**c**) olivine-rich materials in the Beaumont L crater wall, and (**d**) feldspathic materials on the non-mare region.

Similar to *Nectaris* Basin, *Mare Orientale* and the surrounding highlands area is clearly distinguishable on the IBD map of the *Orientale* Basin region. At the wall of the Il'in crater (e.g., point a in Figure 12c), the M3 spectral reflectance curve shows strong absorption at both 1000 nm and 2250 nm (Figure 14a), which indicates that there are some fresh pyroxene-rich materials in the crater. The basalts in *Mare Orientale* and *Lacus Veris* are yellow in color (e.g., point b in Figure 12c), displaying relatively weak absorption features at both 1000 nm and 2250 nm (Figure 14b). The features indicate that the basalts contain weathered pyroxene. In the northwestern rim of the Kopff E crater, there is a small and fresh crater (point c in Figure 12c). The M3 spectrum of point c displays an absorption band at 1 μm but no absorption band at 2000 nm (Figure 14c), which indicates that there possibly are olivine-rich materials here. The surrounding highlands in the composite spectral parameter map are blue-purple in color (e.g., point d in Figure 12c), which suggests that the composition of the highland plagioclase dominates (Figure 14d).

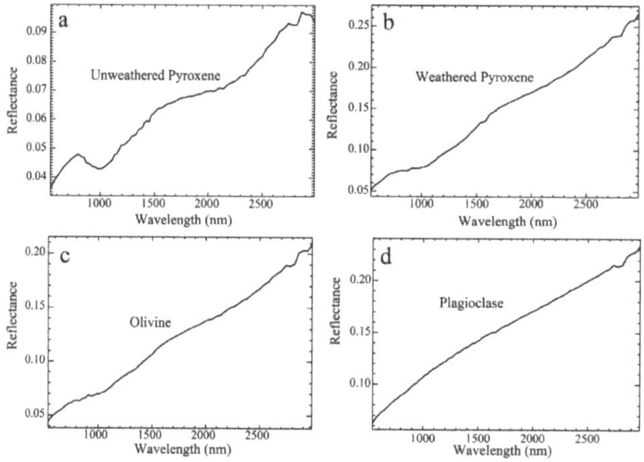

Figure 14. Spectral reflectance curves of the materials in the Orientale Basin: (**a**) unweathered pyroxene-rich materials at Il'in crater wall; (**b**) weathered pyroxene-rich basalt; (**c**) possible olivine rocks in the small and fresh crater wall; (**d**) feldspathic materials on the surrounding highlands.

5. Discussion

The initial goal of this study was to compare the topographical and morphological characteristics of these basins. Topographic profiles traversing their interiors can provide a qualitative assessment of their present topography. Analysis of the topographic profiles of the three basins (Figure 15) show that the average elevation of the Dirichlet–Jackson Basin is higher than the other two basins. In addition, the cavity-like shape in the profiles is consistent with those demonstrated by Byrne [45], who proposed the radial elevation profiles. The topographical analogies of the three basins show the cavity-like shapes of the basins. However, the shape of the Dirichlet–Jackson Basin is not an obvious cavity. This is because there are many craters with large diameters on the surface of Dirichlet–Jackson Basin, which may be because this basin was formed in the pre-Nectarian period. After the formation of the Dirichlet–Jackson Basin, it was subjected to a large number of impacts, severely degrading the surface, and the rings were not easily identified. In addition, the topographies of the *Nectaris* Basin and the *Orientale* Basin are extremely similar, in that they are filled with mare basalts. The rings of the *Nectaris* Basin and *Orientale* Basin are easily distinguishable because the two basins are well preserved. The floors of the two basins are relatively flat, with only a small number of young impact craters.

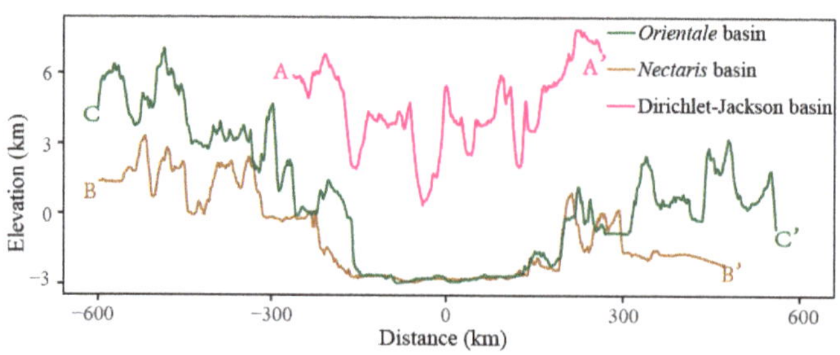

Figure 15. Topographic profile comparison map of the three basins, wherein 0 on the x-axis corresponds to the center of the three basins, negative values on the x-axis represent the direction from the center of the basin to the start of the topographic profile, and positive values on the x-axis represent the direction from the center of the basin to the end of the topographic profile.

We compared the crustal thickness of the three basins based on the GRAIL Crust Thickness Model. The crustal thickness comparison map shows the crustal thickness variations in the three basins (Figure 16). The minimum and maximum crustal thickness positions are located at the centers and the rims of the basins, respectively, which is due to the basin formation process. During the formation of basins, a large amount of material is excavated and redistributed. Thus, the centers of the basins are thinner than the rim. The formation of basins is the main process that leads to variations in lunar crust thickness, which is in line with the results of Johnson et al. [46] from the perspective of simulating the formation of basins. However, the crustal thickness of Dirichlet–Jackson Basin is thicker than the other two basins; the thicker crust corresponds to the higher topography (Figures 15 and 16). *Nectaris* Basin and *Orientale* Basin have thinner crusts, which is probably due to an abnormally large mantle plug [20].

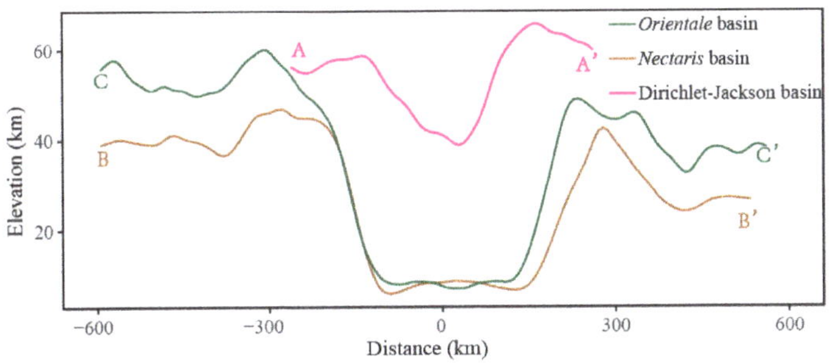

Figure 16. Crustal thickness comparison map of the three basins, wherein 0 on the *x*-axis corresponds to the center of the three basins, negative values on the *x*-axis represent the direction from the center of the basin to the start of the topographic profile, and positive values on the *x*-axis represent the direction from the center of the basin to the end of the topographic profile.

Analysis of the lunar samples showed that the TiO_2 concentrations of lunar mare basalts have large variations, which may be due to the magmatic source and subsequent differentiation processes. According to the TiO_2 contents, mare basalts can be classified as very high titanium (VHT) basalts (>10 wt.% of TiO_2), high-titanium basalts (7.5–10 wt.% of TiO_2), medium-titanium basalts (4.5–7.5 wt.% of TiO_2), low-titanium basalts(1–4.5 wt.% of TiO_2), and very-low-titanium (VLT) (<1 wt.% of TiO_2) [47]. Basalts in *Mare Nectaris* and *Orientale* are in the groups of very-low-titanium basalts and low-titanium basalts (Figure 17). Very-low-titanium and low-titanium basalts (<5 wt.%) are the most widely distributed basalts on the surface of the Moon and may be formed by the partial melting of olivine and orthopyroxene materials [47].

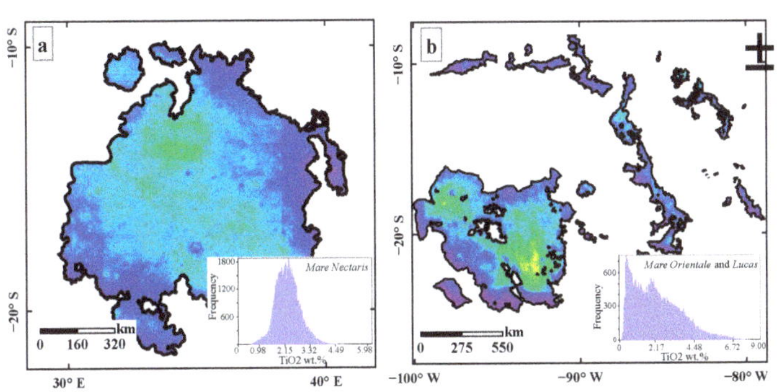

Figure 17. TiO_2 contents of *Mare Nectaris* (**a**), and *Mare Orientale* and *Lucas* within the *Orientale* Basin (**b**).

Large impact events disturb the lunar crust of the target region and excavate materials from the deep parts of the Moon. Maunder (Figure 18a,b) is a complex crater of the Eratosthenian period, with a central peak [1]. The diameter of Maunder crater is approximately 55 km, and this impact event excavated materials to a depth of ~5 km. Topographic analysis of the *Orientale* impact structure suggests that the thickness of the melt sheet in the *Orientale* is approximately 15 km [48]. These materials likely compose the entire wall of Maunder and were brought to the surface from beneath. Thus, Maunder would have both excavated and brought up through structural uplift the deeper stratigraphy of the melt sheet of *Orientale* basin. The Rosse crater in the south of the *Nectaris* basin floor appears to be relatively fresh. The depth and the excavation depth of the Rosse crater are approximately 2.5 km and 1.0 km, respectively (Figure 18c,d). In addition, the maximum thickness of *Mare Nectaris* is over 1.5 km [49]. Thus, the formation of the Rosse crater only excavated the mare basalts. In the M3 composite spectral parameter image of the Maunder and Rosse craters, the yellow/orange colors indicate that the craters are a mixture of pyroxene and olivine-rich materials.

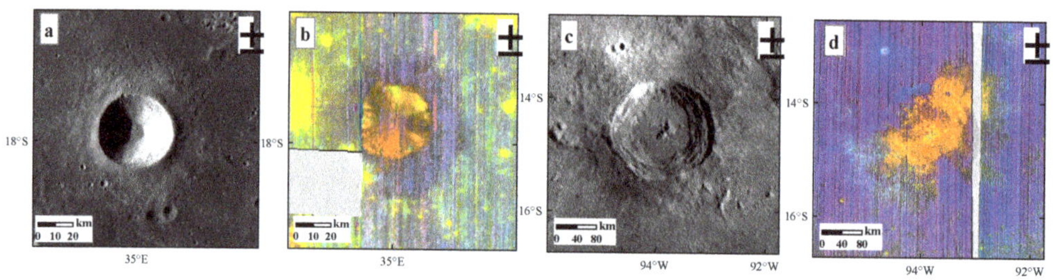

Figure 18. (**a**) Maunder crater in *Orientale* Basin; the base map is an LROC WAC image. (**b**) The M3 composite image displays the mineralogy of Maunder. (**c**) Rosse crater in *Nectaris* basin; the base map is an LROC WAC image. (**d**) The M3 composite image displays the mineralogy of Rosse.

The lunar magma ocean (LMO) hypothesis suggests that as magma oceans begin to solidify, they first crystallize to form olivine, which, due to its higher density, sinks toward the bottom of the magma ocean as it crystallizes and differentiates [50,51]. The impact basin formation is responsible for the exposure of olivine. Large impact events could have penetrated and blasted away the lunar upper crust, excavating deep-seated olivine-rich material [52]. The *Orientale* Basin and the *Nectaris* Basin have thinner crusts [31]. Therefore, it is possible that the olivine-rich materials may have come from the excavation of lunar mantle material during large impact events to form these basins.

6. Conclusions

We systematically studied and compared the characteristics of topography, composition and mineralogy, geological features, and crustal characteristics of Dirichlet–Jackson Basin, *Nectaris* Basin, and *Orientale* Basin using different imaging data sets. The main findings of this paper are presented subsequently.

In terms of the topographical and morphological characteristics of these basins, the topographical analogies are that the basins exhibit cavity-like shapes. However, the shape of the Dirichlet–Jackson Basin is not an obvious cavity. In addition, the surface of Dirichlet–Jackson Basin is severely degraded, and the rings were not easily identified due to the earlier formation of the basin (pre-Nectarian period). The topography of *Nectaris* Basin and *Orientale* Basin are extremely similar, in that they are filled with mare basalts and are well preserved.

The minimum and maximum crustal thickness positions of the three basins are located at the centers and the rims of the basins. The crust–mantle interface in the Dirichlet–Jackson Basin is slightly uplifted, approximately 20 km, which is smaller than in the *Nectaris* Basin and the *Orientale* Basin. The amplitudes of the uplifts in the *Nectaris* Basin and *Orientale* Basin are 35 km and 45 km, respectively.

Analyses of the composition and modeled concentrations of FeO and TiO_2 of lunar basin landforms enable us to investigate the distribution of the concentrations of FeO and TiO_2 in different basin landforms. The concentrations of FeO in the Dirichlet–Jackson Basin are mainly concentrated from ~6 to ~12 wt.%, whereas the TiO_2 abundance is mainly concentrated in the range of 0 to ~1 wt.%. The concentrations of FeO and TiO_2 in the *Nectaris* Basin and the *Orientale* Basin are similar, concentrated between ~3 and ~9 wt.% and 0 and ~5 wt.%, respectively. The concentrations of FeO and TiO_2 in the non-mare formation of the basin and mare area show an expected a bimodal distribution.

The materials within the Dirichlet–Jackson Basin, *Nectaris* Basin, and *Orientale* Basin are different. The Dirichlet–Jackson Basin is mainly composed of plagioclase minerals, whereas *Mare Nectaris* and *Mare Orientale* are mainly mixtures of pyroxene and olivine. The surrounding highland areas of the *Nectaris* Basin and the *Orientale* Basin exhibit a feldspar-dominated composition, which is the original component of the lunar crust. There are some relatively fresh impact craters in the mare region of the *Nectaris* Basin and the *Orientale* Basin, and the walls of the craters contain pyroxene-rich basalt material excavated from the near-surface.

Author Contributions: Conceptualization, J.D. and W.C.; methodology, J.D. and W.C.; software, J.D. and Y.J.; formal analysis, J.D.; writing—original draft preparation, J.D.; writing—review and editing, J.D., W.C. and Y.J.; visualization, J.D. and Y.J.; supervision, J.D. and W.C. All authors have read and agreed to the published version of the manuscript.

Funding: This research was funded by the B–type Strategic Priority Program of the Chinese Academy of Sciences, grant No. XDB41000000, the National Natural Science Foundation of China, No. 42130110, National Natural Science Foundation of China, grant No. 41571388, and the Key projects of national basic work of science and technology, grant No. 2015FY210500.

Acknowledgments: We greatly acknowledge the teams of SELENE Data Archive and the Planetary Data System archive for provide the data used in the study. We are deeply grateful to Mikhail A. Ivanov from V.I. Vernadsky Institute of Geochemistry and Analytical Chemistry, Russian Academy of Sciences for his useful and constructive comments.

Conflicts of Interest: The authors declare no conflict of interest.

References

1. Wilhelms, D.E.; McCauley, J.F.; Trask, N.J. *Geologic History of the Moon*; U.S. Geological Survey Professional Paper: Washington, DC, USA, 1987.
2. Melosh, H.J. *Impact Cratering: A Geologic Process*; Oxford University Press: New York, NY, USA, 1989; pp. 60–184.
3. Wieczorek, M.A.; Phillips, R.J. Lunar Multiring Basins and the Cratering Process. *Icarus* **1999**, *139*, 246–259. [CrossRef]
4. Spudis, P.D.; Gillis, J.J.; Reisse, R.A. Ancient multiring basins on the Moon revealed by Clementine laser altimetry. *Science* **1994**, *266*, 1848–1851. [CrossRef] [PubMed]
5. Miljkovicc, K.; Wieczorek, M.A.; Collins, G.S.; Laneuville, M.; Neumann, G.A.; Melosh, H.J.; Solomon, S.C.; Phillips, R.J.; Smith, D.E.; Zuber, M.T. Asymmetric distribution of lunar impact basins caused by variations in target properties. *Science* **2013**, *342*, 724–726. [CrossRef] [PubMed]
6. Spudis, P.D. *The Geology of Multi-Ring Impact Basins*; Cambridge University Press: Cambridge, UK, 1993.
7. Baker, D.M.; Head, J.W. New morphometric measurements of craters and basins on Mercury and the Moon from MESSENGER and LRO altimetry and image data: An observational framework for evaluating models of peak-ring basin formation. *Planet. Space Sci.* **2013**, *86*, 91–116. [CrossRef]
8. Fassett, C.I.; Head, J.W.; Kadish, S.J.; Mazarico, E.; Neumann, G.A.; Smith, D.E.; Zuber, M.T. Lunar impact basins: Stratigraphy, sequence and ages from superposed impact crater populations measured from Lunar Orbiter Laser Altimeter (LOLA) data. *J. Geophys. Res. Planets* **2012**, *117*, E00H06. [CrossRef]
9. Neumann, G.A.; Zuber, M.T.; Wieczorek, M.A.; Head, J.W.; Baker, D.M.H.; Solomon, S.C.; Smith, D.E.; Lemoine, F.G.; Mazarico, E.; Sabaka, T.J.; et al. Lunar impact basins revealed by Gravity Recovery and Interior Laboratory measurements. *Sci. Adv.* **2015**, *1*, e1500852. [CrossRef]
10. Bussey, D.B.J.; Spudis, P.D. Compositional studies of the Orientale, Humorum, Nectaris, and Crisium lunar basins. *J. Geophys. Res. Planets* **2000**, *105*, 4235–4243. [CrossRef]

11. Bretzfelder, J.M.; Klima, R.L.; Greenhagen, B.T.; Buczkowski, D.L.; Cartwright, S.F.A.; Moriarty, D.P.; Ernst, C.M.; Petro, N.E. Comparative Spectral Analysis of Three Distinct Lunar Basins. In Proceedings of the 50th Lunar and Planetary Science Conference, The Woodlands, TX, USA, 18–22 March 2019.
12. Singh, T.; Srivastava, N. Geology of Grimaldi Basin on the Moon: Evidence for volcanism and tectonism during the Copernican period. *Icarus* **2020**, *351*, 113921. [CrossRef]
13. Stöffler, D.; Ryder, G.; Ivanov, B.A.; Artemieva, N.A.; Cintala, M.J.; Grieve, R.A.F. Cratering History and Lunar Chronology. *Rev. Mineral. Geochem.* **2006**, *60*, 519–596. [CrossRef]
14. Satya Kumar, A.V.; Rajasekhar, R.P.; Tiwari, V.M. Gravity anomalies and crustal structure of the Lunar far side highlands. *Planet. Space Sci.* **2018**, *163*, 106–113. [CrossRef]
15. Pike, R.J.; Spudis, P.D. Basin-ring spacing on the Moon, Mercury, and Mars. *Earth Moon Planets* **1987**, *39*, 129–194. [CrossRef]
16. Howard, K.A.; Wilhelms, D.E.; Scott, D.H. Lunar basin formation and highland stratigraphy. *Rev. Geophys.* **1974**, *12*, 309–327. [CrossRef]
17. Spudis, P.D.; Hawke, B.R.; Lucey, P.G. Geology and deposits of the lunar Nectaris basin. In Proceedings of the 19th Lunar and Planetary Science Conference, Houston, TX, USA, 1 January 1989; pp. 51–59.
18. Head, J.W. Orientale multi-ringed basin interior and implications for the petrogenesis of lunar highland samples. *Earth Moon Planets* **1974**, *11*, 327–356. [CrossRef]
19. Cheek, L.C.; Donaldson Hanna, K.L.; Pieters, C.M.; Head, J.W.; Whitten, J.L. The distribution and purity of anorthosite across the Orientale basin: New perspectives from Moon Mineralogy Mapper data. *J. Geophys. Res. Planets* **2013**, *118*, 1805–1820. [CrossRef]
20. Ishihara, Y.; Goossens, S.; Matsumoto, K.; Noda, H.; Araki, H.; Namiki, N.; Hanada, H.; Iwata, T.; Tazawa, S.; Sasaki, S. Crustal thickness of the Moon: Implications for farside basin structures. *Geophys. Res. Lett.* **2009**, *36*, L19202. [CrossRef]
21. Spudis, P.D.; Hawke, B.R.; Lucey, P. Composition of orientale basin deposits and implications for the lunar basin-forming process. *J. Geophys. Res.* **1984**, *89*, C197. [CrossRef]
22. Yue, Z.; Yang, M.; Jia, M.; Michael, G.; Di, K.; Gou, S.; Liu, J. Refined model age for Orientale Basin derived from zonal crater dating of its ejecta. *Icarus* **2020**, *346*, 113804. [CrossRef]
23. Robinson, M.S.; Brylow, S.M.; Tschimmel, M.; Humm, D.; Lawrence, S.J.; Thomas, P.C.; Denevi, B.W.; Bowman-Cisneros, E.; Zerr, J.; Ravine, M.A.; et al. Lunar Reconnaissance Orbiter Camera (LROC) Instrument Overview. *Space Sci. Rev.* **2010**, *150*, 81–124. [CrossRef]
24. Barker, M.K.; Mazarico, E.; Neumann, G.A.; Zuber, M.T.; Haruyama, J.; Smith, D.E. A new lunar digital elevation model from the Lunar Orbiter Laser Altimeter and SELENE Terrain Camera. *Icarus* **2016**, *273*, 346–355. [CrossRef]
25. Nozette, S.; Rustan, P.; Pleasance, L.; Kordas, J.; Lewis, I.; Park, H.; Priest, R.; Horan, D.; Regeon, P.; Lichtenberg, C.J.S. The Clementine mission to the Moon: Scientific overview. *Science* **1994**, *266*, 1835–1839. [CrossRef]
26. Boardman, J.W.; Pieters, C.M.; Green, R.O.; Lundeen, S.R.; Varanasi, P.; Nettles, J.; Petro, N.; Isaacson, P.; Besse, S.; Taylor, L.A. Measuring moonlight: An overview of the spatial properties, lunar coverage, selenolocation, and related Level 1B products of the Moon Mineralogy Mapper. *J. Geophys. Res. Planets* **2011**, *116*, E00G14. [CrossRef]
27. Green, R.; Pieters, C.; Mouroulis, P.; Eastwood, M.; Boardman, J.; Glavich, T.; Isaacson, P.; Annadurai, M.; Besse, S.; Barr, D. The Moon Mineralogy Mapper (M3) imaging spectrometer for lunar science: Instrument description, calibration, on-orbit measurements, science data calibration and on-orbit validation. *J. Geophys. Res. Planets* **2011**, *116*, E00G19. [CrossRef]
28. Besse, S.; Sunshine, J.M.; Staid, M.I.; Petro, N.E.; Boardman, J.W.; Green, R.O.; Head, J.W.; Isaacson, P.J.; Mustard, J.F.; Pieters, C.M. Compositional variability of the Marius Hills volcanic complex from the Moon Mineralogy Mapper (M3). *J. Geophys. Res. Planets* **2011**, *116*, E00G13. [CrossRef]
29. Besse, S.; Sunshine, J.; Staid, M.; Boardman, J.; Pieters, C.; Guasqui, P.; Malaret, E.; McLaughlin, S.; Yokota, Y.; Li, J.Y. A visible and near-infrared photometric correction for Moon Mineralogy Mapper (M-3). *Icarus* **2013**, *222*, 229–242. [CrossRef]
30. Pieters, C.M.; Boardman, J.; Buratti, B.; Chatterjee, A.; Clark, R.; Glavich, T.; Green, R.; Head, J., III; Isaacson, P.; Malaret, E.J.C.S. The Moon mineralogy mapper (M^3) on chandrayaan-1. *Curr. Sci.* **2009**, *96*, 500–505.
31. Wieczorek, M.A.; Neumann, G.A.; Nimmo, F.; Kiefer, W.S.; Taylor, G.J.; Melosh, H.J.; Phillips, R.J.; Solomon, S.C.; Andrews-Hanna, J.C.; Asmar, S.W.; et al. The Crust of the Moon as Seen by GRAIL. *Science* **2013**, *339*, 671–675. [CrossRef]
32. Kneissl, T.; van Gasselt, S.; Neukum, G. Map-projection-independent crater size-frequency determination in GIS environments-New software tool for ArcGIS. *Planet. Space Sci.* **2011**, *59*, 1243–1254. [CrossRef]
33. Rosenburg, M.A.; Aharonson, O.; Head, J.W.; Kreslavsky, M.A.; Mazarico, E.; Neumann, G.A.; Smith, D.E.; Torrence, M.H.; Zuber, M.T. Global surface slopes and roughness of the Moon from the Lunar Orbiter Laser Altimeter. *J. Geophys. Res.* **2011**, *116*, E02001. [CrossRef]
34. Gillis, J.J.; Jolliff, B.L.; Elphic, R.C. A revised algorithm for calculating TiO_2 from Clementine UVVIS data: A synthesis of rock, soil, and remotely sensed TiO_2 concentrations. *J. Geophys. Res. Planets* **2003**, *108*, 5009. [CrossRef]
35. Lucey, P.G.; Blewett, D.T.; Hawke, B.R. Mapping the FeO and TiO_2 content of the lunar surface with multispectral imagery. *J. Geophys. Res. Planets* **1998**, *103*, 3679–3699. [CrossRef]
36. Lucey, P.G.; Blewett, D.T.; Jolliff, B.L. Lunar iron and titanium abundance algorithms based on final processing of Clementine ultraviolet-visible images. *J. Geophys. Res. Planets* **2000**, *105*, 20297–20305. [CrossRef]
37. Lucey, P.G.; Taylor, G.J.; Malaret, E. Abundance and Distribution of Iron on the Moon. *Science* **1995**, *268*, 1150–1153. [CrossRef] [PubMed]

38. Ivanov, M.A.; Hiesinger, H.; Van Der Bogert, C.H.; Orgel, C.; Pasckert, J.H.; Head, J.W. Geologic History of the Northern Portion of the South Pole-Aitken Basin on the Moon. *J. Geophys. Res. Planets* **2018**, *123*, 2585–2612. [CrossRef]
39. Kusuma, K.N.; Sebastian, N.; Murty, S.V.S. Geochemical and mineralogical analysis of Gruithuisen region on Moon using M3 and DIVINER images. *Planet. Space Sci.* **2012**, *67*, 46–56. [CrossRef]
40. Staid, M.I.; Pieters, C.M.; Besse, S.; Boardman, J.; Dhingra, D.; Green, R.; Head, J.W.; Isaacson, P.; Klima, R.; Kramer, G.; et al. The mineralogy of late stage lunar volcanism as observed by the Moon Mineralogy Mapper on Chandrayaan-1. *J. Geophys. Res.* **2011**, *116*, E00G10. [CrossRef]
41. Mustard, J.F.; Pieters, C.M.; Isaacson, P.J.; Head, J.W.; Besse, S.; Clark, R.N.; Klima, R.L.; Petro, N.E.; Staid, M.I.; Sunshine, J.M.; et al. Compositional diversity and geologic insights of the Aristarchus crater from Moon Mineralogy Mapper data. *J. Geophys. Res.* **2011**, *116*, E00G12. [CrossRef]
42. Liu, J.; Guo, D.; Chen, S.; Sun, Y.; Chen, J.; Wang, X.; Ling, Z.; Zhang, J.; Ding, X. Chinese 1: 2.5 M geologic mapping of the global Moon. In Proceedings of the 47th Lunar and Planetary Science Conference, The Woodlands, TX, USA, 21–25 March 2016; p. 2039.
43. Liu, J.-W.; Liu, J.-Z.; Zhang, L.; Ling, Z.-C.; Chen, J.; Ding, X.-Z.; Han, K.-Y.; Xu, K.-J. New Global Lunar Impact Basins' Map Based on Remote Sensing Interpretation. In Proceedings of the 51st Lunar and Planetary Science Conference, The Woodlands, TX, USA, 16–20 March 2020; p. 1940.
44. Liu, J.; Liu, J.; Yue, Z.; Zhang, L.; Wang, J.; Zhu, K. Characterization and interpretation of the global lunar impact basins based on remote sensing. *Icarus* **2022**, *378*, 114952. [CrossRef]
45. Byrne, C.J. *The Moon's Largest Craters and Basins: Images and Topographic Maps from LRO, GRAIL, and Kaguya*; Springer: Berlin/Heidelberg, Germany, 2016.
46. Johnson, B.C.; Andrews-Hanna, J.C.; Collins, G.S.; Freed, A.M.; Melosh, H.J.; Zuber, M.T. Controls on the Formation of Lunar Multiring Basins. *J. Geophys. Res. Planets* **2018**, *123*, 3035–3050. [CrossRef]
47. Giguere, T.A.; Taylor, G.J.; Hawke, B.R.; Lucey, P.G. The titanium contents of lunar mare basalts. *Meteorit. Planet. Sci.* **2000**, *35*, 193–200. [CrossRef]
48. Vaughan, W.M.; Head, J.W.; Wilson, L.; Hess, P.C. Geology and petrology of enormous volumes of impact melt on the Moon: A case study of the Orientale basin impact melt sea. *Icarus* **2013**, *223*, 749–765. [CrossRef]
49. De Hon, R. Thickness of mare material in the Tranquillitatis and Nectaris basins. In Proceedings of the 5th Lunar and Planetary Science Conference, Houston, TX, USA, 1 January 1974; pp. 53–59.
50. Snyder, G.A.; Taylor, L.A.; Neal, C.R. A chemical-model for generating the sources of mare basalts-Combined equilibrium and fractional crystallization of the lunar magmasphere. *Geochim. Cosmochim. Acta* **1992**, *56*, 3809–3823. [CrossRef]
51. Elkins Tanton, L.T.; Van Orman, J.A.; Hager, B.H.; Grove, T.L. Re-examination of the lunar magma ocean cumulate overturn hypothesis: Melting or mixing is required. *Earth Planet. Sci. Lett.* **2002**, *196*, 239–249. [CrossRef]
52. Yamamoto, S.; Nakamura, R.; Matsunaga, T.; Ogawa, Y.; Ishihara, Y.; Morota, T.; Hirata, N.; Ohtake, M.; Hiroi, T.; Yokota, Y.; et al. Possible mantle origin of olivine around lunar impact basins detected by SELENE. *Nat. Geosci.* **2010**, *3*, 533–536. [CrossRef]

remote sensing

MDPI

Article

A Cartographic Perspective on the Planetary Geologic Mapping Investigation of Ceres

Andrea Naß [1,*,†] and Stephan van Gasselt [2]

1 German Aerospace Center (DLR), Institute for Planetary Research (PF), Rutherfordstr. 2, 12489 Berlin, Germany
2 Geomatics Group, Department of Land Economics, National Chengchi University (NCCU), Taipei City 11605, Taiwan; svg@nccu.edu.tw
* Correspondence: andrea.nass@dlr.de
† Current address: Vulcan Energy Subsurface Solutions GmbH, 76227 Karlsruhe, Germany.

Abstract: The NASA Dawn spacecraft visited asteroid 4 Vesta between 2011 and 2012 and dwarf planet 1 Ceres between 2015 and 2018 to investigate their surfaces through optical and hyperspectral imaging and their composition through gamma-ray and neutron spectroscopy. For the global mapping investigation of both proto-planets, geologic mappers employed Geographic Information System (GIS) software to map 15 quadrangles using optical and hyperspectral data and to produce views of the geologic evolution through individual maps and research papers. While geologic mapping was the core motivation of the mapping investigation, the project never aimed to produce homogeneous and consistent map representations. The chosen mapping approach and its implementation led to a number of inconsistencies regarding cartographic representation, including differential generalization through varying mapping scales, topologic inconsistencies, lack of semantic integrity, and scale consistency, and ultimately, to the management of reusable research data. Ongoing data acquisition during the mapping phase created additional challenges for the homogenization of mapping results and a potential derivation of a global map. This contribution reviews cartographic and data perspectives on the mapping investigation of Ceres and highlights (a) data sources, (b) the cartographic concept, (c) mapping conduct, and (d) dissemination as well as research-data management arrangements. It furthermore discusses decisions and experiences made during mapping and finishes with a set of recommendations from the viewpoint of the cartographic sciences.

Keywords: planetary remote sensing; planetary cartography; cartographic object models; planetary geologic mapping; geoinformation systems; data models; map templates; research data management

1. Introduction and Background

1.1. Background

The planetary geologic exploration of Solar System bodies is a fundamental part of planetary sciences and targets the development of an understanding of planetary evolution under exogenic and endogenic forces. Over recent decades, hundreds of geologic investigations and maps have been published, starting with mapping investigations at small map scales for the Geologic Atlas of the Moon in the early 1960s based on orbital and in situ data [1,2]. Beyond fundamental scientific questions and landing-site characterizations, geologic mapping has become of major importance when analyzing the potential for future mineral exploitation and extraction, e.g., through resource mining [3]. In its most general form, planetary geologic mapping describes an investigation method in which planetary surfaces are classified and mapped according to homogeneous material units that have been formed through a distinct process during a distinct time in geologic history. The aim of such investigations is to understand the geologic formation history of planets, moons, and other small planetary bodies. Map products derived from such investigations thus combine information about surface materials (units), formation processes, as well as relative and absolute timing along with the unit's spatial distribution in its explicitly mapped lateral as well as implicitly displayed vertical directions [4]. Geological maps, therefore, depict the spatio-temporal relationships of planetary surface materials and processes. As such, mapping is one of the most important tools of planetary geology, in which Earth-based principles are applied to other planetary objects in order to break down complex settings and to gain in-depth insights about the origins of units [2,5,6].

As of today, planetary geologic mapping is based on orbital data, mainly as in situ exploration by rovers cannot cover significant areal extents. The majority of planetary geological maps are based on the comparative analysis of optical images, multi- and hyperspectral images, high-energy data, radar data, as well as derivative products developed at ground facilities, such as digital terrain models (DTM). A comprehensive summary of the building blocks of mapping as foundational data products is discussed in Ref. [7]. With the increase in scientific and technological developments since the beginning of sensor-based planetary exploration around the 1960s, the quality and quantity of planetary data increased and mapping could be conducted at much larger map scales and smaller spatial extents.

Planetary geologic maps are commonly created during investigations that are regionally confined to a specific research area, and they are published as image subsets or supplements to research papers [8]. They follow no particular convention, and their layout, contents, and quality differ largely from publication to publication. In a few cases, large map sheets are published as dedicated mapping investigations although outlets are relatively sparse [9]. The majority of planetary geologic maps, however, have been published through coordinated programs headed by the United States Geological Survey (USGS) and funded by NASA or through support from dedicated national funding programs. Those maps and associated reports remain in the public domain and may be downloaded or ordered as analog copies from the survey's data warehouse in which a metadata catalog is maintained. In addition to the USGS mapping program [10,11] and the traditional data archives such as the Planetary Data System [12] and the Planetary Science Archive [13], a number of institutional and community efforts exist to coordinate the collection of spatial information and maps, such as the International Planetary Data Alliance (IPDA) [14], NASA/USGS Mapping and Planetary Spatial Infrastructure Team (MAPSIT) [15], or Planmap [16–18]. Recent efforts have been looking beyond the traditional pipeline, leading to the archival of data and maps, and focusing on investigating topics such as the implementation of data infrastructures with associated research data management in the planetary domain [7,19–24].

Until today, the initial geologic mapping investigations of the proto-planets Vesta and Ceres are the only modern large-scale efforts that have not been led by the USGS but were part of an investigation that was carried out alongside the ongoing data acquisition in orbit around Vesta and Ceres [25–27]. The geologic mapping campaign was developed to provide a systematic and initial characterization of the global and regional geology of both planetary bodies. This investigation aimed to provide context for

1. Other instrument teams onboard Dawn, aiming to determine their geochemistry and mineralogy using a hyperspectral imager and a gamma-ray and neutron detector;
2. The larger planetary-science community for subsequent studies [28,29].

The major aim of the geologic mapping investigation of Ceres was to develop an understanding of the global geology and chronostratigraphy of Ceres through iterative geologic mapping at low and high spatial resolution during the mission's lifetime [29] and depending on the data return from low- and high-altitude orbits. More specifically, the mapping investigation includes the identification of the origin of various surface features and their spatial distribution, the identification and mapping of geologic units, and the identification and characterization of stratigraphic relationships. The mapping approach was different from other mapping investigations as data kept arriving when mapping already started using survey, high-altitude, and lower-resolution data [29].

The detailed geological reasoning has been discussed in more detail in the introductory papers to this Special Issues on the geologic mapping of Ceres and Vesta [28–30] and in individual contributions to the Special Issue.

1.2. Objectives

The background described above provides the motivation for reviewing the mapping investigation at Ceres from a cartographic viewpoint and from the viewpoint of contemporary research-data management needs and to investigate the chosen approach and the potential reuse value. More specifically, this contribution tries to

1. Review and discuss the cartographic object model under the project's boundary constraints, including availability and use of cartographic base data and definition of map sheets;
2. Review mapping conduct and mapping support;
3. Review the approach to disseminate and make the research results available.

The first author of this contribution (A.N.) was involved in the geologic mapping investigation at Ceres to provide support in GIS, cartographic, and data modeling matters.

A discussion for this mapping investigation is timely as it complements the retrospective from the scientific point of view as provided by Ref. [28,29] and might help to discuss or even improve future collaborative mapping investigations. Furthermore, by identifying relevant cartographic requirements and by discussing potentials, a set of recommendations can be developed, which might guide not only human mappers but also machine-based mapping performed through AI-based methods in the near future. As autonomous impact-crater mapping and landform detection and classification have seen new impulses through machine-learning algorithms [31–36], new research on unit delineation, segmentation, and classification is likely soon to spawn also in the planetary domain. Those machine-based methods, however, need to withstand the same criticism when it comes to object models and cartographic realization. And, while past publications have shown the efficiency of such tools, they are usually not validated against manually measured data, which would qualify as the closest we reach a ground truth. This contribution aimed to provide a few first impulses in this direction.

While quadrangle-based mapping has always been conducted in the past (e.g., the USGS Geologic Atlas of the Moon), the steps and the requirements from quadrangle-based mapping to producing a coherent and consistent global cartographic product as an additional product have never been exploited in any detail thus far. This contribution, therefore, aims to identify cartographic and data management aspects that help close this gap. However, for the geologic mapping investigation of Ceres, compilation (or unification [37]) has never been an objective.

Another point of motivation is that there has been an evolving need to make results accessible to the public and peer research by introducing strategies for research data management and data reuse. This investigation attempts to discuss this topic to highlight aspects related to more efficient planning for future reuse of research data. It aims to improve the contextualization of research products in order to derive new findings by linking spatial information more efficiently with each other. A development of this type would also support initiatives such as GoFair [38]; the Committee on Data for Science and Technology [39]; Research Data Alliance [40]; and finally, the FAIR-principles [41].

The concrete objectives of this investigation can be summarized as follows:

1. Creating a higher-level cartographic abstraction of the geological mapping investigation of Ceres to highlight and discuss major cartographic tasks associated with the mapping investigation;
2. Discussing relevant constituents of the cartographic concept and their implementation and potential, ranging from data sources and conducting over-mapping to dissemination and research data management;
3. Providing recommendations based on the lessons learned in this mapping investigation.

1.3. Structure

In the first part of this contribution, the investigation is contextualized by first introducing the motivation and requirements for the mapping investigation in general and by emphasizing the need to review and discuss the cartographic viewpoint. Also here, the main objectives related to this research and the potential impact beyond the limited boundaries of this research paper are highlighted. The methodology (Section 2) describes and discusses the cartographic components within the geologic mapping investigation and how they were integrated into the overall process and within each project phase. This section describes the underlying approach for how the geologic mapping investigation was cartographically conceptualized under given project constraints and how it was logically designed and implemented. The focus here is put on four core elements: (1) cartographic base data, (2) the cartographic design concept (*sensu stricto*), (3) mapping conduct, and (4) research data and dissemination.

As highlighted in the subsequent section, design freedom for the mapping investigation was limited due to overarching project definitions. Consequently, the results (cf. Section 3) will focus on the presentation and discussion of the actual mapping results, i.e., maps and data products, and the potentials rather than on the mapping management and coordination itself. In addition, a discussion about the future reuse of mapping results is given due to the growing relevance of efficient research data management. More importantly, this section will discuss the potential of reaching additional goals and discuss the required effort and feasibility. This may provide a benchmark for future collaborative mapping investigations that might be time-constrained. In the conclusions (Section 4), a synthesis and major recommendations based on this investigation are presented.

2. Data and Methods

2.1. Base Data and Data Management

Dwarf planet Ceres has a dimension of (964.4 km $\times$ 964.2 km $\times$ 891.8 km) $\pm$ 0.2 km and is best represented as a triaxial ellipsoid. For cartographic and mapping purposes, its shape is reduced to a sphere with a radius of 470 km. Its longitude domain is traditionally 180° in the positive east direction with impact crater Kait located at 0.2°S, marking the central meridian λ_0 [42].

Cartographic base data for mapping consist exclusively of data from the instruments onboard the Dawn spacecraft as no prior orbital observations exist. As the mission was still ongoing during the time of the geologic mapping investigation, lower-resolution base data had to be updated to high-resolution data during the mapping phase, which ultimately required a more iterative mapping conduct. The main basis for the geologic mapping investigation was optical data from the dual framing camera instruments (FC), which consisted of a 1024 $\times$ 1024 pixel charge-coupled device (CCD) sensor and a 150 mm focal-length telescope. During Dawn's mapping phase, the spacecraft imaged Ceres from three different orbital altitudes: a 4425 km survey orbit, a 1475 km high-altitude mapping orbit (HAMO), and a 375 km to 410 km low-altitude mapping orbit (LAMO) [43].

In HAMO, the framing camera took about 2490 clear-filter images with a resolution of about 140 m/pixel. The individual frames were selected according to homogeneous illumination conditions and a subset was processed by the FC team to generate a panchromatic near-global image mosaic [44]. Based on this dataset, 15 map sheets using a traditional quadrangle schema were created at a map scale of 1:750,000 [44]. During LAMO, the camera took about 31,300 clear-filter images with an image scale of 35 m/px from which a global image mosaic was created [45]. Again, based on a subset of that data, a Ceres LAMO atlas consisting of 62 map tiles with a map scale of 1:250,000 was compiled and published [46].

Based on the HAMO clear-filter image collection, a global digital terrain model (DTM) was photogrammetrically derived at a scale of 135 m/px (60 pixels per degree) and with a vertical accuracy of about 10 m [46,47]. The heights are calculated as ellipsoidal heights and refer to an oblate spheroid (two-axial ellipsoid) with a semi-major axis of 482 km and a semi-minor axis of 446 km [47]. The digital terrain model not only serves in height and relief estimates but also is used for photogrammetric control for the HAMO mosaic. For the mapping investigation, a global DTM was provided in an equidistant cylindrical projection. Hemispheric DTMs were provided for both polar regions in polar-stereographic projections. DTMs for GIS integration are formatted as 16-bit raster images with raster values representing the height above the reference spheroid with a 1 m binning [47].

While for the geologic mapping investigation, data sources were limited, the decision was made to not systematically include additional single frames with varying illumination conditions in order to balance the limited expected information gain and the expected complexity when adding new data sources. Accordingly, by using the pre-compiled image mosaic, only a small subset of images has effectively been used for mapping [29,44,45].

Hyperspectral information was also not systematically added to the database (Figure 1) but was provided for individual mappers if required in order to limit the complexity of the mapping investigation.

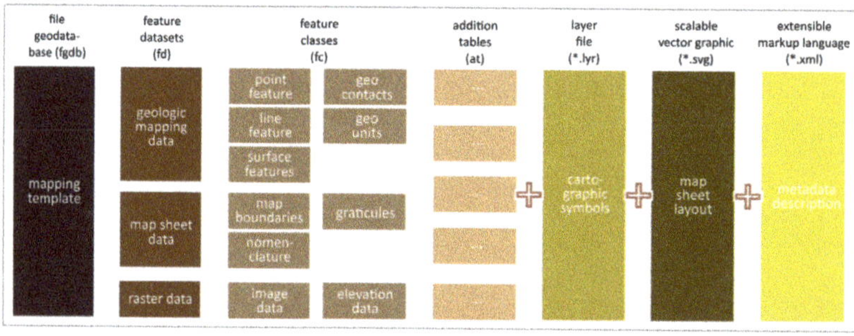

Figure 1. Data model as implemented in ESRI's ArcGIS File-Geodatabase format.

In addition to this base data, feature datasets (FD) were created within ESRI's ArcGIS 10 File-Geodatabase (FGDB) container to host individual feature classes (FC) containing the actual mapping information (see Figure 1). The FD concepts allowed us to maintain a consistent map projection throughout the project for all data that were added into the FD container. Individual feature classes for geologic mapping contained point, line, and surface geometries for geological contacts and unit representation. The image data mosaic, digital terrain mode data, and nomenclature information were added from external sources to form a consistent dataset.

The map layout and database design were kept minimal to not limit mappers through too many constraints which could have led to abandoning the data model due to frustration or conceived technical issues [48]. The original motivation was the implementation of a data model as presented in Ref. [49] which would put more constraints on mappers.

2.2. Cartographic Concept

While not universally defined, a cartographic concept addresses the way data and objects are modeled and put into relationship to each other, and defines how they are cartographically visualized considering the needs of the target audience. With modern cartography being a wide field, a cartographic concept might be implemented in different ways for different purposes. The cartographic concept will, here, in its most general form, be understood as a combination of aspects targeted at managing input data and converting a topic into a comprehensible form using cartographic techniques. The concept, therefore, comprises aspects such as base-data selection, as well as cartographic objects and their relationships, i.e., the digital object model and the cartographic (visualization) model. Due to the limited variety of source data in this project and due to a number of pre-defined project constraints, the cartographic concept could not be developed from bottom up but rather represented a simplified informal top-down approach loosely related to concepts established at the USGS [50]. The target audience as defined in Ref. [29] were instrument teams of the DAWN mission as well as planetary researchers that will be working on geologic topics, i.e., the same community that performs the actual geologic mapping. Consequently, no adaptation regarding the topic's complexity had to be introduced as the target audience would expect to see and work with the same style of research papers and maps that they have been used to.

The following paragraphs therefore concentrate on the cartographic representation and some of the reasoning behind it. This includes specifically the (1) definition of map sheets including subdivision of the surface into smaller sections in order to carry out this project globally within the given project time and (2) specification of map products and map formats, including final map scale, mapping scale, and map projection parameters to provide a scientifically sound and useful basis for mappers. While being part of the cartographic concept, a description of the actual mapping conduct and compilation follows in the next section (Section 2.3). Topics related to the dissemination of mapping results, including support for data homogenization and archiving for future reuse, are covered thereafter (Section 2.4).

The underlying framework for this mapping investigation were losely defined through the approach taken by individuals in USGS-coordinated mapping programs. While certain aspects were implemented for the actual mapping, it did not include the compilation of map sheets, neither in digital nor in analog form.

As base data for the geologic mapping investigation of Ceres, quadrangle maps from the HAMO (1:750,000) and LAMO (1:250,000) observations have been used. They conform to the quadrangle scheme presented in Greeley and Batson [51] for medium-sized bodies as outlined in Roatsch et al. [44,45], and they represent dedicated definitions for printed map sheets. The map scale for the geologic mapping investigation was defined to be 1:500,000 using the LAMO, and to some extent HAMO input data [29]. For polar areas, a map scale of 1:1,000,000 was defined.

All mapping bases used the pre-defined projection for each quadrangle mosaic. For the polar areas between 65° and 90°, an azimuthal stereographic projection was employed; for mid-latitudes, a Lambert conformal conic projection between 21° and 66° was employed with standard parallels at 30° and 58°; and for equatorial latitudes between 225°S and 22°N, a secant normal form of the Mercator projection was employed with standard parallels at 13°S and 13°N (Figure 2). In the latitudinal direction, each tile overlaps by 1°, and this representation was adopted for the geologic mapping investigation as well. All three map projections are conformal and thus maintain true shapes, which renders all impact craters as circular features despite having distortions in lengths and areas with increasing distance from standard parallels and the projection centers. Due to the ellipsoidal shape of Ceres and its simplified representation through a sphere, however, the true shape cannot effectively be maintained.

The mapping database was complemented by a layer for geologic symbols, according to the planetary geologic symbol set as published by the Federal Geographic Data Committee (FGDC), as well as metadata information (Figure 1) [52]. Color definitions for units followed, to a larger degree, the stratigraphic color assignments known from the Moon and also Mars (cf. Figures 3 and 4). Due to the associative character of color hue as a cartographic variable, it is straightforward to immediately detect and spatially correlate similar material units across the map and across map sheets. Material deposited during similar time periods show variations in visually variable color saturation and color intensity and thus allows for direct association of various materials formed during the same time period. Due to the lack of geologic complexity, as mapped in the investigation, unit and color variations are, however, limited.

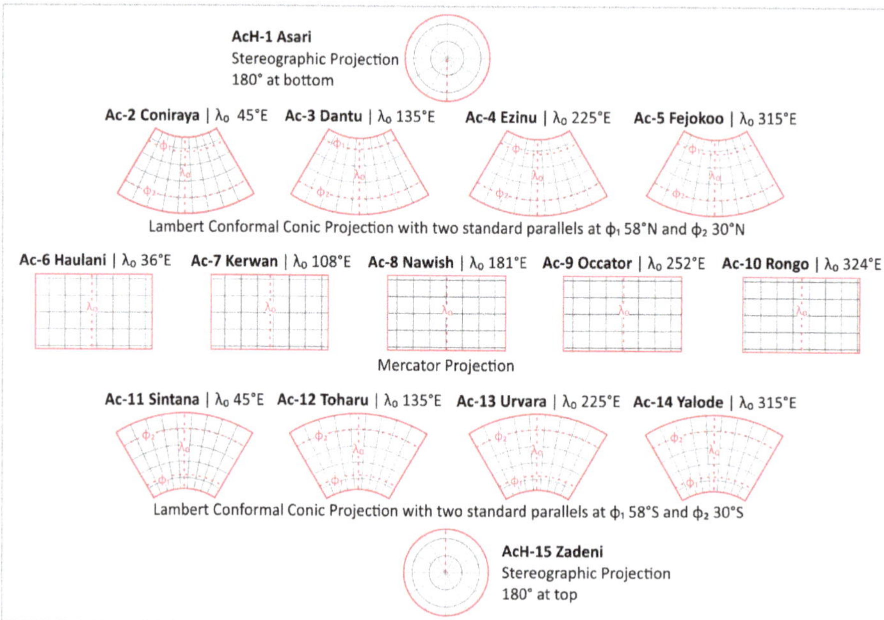

Figure 2. Mapping schema as implemented and applied for the geologic mapping investigation of Ceres (140 m/pixel, 1:1 M scale [45]). Latitude and longitude are adapted from the Kait coordinate system, approved by the IAU.

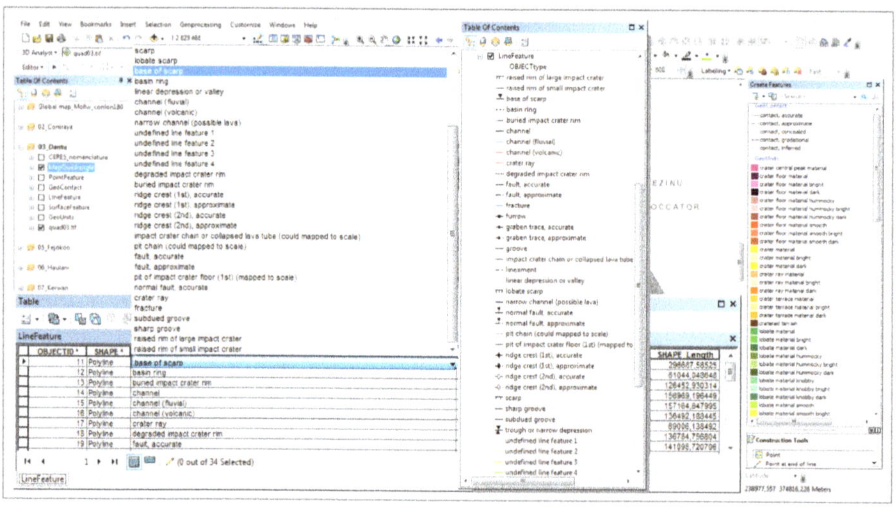

Figure 3. GIS interface with predefined attributes and related symbology.

The map-sheet layer along with symbol configuration contained map-frame definitions and the standard map graticule for print reproduction. For potential printing of map sheets, a map template was created (see Figure 5).

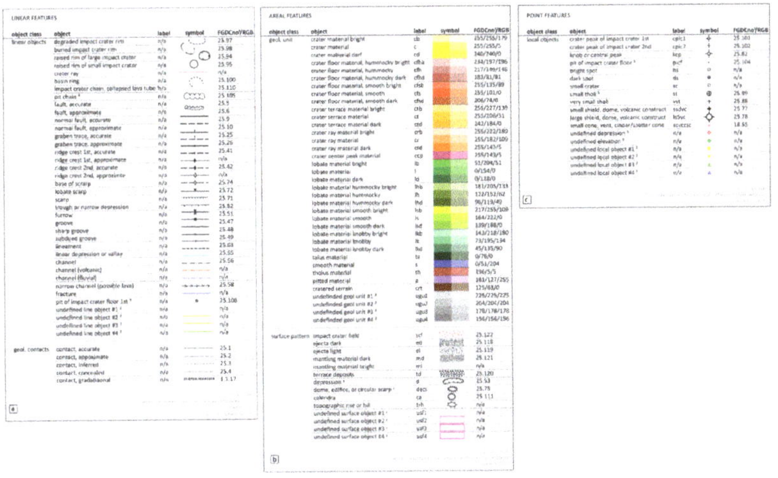

Figure 4. Symbology as used in the geologic mapping investigation. (**a**) Linear features, (**b**) areal features, and (**c**) point features. See also Ref. [53].

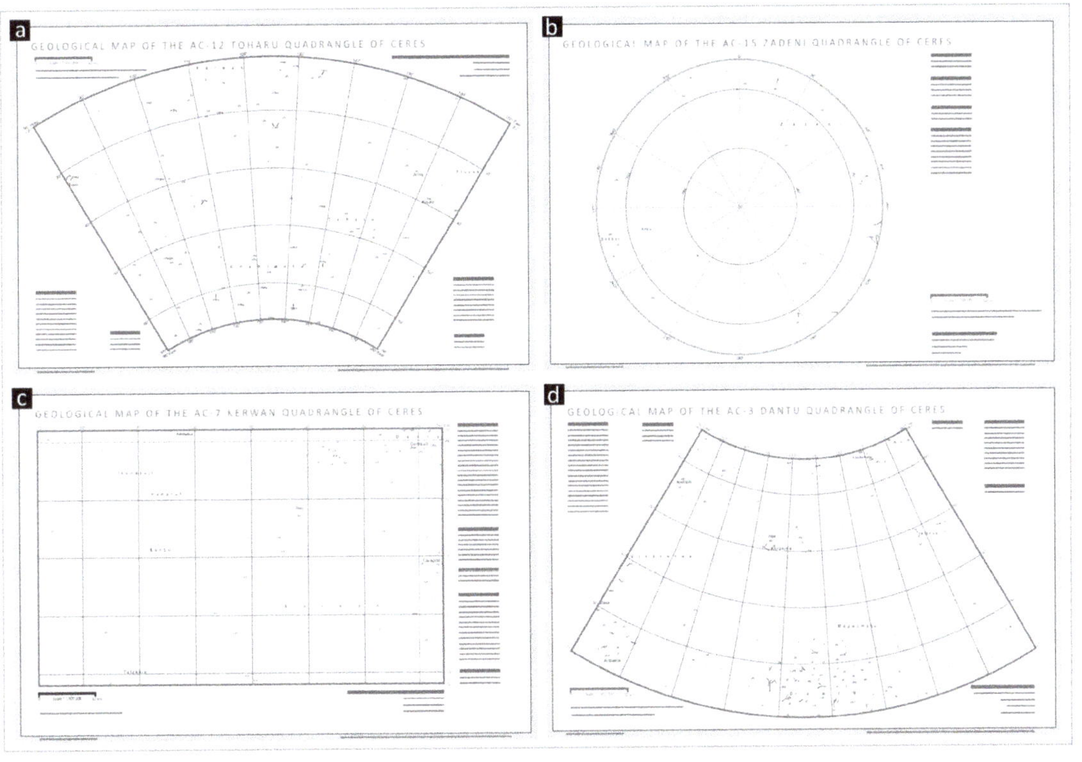

Figure 5. Mapping templates and principle layouts as distributed for the mapping investigation: (**a**) map sheet layout for higher-latitude mapping in the Southern hemisphere in Lambert conformal conic projection, (**b**) map sheet layout for mapping polar latitudes in polar stereographic projection, (**c**) map sheet layout for mapping in equatorial latitudes in Mercator projection, (**d**) map sheet layout for mapping in the Northern hemisphere in Lambert conformal conic projection.

2.3. Mapping Conduct

Geologic mapping is based on differences in materials indicative of different formation mechanisms and times. Compositional characteristics are challenging to distinguish in planetary mapping unless hyperspectral information is used to differentiate units. Differences are marked by traditional geologic criteria, such as lateral discontinuities, texture variations, different reflectances or different erosional levels [6]. Thus, terrain and relief information as well as lighting conditions play a crucial role in the mapping process to distinguish individual units.

Due to the lack of knowledge about a number of material units, a hybrid mapping approach has been taken, which also comprises descriptive terms related to texture properties or other characteristics. This approach is taken frequently for geologic mapping, in particular for historic maps due to lack of better information.

Geologic mapping was conducted on the panchromatic map-projected image as a primary dataset for each quadrangle extent (Figure 6). Depending on the mappers' preferences, units were either defined using geological contacts (contact method) or through geologically homogeneous areas (area method). The intention was to perform a refinement mapping based on the global surveying map by Ref. [54] and to iterate further down from HAMO to LAMO resolution, while the image mosaics were still being produced.

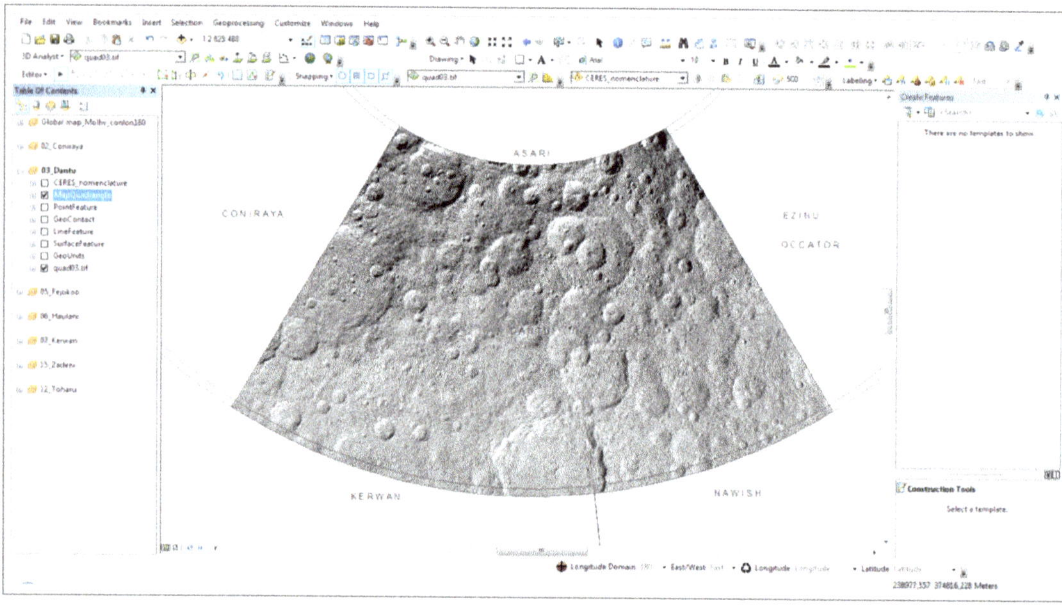

Figure 6. Interface of the GIS-based mapping template.

Mapping scales are commonly set to 3–4 times higher than the map scale in order to create smooth contact lines that account for the actual map scale. It needs to be kept in mind at this point that mapping was performed on screen, which translates to a reduction in effective resolution by a factor of about three due to the difference of screen (about 100 dpi) to paper-print resolution (about 300 dpi). Mappers were expected to communicate with each other for homogenizing quadrangle boundaries in terms of geometry and contents [29].

2.4. Research Data and Dissemination

For this mapping investigation, no concrete dissemination plan has been designed nor was it implemented other than through the production of research papers. While collaborative geologic mapping represented the core of the main investigation, the project did not intend to produce ready-to-print maps for display or as a working basis. Instead, the aims, as stated in Williams et al. [29] and also previously in Williams et al. [28], were primarily to provide a geologic context for other instrument teams and for the community at large to conduct future research. This, accordingly, puts a major constraint on the mapping investigation itself, which requires the mapping to be completed in the shortest amount of time in order to serve as a useful basis for decision-making during the mission's lifetime. Consequently, the reuse of mapping data and detailed map compilations could not be prioritized and any post-processing related to map and research-data generation was not part of the original investigation. In order to provide some degree of reusable data, however, the mapping coordinators agreed on a basic set of metadata that would allow archiving all data products and mapping products along with the most basic pieces of information (see Figure 7).

Geologic mapping results were planned to be published in a Special Issue where each quadrangle-based investigation is represented by a research paper and accompanied by the associated geologic mapping.

label	keyword	standard	value	domain
Planetary Body	TARGET_NAME [1]	PDS	"*Vesta*", "*Ceres*"	coded value (extensible)
Map Title	TITLE [2]	FGDC	"*Geological map of Ceres*"	text
Mapper	FULL_NAME [1]	PDS	"*Andrea Nass*"	coded value (extensible)
Institution	INSTITUTION_NAME [1]	PDS	"*German Aerospace Center*"	coded value (extensible)
Mapping Period	START_DATE [3] FINISH_DATE [3]	n/a	2008-04-13 2008-12-02	
Mapping Keywords	KEYWORDS [2]	FGDC	"*geomorphology*", "*fractures*"	coded value (extensible)
Map Purpose	PURPOSE [4]	n/a	"*Mapping Program*", "*Publication*"	coded value (extensible)
Map Status	STATUS [4]	n/a	"*Complete*", "*In work*", "*Planned*"	coded value
Map Projection Type	MAP_PROJECTION_TYPE [1]	PDS	"*Sinusoidal Equal Area*", "*Polar stereographic*"	coded value (extensible)
or/and	MAP_PROJECTION_NAME [1]	PDS	"*Lambert_Conformal_Conic*", "*Azimuthal_Equidistant*", "*Transverse_Mercator*"	coded value (extensible)
Map Scale	MAP_SCALE [3]	FGDC	"*1:500,000*"	coded value (extensible)
Map Boundary Coordinates	MAXIMUM_LATITUDE [1]	PDS	5.12	value range
	MAXIMUM_LONGITUDE [1]		78.39	value range
	MINIMUM_LATITUDE [1]		-2.03	value range
	MINIMUM_LONGITUDE [1]		94.39	value range
Base Data	IMAGE_ID [1]	PDS	"*.... (image number) .....*"	coded value (extensible)
Additional Data Available	ADD_DATA [3]	FGDC	"*yes*", "*no*", "*N/A*" (default)	coded value
if "yes" >>	ADD_DATA_INFO [3]			
	ADD_DATA_CREATOR [3]		"*morphometric data*", "*age determination analyses*"	coded value (extensible)
			"*Andrea Nass*"	coded value (extensible)

Figure 7. Metadata allocation for users in order to describe a GIS-based mapping product ([1] = PDS keyword, [2] = FGDC keyword, [3] = so far not standardized keyword), [4] = so far optional).

3. Discussion of Results

The geologic mapping investigation of Ceres led to a global geologic image with a scale of 1:10,000,000, HAMO maps at 1:1,000,000 scale, and 15 maps based on LAMO observations with a scale of 1:500,000, as well as research papers for 13 quadrangle investigations [29,55–65]. The global map has not been officially published and is available upon request only [66]. Individual HAMO/LAMO maps have been included as higher-resolution figures in each of the cited research papers; however, no maps have been published as separate map sheets thus far.

After the mapping investigation's termination, data have been made available via the data archive zenodo [67] by the first author. This archive also contains a non-unified version of the global map alongside GIS data packages available for download (Figure 8).

In Sections 3.1–3.3, we highlight three issues in particular: (1) completeness of foundational data, (2) originality of the cartographic concept, and (3) mapping results and reuse.

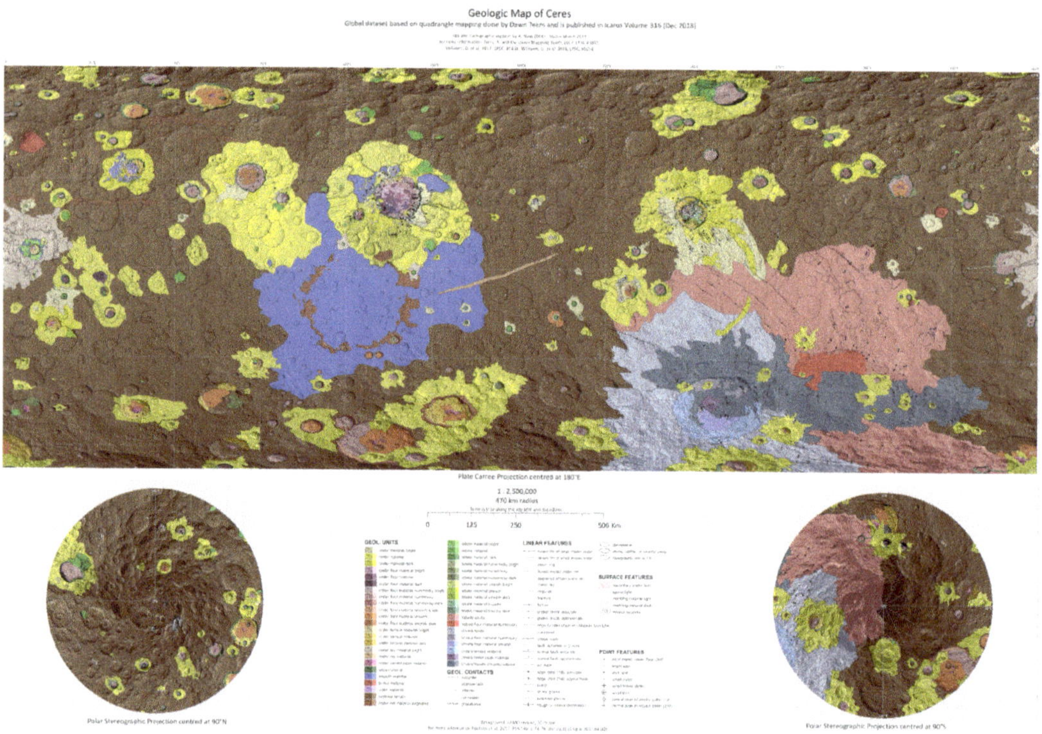

Figure 8. Map sheet of the global geology of Ceres at 1:2,500,000 as compiled from individual quadrangle maps in equidistant cylindrical and polar stereographic map projections [45]. The original data are available at Nass [67]. Note, this is a map sheet merge that was created after the termination of the geologic mapping investigation. This map and the geologic map image derived within the project at a scale 1:10,000,000 are different visualizations.

3.1. Completeness of Foundational Data

The employed data basis (see Section 2.1) consists of a subset of available images that were selected according to homogeneous illumination characteristics within a preceding project [44,45]. For the HAMO FC observations at Ceres, a total number of 2490 images and, for LAMO, a total number of 31,300 images were sent back by the DAWN spacecraft. For the panchromatic image *atlas*, a total number of 386 images for HAMO—largely from one single observation cycle—and about 11,550 images for LAMO have been processed [44,45], translating to roughly 16% of all HAMO and 37% of all LAMO images that have been included into the Ceres image mosaics and map sheets. As geologic mapping largely relies on relief and illumination variations for geologic contact tracing, the decision to not include all available data might raise some concern. In fact, the majority of the available data basis has not been incorporated for mapping in any systematic way.

The effects of varying illumination conditions are known, in particular when it comes to morphometric measurements and to tracing the superposition of units so that views under varying illumination geometries are crucial for detailed mapping. This holds even more as there is no validation available. It is therefore likely, that the selection of a foundational data product that has been created with a different target audience in mind (see the discussion in the next section) has introduced considerable bias and ambiguities that limit the potential of the final product. It has been shown in van Gasselt and Nass [68] that geologic mapping based on remote sensing data introduces a number of uncertainties, and thus by reducing these uncertainties, the integrity of the final product could be improved. Given the lack of geologic complexity in the mapping area, additional datasets might not change the overall picture considerably.

However, since data are readily available, small in size, and straightforward to process, the systematic inclusion of additional data might not have impacted the overall mapping timeline and would have increased the value of the outcome by making the best use of the available data.

3.2. *Originality of a Cartographic Concept*

As outlined in the methodology section, a working cartographic concept requires information about recipients, i.e., the map readers, in order to tailor the cartographic products to the needs of the target audience. By reviewing data products from this project, we do not intend to criticize decisions that are ultimately controlled by other project constraints, overarching boundary conditions, or future plans. Rather, the authors attempt to discuss and highlight the potential that different approaches might have or which could be discussed for future mapping projects. While the projects provide ample material to discuss decisions and take approaches on even the lowest levels, with chosen map scales being one of them, the authors want to focus here on the more overarching issue of borrowing a cartographic concept designed for different target audiences and the consequences.

For the geologic mapping investigation of Ceres, the decision was made to follow the quadrangle definition as proposed by Greeley and Batson [51] and as employed for the so-called *atlasses* of Vesta and Ceres. The original proposition by Greeley and Batson [51] was published at a time that digital cartography was perceived in a traditional map-making way and before the introduction of GIS technology in the planetary sciences. At that time, physical print maps were the only means of communication of cartographic results. And, print maps have also been a major focus when the collection of image map sheets was published for Vesta and Ceres, as discussed in Ref. [44,45,69,70], in particular when it comes to the discussion about map scales and map sheet sizes.

For both Vesta and Ceres, all quadrangle image maps were published and archived in the Asteroid/Dust Subnode of PDS's Small Bodies Node. These include

- Map sheets in the portable document format (PDF)l
- Global and quadrangle-based digital raster image mosaics in the PDS IMG format;
- Raster images in GeoTIFF format ready to use for further processing or analysis.

These main datasets were stored alongside extensive metadata and other data derivatives, such as color products [46,71].

While the image map sheets represent valuable and interesting cartographic products in themselves, it remains to be seen who the target audience will be as it has not been mentioned in any of the associated publications. If the target audience is future researchers, it would be interesting to see how these map sheets are being actually used for investigations, measurements, or future context studies. Until then, the map products must be seen as serving a more intrinsic purpose. On the other hand, the foundational global image mosaics for Vesta and Ceres will likely be used for subsequent studies as they were uploaded to a central archive and are well described through metadata. Diversity of file formats will certainly ensure that future reuse and integration into projects are possible. It would have been exciting to see a publication of these maps within a modern webGIS platform using dynamic zoom functionality and a data layer with contours that are filtered and generalized according to each visualization scale rather than being statically generalized for a single print resolution. Formats and dissemination platforms reflect that conceptual core idea of generating print maps in addition to GIS-ready image products for future reuse.

As the geologic mapping investigation of Ceres (and Vesta) copy the quadrangle scheme into its very own cartographic concept, it would be sensible to assume that the outcome consists of ready-to-use geologic map sheets as well as digital products which are eventually published in a public archive. The target audiences are clearly defined in Williams et al. [28,29]; however, a compilation of print map sheets has not been the aim of that investigation despite forming the core of the cartographic framework by design. Also, the introduction of a GIS workflow along with a GIS data model to help maintain data integrity could not help to create a unified, homogeneous, and consistent output across all quadrangle boundaries.

After all, there is a significant loss of potential as by using the GIS data model in its intended and potentially more constraining form, individual quadrangle mappings could become geometrically, topologically, and semantically consistent, which would have allowed for a unified map compilation and the creation of rich data layers alongside their metadata. Also, the derivation of a unified geologic map as a side product would have been possible. With the combination of digital contours, such geologic maps could become rich in information and they could be integrated into a webGIS platform with dynamic zoom functionality and the potential for overlaying additional information.

3.3. Mapping Results and Reuse

The scientific investigation reports along with individual geologic mappings have been published in the Special Issue "The Geologic Mapping of Ceres" [29]. A compilation of maps and reports has been completed by individual quadrangle mappers. Adjustments of geologic contacts and symbols to homogenize map sheets transitions were encouraged and have been realized between some quadrangles by the mappers themselves. A final geometric and topological adjustment of contacts across quadrangle boundaries and homogenization of results to compile a unified map has not been carried out as no post-processing time was allocated for this project. The mapping project was technically completed within its scheduled time as it was considered complete after all papers were submitted. Individual quadrangle maps have not been submitted to research data repositories in a consistent format thus far, and basic post-processing was conducted outside of the project's timeframe by the first author (see Figures 8 and 9). A number of mappers decided to provide maps as higher-resolution figures associated with their publication.

A dataset comprising a compilation of all individual quadrangle mappings has been uploaded to the archive platform zenodo (zenodo.org, European Organization For Nuclear Research and OpenAIRE [72]). This includes supporting GIS files and a global dataset that, however, are not unified as it would have required discussions with individual mappers about geometric and semantic aspects, which was not achievable after the project's scheduled termination [67].

Given the amount of effort that usually goes into joint discussions, mapping definitions, and the development of data models, a higher level of constraints for achieving a higher consistency and integrity of research data products could have extremely positive effects on data value by only imposing minor limitations. A more restrictive data model with more constraints on geometry and topology, including mechanisms for validation, and a higher degree of consistent generalization, either by enforcing consistent mapping scales or by extensive post-processing, would undoubtedly have reduced the problem of inconsistencies along individual map boundaries. This problem is even more pronounced as mapping quadrangles have overlaps in the latitude direction by design. These problems are not purely geometrical as each geometry relates directly to the mapping contents. Thus, the lack of synchronized contacts among most map boundaries causes non-trivial challenges for post-processing as this is only reasonable in discussion with individual mappers. While the decision was made to give individual mappers the maximum amount of freedom and to not impose any constraints that would limit mappers in their approach, it also causes non-resolvable challenges when it comes to unifying geometric and semantic transitions after a project is terminated. The relevance of this, ultimately, will vary from mapping project to mapping project; however, the imbalance between mapping integrity, efficiency, and freedom might be hard to justify at the end. When considering the intended target audience, i.e., mission instrument teams and future researchers, one has to accept that other instrument teams are likely the group who are least able to resolve such semantic inconsistencies.

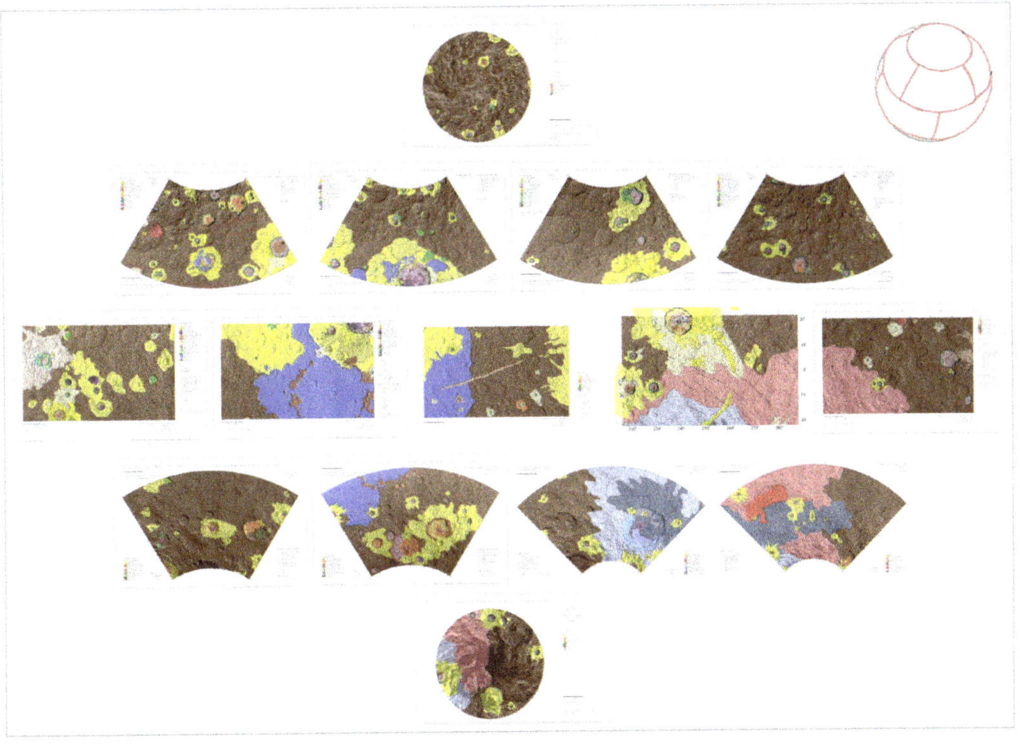

Figure 9. Compilation of the 15 map sheets of the mapping investigation of Ceres as overview figure. For details regarding mapping criteria and map keys refer to the original publications introduced in the Special Issue "The Geologic Mapping of Ceres" [29].

As it was not designed for the mapping investigation, and due to inconsistencies and integrity problems as outlined above, no targeted research data management could be realized. All adjustments and corrections have taken place outside the project's time frame.

4. Summary and Conclusions

4.1. Summary

This geologic mapping investigation focused on a multi-user quadrangle-based mapping approach with individual quadrangles being defined using a map sheet tiling scheme designed for print maps [51]. Mappers were provided with map and mapping templates, and the finalized map contents could optionally be incorporated into individual map-sheet layouts for exporting. The results were published in research papers with maps being simplified and transferred to figures. While map-sheet compilations for print and display take up resources to create, compile, review, and revise, it could have been a secondary aim of such an investigation for which, however, additional time had to be allocated. Furthermore, within the boundaries of the geologic mapping investigation, no resources were committed to producing an archivable and reusable research data product that would provide a transparent and consistent dataset to be exploited by future researchers.

While the aims of this mapping investigation, as stated in Williams et al. [29], have been met with respect to the publication of individual geologic investigations, it could be shown that, by using a more targeted cartographic concept,

1. A global set of HAMO-based geologic quadrangle maps at more conservative map scales could have been produced as digital datasets as well as analog map sheets;
2. A global geologic map at HAMO resolution could have been compiled with consistent geometry, topology, and map semantics by making use of the data model;
3. A consistent research data product could have been designed and archived that would be reusable for future research, in particular for subsequent larger-scale mappings.

Subsequent mapping investigations would then build on these findings and create a more refined mapping investigation using LAMO data (as currently planned to be carried out by the USGS). There are plans currently to remap and compile a global map based on HAMO data and a higher-resolution global geologic map based on LAMO data [73–75]. The LAMO geologic map is planned to be published as a USGS Special Investigation Map (SIM) at a map scale of 1:3,000,000 for the equatorial region (±60°N and S) and 1:1,500,000 scale for the polar regions (>60°N and S) following the standards from the USGS SIM.

4.2. Recommendations

Research Data Management. The awareness of targeted research data management and the need for making research results available for future reuse have been growing significantly over the last decades. Apart from reducing the amount of redundant work, the strategy would also allow for more indirect reviews among peers and for obtaining additional insights through the recombination and derivation of research results. In order to achieve this, not only direct project aims but also strategies for dissemination need to be developed, as it is already requested by various funding agencies. For a mapping investigation similar to the use case reported here, dissemination strategies, data integrity checks, as well as metadata specifications would constitute important points of discussion during the conceptual phase. Furthermore, dissemination strategies would also allow for technical constraints to be adjusted and consistency checks to be enforced, which otherwise would not be implemented. These would be summarized as technical tools for quality control.

Metadata. Targeted metadata are required to make sure not only can data be found in the archives but also that a dataset's lineage and associated researchers are communicated in a transparent and standardized way. Considerable potential has been lost within this geologic mapping investigation although common metadata attributes have been agreed upon. For communication of producers, versions, and other lineage information, it is of utmost importance to attach at least basic metadata information to each map/dataset (cf. Figure 7).

Mapping Constraints. One important tool for quality control is the implementation of targeted mapping constraints. Upon working towards the primary project results, a balance needs to be found between addressing the needs for targeted research data management and leaving wide-open research freedom. While this could lead to a discussion about the role of mapping as a research tool [68], a more pragmatic approach would be to find a balance discussed within the team and constrained through a mapping data model and topology rule checks on contacts and units.

Cartographic Support. Geologic mapping investigations, in particular collaborative ones with an increased level of complexity, depend highly on support from the cartographic and data management domain. This is important not only for the provision of data, templates, and data-support tools but also to increase the level of awareness of the need for such management.

Print Maps vs. webGIS. It feels timely to start discussing how usable as an asset cartographic print maps, i.e., traditional map sheets for display and as a work product, can be assets in the future as efficient tools for exploration. Analog working maps, i.e., charts and plans, might be of limited use nowadays in the planetary domain, in particular as new types of information such as more spatial context cannot be added due to their static character. Printable map sheets certainly remain beautiful products and demonstrate the meticulous work and experience that have gone into the creation of such maps. If the cartographic concept has been well designed, a map sheet compilation is probably a minor additional step to be taken to complete the investigation. It, therefore, seems even more important to discuss the implementation of efficient webGIS solutions which allow for hosting not only the map sheet in digital format but also all cartographic objects along with their semantic value for querying, measuring, and investigations. The provision of data layers served over the internet (e.g., OGC/Web Feature Service (WFS) and OGC/Web Map Services (WMS)) would in addition allow users to access maps using web interfaces or their own systems and start working with such products. In such a way, these maps could become new foundational map products to be used as the basis for future investigation.

Iterative Development. While the geologic mapping investigation consists of distinct phases, the overall conduct is iterative in nature due to new data arriving during the mapping phases and due to changing mapping foci. The project has shown that constant communication between mappers and with data management support would have been needed to arrive at consistent products during the mapping phase.

Planning for Derivative Products. The reviewed investigation has shown that the compilation of mapping quadrangles could have led to a consistent derivative product on the global geology of Ceres compiled at a smaller map scale. Consistent mapping scales, synchronization of unit boundaries at map quadrangle boundaries, and a priori agreement on the terrain descriptors and their exact delineation would have provided a more consistent image. An internal iterative review for each quadrangle with its 4–8 adjacent quadrangles could have effortlessly created a highly consistent set of products, which could then be generalized and simplified systematically to provide input for a global map. This opportunity has been missed as a wider definition of freedom and lack of constraints (including reviews) led to geometric and topologic inconsistencies, which cannot be solved by a cartographer or GIS expert due to a lack of detailed topical knowledge of the mapper's view on the local geology and geomorphology.

Joint Concepts. A joint conceptual understanding and common baseline are needed, not only for the general direction of the mapping investigation but also with respect to the level of detail and the exact criteria for delineation. Varying topical expertise among different mappers can easily lead to different mapping foci, and while similar geologic features might appear across quadrangles, not all mappers have mapped such features. This further begs the question of how far topical mapping rather than quadrangle mapping would constitute an additional step for guaranteeing consistent mapping results.

Author Contributions: Conceptualization, A.N.; methodology, A.N.; writing—original draft preparation, A.N. and S.v.G.; writing—review and editing, A.N. and S.v.G.; visualization, A.N. All authors have read and agreed to the published version of the manuscript.

Funding: This research received no external funding.

Data Availability Statement: The compiled database for the mapping investigation can be found as Geologic Map of Ceres [Dawn Mission] – Global dataset based on the 15 individual quadrangle maps at https://zenodo.org/record/7989803.

Acknowledgments: The authors thank the three anonymous reviewers who have provided comments and suggestions to improve the manuscript.

Conflicts of Interest: The authors declare no conflict of interest.

References

1. USGS. *Geological Atlas of the Moon*; USGS: Reston, VA, USA, 2023. Available online: https://www.lpi.usra.edu/resources/mapcatalog/usgs/ (accessed on 26 August 2023).
2. Wilhelms, D. Geologic Mapping. In *Planetary Mapping*; Greeley, R., Batson, R.M., Eds.; Cambridge Planetary Science Series; Cambridge University Press: Cambridge, UK, 1990; pp. 208–259.
3. Abbud-Madrid, A. Space and Planetary Resources. In *Planetary Geology*; Rossi, A.P., van Gasselt, S., Eds.; Springer Praxis Books; Springer: Cham, Switzerland, 2018; pp. 369–394. [CrossRef]
4. van Gasselt, S.; Neukum, G. Chronostratigraphy. In *Encyclopedia of Astrobiology*; Gargaud, M., Irvine, W.M., Amils, R., Cleaves, H., Pinti, D.L., Quintanilla, J.C., Rouan, D., Spohn, T., Tirard, S., Viso, M., Eds.; Springer: Berlin/Heidelberg, Germany, 2015; pp. 454–466. [CrossRef]
5. Guest, J.; Butterworth, P. *Planetary Geology*; John Wiley and Sons: Hoboken, NJ, USA, 1979; pp. 15–31.
6. Pondrelli, M.; Baker, V.R.; Hauber, E. Geologic Tools. In *Planetary Geology*; Rossi, A.P., van Gasselt, S., Eds.; Springer Praxis Books; Springer: Cham, Switzerland, 2018; pp. 15–31. [CrossRef]
7. Laura, J.R.; Beyer, R.A. Knowledge Inventory of Foundational Data Products in Planetary Science. *Planet. Sci. J.* **2021**, *2*, 18. [CrossRef]
8. Tognon, G.; Pozzobon, R.; Massironi, M.; Ferrari, S. Geologic Mapping and Age Determinations of Tsiolkovskiy Crater. *Remote Sens.* **2021**, *13*, 3619. [CrossRef]
9. Debniak, K.; Mege, D.; Gurgurewicz, J. Geomorphology of Ius Chasma, Valles Marineris, Mars. *J. Maps* **2017**, *13*, 260–269. [CrossRef]
10. Skinner, J.A.; Huff, A.E.; Fortezzo, C.M.; Gaither, T.; Hare, T.M.; Hunter, M.A.; Buban, H. *Planetary Geologic Mapping—Program Status and Future Needs*; Technical Report U.S. Geological Survey Open-File Report 2019–1012; United States Geological Survey: Reston, VA, USA, 2019. [CrossRef]
11. Williams, D. NASA's Planetary Geologic Mapping Program: Overview. *ISPRS—Int. Arch. Photogramm. Remote. Sens. Spat. Inf. Sci.* **2016**, *XLI-B4*, 519–520. [CrossRef]
12. PDS. Planetary Data System. 2023. Available online: https://pds.nasa.gov/ (accessed on 4 January 2023).
13. PSA. Planetary Science Archive (ESA). 2023. Available online: https://www.cosmos.esa.int/web/psa/psa-introduction (accessed on 4 January 2023).
14. The International Planetary Data Alliance (IPDA) Available online: https://ipda.jpl.nasa.gov/ (accessed on 26 August 2023).
15. Radebaugh, J.; Thomson, B.J.; Archinal, B.; Beyer, R.; DellaGuistina, D.; Fassett, C.; Gaddis, L.; Hagerty, J.; Hare, T.; Laura, J.; et al. A roadmap for planetary spatial data infrastructure. In Proceedings of the Lunar and Planetary Science Conference, Woodlands, TX, USA, 18–22 March 2019; p. 1667.
16. Rossi, A.; Massironi, M.; Altieri, F.; van der Bogert, C.; Hiesinger, H.; Mangold, N.; Rothery, D.; Balme, M.; Carli, C.; Pozzobon, R.; et al. PLANMAP: Geological mapping supporting the exploration of the Moon, Mars and Mercury. In Proceedings of the 69th International Astronautical Congress, Bremen, Germany, 1–5 October 2018; p. 47635.
17. Massironi, M.; Altieri, F.; Hiesinger, H.; Mangold, N.; Rothery, D.; Rossi, A.P.; Balme, M.; Carli, C.; Capaccioni, F.; Cremonese, G.; et al. Towards integrated geological maps and 3D geo-models of planetary surfaces: The H2020 PLANetary MAPping project. In Proceedings of the EGU General Assembly Conference Abstracts, Vienna, Austria, 8–13 April 2018; European Geosciences Union (EGU): Munich, Germany, 2018; p. 18106.
18. Semenzato, A.; Massironi, M.; Ferrari, S.; Galluzzi, V.; Rothery, D.A.; Pegg, D.L.; Pozzobon, R.; Marchi, S. An Integrated Geologic Map of the Rembrandt Basin, on Mercury, as a Starting Point for Stratigraphic Analysis. *Remote Sens.* **2020**, *12*, 3213. [CrossRef]
19. Crichton, D.; Padams, J.; Hollins, G.; Hughes, J.S.; Joyner, R.; Law, E.; Cayanan, M. PDS data services initiative: Evolving towards data-driven capabilities to enable planetary science research. In *Lunar and Planetary Science Conference*; LPI Contribution; Lunar and Planetary Institute: Houston, TX, USA, 2020; p. 2754.
20. Erard, S.; Cecconi, B.; Le Sidaner, P.; Chauvin, C.; Rossi, A.P.; Minin, M.; Capria, T.; Ivanovski, S.; Schmitt, B.; Génot, V.; et al. Virtual European Solar and Planetary Access (VESPA): A Planetary Science Virtual Observatory Cornerstone. *Data Sci. J.* **2020**, *19*, 5334. [CrossRef]
21. Law, E.; Cayanan, M.; Crichton, D.; Hollins, G.; Hughes, S.; Padams, J. *Planetary Data System (PDS) Tools and Tool Registry, Proceedings of the EPSC DPS Joint Meeting, Geneva, Switzerland, 15–20 September 2019*; Europlanet/American Astronomical Society: Strasbourg, France. 2019; p. EPSC–DPS2019–170.
22. Baratoux, D.; Chennaoui Aoudjehane, H.; Gibson, R.; Lamali, A.; Reimold, W.; Sapah, M.; Chabou, M.; Habarulema, J.; Jessell, M.; Mogessie, A.; et al. Africa Initiative for Planetary and Space Sciences. *Eos* **2017**. [CrossRef]
23. Europlanet Society. Europlanet Diversity Working Group. 2023. Available online: https://www.europlanet-society.org/diversity-working-group/ (accessed on 18 February 2023).
24. Nass, A.; Asch, K.; van Gasselt, S.; Rossi, A.P.; Besse, S.; Cecconi, B.; Frigeri, A.; Hare, T.; Hargitai, H.; Manaud, N. Facilitating reuse of planetary spatial research data—Conceptualizing an open map repository as part of a Planetary Research Data Infrastructure. *Planet. Space Sci.* **2021**, *204*, 105269. [CrossRef]
25. Sierks, H.; Keller, H.U.; Jaumann, R.; Michalik, H.; Behnke, T.; Bubenhagen, F.; Buettner, I.; Carsenty, U.; Christensen, U.; Enge, R.; et al. The Dawn Framing Camera. *Space Sci. Rev.* **2011**, *163*, 263–327. [CrossRef]

26. De Sanctis, M.C.; Coradini, A.; Ammannito, E.; Filacchione, G.; Capria, M.T.; Fonte, S.; Magni, G.; Barbis, A.; Bini, A.; Dami, M.; et al. The VIR Spectrometer. *Space Sci. Rev.* **2011**, *163*, 329–369. [CrossRef]
27. Prettyman, T.; Feldman, W.; McSween, H.; Dingler, R.; Enemark, D.; Patrick, D.; Storms, S.; Hendricks, J.; Morgenthaler, J.; Pitman, K.; et al. Dawn's Gamma Ray and Neutron Detector. *Space Sci. Rev.* **2011**, *163*, 371–459. [CrossRef]
28. Williams, D.A.; Yingst, R.A.; Garry, W.B. Introduction: The geologic mapping of Vesta. *Icarus* **2014**, *244*, 1–12. [CrossRef]
29. Williams, D.A.; Buczkowski, D.L.; Mest, S.C.; Scully, J.E.C.; Platz, T.; Kneissl, T. Introduction: The geologic mapping of Ceres. *Icarus* **2018**, *316*, 1–13. [CrossRef]
30. Williams, D.A.; Jaumann, R.; McSween, H.Y.; Marchi, S.; Schmedemann, N.; Raymond, C.A.; Russell, C.T. The chronostratigraphy of protoplanet Vesta. *Icarus* **2014**, *244*, 158–165. [CrossRef]
31. Sawabe, Y.; Matsunaga, T.; Rokugawa, S. Automated detection and classification of lunar craters using multiple approaches. *Adv. Space Res.* **2006**, *37*, 21–27. .: 10.1016/j.asr.2005.08.022. [CrossRef]
32. Stepinski, T.; Vilalta, R.; Ghosh, S. Machine Learning Tools for Automatic Mapping of Martian Landforms. *IEEE Intell. Syst.* **2007**, *22*, 100–106. [CrossRef]
33. Urbach, E.R.; Stepinski, T.F. Automatic detection of sub-km craters in high resolution planetary images. *Planet. Space Sci.* **2009**, *57*, 880–887. [CrossRef]
34. Jin, S.; Zhang, T. Automatic detection of impact craters on Mars using a modified adaboosting method. *Planet. Space Sci.* **2014**, *99*, 112–117. [CrossRef]
35. Li, W.; Di, K.; Yue, Z.; Liu, Y.; Sun, S. Automated Detection of Martian Gullies from HiRISE Imagery. *Photogramm. Eng. Remote. Sens.* **2015**, *81*, 913–920. [CrossRef]
36. Jiang, S.; Lian, Z.; Yung, K.L.; Ip, W.H.; Gao, M. Automated Detection of Multitype Landforms on Mars Using a Light-Weight Deep Learning-Based Detector. *IEEE Trans. Aerosp. Electron. Syst.* **2022**, *58*, 5015–5029. [CrossRef]
37. Fortezzo, C.; Spudis, P.D.; Harrel, S.L. Release of the Digital Unified Global Geologic Map of the Moon at 1:5,000,000-Scale. In Proceedings of the 51st Lunar and Planetary Science Conference, Lunar and Planetary Institute, Houston, TX, USA, 15–18 August 2020; p. 2760.
38. GO FAIR. GO FAIR Initiative: Make Your Data & Services FAIR. 2023. Available online: https://www.go-fair.org (accessed on 2 January 2023).
39. CODATA. The Committee on Data for Science and Technology. 2023. Available online: https://codata.org (accessed on 2 January 2023).
40. RDA. Research Data Sharing without Barriers. 2023. Available online: https://www.rd-alliance.org (accessed on 15 February 2023).
41. Wilkinson, M.D.; Dumontier, M.; Aalbersberg, I.J.; Appleton, G.; Axton, M.; Baak, A.; Blomberg, N.; Boiten, J.; Da Silva Santos, L.B.; Bourne, P.E.; et al. The FAIR Guiding Principles for scientific data management and stewardship. *Sci. Data* **2016**, *3*, 160018–160018. [CrossRef] [PubMed]
42. Roatsch, T.; Kersten, E.; Matz, K.; Preusker, F.; Scholten, F.; Jaumann, R.; Raymond, C.A.; Russell, C.T. Ceres Survey Atlas derived from Dawn Framing Camera images. *Planet. Space Sci.* **2016**, *121*, 115–120. [CrossRef]
43. Russell, C.T.; Raymond, C.A. The Dawn Mission to Vesta and Ceres. *Space Sci. Rev.* **2011**, *163*, 3–23. [CrossRef]
44. Roatsch, T.; Kersten, E.; Matz, K.; Preusker, F.; Scholten, F.; Jaumann, R.; Raymond, C.A.; Russell, C.T. High-resolution Ceres High Altitude Mapping Orbit atlas derived from Dawn Framing Camera images. *Planet. Space Sci.* **2016**, *129*, 103–107. [CrossRef]
45. Roatsch, T.; Kersten, E.; Matz, K.; Preusker, F.; Scholten, F.; Jaumann, R.; Raymond, C.A.; Russell, C.T. High-resolution Ceres Low Altitude Mapping Orbit Atlas derived from Dawn Framing Camera images. *Planet. Space Sci.* **2017**, *140*, 74–79. [CrossRef]
46. Roatsch, T.; Kersten, E.; Matz, K.D.; Preusker, F.; Scholten, F.; Elgner, S.; Schroeder, S.; Jaumann, R.; Raymond, C.; Russell, C.T. *DAWN Framing Camera Derived Ceres Global Mosaics V1.0 (DAWN-A-FC2-5-CERESMOSAIC-V1.0)*; NASA Planetary Data System, Asteroid/Dust Subnode; NASA/Planetary Science Institute: Tucon, AZ, USA, 2016..
47. Preusker, F.; Scholten, F.; Matz, K.; Roatsch, T.; Elgner, S.; Jaumann, R.; Joy, S.; Raymond, C.; Russell, C. Shape model and rotational state of dwarf planet Ceres from Dawn FC stereo images. In Proceedings of the European Planetary Science Congress, Nantes, France, 27 September–3 November 2015; p. EPSC2015–186.
48. Nass, A.; the DAWN Mapping Team. One GIS-based Data Structure for Geological Mapping using 15 Map Sheets—Dawn at Ceres. In *Abstracts of the Lunar and Planetary Science Conference XLVIII*; Lunar and Planetary Institute: Houston, TX, USA, 2017; p. 1892.
49. van Gasselt, S.; Nass, A. Planetary mapping—The datamodel's perspective and GIS framework. *Planet. Space Sci.* **2011**, *59*, 1231–1242. [CrossRef]
50. Skinner, J.A.; Huff, A.; Black, S.; Buban, H.; Fortezzo, C.; Gaither, T.; Hare, T.; Hunter, M. *Planetary Geologic Mapping Protocol—2022*; U.S. Geological Survey Standards, Techniques and Methods 11–B13; U.S. Geological Survey: Reston, VI, USA, 2022. [CrossRef]
51. Greeley, R.; Batson, R.M. *Planetary Mapping*; Cambridge University Press: New York, NY, USA, 1990.
52. Nass, A.; van Gasselt, S.; Jaumann, R.; Asche, H. Implementation of cartographic symbols for planetary mapping in geographic information systems. *Planet. Space Sci.* **2011**, *59*, 1255–1264. [CrossRef]

53. Federal Geographic Data Committee FGDC Digital Cartographic Standard for Geologic Map Symbolization. GDC Document Number FGDC-STD-013-2006. Available online: https://ngmdb.usgs.gov/fgdc_gds/geolsymstd.php (accessed on 26 August 2023).
54. Buczkowski, D.L.; Schmidt, B.E.; Williams, D.A.; Mest, S.C.; Scully, J.E.C.; Ermakov, A.I.; Preusker, F.; Schenk, P.; Otto, K.A.; Hiesinger, H.; et al. The geomorphology of Ceres. *Science* **2016**, *353*, aaf4332. [CrossRef]
55. Ruesch, O.; McFadden, L.A.; Williams, D.A.; Hughson, K.H.; Pasckert, J.H.; Scully, J.; Kneissl, T.; Roatsch, T.; Naß, A.; Preusker, F.; et al. Geology of Ceres' North Pole quadrangle with Dawn FC imaging data. *Icarus* **2018**, *316*, 14–27. [CrossRef]
56. Pasckert, J.; Hiesinger, H.; Ruesch, O.; Williams, D.; Nass, A.; Kneissl, T.; Mest, S.; Buczkowski, D.; Scully, J.; Schmedemann, N.; et al. Geologic mapping of the Ac-2 Coniraya quadrangle of Ceres from NASA's Dawn mission: Implications for a heterogeneously composed crust. *Icarus* **2018**, *316*, 28–45. [CrossRef]
57. Scully, J.E.; Buczkowski, D.; Neesemann, A.; Williams, D.; Mest, S.; Raymond, C.; Nass, A.; Hughson, K.; Kneissl, T.; Pasckert, J.; et al. Ceres' Ezinu quadrangle: A heavily cratered region with evidence for localized subsurface water ice and the context of Occator crater. *Icarus* **2018**, *316*, 46–62. [CrossRef]
58. Hughson, K.H.; Russell, C.; Williams, D.; Buczkowski, D.; Mest, S.; Pasckert, J.; Scully, J.; Combe, J.P.; Platz, T.; Ruesch, O.; et al. The Ac-5 (Fejokoo) quadrangle of Ceres: Geologic map and geomorphological evidence for ground ice mediated surface processes. *Icarus* **2018**, *316*, 63–83. [CrossRef]
59. Krohn, K.; Jaumann, R.; Otto, K.; Schulzeck, F.; Neesemann, A.; Nass, A.; Stephan, K.; Tosi, F.; Wagner, R.; Zambon, F.; et al. The unique geomorphology and structural geology of the Haulani crater of dwarf planet Ceres as revealed by geological mapping of equatorial quadrangle Ac-6 Haulani. *Icarus* **2018**, *316*, 84–98. [CrossRef]
60. Williams, D.A.; Kneissl, T.; Neesemann, A.; Mest, S.; Palomba, E.; Platz, T.; Nathues, A.; Longobardo, A.; Scully, J.; Ermakov, A.; et al. The geology of the Kerwan quadrangle of dwarf planet Ceres: Investigating Ceres' oldest, largest impact basin. *Icarus* **2018**, *316*, 99–113. [CrossRef]
61. Frigeri, A.; Schmedemann, N.; Williams, D.; Chemin, Y.; Mirino, M.; Nass, A.; Carrozzo, F.G.; Castillo-Rogez, J.; Buczkowski, D.L.; Scully, J.E.; et al. The geology of the Nawish quadrangle of Ceres: The rim of an ancient basin. *Icarus* **2018**, *316*, 114–127. [CrossRef]
62. Buczkowski, D.; Williams, D.; Scully, J.; Mest, S.; Crown, D.; Schenk, P.; Jaumann, R.; Roatsch, T.; Preusker, F.; Nathues, A.; et al. The geology of the occator quadrangle of dwarf planet Ceres: Floor-fractured craters and other geomorphic evidence of cryomagmatism. *Icarus* **2018**, *316*, 128–139. [CrossRef]
63. Platz, T.; Nathues, A.; Sizemore, H.; Crown, D.; Hoffmann, M.; Schäfer, M.; Schmedemann, N.; Kneissl, T.; Neesemann, A.; Mest, S.; et al. Geological mapping of the Ac-10 Rongo Quadrangle of Ceres. *Icarus* **2018**, *316*, 140–153. [CrossRef]
64. Schulzeck, F.; Krohn, K.; v.d. Gathen, I.; Schmedemann, N.; Stephan, K.; Jaumann, R.; Williams, D.; Wagner, R.; Buczkowski, D.; Mest, S.; et al. Geologic mapping of the Ac-11 Sintana quadrangle: Assessing diverse crater morphologies. *Icarus* **2018**, *316*, 154–166. [CrossRef]
65. Crown, D.A.; Sizemore, H.G.; Yingst, R.A.; Mest, S.C.; Platz, T.; Berman, D.C.; Schmedemann, N.; Buczkowski, D.L.; Williams, D.A.; Roatsch, T.; et al. Geologic mapping of the Urvara and Yalode Quadrangles of Ceres. *Icarus* **2018**, *316*, 167–190. [CrossRef]
66. Williams, D.; Buczkoswki, D.; Crown, D.; Frigeri, A.; Hughson, K.; Kneissl, T.; Krohn, K.; Mest, S.; Pasckert, J.; Platz, T.; et al. DAWN mission final high resolution global geolgic map of Ceres. In Proceedings of the Planetary Geologic Mappers Workshop, London, UK, 1–6 January 2019; LPI Contrib. No. 2154; p. 7003.
67. Nass, A. *Geologic Map of Ceres [Dawn Mission]—Global Dataset Based on the 15 Individual Quadrangle Maps [Data Set]*; Zenodo, CERN: Geneva, Switzerland, 2023. [CrossRef]
68. van Gasselt, S.; Nass, A. A Semantic View on Planetary Mapping—Investigating Limitations and Knowledge Modeling through Contextualization and Composition. *Remote Sens.* **2023**, *15*, 1616. [CrossRef]
69. Roatsch, T.; Kersten, E.; Matz, K.; Preusker, F.; Scholten, F.; Jaumann, R.; Raymond, C. High resolution Vesta High Altitude Mapping Orbit (HAMO) Atlas derived from Dawn framing camera images. *Planet. Space Sci.* **2012**, *73*, 283–286. [CrossRef]
70. Roatsch, T.; Kersten, E.; Matz, K.; Preusker, F.; Scholten, F.; Elgner, S.; Jaumann, R.; Raymond, C.A.; Russell, C.T. High-resolution Vesta Low Altitude Mapping Orbit Atlas derived from Dawn Framing Camera images. *Planet. Space Sci.* **2013**, *85*, 293–298. [CrossRef]
71. Roatsch, T.; Kersten, E.; Matz, K.D.; Preusker, F.; Scholten, F.; Elgner, S.; Schroeder, S.; Jaumann, R.; Raymond, C.; Russell, C.T. *DAWN Framing Camera Derived Vesta Global Mosaics V1.0, DAWN-A-FC2-5-MOSAIC-V1.0*; NASA Planetary Data System, Asteroid/Dust Subnode; NASA/Planetary Science Institute: Tucon, AZ, USA, 2015.
72. European Organization For Nuclear Research. *OpenAIRE*; CERN: Geneva, Switzerland, 2013. [CrossRef]
73. Mest, S.C.; Crown, D.A.; Yingst, R.A.; Berman, D.C.; Williams, D.A.; Buczkowski, D.L.; Scully, J.E.C.; Platz, T.; Hiesinger, H.; Pasckert, J.H.; et al. The HAMO-Based Global Geologic Map and Chronostratigraphy of Ceres. In Proceedings of the 49th Annual Lunar and Planetary Science Conference, Woodlands, TX, USA, 19–23 March 2018; p. 2730.

74. Mest, S.C.; Berman, D.C.; Buczkowski, D.L.; Crown, D.A.; Scully, J.E.C.; Williams, D.A.; Yingst, R.A.; Frigeri, A.; Nass, A.; Neesemann, A.; et al. The Global LAMO-Based Geologic Map of Ceres. In Proceedings of the 2021 Annual Meeting of Planetary Geologic Mappers, Online, 14–15 June 2021; Volume 2610, p. 7035.
75. Mest, S.C.; Crown, D.A.; Williams, D.A.; Buczkowski, D.L.; Scully, J.E.C.; Yingst, R.A.; Berman, D.C.; Frigieri, A.; Nass, A.; Neesemann, A.; et al. The Global Geologic Mapping of Ceres. In Proceedings of the 2022 Annual Meeting of Planetary Geologic Mappers, Online, 20–22 June 2022; Volume 2684, p. 7035.

MDPI AG
Grosspeteranlage 5
4052 Basel
Switzerland
Tel.: +41 61 683 77 34

Remote Sensing Editorial Office
E-mail: remotesensing@mdpi.com
www.mdpi.com/journal/remotesensing

www.ingramcontent.com/pod-product-compliance
Lightning Source LLC
Chambersburg PA
CBHW061618220926
51850CB00005B/534
* 9 7 8 3 7 2 5 8 7 8 5 1 2 *